Ecological Studies

Analysis and Synthesis

Volume 244

Ecological Studies is Springer's premier book series treating all aspects of ecology. These volumes, either authored or edited collections, appear several times each year. They are intended to analyze and synthesize our understanding of natural and managed ecosystems and their constituent organisms and resources at different scales from the biosphere to communities, populations, individual organisms and molecular interactions. Many volumes constitute case studies illustrating and synthesizing ecological principles for an intended audience of scientists, students, environmental managers and policy experts. Recent volumes address biodiversity, global change, landscape ecology, air pollution, ecosystem analysis, microbial ecology, ecophysiology and molecular ecology.

Jean-Baptiste Ramond • Don A. Cowan
Editors

Microbiology of Hot Deserts

Springer

Editors
Jean-Baptiste Ramond
Departamento de Genética Molecular y Microbiología
Pontificia Universidad Católica de Chile
Santiago, Chile

Don A. Cowan
Centre for Microbial Ecology and Genomics - Department of Biochemistry, Genetics and Microbiology
University of Pretoria
Pretoria, South Africa

ISSN 0070-8356 ISSN 2196-971X (electronic)
Ecological Studies
ISBN 978-3-030-98414-4 ISBN 978-3-030-98415-1 (eBook)
https://doi.org/10.1007/978-3-030-98415-1

This Springer imprint is published by the registered company Springer Nature Switzerland AG
The registered company address is: Gewerbestrasse 11, 6330 Cham, Switzerland

Preface

> Parler du désert, ne serait-ce pas, d'abord, se taire, comme lui, et lui rendre hommage non de nos vains bavardages mais de notre silence?
> Théodore Monod

According to the French naturalist and explorer Théodore Monod (1902–2000), who spent most of his career working in the Sahara Desert, the largest hot desert on Earth, the best way to pay tribute to deserts is to remain silent. However, in this text, we have chosen to give voice to probably the most silent of all (desert) organisms, arguably the most important living forms: microorganisms and their communities. The latter are complex assemblages of diverse microorganisms (i.e. viruses, archaea, bacteria, fungi and/or microeukaryotes) that dynamically interact with or against one another and which have colonized almost all available niches on Earth (Merino et al. 2019).

Hot deserts are poly-extreme environments characterized by low water availability, low nutrient contents, high surface UV irradiation and generally high and fluctuating temperatures (Laity 2009; Fig. 1).

As a result of the water deficit and high temperatures, the presence of macroorganisms in hot deserts, particularly higher plants, is greatly reduced in comparison to more temperate ecosystems (Fig. 2). The corollary to this observation is that in desert systems, microbial communities are disproportionately important for their functional capacity and particularly for their primary productivity (e.g. Pointing and Belnap 2012; Makhalanyane et al. 2015; Ramond et al. 2018; Jordaan et al. 2020).

According to the United Nations *Decade for Deserts and the fight against Desertification* 2010–2020 report (http://www.un.org/en/events/desertification_decade/whynow.shtml), drylands cover 41.3% of the Earth's land surface and harbour 35.5% of the human population. Furthermore, with increasing global temperatures, linked to climate change, desert surface area is expected to expand—a process termed desertification. This process will particularly affect developing countries and already vulnerable human and domestic animal populations (Huang et al. 2016; Yao et al. 2020). Models now predict that by the end of the twenty-first century, hot drylands will cover up to 56% of the Earth's

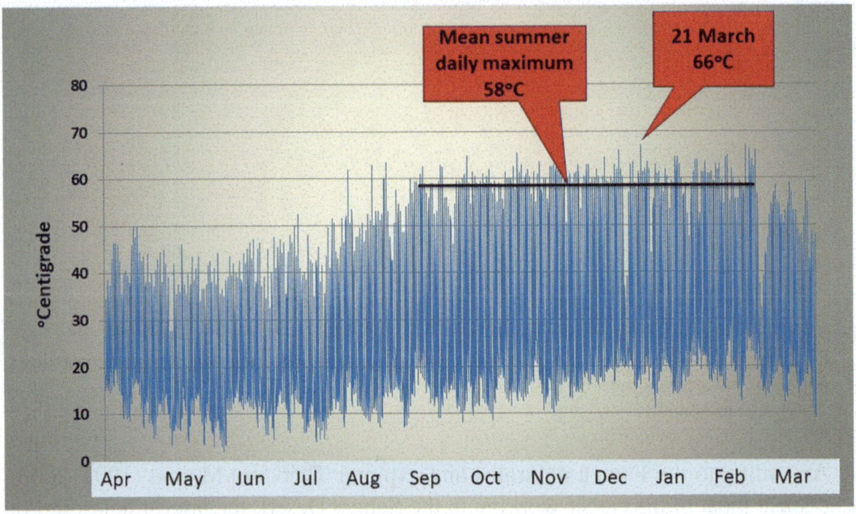

Fig. 1 Annual temperature record for Namib Desert soil (0.5 cm depth). Note that the soil temperature exceeded 60 °C on more than 70 days in the 6-month summer period between October and February

global land surface and be home to more than half of the world's human population (Yao et al. 2020).

While hot deserts/dryland ecosystems exist in all continents apart from Antarctica (the Dry Valleys of East Antarctica are hyperarid cold drylands; Chap. 9) and are expanding at the global scale (Fig. 3; Chap. 12), the microbial ecology of these environments has probably been less studied than more productive ecosystems such as prairies, farmlands and forests. We, therefore, argue that improving our knowledge of all aspects of the microbiomics of deserts is crucial to appropriately comprehend the present, and model the future, functioning of terrestrial Earth.

This text aims to provide readers with a comprehensive introduction to the current body of knowledge of hot desert microbial ecology. Nevertheless, before detailing the biology of hot deserts, it is necessary to describe the physical characteristics of hot deserts. Chapter 1 presents a comprehensive introduction for a wider audience—and particularly for microbial ecologists—to the climate, geomorphology, habitats and soils of hot deserts. Understanding the physical nature of the system is an essential element in designing any macro-scale study of microbial ecology, since the physical, macroenvironmental and chemical properties of the system can (alone or in combination) influence the structure and function of the intrinsic microbial communities (Gombeer et al. 2015; Makhalanyane et al. 2015; Cowan et al. 2020). This chapter also describes how global climate patterns dictate patterns of precipitation and ultimately the origins of drylands. The chapter also details the geomorphological origin of desert landscapes and of their many physical structures (e.g. inselbergs, drainage systems, playas, sand dune systems and pediplains).

Fig. 2 Photographs showing the barren landscapes of the hyperarid Namib Desert. Desert expanses before (**a**) and after (**b**) rainfall, showing the rapid germination of the desert grass *Stipagrostis* spp. (**c**) The Namib Sand Sea, A UNESCO World Heritage Site: a complex structure of mobile sand dunes and stable inter-dune zones. (**d**) Quartz-littered pavements in the Namib gravel desert. *Photo courtesy Don Cowan*

Finally, this chapter describes the formation of soils under arid conditions. The physical characteristics of the *pedoderm* of desert soils, i.e., the first and most 'active' top millimetres and centimetres, are particularly important in the biological context, since this upper horizon harbours the most active biological communities such as biological soil crusts (cf., Chap. 3).

The development of new high-throughput meta'omics' technologies (i.e. metagenomics, -transcriptomics, -proteomics and -metabolomics) and the potential for their integration (e.g. as metaphenomics; Jansson and Hofmockel 2018) has dramatically increased the depth and scope of environmental microbial ecology research. Chapter 2 aims at introducing the reader to the state-of-the-art technical methodologies, detailing the most relevant bioinformatics tools and pipelines, which can be applied to studies of many aspects of desert microbial communities. A comprehensive overview is provided of the short- and long-read sequencing technologies currently used to explore the taxonomic composition (from viruses [cf., Chap. 6] to microeukaryotes) and functional capacities (cf., Chap. 9) of desert microbiomes. Chapter 2 also describes the different approaches used to investigate the functional fraction of microbial communities (e.g. Stable Isotope Probing [SIP], Bio-Orthogonal Non-Canonical Amino Acid Tagging [BONCAT] and metabolomics). We particularly highlight the detailed explanations on how high-resolution imagery (e.g. X-ray microtomography, Catalyzed Reporter Deposition-

Fig. 3 Photographs of deserts from around the world. (**a**) The Atacama Desert in South America. (**b**) Semi-arid/arid shrubland in Australia. (**c**) The southern African Kalahari Desert. (**d**) The Arabian Peninsula Desert. *Photos courtesy of Karen Jordaan (**a** and **b**) and Jean-Baptiste Ramond (**c** and **d**)*

Fluorescence *in situ* Hybridization [CARD-FISH]), spectrometry (e.g. Nanoscale Secondary Ion Mass Spectrometry [NanoSIMS]) and sequencing (Metagenomic plot sampling by sequencing [MaPS-seq]) can resolve desert soil and cryptic microbial communities with great precision; i.e., to the micrometre-scale.

Chapters 3–5 describe the microbial ecology of the most dominant and possibly most important desert niches (biocrusts, open soils and endolithic habitats), detailing the phylogenetic and functional diversities of these microbiomes as well as their potential roles in the broader ecology of desert ecosystems.

Chapter 3 focuses on biological soil crusts (BSC; also known as biocrusts, microbiotic, mycrophytic, cryptogamic or cryptobiotic crusts), phototrophic microbial mats covering the surfaces of many hot and cold desert soils. This chapter discusses the composition of these communities, as well as their global distribution as a function of aridity. These mats, often referred to as the living skin of the Earth, are microbial community diversity and biomass hotspots in otherwise depauperate dryland ecosystems. In addition, BSCs are considered to be crucial features of deserts ecosystems at the landscape scale, as they play an important role in soil stabilization, water retention and primary production. However, the author notes that BSC microbial ecology studies have largely been restricted to four arid regions (the

USA, Europe, Israel and Australia) and highlights the necessity to expand these studies, particularly to Asian and African drylands.

Barren open soils, i.e., soil devoid of any biocrusts, in drylands are the most hostile of the edaphic niches colonized by desert microorganisms, particularly in the most arid deserts such as the hyperarid Atacama and Namib deserts. As discussed in Chap. 4, their microbial ecology remains understudied when compared to the more productive (and visible) BSC niches. This chapter describes the prokaryotic and microeukaryotic composition and the *virosphere* of hot desert edaphic communities, as well as the functional properties of these communities during periods of desiccation and after water events. The current state of knowledge of their '*interactomes*', i.e., the biotic and abiotic interactions identified using co-occurrence networks, is also reviewed.

To evade the extremes of the desert environment, microorganisms colonize cryptic lithic niches (Makhalanyane et al. 2015). Chapter 5 describes in detail the diversity, assembly, adaptive strategies and metabolic capacities of hot desert endolithic microbial communities, i.e. microbial communities within pores (cryptoendolithic) and fissures (chasmoendolithic) of rocks.

The viruses and viral communities of desert ecosystems, which have been little studied (and are little understood; Trubl et al. 2018), are the focus of Chap. 6. This chapter reviews the current state of knowledge of the virosphere of hot desert soils, BSCs, saline water bodies (e.g. playas) and hypoliths. The key role of viruses and phage in affecting community dynamics in other habitats (aquatic and marine, in particular) may not necessarily apply in discrete, particulate low-water activity systems such as desert soils, although the authors note the severe deficit of relevant data. The authors also highlight the current limitations of metaviromics studies, in part due to the general paucity of annotated viral sequences in relevant databases. In this context, the authors identify new avenues for exploring desert virus–host dynamics and also highlight the many questions that remain to be answered in the field of desert viromics.

The roles of desert microbial communities and abiotic processes in biogeochemical cycling of Carbon, Nitrogen and Phosphorus are reviewed in Chap. 7. Here, the low water availability in desert ecosystems has opposing impacts: where limited plant growth increases the relative importance of microbial nutrient cycling, but also limits the capacity of the intrinsic microbial communities to perform such functions. This chapter provides a cogent argument for the contention that microbial communities are the dominant drivers of C, N and P cycling in deserts, and describe the mechanisms by which they participate in both nutrient loss and input processes.

Desert ecosystems typically harbour a wide variety of highly adapted plants (e.g. Eshel et al. 2021). To survive and grow in such hostile environments, such plants have evolved multiple adaptative traits for the acquisition and/or retention of water. There is strong evidence that xerophytes have established intimate relationships with their associated microbiota (e.g. Marasco et al. 2018; Ramond et al. 2019; Peguero-Pina et al. 2020). Desert plant–microbe interactions, particularly those of Plant Growth Promoting (PGP) microorganisms, are discussed in detail in Chap. 8. In the context of global desertification, desert PGP microorganisms may have the potential to be used to improve desert/dryland agricultural practices (Soussi et al. 2016).

Desert microbiology has been at the forefront of Astrobiology. In the era of space exploration and the search for Life Outside of Earth, understanding how Life has adapted to low water availability conditions on Earth represents a valid approach to predicting where to find past or present Life on other *habitable* planets, and may even be used conceptually in future *terraforming* strategies (Friedmann and Ocampo-Friedmann 1995; Warren-Rhodes et al. 2019). Chapter 9 is dedicated to providing an up-to-date review of desert astrobiology studies and expands the scope of this book by including studies from the cold hyperarid Antarctic Dry Valleys. This is justified by the fact that Mars, and many other habitable planetary objects such as the moons of Saturn Enceladus and Titan (Lorenz et al. 1997; McKay et al. 2014), are particularly cold. Chapter 9 summarizes the lessons learned from the search for life on Mars (notably from the Viking, Phoenix and Curiosity Lander and Rover missions) and from microbial ecology work on Earth's deserts and their many edaphic, hypo/endolithic and salt-rich habitats. It further discusses the centrality of water availability, temperature and substrate properties on their microbial ecology; both at the micro- and macro-scales.

Chapters 10 and 11 cover the different adaptative strategies employed by microorganisms to cope with extreme conditions of dryland habitats. After a comprehensive description of the many environmental stresses imposed on hot desert microbial communities (i.e. from lack of water, high salinity, oligotrophy, UV-radiation and elevated temperatures), Chapter 10 reviews the adaptation mechanisms evolved by microorganisms to cope with them at both cellular and community-scale levels. These include strategies such as sporulation, membrane modifications, protein protection and expression of chaperones, DNA repair mechanisms, the salting-in/out strategies, the down-regulation of energy- and water-demanding processes and the upregulation of desiccation-tolerance genes. Community-scale adaptations include complex processes such as EPS/biofilm formation and atmospheric trace gas harvesting. The authors have provided readers with a multifactorial model showing the major molecular and physiological mechanisms for survival of microorganisms in hot deserts.

In water-limited environments, rare rain events represent both a source of 'relief' and stress, where the sudden availability of water stimulates metabolic activities and, provides access to new niches and substrates, but also induces a period of intense competition and the activation of predators and 'pathogens' such as viruses and phage. Chapter 11, therefore, reviews the desert microbial community physiological responses to hydration and desiccation cycles, mainly from BSCs and topsoils.

We hope that this text will provide readers with a comprehensive and useful overview of the current status of research on hot desert microbial ecology and that it may even contribute to the evolution of these studies. We argue that, with the current climate trends, it has never been more important to better understand and preserve these dominant, majestic yet fragile ecosystems which are becoming more extreme and expanding with global climate change (Figs. 2 and 3; Chap. 12).

Acknowledgements J-BR and DAC would like to especially thank the Gobabeb-Namib Research Institute (Namibia) and its present and past members, particularly Mary Seely, Gillian Maggs-Kölling, Eugene Marais and Chef Hendriks who have done so much to support our research on Namib Desert microbial ecology. We would also like to acknowledge the continuous support of the University of Pretoria and of the Pontificia Universidad Católica de Chile.

We would also like to use this introductory preface chapter to sincerely thank the reviewers who agreed to spend their valuable time editing these chapters. In alphabetical order:

- *Anonymous* (many)
- Prof. B. Büdel—University of Kaiserslautern, Germany
- Dr. P. De Maayer—University of the Witwatersrand, South Africa
- Prof. C. Dorador—Universidad de Antofagasta, Chile
- Dr. A. Frossard—Swiss Federal Institute for Forest, Snow and Landscape Research, WSL, Switzerland
- Prof. M.A. Jorquera—Universidad de La Frontera, Temuco, Chile
- Prof. W. Kong—Institute of Tibetan Plateau Research, Chinese Academy of Sciences, China
- Prof. MCY Lau—Institute of Deep-Sea Science and Engineering, Chinese Academy of Sciences, China
- Prof. A. Priemé—University of Copenhagen, Denmark
- Dr. A. Valverde—Spanish National Research Council, Spain

Centre for Microbial Ecology
and Genomics—Department of Biochemistry,
Genetics and Microbiology—University of Pretoria
Pretoria, South Africa

Department of Molecular
Genetics and Microbiology—Pontificia
Universidad Católica de Chile
Santiago, Chile

Jean-Baptiste Ramond

Centre for Microbial Ecology
and Genomics—Department of Biochemistry,
Genetics and Microbiology—University of Pretoria
Pretoria, South Africa

Don A. Cowan

References

Cowan DA, Hopkins DW, Jones BE, Maggs-Kölling G, Majewska R, Ramond JB (2020) Microbiomics of Namib Desert habitats. Extremophiles 24(1):17–29

Eshel G, Araus V, Undurraga S, Soto DC, Moraga C, Montecinos A, Moyano T, Maldonado J, Díaz FP, Varala K, Nelson CW, Contreras-López O, Pal-Gabor H,

Kraiser T, Carrasco-Puga G, Nilo-Poyanco R, Zegar CM, Orellana A, Montecino M, Maass A, Allende ML, DeSalle R, Stevenson DW, González M, Latorre C, Coruzzi GM, Gutiérrez RA (2021) Plant ecological genomics at the limits of life in the Atacama Desert. Proc Natl Acad Sci 118(46):e2101177118
Friedmann EI, Ocampo-Friedmann R (1995) A primitive cyanobacterium as pioneer microorganism for terraforming Mars. Adv Space Res 15(3):243–246
Gombeer S, Ramond JB, Eckardt FD, Seely M, Cowan DA (2015) The influence of surface soil physicochemistry on the edaphic bacterial communities in contrasting terrain types of the Central Namib Desert. Geobiology 13(5):494–505
Huang J, Yu H, Guan X, Wang G, Guo R (2016) Accelerated dryland expansion under climate change. Nat Clim Chang 6:166–171
Jansson JK, Hofmockel KS (2018) The soil microbiome—from metagenomics to metaphenomics. Curr Opin Microbiol 43:162–168
Jordaan K, Lappan R, Dong X, Aitkenhead IJ, Bay SK, Chiri E, Wieler N, Meredith LK, Cowan DA, Chown SL, Greening C (2020) Hydrogen-oxidizing bacteria are abundant in desert soils and strongly stimulated by Hydration. Msystems 5(6)
Laity JJ (2009) Deserts and desert environments, vol 3. John Wiley & Sons, Chichester
Lorenz RD, Lunine JI, McKay CP (1997) Titan under a red giant sun: A new kind of "habitable" moon. Geophys Res Lett 24(22):2905–2908
Makhalanyane TP, Valverde A, Gunnigle E, Frossard A, Ramond JB, Cowan DA (2015) Microbial ecology of hot desert edaphic systems. FEMS Microbiol Rev 39(2):203–221
Marasco R, Mosqueira MJ, Fusi M, Ramond JB, Merlino G, Booth JM, Maggs-Kölling G, Cowan DA, Daffonchio D (2018) Rhizosheath microbial community assembly of sympatric desert speargrasses is independent of the plant host. Microbiome 6(1):1–18
McKay CP, Anbar AD, Porco C, Tsou P (2014) Follow the plume: the habitability of Enceladus. Astrobiology 14(4):352–355
Merino N, Aronson HS, Bojanova DP, Feyhl-Buska J, Wong ML, Zhang S, Giovannelli D (2019) Living at the extremes: extremophiles and the limits of life in a planetary context. Front Microbiol 10:780
Peguero-Pina JJ, Vilagrosa A, Alonso-Forn D, Ferrio JP, Sancho-Knapik D, Gil-Pelegrín E (2020) Living in drylands: Functional adaptations of trees and shrubs to cope with high temperatures and water scarcity. Forests 11(10):1028
Pointing SB, Belnap J (2012) Microbial colonization and controls in dryland systems. Nat Rev Microbiol 10(8):551–562
Ramond JB, Woodborne S, Hall G, Seely M, Cowan DA (2018) Namib Desert primary productivity is driven by cryptic microbial community N-fixation. Sci Rep 8(1):1–9
Ramond JB, Baxter J, Maggs-Kölling G, Martínez-Alvarez L, Read DA, León-Sobrino C, van der Walt AJ, Cowan DA (2019) Microbial ecology of the Namib Desert. In: Model ecosystems in extreme environments. Academic Press, Cambridge, MA, pp 113–143

Soussi A, Ferjani R, Marasco R, Guesmi A, Cherif H, Rolli E, Mapelli F, Ouzari HI, Daffonchio D, Cherif A (2016) Plant-associated microbiomes in arid lands: diversity, ecology and biotechnological potential. Plant Soil 405(1):357–370

Trubl G, Jang HB, Roux S, Emerson JB, Solonenko N, Vik DR, Solden L, Ellenbogen J, Runyon AT, Bolduc B, Woodcroft BJ (2018) Soil viruses are underexplored players in ecosystem carbon processing. MSystems 3(5)

Warren-Rhodes KA, Lee KC, Archer SD, Cabrol N, Ng-Boyle L, Wettergreen D, Zacny K, Pointing SB (2019) Subsurface microbial habitats in an extreme desert Mars-analog environment. Front Microbiol 10:69

Yao J, Liu H, Huang J, Gao Z, Wang G, Li D, Yu H, Chen X (2020) Accelerated dryland expansion regulates future variability in dryland gross primary production. Nat Commun 11(1):1–10

Contents

A Brief Introduction to Hot Desert Environments: Climate, Geomorphology, Habitats, and Soils

1

Frank D. Eckardt, Gillian Maggs-Kölling, Eugene Marais, and Pieter C. de Jager

Abstract

This chapter provides a broad introduction to dryland environments with a focus on desert climate, geomorphology, habitats, and soils that collectively provide opportunities and limitations for microbial life. Desert precipitation is governed by global circulation patterns, which determine the distribution of drylands and associated rainfall gradients. Desert margins especially are subject to pronounced inter-annual variability, and good years may be replaced by a negative departure from mean rainfall, resulting in drought. The depth of drought for any location can be examined using the widely accessible Standardised Precipitation-Evapotranspiration Index (SPEI) and Normalised Difference Vegetation Index (NDVI). The topography of desert surfaces and habitats can be characterised by rocky run-off from uplands and fluvial run-on in channels and fan settings. Other settings may be dominated by mobile sediments such as dunes or shallow groundwater, which promotes evaporation and build-up of salts. Gravel plains, on the other hand, may provide stable surfaces. Desert soils and associated surfaces, the so-called pedoderm, require stability to form and prevail. The abiotic properties of unconsolidated porous media moderate the physical and chemical subsurface environment. Crusts in particular, play a role in regulating fluxes, including water, carbon, and nitrogen, which collectively determine the microbial subsurface environment. Moisture regimes vary in space and time along with geomorphic conditions, the distribution of habitats, and the state of the pedoderm.

F. D. Eckardt (✉)
Environmental and Geographical Science, University of Cape Town, Rondebosch, South Africa
e-mail: frank.eckardt@uct.ac.za

G. Maggs-Kölling · E. Marais
Gobabeb - Namib Research Institute, Walvis Bay, Namibia

P. C. de Jager
Plant and Soil Sciences, University of Pretoria, Hatfield, South Africa

J.-B. Ramond, D. A. Cowan (eds.), *Microbiology of Hot Deserts*, Ecological Studies 244, https://doi.org/10.1007/978-3-030-98415-1_1

Keywords

Dryland precipitation · Geomorphology · Habitat · Soil · Pedoderm

The chapter wishes to introduce hot, arid regions, also referred to as drylands or desert environments. These are defined by a lack of moisture, the result of low precipitation and high evapotranspiration rates (Thornthwaite 1948), and which cover close to half of the earth's surface. Meteorological deserts extend into the polar region and cover large parts of the world's ocean also. For the purpose of this book, we shall focus on hot deserts in terrestrial subtropics. This chapter serves as a broad and general introduction to dryland environments, determined by climate, geomorphology, habitats, and soil conditions. These provide both limitations and opportunities for microbial life at various scales: first globally, in terms of climate, then regionally, due to landscapes and landforms, and subsequently locally, due to habitats and soil processes.

1.1 Desert Climatology

1.1.1 Desert Precipitation

While a number of climatic factors influence microbial activity in deserts, this chapter focuses on precipitation as the dominant limiting resource and driver of desert ecosystems. Hot deserts in terrestrial subtropics are largely determined by the positioning of dry, descending air, associated with high-pressure systems that form part of the global Hadley Cell circulation patterns with quasi-stationary positions over north and south Africa, Australia, Asia, and the west of North and South America (Figs. 1.1 and 1.2). The positions of these high-pressure systems are subject to a seasonal latitudinal shift, giving way to tropical, convective summer rain, especially the Intertropical Convergence Zone towards the equator and frontal, winter rain in Mediterranean latitudes. These pressure constellations generate rainfall gradients that extend from the equatorial regions, with high rainfall regimes, into semi-arid drylands, deserts, and hyper-arid desert cores. For an introduction to desert climatology, refer to Nicholson (2011).

Additional factors play a role in further limiting rain, such as distance from oceans, which are the primary source of moisture, the blocking of moisture bearing, westward flowing air by topographic barriers, and the suppression of convective cloud formation over cool waters, especially ocean upwelling systems along subtropical west coasts. The results are the major arid regions of the world, to be found to the west of North and South America, Southern Africa, the Sahara of North Africa, Arabia, and much of the Middle East, as well as Central Asia and most of Australia. Due to the movement of pressure systems, the core of drylands tends to be hyper-arid, while desert margins are considered semi-arid. The resulting gradients can be steep over short distances, as seen in Fig. 1.3.

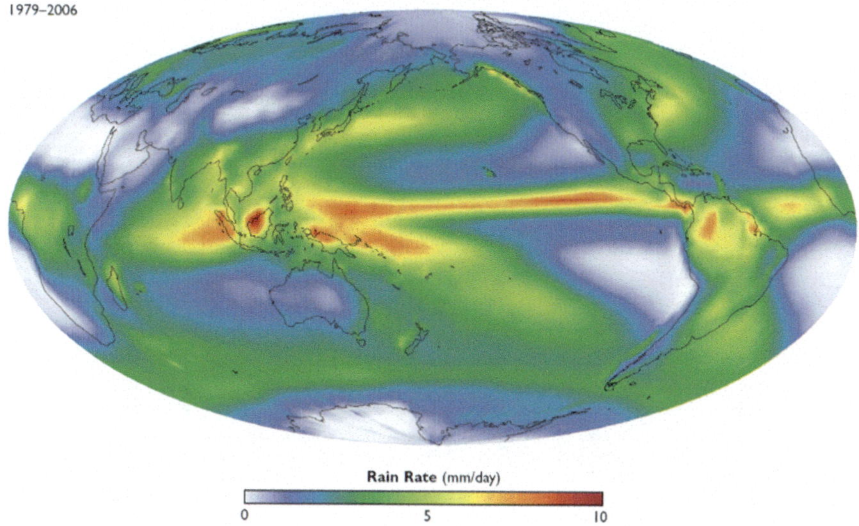

Fig. 1.1 Global long-term rainfall patterns. Note the lack of rain in the subtropics over land and sea (King 2007)

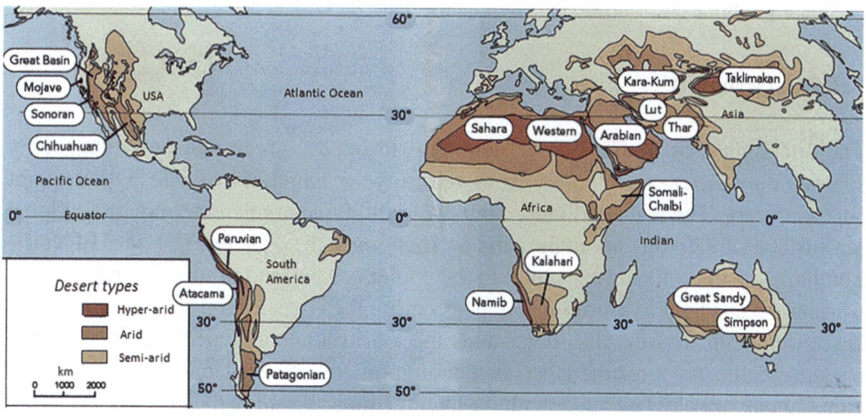

Fig. 1.2 Map of global deserts (Provided by A.S. Goudie)

Opportunities for rain are seasonal but diminish significantly towards the quasi-stationary centres of continental high-pressure systems. This results in rainfall gradients, some of which may be relatively short and dramatic, especially in areas with pronounced topography (Zipser et al. 2006). Since convective rain in the subtropics is also mostly associated with stationary rising air, this often produces localised rain events, making rainfall patterns both patchy and unpredictable in nature. Rains in drylands can be described as infrequent but also often intense, given that the annual total might be associated with a few major events.

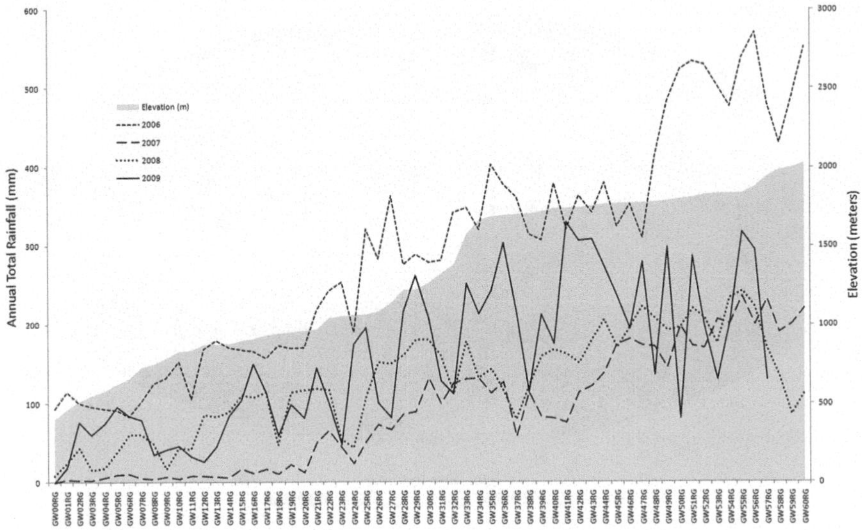

Fig. 1.3 Rainfall gradient in Namibia over a few hundred kilometres from the interior to the coast, between 2006 and 2009. Note the diminishing rainfall towards the hyper-arid coast and the inter-annual variability of rainfall (y-axis on the left). The grey depicts elevation from the interior at 2000 m towards the coasts at 500 m (y-axis on the right) (Eckardt et al. 2013)

1.1.2 Drought

The uncertainty of rainfall is also temporal in nature, given that individual rainy seasons may differ considerably from year to year, which may cause both drought and wet years. The so-called inter-annual variability is largely determined by global sea surface temperature anomalies and oscillations in the equatorial Pacific (El Niño–Southern Oscillation) and beyond. Firstly, cooler oceans tend to suppress rainfall and warmer oceans tend to promote it, and secondly, these associated shifts in rising and descending air masses also determine the positioning and persistence of high-pressure cells. As a result, it is not uncommon for seasonal rainfall yields to be either above or below average, with the average rainfall seldom being met. The mean rainfall, however, establishes a baseline against which to compare individual months and years. It is, furthermore, not uncommon for rainfall patterns at any location to vary in a cyclic manner with several above average years followed by several below average years.

While deserts can be classified according to their average rainfall and evaporation, it is also useful to quantify drought conditions. A meteorological drought would be associated with an anomalously dry month. Several months with a persistent absence of rain would result in below average soil moisture and reduced plant growth, also referred to as agricultural drought. A number of below average dry years would result in a decline of surface waters and drop in groundwater tables, which can be described as a hydrological drought (Fig. 1.4). This cyclic nature is common in all arid regions.

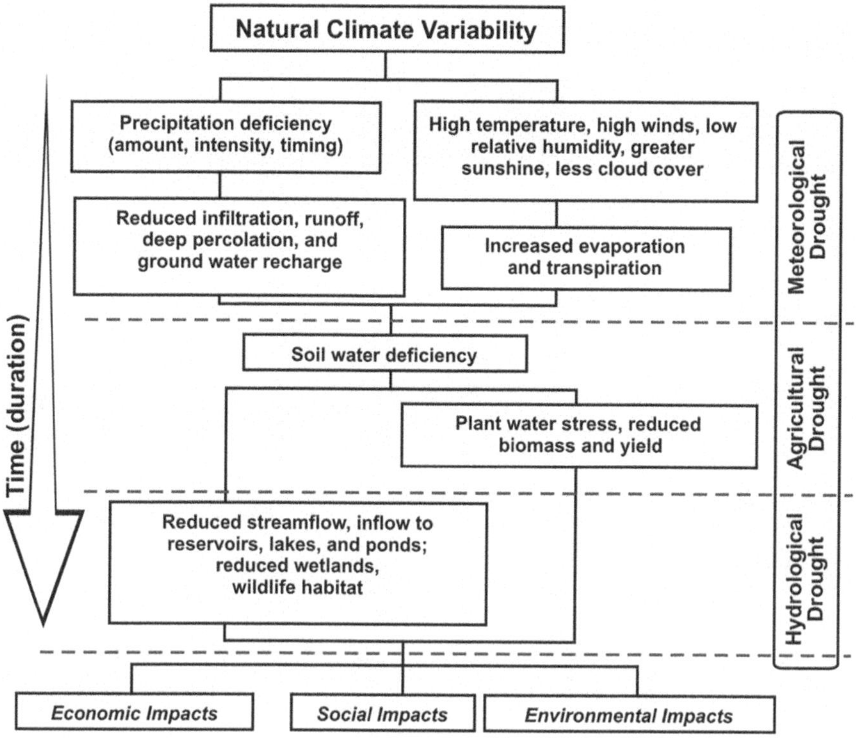

Fig. 1.4 Drought and progression with time, from short-lived meteorological to long-term hydrological drought (source: National Drought Mitigation Centre, University of Nebraska)

Two readily available indices can be used to quantify drought at any location on the globe, namely the Standardised Precipitation-Evapotranspiration Index (SPEI) and Normalised Difference Vegetation Index (NDVI). The unitless SPEI index is available in half degree tiles at a monthly resolution and on a range of timescales from 1 to 48 months. The index makes use of the Thornthwaite (1948) equation and draws on a number of global datasets (Beguería et al. 2014). By resorting to smoothed time series, the different types of drought are identifiable (Fig. 1.5). The resulting 1-month (SPEI-01) product emphasises meteorological drought cycles. The SPEI-03 (3 months) accentuates seasonal short- and medium-term trends, such as soil moisture, and is suitable for the study of agricultural drought. SPEI-12 (1 year) examines annual moisture cycles that would flag hydrological drought conditions. SPEI-48 (4 years) is the coarsest SPEI product and depicts accumulative long-term decadal drought cycles. The SPEI dataset has now been extended to 1901 and provides near real-time results for any location on the earth surface, which is ideally suited for drought monitoring and early-warning purposes. The graphic interface is easy to negotiate, and raw data can be obtained as well.

The satellite-based Moderate Resolution Imaging Spectroradiometer (MODIS) produces a global NDVI (Normalised Difference Vegetation Index) time series. This

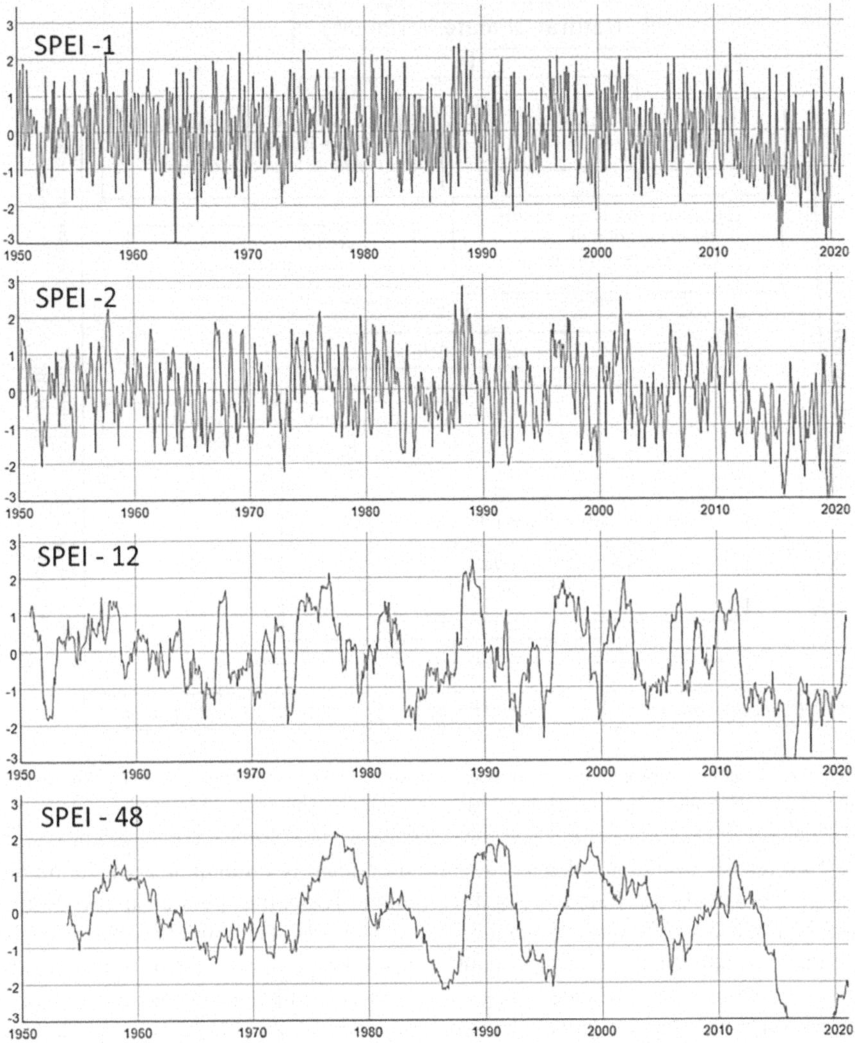

Fig. 1.5 The unitless Standardised Precipitation-Evapotranspiration Index (SPEI) for Bloemfontein South Africa (1954–2020) (source: SPEI online portal https://spei.csic.es/database.html). From top to bottom: SPEI 01 meteorological drought, SPEI-03 agricultural drought, SPEI-12 hydrological drought, SPEI-48 decadal drought. Positive values indicate wetter conditions and negative values indicate the opposite. The recent deep drought is evident in the time series and the result of numerous dry months and years

monitors the state of vegetation, measuring reflected infrared radiation and red absorption as an indicator of plant growth (Tucker 1979). The results can be interrogated through the Global Agriculture Monitoring (GLAM) project portal (Becker-Reshef et al. 2010), which depicts vegetation in 8-day timesteps and

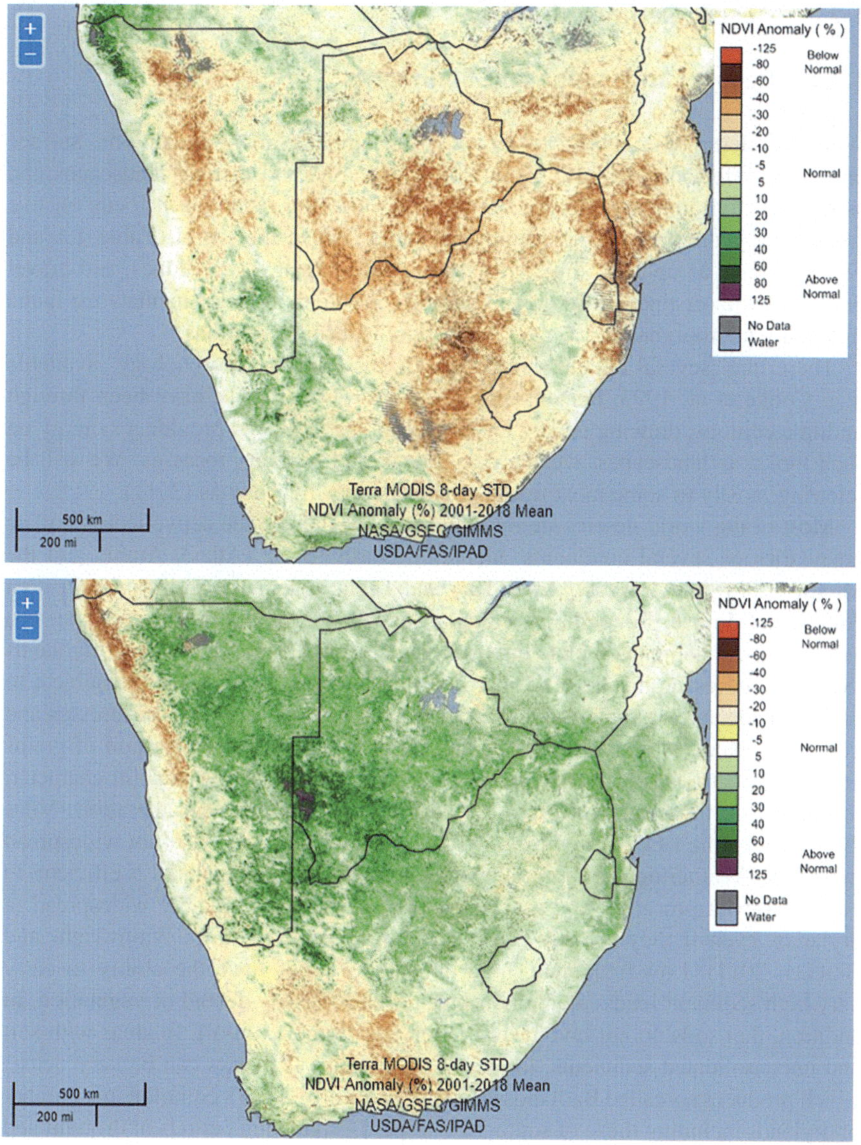

Fig. 1.6 Significant inter-annual variation in chlorophyl response to rainfall for southern Africa, captured by the Normalised Difference Vegetation (NDVI) Anomaly for February 10–17, 2016 (a severe drought year) at the top and March 30–April 06, 2021 (a very wet year) at the bottom (source: Global Agriculture Monitoring (GLAM) Project (https://pekko.geog.umd.edu/usda/test/)

determines the NDVI anomaly at a 250 m resolution since February 2000 for any part of the earth surface (Fig. 1.6). Anomalies for any 8-day time period can be generated.

1.2 Desert Geomorphology

1.2.1 Desert Landscapes and Sediments

While hot deserts are subject to more-or-less similar climatic conditions, surface processes differ considerably, given a host of varied and distinct landscapes and associated environments (Goudie 2002). The following section will focus on five contrasting geomorphic units commonly found around the world (Table 1.1 and Fig. 1.7), such as uplands, drainage systems, saline basins, sand dunes, and gravel plains, all with distinct substrate and abiotic conditions, but also habitats for fauna (including microorganisms; Chap. 4) and flora and soil conditions.

There are several recent textbooks on dryland geomorphology available (e.g. Cooke et al. 1993; Parsons and Abrahams 2009). These have been through multiple editions, drawing on key researchers in the field, and provide a synergy on such topics as landscapes, weathering, fluvial, and aeolian processes. We will be referring mostly to some most recent edited chapters by Thomas (2011).

Most of the world deserts are mountainous and include the active tectonic plate boundaries of several continents, including North Africa (Atlas), Arabia and the Middle East (Sinai, Zagros), Asia (Thar and Takla Makan), as well as the western Americas (Mojave, Sonoran, Atacama) (Rendell 2011). Rocky surfaces are also associated with denudated plate interiors (Karoo and Australia), volcanic highlands (Sahara), and escarpments in Africa. Bedrock outcrops in drylands are subject to limited shade and large diurnal temperature cycles, which may impose stresses and volumetric changes on rock surfaces and result in the gradual separation of grains and rock structures. Weathering in rocky deserts is primarily physical in character, resulting in a fragmentation of rocks, with little or no chemical alteration (Viles 2011). Given the lack of chemical alteration, clay formation is also not widespread in deserts. Weathering may be slow, but the removal of loose material is efficient. In general, fluvial processes and associated landforms are remarkably widespread in drylands, even if they are not active for long periods of time (Wainwright and Bracken 2011). Low-frequency, high-magnitude floods have the ability to carry very high sediment loads in an environment that is largely devoid of vegetation. In addition, thin soils do not favour infiltration but promote run-off. In areas with soft and unconsolidated sediments, the resulting drainage networks can be very dense, which produces so-called Badlands (Howard and Kerby 1983), common in a number of drylands including those of southern Europe. Despite long periods of drought and significant dry seasons, the imprint of fluvial processes in arid landscapes is both pronounced and widespread and often dramatic.

Upland regions and associated fluvial processes tend to generate coarse angular colluvium on slopes. Should flow become unconstrained at the base of slopes, diverging alluvial fans develop, which results in finer and rounder sediment along the toe of fans (Harvey 2011). If flow remains constrained, valley sides or flood plains accumulate terraces through successive episodes of fluvial cutting and filling. Accumulations of fine alluvial sediments may line the valley sides and extend towards river endpoints, including inland basins. It is here where finer material can

Table 1.1 Major desert environmental characteristics

Geomorphology	Substrate	Hydrology	Radiation	Surface Temperature (max, min, and range)	Vegetation (Main source soil organic matter)	Macrofauna	Nutrient dynamics	Substrate processes
Rock outcrops, inselbergs, uplands	Very stable, bare rock, gravel, colluvial or accretion fans/ slopes	Limited surface water retention, rapid infiltration and storage in cracks, aspect, and size/ orogenic effects	Variable (affected considerably by aspect, slope, and surface roughness)	Variable (dependent on colour, aspect, slope, surface roughness, altitude, wind exposure)	Diverse (cremnophilous, refugial, cryptic and specialist species), expect higher diversity and aspect differences	Petrophilous, refugial and herbivorous specialists, niche specialisation, high species to area diversity ratios	Weathering, organic trapping	Weathering, colluvium, accretion, limited pedogenesis
Mixed plains	Stable, desert armour, gravel, sometimes calcrete, gypsum	Rapid surface desiccation, shallow storage of available moisture	High (affected by degree of vegetation cover, surface colour)	High (ameliorated by pebble size, lighter colours, vegetation cover)	Mainly grasses, sparse dwarf shrubs, annuals, geophytes, and succulents	Fossorial, cryptic, vagile and dormant species common, low species to area diversity ratios	Trap dust and nutrients	Aggregation, accretion, pedogenic cementation
Sandy plains, sand sheets, dunes	Unstable to very dynamic sand, Aeolian surface saltation and abrasion	Rapid surface desiccation, porous, store water at depth, water vapour absorption near surface	High (affected by degree of vegetation cover, aspect, sand colour)	Very high due to small grain size (ameliorated by lighter colours, vegetation cover, aspect)	Mainly grass and geophytes, occasional specially adapted succulents or shrubs	Psammophiles, fossorial species, high incidence of omnivores and generalist scavengers, very low species to area diversity ratios	Poor in nutrients and soil organic carbon	Accretion

(continued)

Table 1.1 (continued)

Geomorphology	Substrate	Hydrology	Radiation	Surface Temperature (max, min, and range)	Vegetation (Main source soil organic matter)	Macrofauna	Nutrient dynamics	Substrate processes
Playas, river end points, saline basins	Very stable, sandy, and silty	Porous, moisture often salty and close to the surface/ shallow surface water	Very high, reflective, no shade	High (ameliorated by reflection and evaporative cooling)	Halophytes, specially adapted species	Habitat/Niche specialist (halophiles) and transient spp.	Very poor, high Na Mg content	Aggregation, evaporites
Drainage systems	Mostly stable, sandy, and silty alluvium, downstream changes depending on gradient and frequency and magnitude of discharge	Porous, may hold and retain water at depth (fossil water becomes salty), infrequent and potentially high magnitude floods/ discharge	Variable (high in open channel, ameliorated by vegetation, width to depth ratio of channel, slope, and direction of banks)	Variable (ameliorated by substrate grain/pebble size, substrate colour, vegetation, width to depth ratio of channel, bearing)	Phreatophytes, shrubs and other perennial species, herbaceous opportunists	Linear oasis providing refuge to local, adventive and transient spp., high diversity of poorly adapted spp.	Traps organic materials and other nutrients, higher autochthonous organic deposition	Alluvium, accretion, aggregation, erosion

Fig. 1.7 Examples of desert landscapes and landforms. Significant variation in topography, substrate, habitats, and soils is evident. Depicted are (**a**) rocky uplands and a dry river, Oman, (**b**) sand dune on buried fluvial sediments, Namibia, (**c**) saline pan crusts, USA, (**d**) gravel plain, desert pavement, and vegetation polygons, Namibia, (**e**) incised fluvial sediments in badlands, Spain, (**f**) loess silt deposit, China (credit: F. Eckardt)

be preferentially sorted and deflated by aeolian processes, resulting in the accumulation of coarser material, such as sand, and the loss of fine material such as silt, which provides atmospheric dust. Drainage endpoints also have the tendency to accumulate sand, prone to saltation and creep, which may form a range of dune types, depending on a variety of wind regimes and the amount of sediment supplied (Lancaster 2002). Large sand deposits and active sand seas tend to occupy low-lying continental basins but may also be found along coastal margins close to river mouths, where the distinction between coastal and inland dunes becomes blurred. Large sand seas are particularly prevalent in the Sahara, Southern Africa, Asia, and Australia and to a lesser extent in the Americas (McKee 1979). The associated dust generated from saltating sand and abrasion processes may undergo transport over a variety of distances, accompanied by the dispersion of silt, along with bioavailable nutrients and salts. The deposition of vast quantities of fine-grained silt-sized dust results in the accumulation of loess, which is globally widespread, often covered by vegetation, but also exposed on desert margins and interiors (Pye 1995). Loess found in desert margins is fertile, but easily eroded by wind and water.

Rocky surfaces provide very limited opportunities for retaining water; hence, highland catchments are very quick at converting intense convective rain into flash floods. Upland regions have a limited ability to retain water; however, deep sedimentary valley fills are able to sustain groundwater into the dry season and beyond. Lowland areas, especially, accumulate sediments, but may even store shallow groundwater and occasional surface water. Here, salts may accumulate at discharge points and seeps, around inland basins and playas (Shaw and Bryant 2011), given the high evaporation rates of shallow groundwater tables. Coastal salt pans, or sabkhas as they are also known, are widespread in the Persian Gulf coast and beyond and, despite their proximity to the ocean, often accumulate evaporation products derived from inland subsurface run-off, rather than seawater evaporation (Wood et al. 2002). Salts are only sustained at the surface in the driest of deserts, as any rain quickly results in a solution of most evaporation products. Dry saline surfaces at inland playas and coastal sabkhas are prevalent in the broad valleys of the Americas and Asia or shallow basins of Africa and Australia, as well as associated coastlines. The remaining desert surfaces may be quite unremarkable and include vegetation-covered sand deposits and dunes, or deflated lag gravel deposits. However, here stability may not only result in the accumulation of sediments but also the formation of soils. These varied environments provide a wide range of opportunities for creating habitats and may locally be subject to run-off and run-off which may hinder or facilitate patches of soil formation.

1.3 Desert Habitats

1.3.1 Introduction

The desert landscapes described in the previous section may appear to be barren and devoid of life. However, they provide a range of habitats that are largely determined by the abiotic environment, e.g. climate, topography, and soil properties (Gibson 1996) and support a host of organisms. Desert biota are adapted to survive a scarcity of resources like water and nutrients, of which the spatial and temporal concentration and availability is particularly pronounced in deserts (Shmida et al. 1986), and to exploit brief periods of abundant resources (Seely and Louw 1980). The most defining variable for biodiversity is low, erratic, and unpredictable precipitation that determines water relations and nutrient mobilisation. Understanding the spatial patterns of the biotic environment, and how these change over time, is a key element for characterising nutrient pools in deserts. Biogeochemical cycling is an important driver of both the evolution and persistence of highly adapted, unique, and diverse assemblages of macro- and microorganisms able to flourish in such extreme arid conditions. Habitats are both a result and a feature of landscapes. However, the geomorphic processes introduced in the previous section are the primary mechanism shaping desert habitats. Table 1.1 depicts the relationship between geomorphic units and habitat descriptions listed here.

1.3.2 Uplands, Inselbergs, and Rocky Outcrops

Desert mountain and hill profiles often transition abruptly from the gentle slopes of desert plains (Fig. 1.8). Although sparsely vegetated, these mountainous landforms nevertheless support a much more diverse and abundant biota than the surrounding plains, due to a greater range of microhabitats associated with hydrologic and geohydrologic moisture regimes. Desert mountains host high levels of biodiversity and endemism (Barthlott and Porembski 2000) due to their microhabitat diversity and insularity. Other important influences of mountains on the fauna and flora include size, elevation, topography, orientation, degree of isolation, climatic effects, as well as lithology (Kruckeberg 2002). Spatial complexity of mountainous habitats is particularly noticeable in deserts, primarily due to the influence of microtopographic heterogeneity on moisture regimes. Generally, there is a decrease

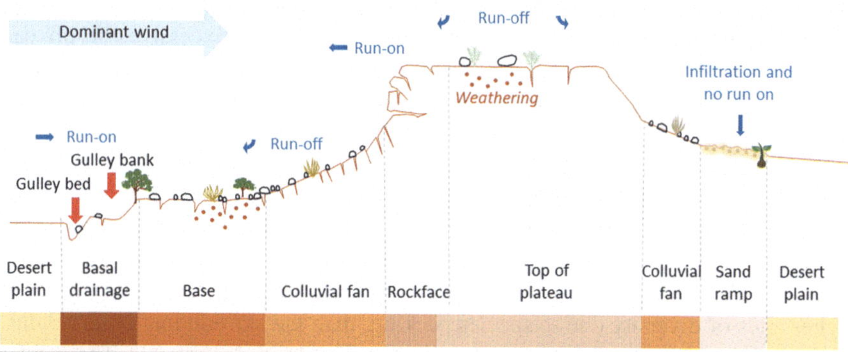

Basal drainage	*Pheatophytes, perennial shrubs, sedges and grasses (ecotone)*
Base	*Desert plains species, pheatophytes (ecotone)*
Colluvial fan	*Phreatophytes, petrophiles, desert plains species, inselberg endemics, cryptic, endogenic and hypolithic species, poikilohydric cryptogams*
Rockface	*Cremnophytes, chasmophytes, petrophiles, refugial species*
Top of plateau	*Chasmophytes, petrophiles, refugial species, inselberg endemics*
Sand ramp	*Psammophiles, geophytes*

Fig. 1.8 Habitats associated with uplands, inselbergs, and rocky outcrops (modified from Shmida et al. 1986)

in the numbers of species of organisms with increasing altitude, although vegetation zonation on desert mountains is not always distinct. Flat mountain tops serve as a sump for moisture, as well as weathered substrate material. As a result, they offer unique habitats for specialised and endemic organisms and are often refugia for relict taxa (Rahbek et al. 2019).

Compact rock surfaces or steep rock faces on the windward side, with high rates of run-off, are less favourable for plant growth, and only hardy lithophytes, including lichens, can flourish. Elsewhere, seedlings of chasmophytes establish opportunistically (Hegazy and Lovett-Doust 2016) in soft materials and moisture that accumulate in fissured rock surfaces and crevices found in sheltered pockets and lee slopes. The microbial ecology of endolithic microbial communities, i.e. those colonising and inhabiting rock pores and fissures, from hot deserts is described in Chap. 5. Pendulous, cremnophilous plants can be found clinging to rock faces (Fig. 1.8). Slope and aspect further influence vegetation, as exposure to the sun and other elements could result in distinctive plant communities.

Colluvial fans on lower, less steep slopes typically have loose boulders and rocky lag material, and accumulate water and soil washed down from steeper slopes (Fig. 1.8). This rockiness enhances species diversity as rocks partition the habitat into more microniches. Biota associated with more mesic environments may occur, but these are often miniaturised. Even in extreme arid areas, these slopes support poikilohydric cryptogams and resurrection-type phanerogams that can respond to irregular moisture pulses.

Run-off from desert slopes provides sufficient moisture to maintain vegetation in run-on piedmont areas, particularly in basal drainage systems, where infiltration often results in high spatial patchiness. Infiltration is enhanced by dense root clusters where shrubs support an understory of grasses and herbs. Excessive run-off can cause erosion that strips surface soils, forms desert pavements, and deepens basal washes (Ward 2009), denuding vegetation cover except for deep-rooted phreatophytes. Sand ramps that may develop on either side of the mountain through aeolian action offer a stable substrate to support psammophilous herbs and geophytes, as well as suitable habitat for fossorial animals.

The high incidence of cracks and crevices, as well as deep hollows formed by weathering, provides abundant shelter for endogenic and hypolithic organisms (Chap. 9). The vegetation, complex topography, and habitat heterogeneity result in higher rates of diversity due to species packing, than the surrounding desert plains, particularly in higher vertebrates such as reptiles, birds, and mammals. The greater abundance of resident or seasonal macrobiota contributes additional types of habitat and nutrient supply, such as accumulations of nest materials and excreta.

1.3.3 Drainage Systems

Ephemeral streams and rivers are the dominant fluvial system in desert and quite common. These accommodate run-off from desert precipitation with a high spatial and temporal variability. Even low precipitation events result in run-off from

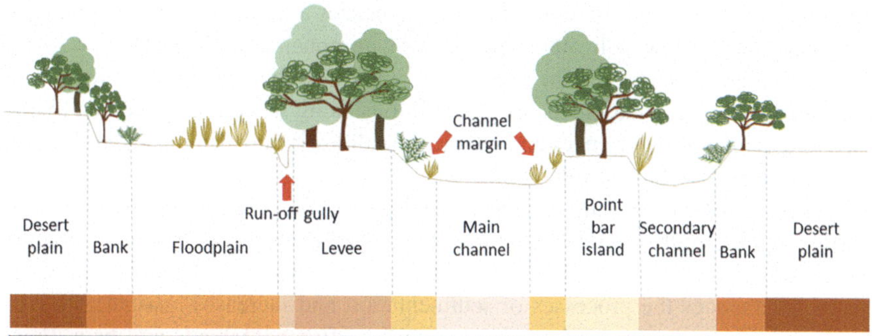

Bank	*Phreatophytes*
Floodplain	*Halophiles, grasses, sedges*
Levee	*Phreatophytes, perennial woody species, adventives*
Channel margin	*Phreatophytes, grasses, sedges, adventives*
Main channel	*Depauperate*
Point bar island	*Phreatophytes, perennial woody species, adventives*
Secondary channel	*Phreatophytes, psammophiles*

Fig. 1.9 Habitats associated with drainage systems

outcrops, and undulating topography (Fig. 1.9). The main catchments for such ephemeral streams often encompass mountainous or rocky areas, where orographic uplift results in higher precipitation, and steep impervious terrain produces more dynamic hydrological regimes. Rain may also originate in more mesic regions before cutting across or terminating in the desert (Shaw and Cooper 2008). The drainage systems often form linear oases in more inhospitable parts of the desert, providing avenues for migration and dispersal of organisms from adjacent, more mesic areas, or refuges for desert organisms during periods of greater stress.

The hydrological regime typically alternates between wet and dry phases, with short periods of seasonally intermittent surface flows, followed by long periods of inactive streambeds, with some subsurface groundwater flow. The fluvial conditions during wet phases are often characterised by punctuated, high-magnitude flood events, followed by periods of low-flow or no-flow conditions with hysteresis and drying out of associated habitats (Von Schiller et al. 2017). Hysteresis heralds the transition to prolonged dry conditions, dominated by arid terrestrial processes. The surface topography along the course of streams and the intensity of wet phases result in considerable variation in stream types, channel geomorphology, vegetation, and flood impacts (Hooke 2016). These features tend to be spatially discontinuous featuring highly dynamic lateral, vertical, and longitudinal connections with their adjacent ecosystems (Hooke et al. 2005).

The hydrological and sediment transport processes associated with episodic flood events are the drivers of geomorphic, vegetation, habitat, and biochemical variation in ephemeral rivers. The episodic floods trigger a switch from terrestrial conditions to aquatic and semi-aquatic ecosystems, although high spatial and temporal variability of flows, and the state of a system, causes a great deal of uncertainty in how a particular flow will affect the state of an ephemeral river (Hooke 2016). Floods may cause changes in the prevalent geomorphic and vegetation structure and associated ecological and biochemical conditions along its course, governed by the threshold values of those episodic hydraulic conditions; thus, similar sized flows in different years may have different effects.

A characteristic of dryland streams is the degree of complexity and irregularity in morphology along their course, coupled with abrupt changes in patterns of riparian vegetation (Sandercock et al. 2007). The general effect of vegetation in these channels enhances the processes of sedimentation and increases resistance to erosion, particularly the establishment and stability of embankments and bars. Point bars, which may develop into in-stream islands (Fig. 1.9), and slack water areas are major sediment and nutrient storage environments (Ringrose et al. 2018). Alluvial sediments at bars and islands may vary considerably in grain size, with less coarse sediments along riverbanks and levees and accumulation of fine-grained clay in slackwater deposits, such as floodplains (Sandercock et al. 2007). Such variations in grain size have considerable effects on the hydrological and nutrient dynamics of ephemeral river habitat assemblages (Von Schiller et al. 2017).

The primary producers associated with desert drainage systems are predominantly phreatophytes that commonly share similar traits for coping with rapid changes in water table levels and long periods of desiccation (Sabater et al. 2017). These traits include morphological characteristics such as load-spreading as well as deep root systems to stabilise banks and access moisture and nutrients at different levels. Shrubs and trees closer to the main discharge channel often have distinct traits such as stem flexibility, or a reduced number of stems and low-level branching, to cope with high-intensity dynamic flow (Shaw and Cooper 2008). Vegetation on more infrequently flooded point bar islands or river embankments often have a greater stem and branch cross-sectional area (blockage ratio) closer to the ground (Sabater et al. 2017). Most of these phreatophytes are likely to have adaptations to contend with other physiological stresses such as evapotranspiration, changes in salinity, and high surface temperatures (Sabater et al. 2017). Slackwater habitats, or flood-out areas, are dominated by sedges and grasses, with occasional salt-tolerant phreatophytes. They experience the lowest disturbance, but greatest variation in hydrological and biochemical stress, often with extended periods of either slightly saline, desiccated hardpan playa or desert plain-like conditions, with high diurnal temperature variation, or waterlogged, semi-aquatic wetland conditions.

This results in highly heterogeneous systems with spatially complex but consistently predictable patterns in the structure and composition of perennial woody plants along ephemeral stream channels, and grass-dominated communities on flood-out peripheries. The perennial shrub and tree species associated with these habitats contribute consistent primary production, organic matter, nutrient storage, and an abundance of structural diversity and shelter that sustains a diversity of fauna not usually associated with desert ecosystems.

1.3.4 Saline Basins, Playas, and River End Points

Inland saline habitats in deserts typically form within high-evaporation basins and lowlands in proximity to shallow groundwater (Fig. 1.10). The high salinity, extremely toxic to most plants, in combination with high temperatures, exerts a strong influence over the vegetation found in these habitats. Xerohalophytes are specifically adapted to thrive in desert areas with high salt concentrations (Gibson 1996). This is a heterogenous assemblage of plants that have evolved structural, phenological, physiological, and biochemical mechanisms for salt tolerance or salt avoidance (Ward 2009), e.g. some accumulate salt in specific organs and are often succulent, while others excrete salt.

Hypersaline, saline, and subsaline zones can be distinguished (Fig. 1.10). There can be a gradual reduction in salinity away from the point sources, the lowest elevations in the basin, due to increasing depth of subterranean water. Plant diversity and abundance increase away from the point source (Halis et al. 2012), and irregular bands of progressively denser plant growth can be observed. Vegetation is absent where soil salinity is very high, particularly where salt crusts are visible on the surface. The point source is often surrounded by an unvegetated expanse prone to occasional or seasonal run-on flooding and waterlogged conditions. These areas are typically inhabited by salt-tolerant extremophiles and occasionally visited by highly mobile transient visitors attracted by moisture. If the water table is high and close to the surface, the lowest areas will be the most saline, but the same adaptions apply.

The inner edge of vegetation, at the outer margin of flooding, where the salt concentration is lower, is usually composed of rhizomatous grasses with shallow

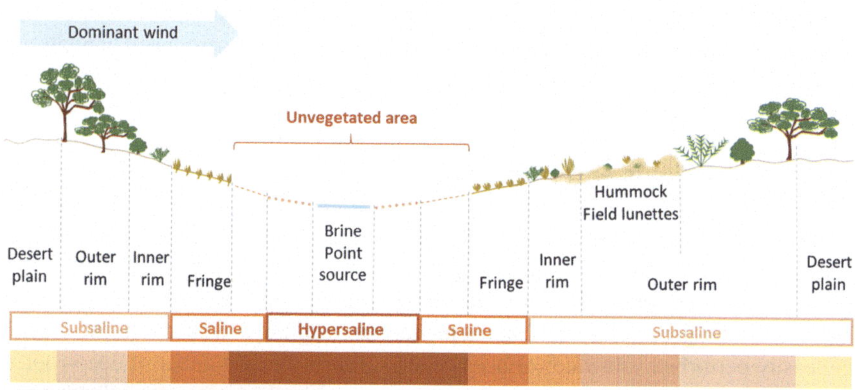

Brine point source and surrounds	*Unvegetated area, halophile aquatic and subaquatic species*
Fringe	*Halophytic grasses and sedges, grazers*
Inner rim	*Halophytic chamaephytes*
Hummock field lunettes	*Halophytic tussock grasses, psammophiles, endogenic species*
Outer rim	*Halophytic to salt-tolerant tussock grasses, shrubs, succulents, phreatophytes (ecotone)*

Fig. 1.10 Habitats associated with saline basins, playas, and river endpoints

roots that form dense mats and sedges. This distinctive inner fringe of vegetation morphs into the outer zones. Halophytic chamaephytes, represented by various perennial herbs and subshrubs, thrive in the inner subsaline zone. On the outer edge, a denser zone of shrubby vegetation, tussock grasses, and scattered woody phreatophytes is found (Fig. 1.10). Halophytic and non-halophytic species can occur together in this zone.

Sand can cover all low-growing vegetation (Danin 1996), while shrubs in the outer edge trap airborne sand grains on the leeward side of the depression to form hummocks (nebkas/hillocks). Unique microhabitats are created in the accumulated sand and hummock fields that are distinctive features around saline depressions. Over prolonged arid periods these hummock fields may grow into substantial lunette dune systems.

In less saline systems, such as freshwater playas and the endpoints of ephemeral rivers in endorheic basins or against topographical obstructions, very similar zones of progressively denser and more diverse vegetation develop. The inner unvegetated mudflats, subjected to occasional flooding, are fringed by a zone of shallow-rooted vegetation. This inner margin transforms into an outer zone of progressively higher and more diverse shrubs and phreatophytes. Waterlogged areas at endpoints of perennial or semi-perennial rivers can support a few dominant species of monocot perennials like reeds and sedges that form dense stands inhibiting the germination of other species. A special set of habitat types occur at oases, where subsurface flow resurfaces due to geological phenomena such as dykes. At such locations submerged and emergent macrophytes and isolated stands of dense relict or adventive phreatophytes may be found. Many oases are today associated with human habitation, and natural vegetation has been replaced by cultivated plants or weeds. More cosmopolitan, wide-ranging species occur in salt marshes and other saline coastal habitats (Shmida et al. 1986). However, these ecosystems and associated habitats are not representative of typical ecological processes in hot deserts.

1.3.5 Sand Plains and Sand Dunes

Aeolian sand plains and dune systems (Fig. 1.11) represent only some 20% of deserts (Pye and Tsoar 1990). These sand biotopes are dominated by a low diversity of psammophytes, predominantly grasses (Bowers 1982). Where water relations improve, some geophytes and chamaephytes that can cope with the nutrient-poor and easily perturbed substrate surfaces occur, with a limited number of deep-rooted perennial phreatophyte species where local conditions allow access to subterranean water. However, even relatively modest increases in rainfall have a positive feedback effect on vegetation and psammophilous biota (Boyer 1989).

Aeolian sand substrates have poor water-holding capacity, rapidly absorb water, are nutrient deficient, have low levels of organic content, and are usually composed of well-sorted, fine-grained particles without much clay content. Most psammophilous plant species require rainfall for germination and seedling establishment; thus, there is a direct correlation between vegetation and rainfall. In sand

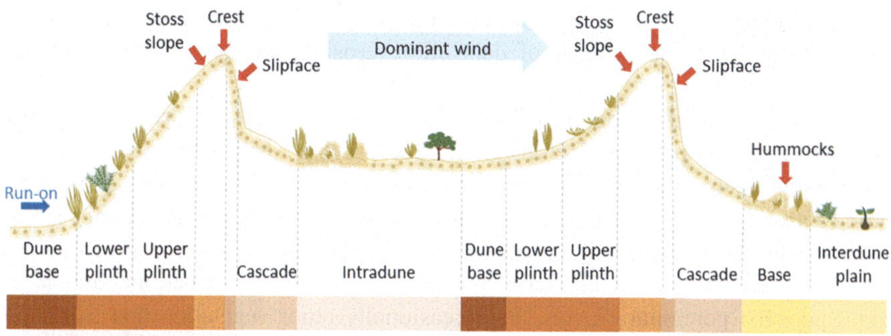

Dune base	*Psammophiles (dominant), occasional pheatophytes, endogenic species (ecotone)*
Plinth	*Psammophiles, endogenic species*
Stoss slope	*Depauperate*
Slipface	*Endogenic, psammophile specialists*
Cascade	*Psammophiles (dune specialists), endogenic species*
Intradune	*Psammophiles, occasional perennial phreatophytes, endogenic species*
Base/Hummocks	*Psammophiles, chamaephytes, endogenic species*
Interdune plain	*Psammophiles, geophytes, grasses, annuals and amphiphytes, endogenic species*

Fig. 1.11 Habitats associated with sand plains and sand dunes (modified from Robinson and Seely 1980)

biotopes, vegetation also contributes organic matter and assists in stabilising surfaces that promote the development of microbiotic crusts (Danin 1996), while increased aeolian activity and other forms of mechanical erosion such as bioturbation or excessive trampling have an inverse correlation with vegetation cover and surface stability (Danin 1996). Non-rainfall moisture, such as dew and fog, only has a notable effect on macrobiotic life once vegetation has established through rainfall.

Short-lived, episodic wet conditions often transform stable sand surfaces such as sand sheets, sandy plains, and dune bases into lush stands of therophytes (Seely and Louw 1980), sprinkled with the flowers and leaves of cryptic geophytes that emerge when conditions are favourable. Even during such highly productive episodes, the species composition typically consists of monocultures of relatively few species. These temporal changes may, however, trigger large-scale emergence of the reproductive stages or migration of a surprisingly high diversity of herbivores and secondary consumers. As is the case with perennial vegetation, therophyte abundance, species richness, and cover are positively correlated with rainfall.

The surface topography of sand biotopes is affected by soil moisture and vegetation cover. Sand sheets and plains are usually uniform and relatively flat. Perennial vegetation occurs at distinct locations where water relations and surface stability allow colonisation, but plant cover is mostly too low to define communities. Where it does occur, low-growing perennial vegetation captures wind-blown sand and increases structural complexity through hummock, coppice, or nebkha formation. The canopy density, height of the canopy above the surface, and number of stems impose changes on the velocity and direction of wind and associated sand movement, which determine patterns of deposition, erosion, and growth in hummock fields.

In contrast, sand dune habitats (Fig. 1.11) have a typical structure that imposes order and zonation patterns (Robinson and Seely 1980; Seely 1990) on associated psammophilous macrobiota. The transition area at the dune base is often defined by more robust stands of perennial grasses and occasional phreatophytes (Danin 1996; Jürgens et al. 1997) exploiting the internal gravity flow of moisture absorbed by the dune sand. Annual grasses and herbs may also emerge at this transitional area after rain. Specialist perennial grasses, and occasionally other vegetation that can cope with erosion and abrasion by wind-blown sand, may colonise the plinth, becoming progressively sparser higher up. The same perennials are often also adapted to recover from burial and occur in the lee, at the base of the cascade, where they are subjected to sand cascades and slumping. The root systems of these grasses often have well-developed rhizosheaths (Marasco et al. 2018) to cope with the very low nutrient content (Buckley et al. 1986) in desert dunes. These perennials, augmented by therophytes, some geophytes, and occasional phreatophytes, may also be found further down the base or in depressions between successive crests of complex or braided dunes. Only transient psammophiles occur on the highly dynamic and severe abrasion zone near the crest on the stoss slope, though specialised biota may exploit the slipfaces and the cascades and slumping features in the lee of a dune crest. This zonation becomes more pronounced and distinctly colonised with increased rainfall, ascribed to differential stresses in the internal moisture regime of dunes, surface stability, interspecies competition, and abiotic stressors (Yeaton 1988).

1.3.6 Desert Plains and Pediplains

Desert plains are the most common biotope of deserts (Fig. 1.12). These extensive landscapes are usually sparsely vegetated with low-growing, arid-adapted, long-lived dwarf shrubs and subshrubs (xerophilous chamaephytes). The substrate is relatively level and stable, commonly protected by desert pavement. Desert or stone pavement typically consists of a dense cover of larger pebbles and stones embedded in a finer matrix, forming a coarse mosaic on the surface (Cook et al. 1993), restricting moisture penetration and consequently further limiting plant productivity. The shallow stony soils frequently have a subsurface rock-like horizon, or basement, which is water-repellent and which plant roots find difficult to penetrate (Gibson 1996; Walter 1983).

Grasses are the dominant plants across desert plains where summer rainfall prevails. Some grasses may be annual or perennial (amphiphytic) depending on local conditions. Perennial grasses can remain dormant for many years until favourable conditions return. In contrast, annual grasses and herbs (therophytes) live for one growing season and survive as seeds until the following growing season. Usually occurring in mixed-species groups, these ephemerals can transform plains into fleeting floral carpets after a moisture pulse. Usually, the ephemeral vegetation

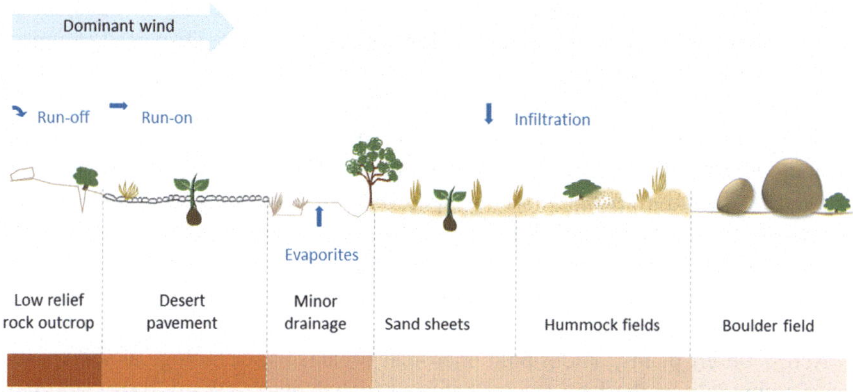

Low relief rock outcrop	*Petrophiles, chasmophytes, pheatophytes, cryptic species*
Desert pavement	*Petrophiles, therophytes, geophytes, epilithic, hypolithic, and cryptic species*
Minor drainage	*Phreatophytes, endolithic species*
Sand sheets and hummock fields	*Psammophiles, geophytes, endogenic and cryptic species*
Boulder field	*Petrophiles (epipetric, hypopetric, endopetric)*

Fig. 1.12 Habitats associated with desert plains and pediplains

is unevenly distributed in rivulets and depressions, while a patchy occurrence may be due to very localised rain showers. Features differ between summer and winter annuals, e.g. plant height and leaf morphology (Ward 2009), as well as species composition.

Where they occur, boulder fields and localised rock outcrops (Fig. 1.12) provide moister, cooler microclimates for plants growing around and between the boulders and in rock crevices. Low-relief bare rock surfaces are kept clear of sand and gravel by the wind and support low-growing plant life, often cryptic and succulent, rooting within crevices. Some rock types with the capacity to hold water at shallow depths, such as sandstones or exfoliating granite, support a different flora than the surrounds (Gibson 1996).

Geophytes, many of which produce leaves and flowers at different times (hysteranthous), are a common lifeform in desert plains. In sandier areas, geophytes may have contractile roots drawing the underground organs deeper into the substrate, a strategy triggered by erosion, predation, and soil disturbance. Extended flats of gravel plains accumulate wind-borne materials (Hegazy and Lovett-Doust 2016) in surface sheets. As the sheet becomes deeper, more plant species appear, and support psammophiles that can resist sand-blasting or endogenous faunal species.

Dense, dome-shaped chamaephytes are found in wind-swept areas, with canopies close to the soil surface. This growth form minimises direct exposure of the foliage to sunlight (Hegazy and Lovett-Doust 2016), traps organic debris, and inorganic particles and provides moister shaded microclimates for other taxa. In areas with greater sand movement, shifting sand is often deposited where grass tussocks or the lower branches of woody and succulent chamaephytes inhibit aeolian transport to form wind-heaped hummocks around and downwind of plants. Dispersed summer-active herbaceous, geophytic, and low caudiform vines escape the strong dry winds associated with hot deserts by hugging the surface.

Due to their low-relief and poorly developed topography, shallow drainage channels are a notable feature of most desert plains. These channels are often visible as linear vegetative patterns along their extent. Chamaephytes and woody phreatophytes are commonly found along their edge, where subsurface moisture can be directly tapped. Plant growth within the channel is often constrained by exposed evaporites and episodic water flows and may be absent from channels with frequent or high-energy flows.

Across desert plains, autotrophic microbial organisms, or biological soil crusts, occur commonly in the bare spaces between higher plants (Ward 2009). As they increase infiltration and nitrogen fixation, reduce erosion, and contribute to soil organic matter, they contribute positive local effects on plains with sparse vegetation.

1.4 Desert Soils

1.4.1 Semi-arid and Arid Soils

Given the lack of water and leaching, very shallow or absent weathering fronts, limited chemical alteration, and clay production, very little soil formation takes place in most drylands. Soil formation is more pronounced on desert margins and stable surfaces away from steep uplands and dunes. Several soil types can nevertheless be found (Dunkerley and Brown 1997) and include entisols, which are widespread, but are merely unaltered sediments with little sign of post-sedimentary modification, occurring at the heart of continental desert regions, such as the Sahara and in Australia. Aridisols are slightly more evolved and depict very limited leaching or evaporation, and can promote the minor accumulations of calcium, silica, gypsum, and salts in their profile. Mollisols are associated with plains and grasslands, particularly in North America and Central Asia. They may feature a thin organic layer, along with calcium accumulations, and are fertile, but also prone to deflation, if stripped of their vegetation cover. Mollisols in part produced the dustbowl, during the great depression in the USA (Lee and Gill 2015). Alfisols, the most evolved dryland soils, are located in the semi-arid margins, towards temperate and equatorial regions in Europe, South America, Africa, and Australia. They are moderately leached, featuring a clay-enriched subsoil, which also leads to a higher productivity, but are also prone to erosion by water (Cogle et al. 2002). Opportunities for microbes in desert soils exist especially in low-lying, run-on areas which may accumulate sediments, weathering products, nutrients, and water and with time may form a subsurface habitat.

1.4.2 Introduction to Unconsolidated Porous Media

There are several advantages for microbial life below the surface (Table 1.2), since temperature fluctuations are less pronounced and water in its liquid form may be available and has the ability to retain nutrients. Especially an unconsolidated porous subsurface media and associated pore geometry provide biochemical reactions and interactions, which shape the microbial habitat while providing shielding from UV radiation. It has been shown that pore geometry, access to organic material, and microbial diversity are linked (Rabbi et al. 2016). However, prolonged water evaporation may promote salt accumulation, which can create more extreme conditions in shallow groundwater settings also (Fig. 1.10). Abiotic properties of unconsolidated porous media are found in Table 1.2.

Table 1.2 Fundamental abiotic properties of unconsolidated porous media which moderate the subsurface environment

	Property	Consequence for microbial life
Physical	Low UV radiation transmittance	Protects against solar radiation. UVA and UVB radiation penetration varies between minerals; however, it is typically less than 0.5 mm (Carrier et al. 2019)
	Poor conductor of heat	The temperature regime below the surface is less extreme than at the surface (Hillel 2004)
	Retains water either through surface adsorption or capillarity This ability is enhanced with decrease in particle size (increase in surface area)	The propensity of water to change phase decreases compared to free water. More energy is needed to evaporate water from a porous medium If a sufficient number of micropores are present, the medium will exhibit the ability to retain water against gravity (Hillel 2004)
Chemical	All mineral surfaces carry charge Variable or pH dependable charges are more common; however, a select group of 2:1 phyllosilicates additionally have permanent negative charge in their crystal structures The more finely divided the material is, the greater the sum surface charge per mass of material	Chemical reactivity is imparted to the medium This, in turn, results in mineral surfaces that: 1. Interact with polar constituents, for example, cations, anions, water, and other mineral particles (Essington 2015) 2. Induce surface precipitation 3. Interact, transform, and protect organic compounds (Kleber et al. 2005, Barré et al. 2014) The net effect of abovementioned is the ability to accumulate life essential elements

1.4.3 The Desert Pedoderm

Desert sediments and soils provide a critical but marginal microbial habitat. We will focus on the "pedoderm", a term originally proposed by Brewer et al. (1970), in reference to the top few millimetres or centimetres of the terrestrial surface. It defines a distinct microenvironment that is physically, chemically, and mineralogically diverse (Fey et al. 2006). The pedoderm pore geometry regulates gaseous exchange as well as water and nutrient acquisition, storage, and movement (Young and Crawford 2004; Rabbi et al. 2016) and requires relatively stable conditions to form and prevail. The following section will examine the physical, chemical and biological characteristics associated with the desert pedoderm, which vary significantly across landscapes and landforms and would be more developed on stable run-on surfaces (Burket et al. 2012) (second column of Table 1.3).

The bare single-grained pedoderm (S) (Table 1.3 and Fig. 1.13b) is associated with a high content of sand-sized particles found in dunes and sandy plains (Fig. 1.11). Due to high rates of mobility, no crusting or aggregation are observed. The formation of WSA requires colloidal organic material, iron oxides, and secondary clay minerals to bond larger particles into stable units (Tisdall and Oades 1982; Fernández-Ugalde et al. 2013). As a result, WSA are more common on dryland margins. Clay minerals, associated with such intermediate stages of weathering (so-called 2:1 clays), include smectite, vermiculite, and chlorite mineral groups (Essington 2015). However, smectite and vermiculite are also particularly susceptible to dispersion by cations like sodium (Na^+), because of their high cation exchange capacity (CEC) (Itami and Kyuma 1995; Ruiz-Vera and Wu 2006) and hence prone to decay. Water-unstable aggregates (WUA) with a high silt content lack the cohesion of clay and iron oxides (Shainberg et al. 2003; Warrington et al. 2009) and are common in drylands.

Salt crusts (SC) occur, where net evaporation is very high and often results in the formation of physical crusts (PC). These may entail a biological component as well as hydrated salts. Desert pavements (DP) on the other hand offer among the hardest protection layers, the result of silt deflation and pebble armouring at the surface. (Fig. 1.8). Clearly, factors which support a stable pedoderm are patchy in dryland areas.

Biological soil crust (BSC) can also establish on surfaces and decrease water infiltration. The microbiology of BSCs is described in Chap. 3. This is attributed to surface densification and the entrapment of silt-sized particles over coarser underlying material (Felde et al. 2014) and the development of hydrophobicity when polysaccharides, excreted by cyanobacteria, dry out (Mazor et al. 1996; Eldridge et al. 2000). The flora of BSC can be diverse ranging from cyanobacterial dominated crusts (BSC_{cyan}), crusts with multiple species (BSC_{muli}), to algal dominated BSC (BSC_{alg}). Additional pedoderm classes are worthy of note and include those with rock mulch (RM) trapped between large rocks, exposed cemented horizons (CEM), and erosion pavements (EP), the result of denudation (Table 1.3).

Table 1.3 Pedoderm classification of arid environments and their possible genesis

Pedoderm classes of Burket et al. (2012)		Possible genesis
1. Characteristics the result of abiotic processes	*Bare single-grained soil (S)*. The pedoderm consists of particle in the sand fraction range and exhibit negligible particle aggregation.	The result of limited weathering and/or the net removal of finer particles by wind or /and water. Sand plains and sand dunes (Fig. 1.11)
	Water-unstable aggregates (WUA). Aggregates at the soil surface which easily disintegrate upon wetting and/or raindrop impact	Aggregates lack strong bonding agents. Associated with high silt and low organic material content. Slaking susceptibility increases, with increasing Na^{+} content
	Water-stable aggregates (WSA). Well-formed or distinct aggregates at the soil surface which does not slake upon wetting	Combination of clay, iron oxides, and organic material results in aggregates that exhibit resistance against physical and chemical dispersion. The dominance of flocculating cations (e.g. Ca^{2+}) rather than dispersive cations is also a feature
	Rock mulch with stable soil (RM). Material is trapped and protected by closely spaced and partially embedded rock fragments	Rough surfaces on windward slopes improve dust and nutrient acquisition run-off will result in particles sorting, possibly favouring the removal of colloidal-sized particles. RM is most likely associated with uplands, inselbergs, and rocky outcrops (Fig. 1.8)
	Physical crust (PC). Usually platy or massive, no substantial biological component	The disintegration of especially WUA that leads to in situ structural crust development in areas where run-on occurs. Aggregate breakdown driven by raindrop kinetic energy transfer and wetting rate (mm h^{-1}) (Agassi et al. 1981; Shainberg et al. 2003)
	Cemented horizon exposed at surface (CEM)	Areas that experience run-off, erosion exposing petrogypsic, petrocalcic, petroduric horizons
	Salt crust (SC) of fine to extremely coarse evaporite crystals or visible whitening	Areas experience where shallow water table is present.

(continued)

Table 1.3 (continued)

Pedoderm classes of Burket et al. (2012)		Possible genesis
	on the soil surface. May include biological components	Saline basins, playas, and river endpoints (Fig 1.10)
	Erosion pavement (EP). The erosion of finer soil material left rock fragments that occur in a dense uniform pavement. Individual fragments may be displaced during run-off events	Run-off areas that experience net loss of material
	Desert pavement (DP). A concentration of closely packed and varnished rock fragments at the soil surface, embedded in a vesicular crust	Vesicular crust develops as a result of dust acquisition; however, platy structure is also associated with it. Low hydraulic conductivities are the norm because of these features (Turk and Graham 2011). DP is associated with desert plains and pediplains (Fig. 1.12)
2. Characteristics the result of both abiotic processes and biological soil crusts (BSC) formation	*Physical–biological crust (PBC)*. Few cyanobacterial sheaths with no darkening from cyanobacteria	Currently unclear if nutrient and fine particle enrichment associated with BSC is solely the result of particle acquisition and entanglement. Or if BSC establishes preferentially where nutrient and fine particle enrichment in the landscape occurs. Then afterward, BSC aid in preventing fine particle export However, with increased dominance, the stronger will the influence of BSC be on the pedoderm properties.
3. Characteristics predominantly the result of BSC	*Biological soil crust dominated by cyanobacteria* (BSC_{cyan}). Dense cyanobacterial sheaths that form a smooth or dimpled crust of variable darkness. It can include algae, lichen, moss	
	Biological soil crust with multiple structural groups (BSC_{multi}). Two or more structural groups (cyanobacteria, algae, moss, lichen) form a rugose, pinnacled, or rolling crust	
	Biological crust dominated by algae (BSC_{alg}). Rubbery algal crust, with or without lichen showing *cracking or curling*	
4. Characteristics imparted by vegetation	*Duff (D)*. Non-decomposed to fully decomposed plant and organic matter located above the A horizon (a patchy or continuous O horizon)	This pedoderm class will occur where extensive vegetation can be supported, for example, drainage systems (Fig. 1.9)

Adapted from Burket et al. (2012)

Fig. 1.13 A well-aggregated soil (**a**) versus a sandy, single-grained soil (**b**). The details of these pedoderm classes can be found in Table 1.3. A surface crust that developed because of a lack of basal cover, leaving the soil surface exposed to raindrop impact and surface crust formation (**c**). The layered nature of a surface crust (**d**) (credit: P de Jager)

1.4.4 Impeded Infiltration

The ability for soils to absorb water is highly variable at the slope scale. With the losses from run-off areas, run-on areas become subsequently enriched, with particle-bonded nutrients, such as phosphate and dissolved constituents, inorganic ammonium, nitrate, and sulphate. This further increases the spatial heterogeneity of nutrient distribution. The additional process of densification can accentuate the heterogeneity of spatial water and nutrient distribution further, which is a common and natural process in deserts. In fact, many of the characteristics of the pedoderm classes in Table 1.3 are prone to densification. The exceptions are single-grained, rock mulch, and erosion pavements. For example, if void spaces of rock mulch or erosion pavements are filled in with finer material, the porosity of the inter-rock space decreases and in essence densifies. This is best explained with the basic equation for bulk density (ρ_b):

$$_{b} = M_{solid}/V_{total} \tag{1.1}$$

where M_{solids} is the mass of solids (kg) and V_{total} is the volume the solid particles occupy (m^3) (Hillel 2004)

Surface and soil densification caused by soil compaction (and causing soil degradation) is common in anthropogenically impacted environments (e.g. in urban environments and agriculture). This is the result of a static and/or a dynamic force applied to the surface (e.g. trampling, vehicles), which results in the packing of the particles in a smaller space (V_{total} decrease). However, the bulk density of a porous medium can also be decreased if the numerator (M_{solids}) increases, as in the case of void spaces that are filled with finer particles. The natural densification of desert surfaces is most likely a combination of M_{solid} increase and denominator (V_{total}) decrease.

Apart from direct densification, the layering of material at the surface may further impede water infiltration. The abrupt widening of conductive pores in a porous medium is as disruptive to water movement as the sudden narrowing of pores. An example of abrupt widening of pores can be found in vesicular subsurface structures associated with desert pavement. As a result, decreased water infiltration, with increasing vesiculation, is commonly reported (Brown and Dunkerley 1996). Finer material that abruptly transitions to coarser material represents another example where the continuity of pores is often disrupted. More detail on the impact of pore geometry changes and water conductivity can be found elsewhere (Hillel 2004). The consequence of this for run-on areas is that under desert conditions this can concentrate water-soluble constituents at or near the surface.

1.4.5 Organic and Inorganic Carbon in Desert Soils

Organic carbon, apart from being an important energy source for heterotrophic organisms, also contributes to aggregate stability (Tisdall and Oades 1982). Deserts have lower levels of organic material (above and below ground) compared to other regions, illustrated by the southern African pedoderms, with lower organic carbon content of the driest country in the region (Namibia) relative to wetter countries (Table 1.4).

Organic matter decomposition in deserts is high and rates are comparable to tropical climates have been reported (Steinberger and Whitford 1988). This has been linked to photo-oxidation by UVB radiation resulting in fast surface litter decomposition rates and the creation of a direct CO_2 release to the atmosphere and the circumvention of soil and microbes (Austin and Vivanco 2006).

The nature of organic material in desert soils is also variable, since organic compounds derived from plants differ from that of microbes. For example, plants impart lignin which has a slow decomposition rate because of its structural complexity and an abundance of aromatic carbon (Thevenot et al. 2010). Plant

Table 1.4 Pedoderm organic carbon content (%)[a] of various Southern African countries from the hyper-arid west to the semi-arid interior and tropical east

	Country				
	Namibia	Botswana	Zambia	Zimbabwe	Mozambique
Climate[b]	Largely warm desert (BWh) with warm semi-arid (BSh) in northeast	Largely warm semi-arid with warm desert in southwest	Predominantly humid subtropical (Cwa)	Semi-arid to humid subtropical	Tropical savannah (Aw), in the south, inland is warm semi-arid
n	117	89	101	47	104
Mean	0.39	0.65	1.05	0.96	0.92
Median	0.29	0.55	0.98	0.82	0.81
Standard deviation	0.47	0.48	0.49	0.54	0.49
95% Confidence interval[c]	0.31–0.48	0.55–0.75	0.95–1.14	0.80–1.12	0.83–1.02

[a]Oxidisable carbon determined by dichromate method of Walkley-Black (Nelson and Sommers 1996). This method reasonably isolates lower oxidation state organic carbon from inorganic carbon (carbonates), which is carbon at its highest oxidation state, C(+4)
[b]Based on Köppen-Geiger climate classes reported by Beck et al. (2018)
[c]The 95% confidence interval for a Student's *t*-distribution, chosen because of relatively small samples for some countries

polysaccharides also differ from microbial polysaccharides, with the latter lacking pentoses (Essington 2015).

Carbon accumulation in the form of CaCO3 can be 10–17 time greater than organic carbon accumulation in desert soils (Díaz-Hernández et al. 2003), particularly given their high accumulation in inorganic carbonate ion CO_3^{2-}. Globally, areas receiving <250 mm of precipitation have the highest soil inorganic carbon (SIC) with accumulation exceeding 32 kg C m^{-2} (Zamanian et al. 2016). Arid environments therefore appear to strongly drive the oxidation of organic carbon and the accumulation of inorganic carbon.

1.4.6 Nitrogen Dynamics in Desert Soils

Nitrogen is critical for life, and understanding the environmental processes that direct its movement is important. In natural systems, atmospheric N_2 enters the pedoderm either through abiotic processes, for example N deposition, or through microbiological N fixation. The fraction not assimilated by living organisms are subjected to oxidation to nitrate (NO_3^-), the most leachable form of N. N biogeochemical cycling in hot deserts is further detailed in Chap. 7.

Subsurface accumulation of nitrate (NO_3^-) is commonly reported for arid environments of North America (Walvoord et al. 2003; Graham et al. 2008), Namibia (Stone and Edmunds 2014), and Australia (Barnes et al. 1990). Graham

Table 1.5 Pedoderm N content (%)[a] of various southern African countries from the hyper-arid west to the semi-arid interior and tropical east

	Country				
	Namibia	Botswana	Zambia	Zimbabwe	Mozambique
Climate[b]	Largely warm desert (BWh) with warm semi-arid (BSh) in northeast	Largely warm semi-arid with warm desert in southwest	Predominantly humid subtropical (Cwa)	Semi-arid to humid subtropical	Tropical savannah (Aw), in the south, inland is warm semi-arid
n	117	89	101	47	104
Mean	0.05	0.06	0.09	0.09	0.08
Median	0.03	0.04	0.08	0.08	0.07
Standard deviation	0.08	0.06	0.04	0.07	0.04
95% Confidence interval[c]	0.03–0.06	0.06–0.07	0.08–0.10	0.08–0.10	0.07–0.09

[a]Catalysed high temperature combustion (Dumas method) (Bremner 1996)
[b]Based on Köppen-Geiger climate classes reported by Beck et al. (2018)
[c]The 95% confidence interval for a Student's *t*-distribution, chosen because of relatively small samples for some countries

et al. (2008) reported nitrate levels equivalent to 8.9–12.7 tons N ha^{-1} at fairly shallow depths (<1 m) below desert pavement in the Mojave Desert, California, and attributed this to NO_3^- entering the soil in areas of the landscape where appreciable infiltration occurs. NO_3^- then moves and accumulates under desert pavement because of its exceedingly low hydraulic conductivity that prevents water from further transporting it. Walvoord et al. (2003) also attributed NO_3^- subsoil accumulation in desert soils to the fact that it is moved beyond biological reach.

Returning to the pedoderm, the alkalinity of arid surfaces increases the propensity of nitrogen from biological origin to volatilise. Westerman and Tucker (1978) reported that 70% of NH_4 applied to desert soils is lost to the atmosphere. This was confirmed by Peterjohn and Schlesinger (1990) who reported that up to 77% of N inputs was lost in the dry Southwestern USA. It must be, however, noted that this later study was published before the quantification of subsurface nitrate in these areas.

Similar to organic carbon, when the pedoderm N content of southern Africa is compared, Namibia again exhibits the lowest levels with a confidence interval stretching over a lower range compared to other countries (Table 1.5). Michalski et al. (2004) estimated NO_3^- atmospheric deposition for deserts of the southern hemisphere to range from 21 to 84 mg N m^{-2} a^{-1} (0.21–0.84 kg N ha^{-1} a^{-1}) with a reported mean of 0.43 kg N ha^{-1} a^{-1}. The N content confidence interval for Namibia equates to 47.5–88.8 kg N ha^{-1}, assuming a 1 cm depth and bulk density of 1400 kg m^{-3}. The deposition rates appear minor when directly compared to the pedoderm N content. However, in theory, it will take only 226–423 years to reach the measured surface N levels based on the conservative deposition estimate (0.21 kg N ha^{-1} a^{-1}) and 108–202 years based on the mean deposition rate. If the biological N fixation is taken as double the conservative atmospheric deposition rate (0.42 kg N ha^{-1} a^{-1}), as estimated by Peterjohn and Schlesinger (1990), the rate at which the pedoderm acquires N totals 0.63 kg N ha^{-1} a^{-1}. The time frame to accumulate current N pedoderm levels then decreases to 75–141 years. If ammonium deposition, or any other N additions, is added, it will require even less time to reach the measured pedoderm N levels. These rudimentary calculations highlight the transient nature of N in the Namibian pedoderm.

1.5 Conclusion

Globally rainfall is not equally distributed in both space and time and the varied surface geomorphology does not equally store and channel the resulting surface water and subsurface moisture. Hence. opportunities for desert microbes are not equal either. At the local scale, the most defining variables for microbial ecology are moisture and substrate stability, while microbial species richness is defined by microhabitat moisture regimes, stability, and diversity of nutrient sources. The dominant climate, geomorphology, and soil type are primary determinants of microbial ecology in deserts, while macrobiotic colonisation, diversity, population densities, endemism, life histories, and rate of species turnover are dependent on nutrient pools and hydrology of shallow surface soils associated with desert habitats. Low and erratic rainfall accentuates relatively small differences in moisture retention and nutrient mobilisation, which have cascading effects on the suitability for colonisation and development of microbial communities. The relatively high biotic diversity and microtopographic heterogeneity typical of desert mountains and hills result in complex microbial mosaics at minute spatial scales, which contrasts with more homogenous microbial communities associated with extensive desert plains. Similarly, switches between dry terrestrial conditions and aquatic and semi-aquatic settings are associated with ephemeral streams and drainage. However, at the microbial scale, other desert habitats experience similar changes after rainfall events, though for briefer periods. One of the more significant influences of ephemeral rivers on microbial ecology is the introduction of propagules for colonisation during flood events, though drainage systems also homogenise propagules to directionally disperse downstream. Wind and opportune settling of propagules are the more likely dispersal agents in other desert habitats. Microbial communities are more depauperate, and often more specialised, where the salient features of the system enhance desiccation such as in saline habitats, impenetrable clay or rock surfaces, or expanses of sand dominated by a homogenous, unstable, and nutrient-poor substrate with rapid infiltration and poor water-holding capacity. At a microbial scale, habitat islands such as vegetation that impart stability, nutrients, and organic matter (Danin 1996) are important for promoting diversity and providing ecological refuges (Pointing and Belnap 2012; Chaps. 3, 5, and 8). Soil pore spaces many opportunities if and where they are present. In general, soils are more mature on desert margins than in interiors. Run-off and run-on settings determine the accumulation and accretion of material at the local scale of geomorphic contexts and habitats. The pedoderm regulates the exchange between the atmosphere and subsurface soil and is dependent among other factors on crusts, cover, and compaction—all of which translate into variable carbon and nitrogen levels in the subsurface environment. In conclusion, opportunities for microbes vary globally as a function of spatial and temporal rainfall patterns and local moisture availability, which is modified by geomorphic conditions and the resulting habitats. At the scale of soil samples, pedoderm development and soil pore geometry act as additional fine-scale controles. Integrating and interpreting microbial field observations, across such a varied scales and settings, remains a formidable challenge.

Acknowledgements We would like to thank Leandri Wessels for preparing Figs. 1.8, 1.9, 1.10, 1.11, and 1.12.

References

Agassi M, Shainberg I, Morin J (1981) Effect of electrolyte concentration and soil sodicity on infiltration rate and crust formation. Soil Sci Soc Am J 45:848–851

Austin A, Vivanco L (2006) Plant litter decomposition in a semi-arid ecosystem controlled by photodegradation. Nature 442:555–558

Barnes CJ, Jacobson G, Smith GD (1990) The origin of high-nitrate ground waters in the Australian arid zone. J Hydrol 137:181–197

Barré P, Fernandez-Ugalde O, Virto I, Velde B, Chenu C (2014) Impact of phyllosilicate mineralogy on organic carbon stabilization in soils: incomplete knowledge and exciting prospects. Geoderma 235–236:382–395

Barthlott W, Porembski S (2000) Why study inselbergs? In: Porembski S, Barthlott W (eds) Inselbergs: biotic diversity of isolated rock outcrops in tropical and temperate regions. Springer, Berlin, pp 1–6

Beck H, Zimmermann N, McVicar T, Vergopolan N, Berg A, Wood EF (2018) Present and future Köppen-Geiger climate classification maps at 1-km resolution. Sci Data 5:180214

Becker-Reshef I, Justice C, Sullivan M, Vermote E, Tucker C, Anyamba A, Small J, Pak E, Masuoka E, Schmaltz J, Hansen M (2010) Monitoring global croplands with coarse resolution earth observations: the Global Agriculture Monitoring (GLAM) project. Remote Sens 2(6): 1589–1609

Beguería S, Vicente-Serrano SM, Reig F, Latorre B (2014) Standardized precipitation evapotranspiration index (SPEI) revisited: parameter fitting, evapotranspiration models, tools, datasets and drought monitoring. Int J Climatol 34(10):3001–3023

Bowers JE (1982) The plant ecology of inland dunes in western North America. J Arid Environ 5(3):199–220

Boyer DC (1989) Some characteristics of plant communities of three dunes situated across a climatic gradient in the Namib Desert. Madoqua 16(2):141–148

Bremner JM (1996) Nitrogen-total. In: Sparks DL, Page A, Helmke PR, Loeppert R, Soltanpour PN, Tabatabai MA, Johnston CT, Sumner ME (eds) Methods of soil analysis, part 3 : chemical method. Soil Science Society of America Book Soil Science Society of America Book Series. Soil Science Society of America Inc and American Society of Agronomy, Madison, WI, pp 1085–1121

Brewer R, Crook KAW, Speight JG (1970) Proposal for soil stratigraphic units in the Australian Stratigraphic Code. Aust J Earth Sci 17:103–111

Brown KJ, Dunkerley DL (1996) The influence of hillslope gradient, regolith texture, stone size and stone position on the presence of a vesicular layer and related aspects of hillslope hydrologic processes: a case study from the Australian arid zone. Catena 26(1-2):71–84

Buckley RC, Chen W, Liu Y, Zhu Z (1986) Characteristics of the Tengger dunefield, north-central China, and comparison with the central Australian dunefields. J Arid Environ 10(2):97–101

Burket LM, Bestelmeyer BT, Tugel AJ (2012) A field guide to pedoderm and pattern classes. Version 2.3. USDA-ARS Jornada Experimental Range, Las Cruces, NM

Carrier BL, Abbey W, Beegle LW, Bhartia R, Liu Y (2019) Attenuation of ultraviolet radiation in rocks and minerals: implications for mars science. J Geophys Res Planet 124:2599–2612

Cogle AL, Rao KPC, Yule DF, Smith GD, George PJ, Srinivasan ST, Jangawad L (2002) Soil management for Alfisols in the semiarid tropics: erosion, enrichment ratios and runoff. Soil Use Manag 18(1):10–17

Cooke RU, Warren A, Goudie AS (1993) Desert geomorphology. University College London, London

Danin A (1996) Plants of desert dunes. Springer, Berlin

Díaz-Hernández JL, Fernández EB, González JL (2003) Organic and inorganic carbon in soils of semiarid regions: a case study from the Guadix–Baza basin (Southeast Spain). Geoderma 114: 65–80

Dunkerley DL, Brown KJ (1997) Desert soils. In: Thomas DSG (ed) Arid zone geomorphology: process, form and change in drylands, 2nd edn. Wiley-Blackwell, Chichester, UK, pp 55–68

Eckardt FD, Soderberg K, Coop LJ, Muller AA, Vickery KJ, Grandin RD, Jack C, Kapalanga TS, Henschel J (2013) The nature of moisture at Gobabeb, in the central Namib Desert. J Arid Environ 93:7–19

Eldridge DJ, Zaady E, Shachak M (2000) Infiltration through three contrasting biological soil crusts in patterned landscapes in the Negev, Israel. Catena 40(3):323–336

Essington ME (2015) Soil and water chemistry: an integrative approach, 2nd edn. CRC Press, Boca Raton Florida

Felde VJMNL, Peth S, Uteau-Puschmann D, Drahorad S (2014) Soil microstructure as an under-explored feature of biological soil crust hydrological properties: case study from the NW Negev Desert. Biodivers Conserv 23:1687–1708

Fernández-Ugalde O, Barré P, Hubert F, Virto I, Girardin C, Ferrage E, Caner L, Chenu C (2013) Clay mineralogy differs qualitatively in aggregate-size classes: clay-mineral-based evidence for aggregate hierarchy in temperate soils. Eur J Soil Sci 64:410–422

Fey MV, Mills AJ, Yaalon DH (2006) The alternative meaning of pedoderm and its use for soil surface characterization. Geoderma 133(3–4):474–477

Gibson AC (1996) Structure-function relations of warm desert plants. Springer, Berlin

Goudie AS (2002) Great warm deserts of the world: landscapes and evolution. Geomorphological landscapes of the world 1. Oxford University Press, Oxford

Graham RC, Hirmas DR, Wood YA, Amrhein C (2008) Large near-surface nitrate pools in soils capped by desert pavement in the Mojave Desert, California. Geology 36(3):259–262

Halis Y, Benhaddya ML, Bensaha H, Mayouf R, Lahcini A, Belhamra M (2012) Diversity of halophyte desert vegetation of the different saline habitats in the Valley of Oued Righ, Low Sahara Basin, Algeria. Res J Environ Earth Sci 4(3):308–315

Harvey A (2011) Dryland alluvial fans. In: Thomas DSG (ed) Arid zone geomorphology: process, form and change in drylands. Wiley-Blackwell, Chichester, UK, pp 333–371

Hegazy A, Lovett-Doust J (2016) Plant ecology of the Middle East. Oxford University Press, Oxford

Hillel D (2004) Introduction to soil physics. Elsevier, San Diego, CA

Hooke JM (2016) Morphological impacts of flow events of varying magnitude on ephemeral channels in a semiarid region. Geomorphology 252:128–143

Hooke JM, Brookes CJ, Duane W, Mant JM (2005) A simulation model of morphological, vegetation and sediment changes in ephemeral streams. Earth Surf Process Landforms 30: 845–866

Howard AD, Kerby G (1983) Channel changes in Badlands. Geol Soc Am Bull 94(6):739–752
Itami K, Kyuma K (1995) Dispersion behavior of soils from reclaimed lands with poor soil physical properties and their characteristics with special reference to clay mineralogy. Soil Sci Plant Nutr 41(1):45–54
Jürgens N, Burke A, Seely MK, Jacobson KM (1997) Desert. In: Cowling RM, Richardson DM, Pierce SM (eds) Vegetation of Southern Africa. Cambridge University Press, Cambridge, pp 189–214
King MD (2007) Our changing planet. Cambridge University Press, Cambridge
Kleber M, Mikutta R, Torn MS, Jahn R (2005) Poorly crystalline mineral phases protect organic matter. Eur J Soil Sci 56:717–725
Kruckeberg AR (2002) Geology and plant life. University of Washington Press, Seattle
Lancaster N (2002) Geomorphology of desert dunes. Routledge physical environment. Routledge, London
Lee JA, Gill TE (2015) Multiple causes of wind erosion in the Dust Bowl. Aeolian Res 19:15–36
Marasco R, Mosqueira MJ, Fusi M, Ramond J-B, Merlino G, Booth JM, Maggs-Kölling G, Cowan DA, Daffonchio D (2018) Rhizosheath microbial community assembly of sympatric desert speargrasses is independent of the plant host. Microbiome 6:215
Mazor G, Kidron GJ, Vonshak A, Abeliovich A (1996) The role of cyanobacterial exopolysaccharides in structuring desert microbial crusts. FEMS MicrobiolEcol 21(2):121–130
McKee ED (ed) (1979) A study of global sand seas. US Geological Survey Professional Paper 1052. US Government Printing Office, Washington, DC
Michalski G, Böhlke JK, Thiemens M (2004) Long term atmospheric deposition as the source of nitrate and other salts in the Atacama Desert, Chile: new evidence from mass-independent oxygen isotopic compositions. Geochim Cosmochim Acta 68(20):4023–4038
Nelson DW, Sommers LE (1996) Total carbon, organic carbon and organic matter. In: Sparks DL, Page A, Helmke PR, Loeppert R, Soltanpour PN, Tabatabai MA, Johnston CT, Sumner ME (eds) Methods of soil analysis, part 3: chemical method. Soil Science Society of America Book Series. Soil Science Society of America Inc and American Society of Agronomy, Madison, WI, pp 961–1010
Nicholson SE (2011) Dryland climatology. Cambridge University Press, Cambridge, UK
Parsons AJ, Abrahams AD (eds) (2009) Geomorphology of desert environments, 2nd edn. Springer, Heidelberg, Germany
Peterjohn WT, Schlesinger WH (1990) Nitrogen loss from deserts in the southwestern United States. Biogeochemistry 10:67–79
Pointing SB, Belnap J (2012) Microbial colonization and controls in dryland systems. Nat Rev Microbiol 10(8):551–562
Pye K (1995) The nature, origin and accumulation of loess. Quaternary Sci Rev 14(7-8):653–667
Pye K, Tsoar H (1990) Aeolian sand and sand dunes. Springer, Berlin
Rabbi S, Daniel H, Lockwood P, Macdonald C, Pereg L, Tighe M, Wilson BR, Young IM (2016) Physical soil architectural traits are functionally linked to carbon decomposition and bacterial diversity. Sci Rep 6:33012
Rahbek C, Borregaard MK, Antonelli A, Colwell RK, Holt BG, Nogues-Bravo D, Rassmussen CMØ, Richardson K, Rosing MT, Whittaker RJ, Fjeldså J (2019) Building mountain biodiversity: geological and evolutionary processes. Science 365(6458):1114–1119
Rendell H (2011) Tectonic frameworks. In: Thomas DSG (ed) Arid zone geomorphology: process, form and change in drylands. Wiley-Blackwell, Chichester, UK, pp 17–25
Ringrose S, Seely M, Matheson W, Cassidy L, Kemosidile T, Diskin S, Coetzee S (2018) Nature and possible origins of hyper-arid floodplain islands: exemplified by the Kuiseb river, Namibia. Trans R Soc S Afr 73(2):143–157
Robinson MD, Seely MK (1980) Physical and biotic environments of the southern Namib dune ecosystem. J Arid Environ 3:183–203
Ruiz-Vera VM, Wu L (2006) Influence of sodicity, clay mineralogy, prewetting rate, and their interaction on aggregate stability. Soil Sci Soc Am J 70:1825–1833

Sabater S, Timoner X, Bornette G, De Wilde M, Stromberg JC, Stella JC (2017) The biota of intermittent rivers and ephemeral streams: algae and vascular plants. In: Datry T, Bonada N, Boulton A (eds) Intermittent rivers and ephemeral streams. Academic, London, pp 189–216
Sandercock PJ, Hooke JM, Mant JM (2007) Vegetation in dryland river channels and its interaction with fluvial processes. Prog Phys Geogr Earth Environ 31(2):107–129
Seely MK (1990) Patterns of plant establishment on a linear desert dune. Isr J Bot 39:443–451
Seely MK, Louw GN (1980) First approximation of the effects of rainfall on the ecology and energetics of a Namib Desert dune ecosystem. J Arid Environ 3:25–54
Shainberg I, Mamedov AI, Levy GJ (2003) Role of wetting rate and rainfall energy in seal formation and erosion. Soil Sci 168(1):54–62
Shaw PA, Bryant RG (2011) Pans, playas and salt lakes. In: Thomas DSG (ed) Arid zone geomorphology: process, form and change in drylands. Wiley, Chichester, UK, pp 373–401
Shaw JR, Cooper DJ (2008) Linkages among watersheds, stream reaches, and riparian vegetation in dryland ephemeral stream networks. J Hydrol 350:68–82
Shmida A, Evanari M, Noy-Meir I (1986) Hot desert ecosystems: an integrated view. In: Evenari M, Noy-Meir I, Goodall DW (eds) Ecosystems of the world 12B, hot deserts and arid shrublands, B. Elsevier, Amsterdam, pp 379–387
Steinberger Y, Whitford WG (1988) Decomposition process in Negev ecosystems. Oecologia 75: 61–66
Stone AEC, Edmunds WM (2014) Naturally high nitrate in unsaturated zone sand dunes above the Stampriet Basin, Namibia. J Arid Environ 105:41–51
Thevenot M, Dignac MF, Rumpel C (2010) Fate of lignins in soils: a review. Soil Biol Biochem 42(8):1200–1211
Thomas DSG (ed) (2011) Arid zone geomorphology: process, form and change in drylands, 3rd edn. Wiley-Blackwell, Chichester, UK
Thornthwaite CW (1948) An approach toward a rational classification of climate. Geogr Rev 38(1): 55–94
Tisdall JM, Oades JM (1982) Organic matter and water-stable aggregates in soils. J Soil Sci 33: 141–163
Tucker CJ (1979) Red and photographic infrared linear combinations for monitoring vegetation. Rem Sens Environ 8(2):127–150
Turk JK, Graham RC (2011) Distribution and properties of vesicular horizons in the western United States. Soil Sci Soc Am J 75:1449–1461
Viles HA (2011) Weathering systems. In: Thomas DSG (ed) Arid zone geomorphology: process, form and change in drylands. Wiley-Blackwell, Chichester, pp 83–100
Von Schiller D, Bernal S, Dahm CN, Martí E (2017) Nutrient and organic matter dynamics in intermittent rivers and ephemeral streams. In: Datry T, Bonada N, Boulton A (eds) Intermittent rivers and ephemeral streams. Academic, London, pp 135–160
Wainwright J, Bracken LJ (2011) Runoff generation, overland flow and erosion on hillslopes. In: Thomas DSG (ed) Arid zone geomorphology: process, form and change in drylands. Wiley-Blackwell, Chichester, pp 235–267
Walter H (1983) Vegetation of the earth and the ecological systems of the geo-biosphere. Springer, Berlin, pp 113–147
Walvoord MA, Phillips FM, Stonestrom DA, Evans RD, Hartsough PC, Newman BD, Striegl RG (2003) A reservoir of nitrate beneath desert soils. Science 302:1021–1024
Ward D (2009) The biology of deserts. Oxford University Press, Oxford
Warrington DN, Mamedov AI, Bhardway AK, Levy GJ (2009) Primary particle size distribution of eroded material affected by degree of aggregate slaking and seal development. Eur J Soil Sci 60: 84–93
Westerman RL, Tucker TC (1978) Factors affecting denitrification in a Sonoran Desert soil. Soil Sci Soc Am J 42:596–599
Wood WW, Sanford WE, Habshi ARSA (2002) Source of solutes to the coastal sabkha of Abu Dhabi. Geol Soc Am Bull 114(3):259–268

Yeaton RI (1988) Structure and function of the Namib dune grasslands: characteristics of the environmental gradients and species distributions. J Ecol 76(3):744–758
Young IM, Crawford JW (2004) Interactions and self-organization in the soil-microbe complex. Science 304(5677):1634–1637
Zamanian K, Pustovoytov K, Kuzyakov Y (2016) Pedogenic carbonates: forms and formation processes. Earth Sci Rev 157:1–17
Zipser EJ, Cecil DJ, Liu C, Nesbitt SW, Yorty DP (2006) Where are the most intense thunderstorms on earth? Bull Am Meteorol Soc 87:1057–1071

Novel Methods for Studying the Structure and Function of Hot Desert Microorganisms and Their Communities

2

Marc W. Van Goethem, Estelle Couradeau, and Trent R. Northen

Abstract

Since the 1980s, the quantity of sequence data has boomed by seven orders of magnitude. These successive technical revolutions, among others, have allowed us to begin shedding light on complex microbial communities. Here, we consolidate literature on the most recent and innovative methods that could be applied to the study of microbial consortia from hot desert environments. While a variety of soil niches constitute the habitats for hot desert microflora and fauna, we focus specifically on biological soil crusts (biocrusts). Biocrusts are commonly found in hot desert environments around the globe and are a powerful testbed for new approaches due to their experimental tractability. Within this chapter, we focus on several avenues for examining the ecology of soil microbial communities, namely through (1) DNA sequencing of entire microbial communities, (2) identifying active community members, (3) using metabolomics to explore microbial nutrient release and uptake, and (4) applying high-resolution imaging of communities. Finally, we offer considerations about the utility of each approach for hot desert

M. W. Van Goethem
Environmental Genomics and Systems Biology Division, Lawrence Berkeley National Laboratory, Berkeley, CA, USA

E. Couradeau
Department of Ecosystem Science and Management, The Pennsylvania State University, State College, PA, USA

T. R. Northen (✉)
Environmental Genomics and Systems Biology Division, Lawrence Berkeley National Laboratory, Berkeley, CA, USA

DOE Joint Genome Institute, Integrative Genomics Building, Lawrence Berkeley National Laboratory, Berkeley, CA, USA
e-mail: trnorthen@lbl.gov

J.-B. Ramond, D. A. Cowan (eds.), *Microbiology of Hot Deserts*, Ecological Studies 244, https://doi.org/10.1007/978-3-030-98415-1_2

ecology and offer potential future lines of inquiry for research within the field of hot desert microbiology.

Keywords

Biocrust · Arid lands · Metagenomics · Active microbes · Sequencing · Metabolomics · Imaging

2.1 Introduction

Our depth and breadth of knowledge about hot desert microbial communities and their ecology continues to expand in the twenty-first century. Accordingly, the number of novel methodologies that are available to researchers grows even more rapidly. In this book chapter, we consolidate and discuss novel approaches that could significantly enhance our understanding of hot desert microbial ecology in the future (Fig. 2.1, Table 2.1). We use examples from other natural environments that have benefited from adopting new approaches and draw on inspiration from studies in the broader field of microbial ecology.

Historically, scientists have cultured microorganisms from samples of interest to study the physiology and morphology of novel species. Cultivation-based approaches remain vital to document antibiotic resistance of pathogenic microbes within the human microbiome (Lagier et al. 2018). Moreover, microbial cultivation has revealed that many soil bacteria, fungi, and archaea contribute important ecosystem services across the globe (Delgado-Baquerizo et al. 2018). Studying microbes from culture is undoubtedly the gold-standard for identifying and characterizing

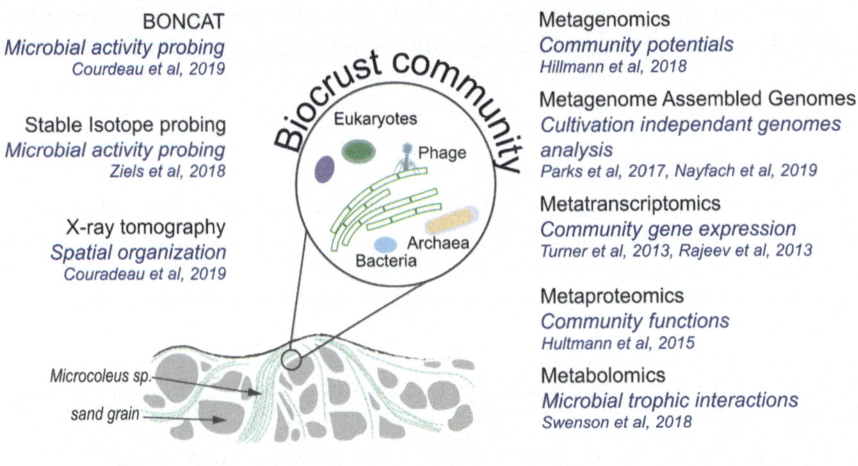

Fig. 2.1 Numerous novel tools are available for study of the complex soil microbial communities. Here we highlight major technical branches that can be used to explore community structure, activity, or physical organization

Table 2.1 Technologies with their application, strengths, weaknesses, and reference examples

Technique	Useful for?	Positives	Challenges	References
Culturing microbes in the laboratory	Identify new species and observe morphology	Can explore physiology and extract DNA directly	Slow and difficult to determine cultivation conditions	Imachi et al. (2020)
16S rRNA and ITS sequencing	Describe community composition	Efficient and inexpensive	No direct information on microbial function	Smets et al. (2016)
Exact Sequence Variants (ESV)	Estimate taxonomic diversity	Based on nucleotide differences (accurate)	May overestimate diversity	Callahan et al. (2016)
Operational Taxonomic Units (OTU)	Estimate taxonomic diversity	Clusters similar sequences (routine, rapid)	Valid lineages may be discarded by clustering	Schloss et al. (2009)
Shotgun Metagenomics	Describe genetic potential	Access sequences from most community members	Biased toward abundant taxa	Handelsman (2004)
Short-read metagenomic sequencing	Capturing genetic potential of multiple organisms in a sample	Offers both taxonomic and function genetic information	Excludes rare taxa, amplification biases	Van Goethem et al. (2019)
Long-read metagenomic sequencing	Access very long gene clusters and repeat regions	Produces very long contigs that can be annotated for novel functions	High error rates in raw reads	Hiraoka et al. (2019)
Metagenome-assembled genomes	Reconstruct genomes from complex samples	Genomes of uncultured microbes	Cannot yet recreate eukaryotic genomes	Moss et al. (2020), Nayfach et al. (2019)
Metatranscriptomics	Determine microbial activity	Insights into microbial transcription	RNA molecules have short half-lives	León-Sobrino et al. (2019)
tRNA-seq	Estimate tRNA abundance at single gene resolution	Generates taxon-specific insights into in situ dynamics of tRNA gene expression	High biomass requirement (up to 10 μg total RNA)	Schwartz et al. (2018)
scRNA-seq	High-resolution expression profiles of individual cells in a microenvironment	Specialized cell functions can be identified	High levels of sequence amplification results in uneven coverage	Stubbington et al. (2017)

(continued)

Table 2.1 (continued)

Technique	Useful for?	Positives	Challenges	References
cDNA Microarray	Access community transcripts rapidly	Measure expression levels of multiple genes simultaneously	May obscure differences in gene expression among individual cells	Rick et al. (2001)
GeoChip	Linking transcripts to functions and taxonomy by functional gene array	Direct linkage between ecosystem functions and microbial genes	Difficulty in comparing microarray data among laboratories	Tu et al. (2014)
Metaproteomics	Data regarding microbial functional outputs is gleaned	Translational products indicate community phenotypes	Obtaining sufficient yield from soil samples is challenging	Hultman et al. (2015)
Metabolomics	Study of all metabolites produced by the community	Potential for determining gene function and monitoring response to change	Minerals can limit metabolite extraction, metabolite identification is often difficult esp. given limited number of authentic chemical standards	Allen et al. (2003), Swenson et al. (2015)
Targeted metabolomics	Measure chemically characterized and annotated metabolites	Accurate measurements of predetermined sets of metabolites of interest	Stringent experimental design is needed to prevent intrinsic variability—this is a very vague statement. How about: Ignores much of the chemical diversity in typical samples	Roberts et al. (2012)
Untargeted metabolomics	Investigate metabolic phenotype of a complex community	Metabolites are informative as to the biochemical activity of the community	Difficulty in confidently identifying many metabolites	Swenson et al. (2018)
Linking multi-omics datasets	Identifying links between complex datasets	Applicable to relative abundance-type data	Requires paired data (i.e., multi-omics)	Morton et al. (2019)
Fluorescence-activated cell sorting (FACS)	Sort heterogeneous cell mixtures into meaningful categories	Wide range of applications to recover labeled cells from a complex mixture	Not all cells are sorted efficiently (cell size and structure)	Hatzenpichler and Orphan (2015)

DNA-SIP (Stable isotope probing)	Enrich DNA of active microbes through heavy isotope incorporation	Combination with shotgun sequencing can reveal novel enzymes and operons	Relies on cell division occurring in the presence of the labeled substrate	Ziels et al. (2018)
Stable isotope analyses	Trace the flow of nutrients through trophic levels	Determine producers in soil systems	May require a lot of sample material to obtain representative replicates	Ramond et al. (2018)
BONCAT-FACS-seq	Active cells in a population are identified and sequenced	Translationally active cells can be sequenced to identify active taxa	Cells are inadvertently stimulated by wetting	Couradeau et al. (2019)
X-ray tomography	Non-destructive visualization of the 3-D geometry of a sample	Provides accurate quantification of microstructural features	Data produced are very large (this impacts storage and retrieval of files)	Couradeau et al. (2018)
FISH & FACS (Fluorescence in situ hybridization-Fluorescence-activated cell sorting)	Targeted enrichment of microbial clades from complex samples	Retrieval of uncultivated bacterial clades from a sample	Requires separation of cells from soil or sediment samples	Grieb et al. (2020)
CARD-FISH (Catalyzed Reporter Deposition-Fluorescence in situ hybridization)	Insights into environmental microbiology at single-cell resolution	Detection of RNA and genes encoded on chromosomes of microbes	Technique produces radicals, and fixing cells complicates sequencing due to chemical cross-linking DNA	Bao et al. (2014)
NanoSIMS (Nanoscale secondary ion mass spec)	Precise measurements of isotopes to link phylogeny and metabolism	High sensitivity and spatial resolution (micrometer)	Sample preparation is complex for biological materials	Dekas et al. (2016)
MaPS-seq	Sequence populations within their appropriate spatial scale	Link microbes to their niches, observe interactions	Not adapted for soil microbiomes currently	Sheth et al. (2019)

novel species, however we know that many (if not most) microbes are not culturable on media in the laboratory.

Estimates regarding the proportion of culturable microbes range from <1% to as high as 50% (Pham and Kim 2012; Nunes da Rocha et al. 2015), but remains a hotly debated topic (Martiny 2019; Steen et al. 2019). Recent studies have applied innovative methods to successfully isolate previously unculturable microbes (Cross et al. 2019). For instance, *Abditibacterium utsteinense*, the first cultured representative of the candidate FBP phylum, was isolated from Antarctic soils by limiting carbon sources present in the media which excluded many copiotrophic microbes (Tahon et al. 2018). Multiple novel *Planctomycetes* species ($n = 79$) were characterized from complex environmental samples by enriching culturing conditions for their slow growth and metabolic preference for *N*-acetylglucosamine (Wiegand et al. 2020). The first representative of an Asgard archaeon was cultured after a decade-long isolation experiment from marine sediment (Imachi et al. 2020). Culturing "unculturable" microbes thus remains an exciting research topic and particularly in polyextreme environments as novel species may have interesting activities and/or biotechnological applications (Browne et al. 2016).

An alternative means of tapping into the microbial diversity of soil samples has been to rely on culture-independent techniques. Historically, techniques such as 16S ribosomal RNA (rRNA) gene profiling through Denaturing Gradient Gel Electrophoresis (DGGE (Muyzer et al. 1993)) and Terminal Restriction Fragment Length Polymorphisms (T-RFLP (Liu et al. 1997)) have provided important initial insights into the uncataloged microbial world that had remained elusive previously (Osborn et al. 2000). These and others techniques, such as Amplified Ribosomal DNA Restriction Analysis (ARDRA (Vaneechoutte et al. 1992)), gave a coarse overview of the community structure of a sample. Scientists could now document the constituents of whole microbial communities rapidly, which allowed for the descriptions of novel phyla from a range of environmental samples.

In parallel, the costs of sequencing continued (and continue) to decrease and the development of Next-Generation Sequencing (NGS) gave researchers unprecedented access to the actual gene content of uncultured microbes. Although 16S rRNA gene sequencing remains a valuable means of cataloging novel microbial diversity today (Smets et al. 2016), much more effort is being directed toward sequencing entire microbial communities through shotgun metagenomics (Hillmann et al. 2018). By extracting genomic DNA from a complex microbial community it becomes possible to explore the metagenome, which can be viewed as the sum of the genomes present within a complex sample (Handelsman 2004). Metagenomic sequencing has since developed into a household technique that is being applied to many microbial ecology studies (Quince et al. 2017). Importantly, studying the genetic potential of an organism gives no insights into actual functional outputs nor the activity of the organism at the time of sampling. This drawback means that inferring function directly from genetic potential is inaccurate, and it is cogent to link DNA sequence information with other methods that give access to functional activity or evidence of actual metabolic output.

Fig. 2.2 Biological soil crust (biocrust) are organosedimentary assemblages that form intimate associations with the soil surface. Biocrust are initially colonized by filamentous *Cyanobacteria* which recruit heterotrophic bacteria. Photo credit: MWVG

Throughout the chapter, we use biological soil crust communities (referred to as biocrust; Fig. 2.2) as model systems that could be interrogated using the methods described in this chapter. Biocrusts are common organosedimentary features of hot desert surface soils (Garcia-Pichel et al. 2013). Biocrust are microbial ecosystems encompassing a few well-defined metabolic guilds that are responsible for major services such as soil stability, hydrology, and fertility (Belnap 2006), making them ideal candidates upon which novel approaches could be benchmarked. A detailed description of the microbial ecology of desert biocrusts is provided in Chap. 3 of this book. Moreover, biocrust are easily transferable to the laboratory and much is known regarding their ecology across the globe (Belnap et al. 2016), for instance, that the cyanobacterium *Microcoleus* sp. is key to biocrust formation (Garcia-Pichel and Wojciechowski 2009). Within the scope of the proposed techniques we highlight both potential benefits and pitfalls, as well as possible outcomes that would greatly improve our knowledge of biocrusts and other hot desert microbial communities.

As discussed in other chapters of this book (notably in "Introduction to Microbiology of Hot Deserts (Preface)", Chaps. 1, 7, and 12) and elsewhere (Makhalanyane et al. 2015), hot desert environments suffer from imposing physical and climatic extremes that render microbial communities relatively simple in terms of their trophic structure compared to other soil ecosystems such as forest soils, which are more phylogenetically diverse (Schindlbacher et al. 2011). A simpler community structure may have direct implications on the utility of the approaches presented within this chapter. Where possible we have highlighted the drawbacks of studying microorganisms optimized for growth and survival in multi-stress environments. Notably, hot deserts are characterized by their water scarcity, high temperatures, high levels of incident ultra-violet (UV) irradiation receipt, and low nutrient status, among other stresses that are discussed elsewhere (Belnap et al. 2016).

It is noteworthy that the half-lives of nucleic acids and proteins decrease as temperatures increase, although numerous mechanisms are used by cells to contend with denaturation and degradation (Grosjean and Oshima 2007). In parallel researchers should aim to preserve these cellular components by snap freezing samples immediately after collection or by storing their sample material in stabilization solutions that cease cellular activity (such as RNAlater® or LifeGuard™). Thus, high temperatures may reduce the utility of techniques that rely on these molecules for quantifying microbial community structure or activity, such as metatranscriptomics and proteomics. Other environmental challenges to consider in hot deserts are generally poor soil nutrient status, low water activity, unpredictable dry-wetting cycles, the reliance of microbes on sporadic moisture events for their activity (inundating a soil sample with a liquid inadvertently stimulates cells), and the desiccated nature of desert soils which causes high levels of cellular adsorption into the soil matrix (Felde et al. 2014).

2.2 Multi-omics and Bioinformatic Advances to Explore Community Structure and Functional Potential

Metagenomics is the study of the microbial genomes present within a complex sample (Handelsman 2004). This now routine sequence-based approach offers insight into both the taxonomic and functional attributes of the microbial communities while circumventing the need to isolate and cultivate individual community members. Metagenomic sequencing allows researchers to chronicle the myriad of taxa from their sample of interest through nucleic acid sequencing and analysis. Marker gene surveys, notably through amplicon sequencing, can provide similar insights but are generally focused on genes of interest, e.g. the 16S ribosomal RNA (rRNA) gene for bacterial and archaeal phylogeny, the internal transcribed spacer (ITS) gene region for fungal taxa, or the *nifH* gene for nitrogen fixation potential. Phylogenetic marker genes can also be retrieved from unassembled or assembled metagenomes, and the analysis of community diversity from metagenomes is comparable to the amplicon sequence analysis described below.

Researchers frequently make use of operational taxonomic units (OTUs), and more recently exact sequence variants (ESVs) (Callahan et al. 2016), to classify the microbes within a community against reference databases of known taxonomy. Two prominent open-source software platforms were initially developed for the analysis of 16S rRNA gene type sequence information, namely mothur (Schloss et al. 2009) and QIIME (Caporaso et al. 2010), and note that QIIME 2 has recently been released (Bolyen et al. 2019). Broadly, these programs compute a series of steps that take an input of raw sequence data through denoising and demultiplexing, to sequence alignment and clustering at a predetermined threshold followed by chimera filtering and finally to sequence classification. Ultimately, an OTU table can be produced that captures the relative abundances of microorganisms present within and across multiple samples of interest. ESV algorithms such as DADA2 (Callahan et al. 2016) produce higher-resolution analogs of OTU tables by detecting single nucleotide differences among sequences and records their frequency.

OTU or ESV tables can then be used to identify ecologically relevant microbes that are significantly over- or under-represented within a treatment or niche. It is also common to identify "keystone" species by analyzing co-occurrence-based networks that compare the relative abundances of individual taxa and their prevalence across samples. For example, *Cyanobacteria* have been identified as keystone taxa in hot desert hypoliths (Valverde et al. 2015; Van Goethem et al. 2017), which is entirely consistent with their roles as primary producers in these desert niches. Similarly, co-occurrence-based analyses of biocrust communities provided insights into the microbes that significantly co-occur with *M. vaginatus* (Nunes da Rocha et al. 2015).

Many studies have performed network analysis tests using Pearson correlations (product moment parametric test) or Spearman's correlations (rank-order non-parametric test) of relative abundance data (often referred to as compositional data). However, these tests have been shown to be statistically invalid for data such as OTU or ESV tables (Morton et al. 2019). This is primarily because the relative abundance of each OTU or ESV in the table is directly influenced by the proportion of every other OTU. In other words, as the relative abundance of OTU_i increases, the proportion of OTU_j in the community will decrease regardless of whether it is changing (increasing or decreasing) or remaining unchanged. Both Spearman and Pearson correlations were designed and are applicable for absolute abundance data. For testing OTU or ESV co-occurrences among microbes from relative abundance data we advise the use of dedicated statistical tools such as SparCC (Sparse Correlations for Compositional data (Friedman and Alm 2012); or SPIEC-EASI (SParse InversE Covariance Estimation for Ecological Association Inference (Kurtz et al. 2015), which are not influenced by proportionality.

Extracting full-length 16S ribosomal RNA (rRNA) genes from metagenomes can be challenging yet is required to determine the bulk catalog of prokaryotes that coexist within a community. Bioinformatic tools including SortMeRNA (Kopylova et al. 2012) or RNAmmer (Lagesen et al. 2007) can extract ribosomal genes from unassembled, quality-corrected metagenomes. An additional worthwhile goal is often to assemble the short, noisy "shotgun" sequences into longer contiguous sequences, referred to as contigs. The steps involved in quality controlling raw metagenomic reads through to assembly and annotation can be found elsewhere (Breitwieser et al. 2019). There are many prominent bioinformatic tools designed to assemble metagenomic sequences. Naturally, all tools differ in their implementation and utility, but some noteworthy examples for short-read Illumina sequence assembly include metaSPAdes (Nurk et al. 2017), MEGAHIT (Li et al. 2015), MetaVelvet (Namiki et al. 2012), and IDBA-UD (Peng et al. 2012).

Another common aim in a metagenomic sequence analysis pipeline is to reconstruct whole microbial genomes, referred to as metagenome-assembled genomes (MAGs) (Bowers et al. 2017). This is made possible by computing differences in sequence coverage, G+C content, and tetranucleotide frequencies across contigs, and binning similar contigs to generate draft genomes of uncultured microbes. Prominent genome binning algorithms include CONCOCT (Alneberg et al. 2014), MetaBAT 2 (Kang et al. 2019), and MaxBin (Wu et al. 2015), while platforms such as Anvi'o (Eren et al. 2015) can make visualizing variations in metagenomic and genomic sequence data feasible. Critically, MAGs can be representatives of dominant or rare (low abundance) microbes making it possible to glean insights into the metabolism of thousands of novel microorganisms with no known cultured representatives (Nayfach et al. 2019). This type of genetic information has the potential to explain, for example, how microbes cope with the extreme environmental stresses of hot deserts. Beyond functional adaptations, such as coding for heat shock proteins or adjusting tRNA secondary structure, warm-adapted microbes also reduce the size of their genomes through generations of streamlining (Sabath et al. 2013). This feature may make reconstituting MAGs from metagenomic sequence information easier as fewer contigs may be required to complete assembly of the entire genome. However, the higher G + C content across warm-adapted genomes (Jegousse et al. 2017) could lead to extraction and amplification biases and may complicate downstream sequence assembly across GC-rich areas of the genome.

Metagenomic sequencing has provided insights into the symbioses of soil microbes and their plant hosts (Finkel et al. 2016; Fonseca-García et al. 2016; Marasco et al. 2018; see also Chap. 8). Intriguing recent findings have shown that opposing selection processes govern microbial assembly in the rhizosphere of cacti (*Myrtillocactus geometrizans* and *Opuntia robusta*), where host-specific patterns emerge where fungi interact within the host stem (Fonseca-García et al. 2016), compared to the rhizosheath of speargrass in which sand-driven selection supersedes microbial recruitment by the plant (Marasco et al. 2018). Regardless of the mechanisms governing assembly, the microbial colonizers of the rhizosphere and phyllosphere of desert plants show genomic adaptations to these discrete niches. For example, halophilic and alkaliphilic bacteria were selected for in the phyllosphere of the salt-secreting tree *Tamarix aphylla* (Finkel et al. 2016), which was supported to the reconstruction of metagenome-assembled genomes (MAGs) that encode these traits.

As sequencing technologies continue to improve in accuracy and fidelity, so do the bioinformatic tools required to analyze their outputs. New sequencing technologies have become available and include those developed by Pacific Biosciences (i.e., Sequel II which uses Single-Molecule Real-Time [SMRT] sequencing), and Oxford Nanopore (i.e., Plongle). Although much more commonly used for whole genome sequencing (Rhoads and Au 2015), long-read sequencers can yield enough coverage to produce entire metagenomes. PacBio metagenomes have recently been used to describe dominant taxonomic groups by leveraging the remarkable range of sequence length and coverage that is attainable from ultra-long reads and contigs. Recent studies have made use of PacBio circular consensus sequencing (CCS) to reconstitute MAGs from complex metagenomes. CCS can produce high-quality reads of ~2.5 kbp and more than 99% accuracy (Frank et al. 2016). Two CCS metagenomes from Lake Biwa, Japan, provided sufficient sequencing depth to reconstruct 19 draft genomes of uncultivated prokaryotic groups (Hiraoka et al. 2019). Similarly, nanopore sequencing was successfully used to reconstitute multiple MAGs into single, circular contigs from the human gut (Moss et al. 2020). Long-read metagenomic sequencing of hot desert microbial communities will certainly provide novel insights into previously unexplored microbial diversity. De novo assembly of long-read sequence information requires the use of specific assemblers such as Canu (Koren et al. 2017) or metaFlye (Kolmogorov et al. 2019).

Viruses are the most abundant bioactive entities in the world. Bacteriophages, viruses that infect and replicate within a bacterial host, are integral components of the soil biosphere (Emerson et al. 2018). However, we still lack a lot of information about the roles of bacteriophages (also commonly referred to simply as phages) in hot deserts, despite major advances in the field recently (Adriaenssens et al. 2015; Zablocki et al. 2016). Chapter 6 provides with an up-to-date description of the knowledge existing on hot desert viromics. There are numerous factors that have contributed to this lag in research and include, but are not limited to, (1) difficulties isolating viable phages through soil resuspension (Trubl et al. 2016), (2) the lack of a universal marker gene for viruses, (3) very small genome sizes, and (4) high mutation rates. Other major issues in the field of viral ecology include those related to host DNA contamination during nucleic acid isolation and taxonomically classifying viral sequences or genomes (this remains a massive undertaking that is continually being tackled by the ICTV [International Committee on Taxonomy of Viruses] (King et al. 2018)). Critically, no universal phylogenetic marker gene exists for viruses, such as the 16S rRNA gene for most prokaryotes, or the ITS region for fungi. However, there have been significant bioinformatic advances aimed at describing uncultivated DNA viruses from complex community metagenomes (Schulz et al. 2020) based on Hidden Markov Models (Schulz et al. 2018) or protein-sharing networks (Bolduc et al. 2017).

An example of a pipeline to describe virus diversity in a sample could be to use VirSorter (Roux et al. 2015) in conjunction with vConTACT2 (Jang et al. 2019) or an equivalent protein-clustering tool. After metagenome assembly, contigs containing viral sequences can be identified based on significant sequence similarity to known viral genes (VirSorter), and then both taxonomy and function of these sequences can be inferred based on shared protein content with all known viral diversity within the RefSeq Viral Database using vConTACT2 (Brister et al. 2014). This type of approach has recently been applied to hot desert biocrust communities for the first time (Van Goethem et al. 2019) and certainly has the potential to increase understanding of virus–host dynamics in other soil systems. Here the authors used biocrust metagenomes exposed to a dramatic wetting event to investigate the relationship between a *Firmicutes* bloom and their bacteriophages which they link in silico. Alternatively, pre-processing of the sample could exclude contaminating host DNA (DNase treatment), followed by the direct sequencing of the viral DNA that was present in viral capsids which constitutes the virome (Zablocki et al. 2014). An extension of this work could be to focus on the transcripts produced by phages and their hosts, which could link phage activity to the response of microbial host to infection.

Metatranscriptomics, the collection and sequencing of microbial community transcripts, can provide a glimpse into which functional processes occur at the time of sample collection (Urich et al. 2008). For example, a metatranscriptomic analysis of hot desert hypoliths showed that chemoautotrophic processes are prioritized over photoautotrophic carbon fixation across daily cycles despite the phylogenetic dominance of *Cyanobacteria* (León-Sobrino et al. 2019). Metatranscriptomes can be assembled in the same manner as metagenomes, and combining the two approaches together can shift questions from "Which microbes are present across treatments?" to "Which microbes show the most dramatic responses to a treatment?"

High-throughput transfer RNA sequencing (tRNA-seq) has recently been used on the gut microbiome (Schwartz et al. 2018). Here, sequencing tRNA transcripts gives insights into the physiological states of microbial populations within the context of protein synthesis. Since tRNA molecules undergo a series of post-transcriptional modifications it becomes possible to identify in situ tRNA gene expression dynamics. Variations in tRNA modifications could reflect microbial responses to different environmental effects. Single-cell RNA sequencing (scRNA-seq) allows for the interrogation of individual cell populations at ultra-high resolution by measuring gene expression in single cells through massively parallel RNA sequencing (Stubbington et al. 2017). Currently, scRNA-seq is being used almost exclusively in the human health sector, where studying cirrhosis in individual liver cells becomes possible (Ramachandran et al. 2019), but it may be suited to studying soil communities owing to the presence of different cell types and morphologies. The technique brings together single-cell sorting (potentially through fluorescence-activated cell sorting [FACS]) and genome-wide quantification of RNA transcripts to quantify expression.

Microarrays offer a well-established technique with which to gauge microbial transcriptional activity (Rick et al. 2001). In environmental studies, GeoChip is a commonly used technique that makes use of DNA probes to target key genes involved in biogeochemical cycles (Carbon, Nitrogen, Phosphorus), stress response and phylogeny (gyrase B; *gyrB*) (Tu et al. 2014). The recently developed GeoChip 5.0 covers more than 1500 functional genes and sheds light onto the prevalence of functional processes and identifies the microorganisms that drive those processes (Tu et al. 2014). Some drawbacks of relying on transcripts to capture microbial functionality include the instability of mRNA molecules which could compromise sample integrity, and the difficultly in resolving nuanced differences among cell populations as transcripts are typically extracted from populations of cells within a complex sample.

Metaproteomics moves even further along the central dogma of biology than either metagenomics or metatranscriptomics and measures the expressed proteins from a microbial community. An example pipeline could be to use isobaric tags for relative and absolute quantitation (iTRAQ) to determine the amount of protein from multiple samples (Aggarwal et al. 2006). The iTRAQ quantitation relies on covalent labeling on the *N*-terminus and side chain amines of peptides derived from protein digestion. The digestions use tags of varying mass to label all peptides from the sample after which LC-MS/MS is performed to produce fragmentation spectra. The fragmentation data can then be queried against reference databases to assign identities to the labeled peptides (Karp et al. 2010). Knowledge of which proteins are expressed can be particularly powerful in interpreting community activities and provides direct measures of who is doing what within the community. The analysis of the major catalytic domains within a sample is an ideal parallel to metagenomic sequence information as it links the genotype with the phenotype in situ. Metaproteomics is sensitive enough to reveal changes in the expressed protein content which may reflect the microbial community response to a stressor. Some examples include its utilization to detect the expression of cold shock proteins during changing environmental conditions such as the shift from frozen to thawed permafrost (Hultman et al. 2015) or to identify the most active litter decomposing enzymes in forest soil (Schneider et al. 2012). Although metaproteomics offers much promise in our quest to understand microbial function, its application lags considerably behind nucleic acid based sequencing owing primarily to technological hurdles, although recent developments allow for the sequential extraction of DNA, RNA, and proteins from soil samples (Gunnigle et al. 2014). Metaproteomics also suffers from the complex nature of soils, notably heterogeneity and nestedness, as well as the potential for salts and humic acids to interfere with analyses and extraction (Wilmes et al. 2015). Notwithstanding these drawbacks, metaproteomic analyses should become a central pillar of multi-omics studies in the foreseeable future as improvements in methodological and computational techniques improve measurements of peptide masses (i.e., through shotgun LC-MS/MS [liquid chromatography tandem mass spectrometry]) and protein identification.

Finally, newly developed tools like mmvec (microbe-metabolite vectors (Morton et al. 2019)) could be used to explore correlations among features from different multi-omic datasets through neural networks, such as linking community membership with the metabolites or proteins produced by the community. The use of multi-omics approaches increases annually as costs and instrument capabilities improve. Major advances have already been made in the cold biosphere by linking sequence data with transcripts and protein information (Hultman et al. 2015). This could serve as a template for hot desert research. Importantly, regardless of which sequencing platform or analysis pipeline is used, nucleic acids alone do not provide any insight into which taxa are active at the time of sampling or DNA isolation. In the following section, we highlight breakthroughs in techniques that give access to the active fraction of microbes in complex communities.

2.3 Approaches to Probe Active Members

A critical endeavor in microbial ecology research has been to identify the spectrum of active members within a complex community. This information is important in the study of hot desert microbial communities as microbial activity depends on rainfall or fog events, as deserts are water pulse system-driven environments (Cowan et al. 2019). There are many approaches that discriminate between active and inactive community members in a sample. The techniques listed here all have the potential to be coupled to genome-resolved metagenomics to answer questions about the timing of microbial activity and the metabolic processes that occur in situ.

Soil isotopic analyses can complement metagenome-informed functional analyses based on gene frequencies or metatranscriptomic sequence data. Common applications of soil isotopic ratios are to explore Carbon or Nitrogen natural isotopes (i.e., those occurring in nature) or by spiking in a labeled isotope and monitoring microbial uptake. Soil isotopic ratios can then be calculated by measuring the depletion or enrichment of natural abundance data against conventional standards such as VPDB (Vienna Pee Dee Belemnite) for $\delta^{13}C$ and N_2 gas for $\delta^{15}N$. Soil isotopic signatures are thus measures of the ratio of stable isotopes, such as ^{13}C:^{12}C, in parts per mille (‰). From these data it becomes possible to understand the prevalence of carbon and nitrogen transfer in situ as both $\delta^{13}C$ and $\delta^{15}N$ vary temporally and serve as indicators of productivity and abiotic transfer between reservoirs (Boschker and Middelburg 2002). Natural isotope analysis has shown microbial preferences for nitrogen fixation followed by nitrate assimilation in cyanobacteria-dominated hypoliths in the Namib Desert (Ramond et al. 2018), and this type of information helps to explain the fate of soil C and N in hot deserts. Moreover, these metabolic predictions can be compared to the isotopic signatures that are preserved in organic matter throughout geological time. For instance, it was recently suggested that biocrust analogs might have been major contributors of landmass nitrogen cycle during the Precambrian era (Thomazo et al. 2018).

Stable isotope probing (SIP) makes it possible to measure microbial growth rates using ^{18}O-labeled H_2O, referred to as [^{18}O] water. The ^{18}O becomes incorporated into newly synthesized DNA during replication (Hungate et al. 2015). This isotopic tracer assimilation allows for a quantitative measure of isotopic enrichment (termed qSIP) as the extracted DNA can then be centrifuged in an isopycnic solution and all density fractions are then sequenced separately. DNA Stable Isotope Probing (DNA-SIP) relies on the assumption that active microbes incorporate the isotopically labeled substrate that is unavailable to inactive microbes (those that are dormant, sporulated, or moribund). Incubating microbial communities with isotopically labeled compounds, such as ^{13}C, ^{15}N, or ^{18}O, makes it possible to fractionate active microbes that utilized the substrate from the inactive microbes that did not take up the labels (Ziels et al. 2018).

Separating ^{13}C "heavy" and ^{12}C "light" DNA fractions along cesium chloride density gradients physically fractionates the high-density DNA fractions that have been isotopically labeled from the inactive, unlabeled fraction of DNA (Malmstrom and Eloe-Fadrosh 2019). Sequencing these fractions can then reveal which microbes were active or inactive at the time of the substrate addition. DNA-SIP has provided insights as to which microbes degrade hydrocarbons and toluene following oil spills (Winderl et al. 2010; Gutierrez et al. 2013). An application of DNA-SIP in hot deserts could be to investigate which microbes become active during plant litter degradation or how physical trampling influences the activity of biocrust community members (Pepe-Ranney et al. 2016a). Exciting recent research has shown the widespread capacity of desert soil microbes to scavenge atmospheric trace gases to supplement their energy needs (Ji et al. 2017). Here, the innovative combination of gas chromatography, radiolabeled carbon ($^{14}CO_2$) incubations, and metagenomic sequencing provided evidence that atmospheric trace gases (H_2, CO, CO_2) are viable sources of energy in oligotrophic soils (Ji et al. 2017; Bay et al. 2018). Future research could reveal just how widespread these mechanisms of nutrient acquisition are in hot deserts by interrogating various soil biotopes.

A recently developed approach to gain access to active microbes is using Bio-Orthogonal Non-Canonical Amino Acid Tagging (BONCAT). BONCAT is a method of labeling and then separating active microbes based on labeled peptide uptake (such as methionine or homopropargylglycine [HPG]). The technique which was originally designed for use on mammalian cells (Dieterich et al. 2006) was adapted for marine microbial communities (Hatzenpichler and Orphan 2015) and has recently been used on temperate soils for the first time (Couradeau et al. 2019). BONCAT hinges upon the microbial uptake and incorporation of a modified amino acid bearing an additional alkyne residue that can be combined to a fluorescent probe bearing an azide group ($-N_3$) post-incubation. The introduced modified amino acids are only consumed by active cells which then incorporate labeled amino acids into peptides during translation, they are rendered fluorescent post-incubation through the cycloaddition of a fluorescent probe to the modified amino acid. Then, using Fluorescence-Activated Cell Sorting (FACS) it is possible to identify which cells consumed the labeled amino acid based on their fluorescence after a click chemistry reaction (this only occurs in the BONCAT-positive cells; Fig. 2.3). The sorted cells can then be sequenced to document which members constitute active or inactive populations (BONCAT-FACS-seq). This approach has demonstrated that both marine and temperate soil ecosystems contain a very high fraction of microbes that are active simultaneously (Hatzenpichler and Orphan 2015; Couradeau et al. 2019). It is important to note that BONCAT-FACS-seq assumes an equal uptake of the substrate by all active microbes, that there is no bias among isogenic cells regarding the labeled methionine, and that the addition of the substrate does not initiate activity of previously dormant cells.

2.4 Metabolomics for Studying Nutrient Exchange

Microbial metabolites serve as growth substrates and chemical effectors that can also mitigate environmental stresses in hot deserts (e.g., by producing sunscreen pigments, compatible solutes, quorum quenchers, or free-radical scavengers). Metabolomic techniques allow the characterization of both the intracellular (endometabolomics) and extracellular (exometabolomics) metabolite pools, which enables direct examination of the chemical ecology of microbes. Exometabolomic profiling methods (also referred to as "metabolic footprinting" (Allen et al. 2003, Kell et al. 2005)) have been used to characterize environmental metabolites in soils (Warren 2016; Swenson et al. 2018), root exudates (Strehmel et al. 2014; Zhalnina

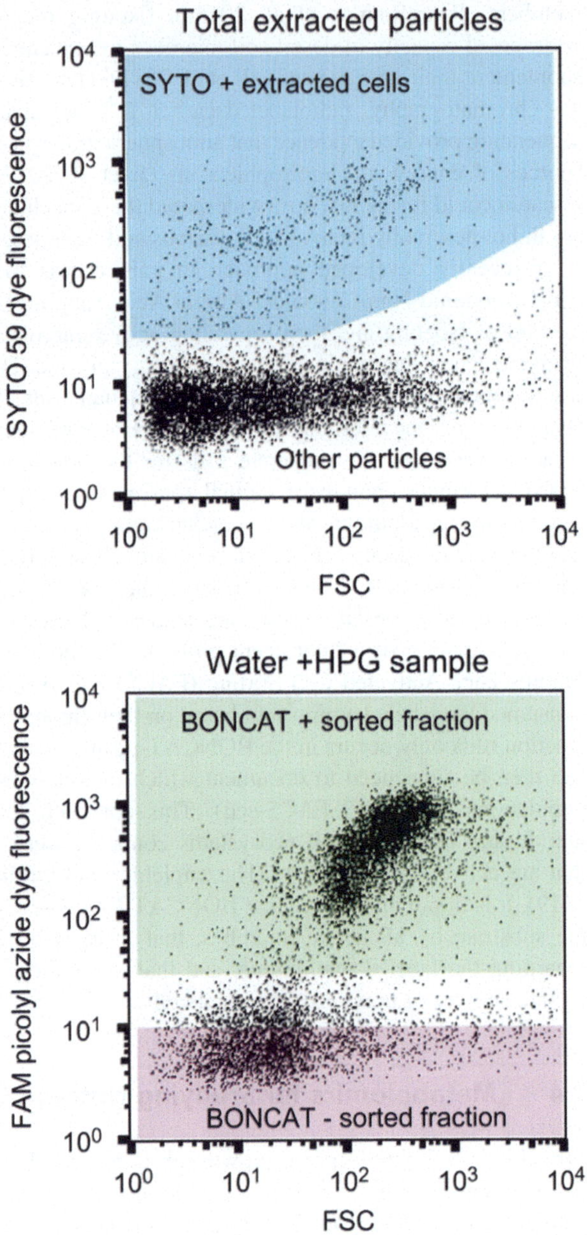

Fig. 2.3 Bio-orthogonal non-canonical amino acid tagging (BONCAT) coupled with fluorescence-activated cell sorting (FACS) allows for the sequential separation of cells from background particles (top panel) based on DNA dye staining fluorescence (SYTO59, shown in the blue portion of the panel). Next, the BONCAT + cells (green panel), i.e. those that are translationally active, are then separated from the translationally inactive cells (BONCAT -, pink panel) based on FAM picolyl azide dye fluorescence. Figure modified from Couradeau et al. (2019)

et al. 2018), and their transformations by microbes (Allen et al. 2003; Baran et al. 2015).

Central to these approaches is the ability to quantitatively measure and identify metabolites from complex mixtures. This is typically accomplished by integrating

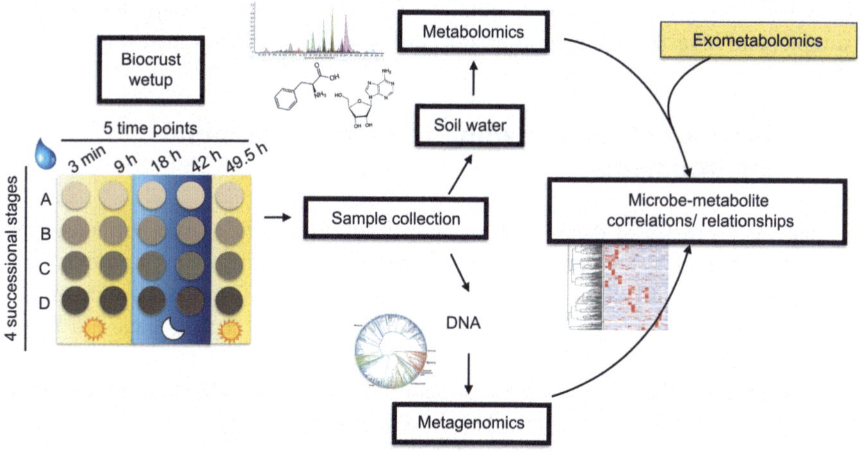

Fig. 2.4 A potential workflow to link metagenomic and metabolomic information modified from Swenson et al. (2018). Spearman rank correlations were used to determine which soil metabolites identify by liquid chromatography/mass spectrometry matched predicted functional genes identified using shotgun metagenomic analysis

chromatographic separation with tandem mass spectrometry-based analysis. Both liquid and gas chromatography are routinely integrated with mass spectrometry (LC-MS and GC-MS, respectively) and are excellent candidates for soil metabolite analysis. LC-MS is better suited for analysis of novel metabolites as well as those that are thermally labile, those with low volatility (even when derivatized), or large molecular weight metabolites. GC-MS is typically better suited for volatile metabolites and difficult to resolve isomers (e.g., monosaccharides). The electron ionization used in GC-MS results in rich fragmentation spectra for comparisons with spectra from authentic standards. While the electrospray ionization used for LC-MS does not typically result in as extensive fragmentation, tandem mass spectrometry provides additional information (in addition to exact mass, retention time, and isotope distribution) for comparison with spectral libraries and authentic standards (Liu and Locasale 2017) to aid in downstream metabolite identification.

In metabolomics there are two major data analysis approaches: "targeted" analysis of specific compounds of interest (Roberts et al. 2012) and "untargeted" global analysis (Alonso et al. 2015). In the first case, data corresponding to the compounds of interest are extracted and analyzed at the exclusion of the remaining compounds. In the second case, statistical methods are used to first identify features of interest (e.g., those altered between conditions) and then efforts are made to identify these compounds.

Linking metabolites to genomic information remains a key goal in metabolomics (Fig. 2.4). Tools such as MAGI (Metabolite Annotation and Gene Integration) achieve this by using biochemical reaction networks to reach a consensus between metabolite identifications and gene annotations (Erbilgin et al. 2019). This moves beyond analyzing the translational outputs of communities, by metaproteomics, for example, and investigates the end products of complex metabolic pathways.

2.5 High-Resolution Imaging to Resolve Microbial Environments

The habitat geometry of desert soils can constrain the ability of microorganisms to access key resources for their cellular activity such as liquid water, nutrients, and light. Imaging hot desert microorganisms in the context of their soil microniches might help us spatialize the information retrieved from molecular data and better predict microbe–microbe interactions in their environment.

X-ray tomography can reveal soil structure and help us understand the physical dynamics that occur within a sample by producing a cross-sectional image of the sample in a non-invasive fashion. In a recent analysis of biocrust samples (Couradeau et al. 2018), the authors used synchrotron-based X-ray microtomography (using hard X-rays) to answer whether exopolysaccharide (EPS) sheaths produced by *Cyanobacteria* in biocrust retained more moisture than the surrounding soil environment that was non-crusted (Fig. 2.5). They provided clear evidence of decreasing moisture gradients away from the cyanobacterial sheath in biocrust, which indicates stability in the cyanobacterial bundle volume as microbes retained their water content. Answering how the dry-down of non-crusted soils occurs is a valuable avenue for future research and microscale X-ray tomography could shed light on water redistribution across different soil types (Felde et al. 2014). The intensive computing requirement in terms of image segmentation and data analysis might limit the use of such techniques, however new avenues of image analysis using machine learning techniques seem to have yielded very promising results and might allow more systematic studies of these type of data (O'Mara et al. 2017).

Catalyzed reporter deposition-fluorescence in situ hybridization (CARD-FISH) is a powerful technique that can be easily coupled to confocal microscopy to visualize the localization of targeted microbes in a complex sample. The CARD-FISH method relies on the hybridization of a mono-labeled oligonucleotide probe that specifically targets a group of organisms, it can be as general as probing all bacteria and archaea or as specific as targeting a single species of interest. The signal amplification performed during the CARD-FISH allows the localization of single cells in the context of the soil (Schmidt and Eickhorst 2014). Such a powerful technique could be used with probes targeting specific taxonomic groups, such as UCYN-A732 for unicellular cyanobacteria (Krupke et al. 2013), to resolve the distribution of microbial functional labors at the relevant spatial scale. A drawback of the CARD-FISH approach is the hydrogen peroxide produced by the radical reaction that can damage cellular DNA.

An alternative to CARD-FISH could be the combination of FISH and fluorescence-activated cell sorting (FACS) coupled with metagenomic sequencing (FISH & FACS mini-metagenomes) which has the scope to enrich for taxonomic

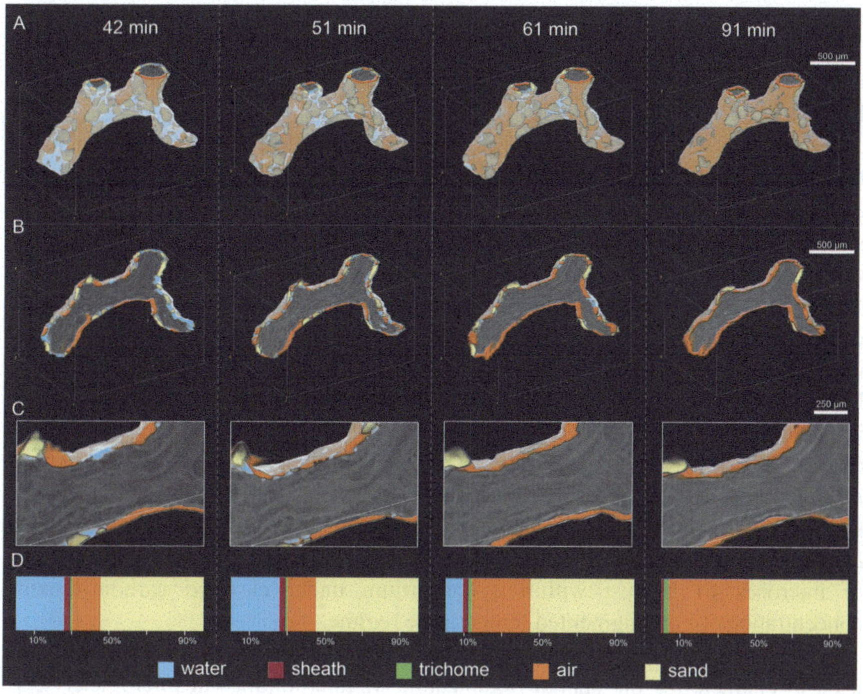

Fig. 2.5 In situ X-ray tomography is a powerful tool for visualizing the physical properties of sample material. Here, Couradeau et al. (2018) visualized how the rate of dehydration around biocrust bundles compared to the surrounding soil. Over a 91-minute period the bundles remained more hydrated than the soil as sheath exopolysaccharides retained moisture. Figure adapted from Couradeau et al. (2018). In the panels, (A) represents the full bundle visualization, (B) shows a cross section of the bundle while (C) indicates a magnified view of the bundle. Finally, (D) indicates the percentage of water, sheath, trichome, air and sand present in each sample

clades from the environment in a targeted fashion (Grieb et al. 2020). Unlike CARD-FISH the FISH & FACS approach relies on a two-step hybridization chain reaction FISH (HCR-FISH). Here, two fluorescently-labeled hairpin oligonucleotides and a chain reaction with an initiator sequence amplify the fluorescent signal to comparable intensities observed with CARD-FISH, yet is radical-free and does not require cells to be fixed thus aiding genome amplification (Yamaguchi et al. 2015).

Fluorescence-activated cell sorting (FACS) and laser-assisted microdissection are two techniques that are sensitive enough to capture individual microscopic particles for sequencing. In hot deserts, the targets of these methodologies could be applied to individual sand grains from different soil biotopes, for example. For biocrust research specifically, these techniques could reveal the community composition of soil particles that are physically associated with *Microcoleus* bundles (Garcia-Pichel and Wojciechowski 2009). This has long been a question in biocrust ecology and could provide insight into the diazotroph populations that are thought to support the nitrogen needs of *Microcoleus* sp. (Pepe-Ranney et al. 2016b). Critically, these micron-scale measurements could lead research away from the confounding effects of habitat heterogeneity (Armitage and Jones 2019) by offering the potential to measure cell–cell interactions over the appropriate spatial scale (Ladau and Eloe-Fadrosh 2019).

Nanoscale secondary ion mass spectrometry (NanoSIMS) is a powerful technique that has been applied to both non-soil ecosystems, particularly marine environments and terrestrial hot springs (McNichol et al. 2018), as well as temperate soil systems such as the rhizosphere of wheat plants (Vidal et al. 2018). The approach allows researchers to directly measure the activity of prokaryotes at the single-cell level via the incorporation of isotopically labeled substrates (Singer et al. 2017). NanoSIMS offers researchers the ability to quantify the anabolic activity of cells at 50 nm resolution by measuring both the elemental and isotopic composition of the sample. Moreover, when coupled with fluorescent in situ hybridization (FISH-NanoSIMS) it is possible to directly test for anabolic activity within the context of spatial associations between microbes. The technique is sufficiently powerful to discriminate activity between closely related microbes (those with high phylogenetic similarities) that occupy unique spatial niches (Dekas et al. 2016). Within the context of hot desert ecosystems, it would be possible, for example, to measure the activity of microbes of interest within a consortium under elevated carbon dioxide concentrations or an augmented temperature regime.

Metagenomic plot sampling by sequencing (MaPS-seq (Sheth et al. 2019)) can be used to characterize the spatial organization of microbiomes at micrometer-scale resolution. This technique is ideal for measuring community structure of intact microbiome samples (i.e., soil, biocrust, or lithic material). Initially, the sample of interest is immobilized within a gel matrix and then cryofractured into particles. Neighboring microbial taxa within the particles are then identified by droplet-based encapsulation, barcoded 16S rRNA gene amplification, and deep sequencing. The application of MaPS-seq on mammalian gut samples revealed a heterogeneous distribution of microbes which likely corresponds with their adaptations to discrete niches. These spatial metagenomes could be very useful in hot deserts as patches of activity are non-uniform and could reveal how microbes are intrinsically associated with soil particles, and with each other.

2.6 Conclusion and Future Perspectives

The approaches described in this chapter highlight technologies that can propel our investigations in microbial ecology. Clearly a major challenge is the integration of techniques to understand the ecology of soil microbes across relevant spatial and temporal scales. For example, metabolomics and proteomics provide important insights into in situ activities but are currently not sensitive enough for detailed analysis of micron sized samples that are accessible through sequencing and imaging approaches.

Thus, it is important to consider aspects such as spatial and temporal scales, as well as sample heterogeneity which can plague the interpretation of data generated

(Armitage and Jones 2019). Great caution is needed in extrapolating findings made at the local-scale to larger-scale processes. Factors that drive observations on the local scale may not be relevant on landscape scales, where larger dynamics such as climate regime become more important (Ladau and Eloe-Fadrosh 2019).

Acknowledgements This work was supported by funds provided by the Office of Science Early Career Research Program, Office of Biological and Environmental Research, of the U.-S. Department of Energy under contract no. DE-AC02-05CH11231 to Lawrence Berkeley National Laboratory.

References

Adriaenssens EM, Van Zyl L, De Maayer P, Rubagotti E, Rybicki E, Tuffin M, Cowan DA (2015) Metagenomic analysis of the viral community in Namib Desert hypoliths. Environ Microbiol 17:480–495

Aggarwal K, Choe LH, Lee KH (2006) Shotgun proteomics using the iTRAQ isobaric tags. Brief Funct Genomics 5:112–120

Allen J, Davey HM, Broadhurst D, Heald JK, Rowland JJ, Oliver SG, Kell DB (2003) High-throughput classification of yeast mutants for functional genomics using metabolic footprinting. Nat Biotechnol 21:692

Alneberg J, Bjarnason BS, de Bruijn I, Schirmer M, Quick J, Ijaz UZ, Lahti L, Loman NJ, Andersson AF, Quince C (2014) Binning metagenomic contigs by coverage and composition. Nat Methods 11:1144–1146

Alonso A, Marsal S, Julià A (2015) Analytical methods in untargeted metabolomics: state of the art in 2015. Front Bioeng Biotechnol 3:23

Armitage DW, Jones SE (2019) How sample heterogeneity can obscure the signal of microbial interactions. ISME J 13(11):2639–2646

Bao Z et al (2014) Metaproteomic identification of diazotrophic methanotrophs and their localization in root tissues of field-grown rice plants. Appl Environ Microbiol 80(16):5043–5052

Baran R, Brodie EL, Mayberry-Lewis J, Hummel E, Da Rocha UN, Chakraborty R, Bowen BP, Karaoz U, Cadillo-Quiroz H, Garcia-Pichel F (2015) Exometabolite niche partitioning among sympatric soil bacteria. Nat Commun 6:8289

Bay S, Ferrari B, Greening C (2018) Life without water: how do bacteria generate biomass in desert ecosystems? Microbiol Aust 39:28–32

Belnap J (2006) The potential roles of biological soil crusts in dryland hydrologic cycles. Hydrol Process Int J 20:3159–3178

Belnap J, Weber B, Büdel B (2016) Biological soil crusts as an organizing principle in drylands. In: Biological soil crusts: an organizing principle in drylands. Springer, Cham, pp 3–13

Bolduc B, Jang HB, Doulcier G, You Z-Q, Roux S, Sullivan MB (2017) vConTACT: an iVirus tool to classify double-stranded DNA viruses that infect Archaea and Bacteria. PeerJ 5:e3243

Bolyen E, Rideout JR, Dillon MR, Bokulich NA, Abnet CC, Al-Ghalith GA, Alexander H, Alm EJ, Arumugam M, Asnicar F (2019) Reproducible, interactive, scalable and extensible microbiome data science using QIIME 2. Nat Biotechnol 37:852–857

Boschker H, Middelburg J (2002) Stable isotopes and biomarkers in microbial ecology. FEMS Microbiol Ecol 40:85–95

Bowers RM, Kyrpides NC, Stepanauskas R, Harmon-Smith M, Doud D, Reddy T, Schulz F, Jarett J, Rivers AR, Eloe-Fadrosh EA (2017) Minimum information about a single amplified genome (MISAG) and a metagenome-assembled genome (MIMAG) of bacteria and archaea. Nat Biotechnol 35:725

Breitwieser FP, Lu J, Salzberg SL (2019) A review of methods and databases for metagenomic classification and assembly. Brief Bioinform 20:1125–1136

Brister JR, Ako-Adjei D, Bao Y, Blinkova O (2014) NCBI viral genomes resource. Nucleic Acids Res 43:D571–D577

Browne HP, Forster SC, Anonye BO, Kumar N, Neville BA, Stares MD, Goulding D, Lawley TD (2016) Culturing of 'unculturable' human microbiota reveals novel taxa and extensive sporulation. Nature 533:543–546

Callahan BJ, McMurdie PJ, Rosen MJ, Han AW, Johnson AJA, Holmes SP (2016) DADA2: high-resolution sample inference from Illumina amplicon data. Nat Methods 13:581

Caporaso JG, Kuczynski J, Stombaugh J, Bittinger K, Bushman FD, Costello EK, Fierer N, Pena AG, Goodrich JK, Gordon JI (2010) QIIME allows analysis of high-throughput community sequencing data. Nat Methods 7:335

Couradeau E, Felde VJMNL, Parkinson D, Uteau D, Rochet A, Cuellar C, Winegar G, Peth S, Northen TR, Garcia-Pichel F (2018) In situ X-ray tomography imaging of soil water and cyanobacteria from biological soil crusts undergoing desiccation. Front Environ Sci 6:65

Couradeau E, Sasse J, Goudeau D, Nath N, Hazen TC, Bowen BP, Chakraborty R, Malmstrom RR, Northen TR (2019) Probing the active fraction of soil microbiomes using BONCAT-FACS. Nat Commun 10:1–10

Cowan DA, Hopkins D, Jones B, Maggs-Kölling G, Majewska R, Ramond J-B (2019) Microbiomics of Namib Desert habitats. Extremophiles 1–13

Cross KL, Campbell JH, Balachandran M, Campbell AG, Cooper SJ, Griffen A, Heaton M, Joshi S, Klingeman D, Leys E (2019) Targeted isolation and cultivation of uncultivated bacteria by reverse genomics. Nat Biotechnol 37:1314–1321

Dekas AE, Connon SA, Chadwick GL, Trembath-Reichert E, Orphan VJ (2016) Activity and interactions of methane seep microorganisms assessed by parallel transcription and FISH-NanoSIMS analyses. ISME J 10:678

Delgado-Baquerizo M, Oliverio AM, Brewer TE, Benavent-González A, Eldridge DJ, Bardgett RD, Maestre FT, Singh BK, Fierer N (2018) A global atlas of the dominant bacteria found in soil. Science 359:320–325

Dieterich DC, Link AJ, Graumann J, Tirrell DA, Schuman EM (2006) Selective identification of newly synthesized proteins in mammalian cells using bioorthogonal noncanonical amino acid tagging (BONCAT). Proc Natl Acad Sci USA 103:9482–9487

Emerson JB, Roux S, Brum JR, Bolduc B, Woodcroft BJ, Jang HB, Singleton CM, Solden LM, Naas AE, Boyd JA (2018) Host-linked soil viral ecology along a permafrost thaw gradient. Nat Microbiol 3:870

Erbilgin O, Rübel O, Louie KB, Trinh M, Raad MD, Wildish T, Udwary D, Hoover C, Deutsch S, Northen TR (2019) MAGI: a method for metabolite annotation and gene integration. ACS Chem Biol 14:704–714

Eren AM, Esen ÖC, Quince C, Vineis JH, Morrison HG, Sogin ML, Delmont TO (2015) Anvi'o: an advanced analysis and visualization platform for 'omics data. PeerJ 3:e1319

Felde VJMNL, Peth S, Uteau-Puschmann D, Drahorad S, Felix-Henningsen P (2014) Soil microstructure as an under-explored feature of biological soil crust hydrological properties: case study from the NW Negev Desert. Biodivers Conserv 23:1687–1708

Finkel OM, Delmont TO, Post AF, Belkin S (2016) Metagenomic signatures of bacterial adaptation to life in the phyllosphere of a salt-secreting desert tree. Appl Environ Microbiol 82:2854–2861

Fonseca-García C, Coleman-Derr D, Garrido E, Visel A, Tringe SG, Partida-Martínez LP (2016) The cacti microbiome: interplay between habitat-filtering and host-specificity. Front Microbiol 7:150

Frank JA, Pan Y, Tooming-Klunderud A, Eijsink VG, McHardy AC, Nederbragt AJ, Pope PB (2016) Improved metagenome assemblies and taxonomic binning using long-read circular consensus sequence data. Sci Rep 6:25373

Friedman J, Alm EJ (2012) Inferring correlation networks from genomic survey data. PLoS Comput Biol 8:e1002687

Garcia-Pichel F, Wojciechowski MF (2009) The evolution of a capacity to build supra-cellular ropes enabled filamentous cyanobacteria to colonize highly erodible substrates. PLoS One 4: e7801

Garcia-Pichel F, Loza V, Marusenko Y, Mateo P, Potrafka RM (2013) Temperature drives the continental-scale distribution of key microbes in topsoil communities. Science 340:1574–1577

Grieb A, Bowers RM, Oggerin M, Goudeau D, Lee J, Malmstrom RR, Woyke T, Fuchs BM (2020) A pipeline for targeted metagenomics of environmental bacteria. Microbiome 8:1–17

Grosjean H, Oshima T (2007) How nucleic acids cope with high temperature. In: Physiology and biochemistry of extremophiles. American Society of Microbiology, Washington, DC, pp 39–56

Gunnigle E, Ramond J-B, Frossard A, Seeley M, Cowan D (2014) A sequential co-extraction method for DNA, RNA and protein recovery from soil for future system-based approaches. J Microbiol Methods 103:118–123

Gutierrez T, Singleton DR, Berry D, Yang T, Aitken MD, Teske A (2013) Hydrocarbon-degrading bacteria enriched by the Deepwater Horizon oil spill identified by cultivation and DNA-SIP. ISME J 7:2091

Handelsman J (2004) Metagenomics: application of genomics to uncultured microorganisms. Microbiol Mol Biol Rev 68:669–685

Hatzenpichler R, Orphan VJ (2015) Detection of protein-synthesizing microorganisms in the environment via bioorthogonal noncanonical amino acid tagging (BONCAT). In: Hydrocarbon and lipid microbiology protocols. Springer, Heidelberg, pp 145–157

Hillmann B, Al-Ghalith GA, Shields-Cutler RR, Zhu Q, Gohl DM, Beckman KB, Knight R, Knights D (2018) Evaluating the information content of shallow shotgun metagenomics. Msystems 3:e00069–e00018

Hiraoka S, Okazaki Y, Anda M, Toyoda A, Nakano S-i, Iwasaki W (2019) Metaepigenomic analysis reveals the unexplored diversity of DNA methylation in an environmental prokaryotic community. Nat Commun 10:159

Hultman J, Waldrop MP, Mackelprang R, David MM, McFarland J, Blazewicz SJ, Harden J, Turetsky MR, McGuire AD, Shah MB (2015) Multi-omics of permafrost, active layer and thermokarst bog soil microbiomes. Nature 521:208

Hungate BA, Mau RL, Schwartz E, Caporaso JG, Dijkstra P, van Gestel N, Koch BJ, Liu CM, McHugh TA, Marks JC (2015) Quantitative microbial ecology through stable isotope probing. Appl Environ Microbiol 81:7570–7581

Imachi H, Nobu MK, Nakahara N, Morono Y, Ogawara M, Takaki Y, Takano Y, Uematsu K, Ikuta T, Ito M (2020) Isolation of an archaeon at the prokaryote–eukaryote interface. Nature 1–7

Jang HB, Bolduc B, Zablocki O, Kuhn J, Roux S, Adriaenssens E, Brister JR, Kropinski A, Krupovic M, Turner D (2019) Gene sharing networks to automate genome-based prokaryotic viral taxonomy. BioRxiv. https://doi.org/10.1101/533240

Jegousse C, Yang Y, Zhan J, Wang J, Zhou Y (2017) Structural signatures of thermal adaptation of bacterial ribosomal RNA, transfer RNA, and messenger RNA. PLoS One 12:e0184722

Ji M, Greening C, Vanwonterghem I, Carere CR, Bay SK, Steen JA, Montgomery K, Lines T, Beardall J, van Dorst J (2017) Atmospheric trace gases support primary production in Antarctic desert surface soil. Nature 552:400

Kang D, Li F, Kirton ES, Thomas A, Egan RS, An H, Wang Z (2019) MetaBAT 2: an adaptive binning algorithm for robust and efficient genome reconstruction from metagenome assemblies. PeerJ Preprints 7:e27522v27521

Karp NA, Huber W, Sadowski PG, Charles PD, Hester SV, Lilley KS (2010) Addressing accuracy and precision issues in iTRAQ quantitation. Mol Cell Proteomics 9:1885–1897

Kell DB, Brown M, Davey HM, Dunn WB, Spasic I, Oliver SG (2005) Metabolic footprinting and systems biology: the medium is the message. Nat Rev Microbiol 3:557

King AM, Lefkowitz EJ, Mushegian AR, Adams MJ, Dutilh BE, Gorbalenya AE, Harrach B, Harrison RL, Junglen S, Knowles NJ (2018) Changes to taxonomy and the International Code of Virus Classification and Nomenclature ratified by the International Committee on Taxonomy of Viruses (2018). Arch Virol 163:2601–2631

Kolmogorov M, Rayko M, Yuan J, Polevikov E, Pevzner P (2019) metaFlye: scalable long-read metagenome assembly using repeat graphs. bioRxiv. https://doi.org/10.1101/637637

Kopylova E, Noé L, Touzet H (2012) SortMeRNA: fast and accurate filtering of ribosomal RNAs in metatranscriptomic data. Bioinformatics 28:3211–3217

Koren S, Walenz BP, Berlin K, Miller JR, Bergman NH, Phillippy AM (2017) Canu: scalable and accurate long-read assembly via adaptive k-mer weighting and repeat separation. Genome Res 27:722–736

Krupke A, Musat N, LaRoche J, Mohr W, Fuchs BM, Amann RI, Kuypers MM, Foster RA (2013) In situ identification and N2 and C fixation rates of uncultivated cyanobacteria populations. Syst Appl Microbiol 36:259–271

Kurtz ZD, Müller CL, Miraldi ER, Littman DR, Blaser MJ, Bonneau RA (2015) Sparse and compositionally robust inference of microbial ecological networks. PLoS Comput Biol 11: e1004226

Ladau J, Eloe-Fadrosh EA (2019) Spatial, temporal, and phylogenetic scales of microbial ecology. Trends Microbiol 27(8):662–669

Lagesen K, Hallin P, Rødland EA, Stærfeldt H-H, Rognes T, Ussery DW (2007) RNAmmer: consistent and rapid annotation of ribosomal RNA genes. Nucleic Acids Res 35:3100–3108

Lagier J-C, Dubourg G, Million M, Cadoret F, Bilen M, Fenollar F, Levasseur A, Rolain J-M, Fournier P-E, Raoult D (2018) Culturing the human microbiota and culturomics. Nat Rev Microbiol 16:540–550

León-Sobrino C, Ramond J-B, Maggs-Kölling G, Cowan DA (2019) Nutrient acquisition, rather than stress response over diel cycles, drives microbial transcription in a hyper-arid Namib Desert soil. Front Microbiol 10:1054

Li D, Liu C-M, Luo R, Sadakane K, Lam T-W (2015) MEGAHIT: an ultra-fast single-node solution for large and complex metagenomics assembly via succinct de Bruijn graph. Bioinformatics 31: 1674–1676

Liu X, Locasale JW (2017) Metabolomics: a primer. Trends Biochem Sci 42:274–284

Liu W-T, Marsh TL, Cheng H, Forney LJ (1997) Characterization of microbial diversity by determining terminal restriction fragment length polymorphisms of genes encoding 16S rRNA. Appl Environ Microbiol 63:4516–4522

Makhalanyane TP, Valverde A, Gunnigle E, Frossard A, Ramond J-B, Cowan DA (2015) Microbial ecology of hot desert edaphic systems. FEMS Microbiol Rev 39:203–221

Malmstrom RR, Eloe-Fadrosh EA (2019) Advancing genome-resolved metagenomics beyond the shotgun. MSystems 4:e00118–e00119

Marasco R, Mosqueira MJ, Fusi M, Ramond J-B, Merlino G, Booth JM, Maggs-Kölling G, Cowan DA, Daffonchio D (2018) Rhizosheath microbial community assembly of sympatric desert speargrasses is independent of the plant host. Microbiome 6:215

Martiny AC (2019) High proportions of bacteria are culturable across major biomes. ISME J 13(8): 2125–2128

McNichol J, Stryhanyuk H, Sylva SP, Thomas F, Musat N, Seewald JS, Sievert SM (2018) Primary productivity below the seafloor at deep-sea hot springs. Proc Natl Acad Sci USA 115:6756–6761

Morton JT, Aksenov AA, Nothias LF, Foulds JR, Quinn RA, Badri MH, Swenson TL, Van Goethem MW, Northen TR, Vazquez-Baeza Y (2019) Learning representations of microbe–metabolite interactions. Nat Methods 16:1306–1314

Moss EL, Maghini DG, Bhatt AS (2020) Complete, closed bacterial genomes from microbiomes using nanopore sequencing. Nat Biotechnol 1–7

Muyzer G, De Waal EC, Uitterlinden AG (1993) Profiling of complex microbial populations by denaturing gradient gel electrophoresis analysis of polymerase chain reaction-amplified genes coding for 16S rRNA. Appl Environ Microbiol 59:695–700

Namiki T, Hachiya T, Tanaka H, Sakakibara Y (2012) MetaVelvet: an extension of Velvet assembler to de novo metagenome assembly from short sequence reads. Nucleic Acids Res 40:e155

Nayfach S, Shi ZJ, Seshadri R, Pollard KS, Kyrpides NC (2019) New insights from uncultivated genomes of the global human gut microbiome. Nature 568:505

Nunes da Rocha U, Cadillo-Quiroz H, Karaoz U, Rajeev L, Klitgord N, Dunn S, Truong V, Buenrostro M, Bowen BP, Garcia-Pichel F (2015) Isolation of a significant fraction of non-phototroph diversity from a desert biological soil crust. Front Microbiol 6:277

Nurk S, Meleshko D, Korobeynikov A, Pevzner PA (2017) metaSPAdes: a new versatile metagenomic assembler. Genome Res 27:824–834

O'Mara A, King AE, Vickers JC, Kirkcaldie MT (2017) ImageSURF: an ImageJ plugin for batch pixel-based image segmentation using random forests. J Open Res Softw 5:1–7

Osborn AM, Moore ER, Timmis KN (2000) An evaluation of terminal-restriction fragment length polymorphism (T-RFLP) analysis for the study of microbial community structure and dynamics. Environ Microbiol 2:39–50

Peng Y, Leung HC, Yiu S-M, Chin FY (2012) IDBA-UD: a de novo assembler for single-cell and metagenomic sequencing data with highly uneven depth. Bioinformatics 28:1420–1428

Pepe-Ranney C, Campbell AN, Koechli CN, Berthrong S, Buckley DH (2016a) Unearthing the ecology of soil microorganisms using a high resolution DNA-SIP approach to explore cellulose and xylose metabolism in soil. Front Microbiol 7:703

Pepe-Ranney C, Koechli C, Potrafka R, Andam C, Eggleston E, Garcia-Pichel F, Buckley DH (2016b) Non-cyanobacterial diazotrophs mediate dinitrogen fixation in biological soil crusts during early crust formation. ISME J 10:287

Pham VH, Kim J (2012) Cultivation of unculturable soil bacteria. Trends Biotechnol 30:475–484

Quince C, Walker AW, Simpson JT, Loman NJ, Segata N (2017) Shotgun metagenomics, from sampling to analysis. Nat Biotechnol 35:833

Ramachandran P, Dobie R, Wilson-Kanamori J, Dora E, Henderson B, Luu N, Portman J, Matchett K, Brice M, Marwick J (2019) Resolving the fibrotic niche of human liver cirrhosis at single-cell level. Nature 575:512–518

Ramond J-B, Woodborne S, Hall G, Seely M, Cowan DA (2018) Namib Desert primary productivity is driven by cryptic microbial community N-fixation. Sci Rep 8:6921

Rhoads A, Au KF (2015) PacBio sequencing and its applications. Genomics Proteomics Bioinformatics 13:278–289

Rick WY, Wang T, Bedzyk L, Croker KM (2001) Applications of DNA microarrays in microbial systems. J Microbiol Methods 47:257–272

Roberts LD, Souza AL, Gerszten RE, Clish CB (2012) Targeted metabolomics. Curr Protoc Mol Biol 98:30.32.31–30.32.24

Roux S, Enault F, Hurwitz BL, Sullivan MB (2015) VirSorter: mining viral signal from microbial genomic data. PeerJ 3:e985

Sabath N, Ferrada E, Barve A, Wagner A (2013) Growth temperature and genome size in bacteria are negatively correlated, suggesting genomic streamlining during thermal adaptation. Genome Biol Evol 5:966–977

Schindlbacher A, Rodler A, Kuffner M, Kitzler B, Sessitsch A, Zechmeister-Boltenstern S (2011) Experimental warming effects on the microbial community of a temperate mountain forest soil. Soil Biol Biochem 43:1417–1425

Schloss PD, Westcott SL, Ryabin T, Hall JR, Hartmann M, Hollister EB, Lesniewski RA, Oakley BB, Parks DH, Robinson CJ (2009) Introducing mothur: open-source, platform-independent, community-supported software for describing and comparing microbial communities. Appl Environ Microbiol 75:7537–7541

Schmidt H, Eickhorst T (2014) Detection and quantification of native microbial populations on soil-grown rice roots by catalyzed reporter deposition-fluorescence in situ hybridization. FEMS Microbiol Ecol 87:390–402

Schneider T, Keiblinger KM, Schmid E, Sterflinger-Gleixner K, Ellersdorfer G, Roschitzki B, Richter A, Eberl L, Zechmeister-Boltenstern S, Riedel K (2012) Who is who in litter decomposition? Metaproteomics reveals major microbial players and their biogeochemical functions. ISME J 6:1749

Schulz F, Alteio L, Goudeau D, Ryan EM, Feiqiao BY, Malmstrom RR, Blanchard J, Woyke T (2018) Hidden diversity of soil giant viruses. Nat Commun 9:4881

Schulz F, Roux S, Paez-Espino D, Jungbluth S, Walsh D, Denef VJ, McMahon KD, Konstantinidis KT, Eloe-Fadrosh EA, Kyrpides N (2020) Giant virus diversity and host interactions through global metagenomics. Nature 1–7

Schwartz MH, Wang H, Pan JN, Clark WC, Cui S, Eckwahl MJ, Pan DW, Parisien M, Owens SM, Cheng BL (2018) Microbiome characterization by high-throughput transfer RNA sequencing and modification analysis. Nat Commun 9:5353

Sheth RU, Li M, Jiang W, Sims PA, Leong KW, Wang HH (2019) Spatial metagenomic characterization of microbial biogeography in the gut. Nat Biotechnol 37:877–883

Singer E, Wagner M, Woyke T (2017) Capturing the genetic makeup of the active microbiome in situ. ISME J 11:1949–1963

Smets W, Leff JW, Bradford MA, McCulley RL, Lebeer S, Fierer N (2016) A method for simultaneous measurement of soil bacterial abundances and community composition via 16S rRNA gene sequencing. Soil Biol Biochem 96:145–151

Steen AD, Crits-Christoph A, Carini P, DeAngelis KM, Fierer N, Lloyd KG, Thrash JC (2019) High proportions of bacteria and archaea across most biomes remain uncultured. ISME J 13: 3126–3130

Strehmel N, Böttcher C, Schmidt S, Scheel D (2014) Profiling of secondary metabolites in root exudates of Arabidopsis thaliana. Phytochemistry 108:35–46

Stubbington MJ, Rozenblatt-Rosen O, Regev A, Teichmann SA (2017) Single-cell transcriptomics to explore the immune system in health and disease. Science 358:58–63

Swenson TL et al (2015) Untargeted soil metabolomics methods for analysis of extractable organic matter. Soil Biol Biochem 80:189–198

Swenson TL, Karaoz U, Swenson JM, Bowen BP, Northen TR (2018) Linking soil biology and chemistry in biological soil crust using isolate exometabolomics. Nat Commun 9:19

Tahon G, Tytgat B, Lebbe L, Carlier A, Willems A (2018) Abditibacterium utsteinense sp. nov., the first cultivated member of candidate phylum FBP, isolated from ice-free Antarctic soil samples. Syst Appl Microbiol 41:279–290

Thomazo C, Couradeau E, Garcia-Pichel F (2018) Possible nitrogen fertilization of the early Earth Ocean by microbial continental ecosystems. Nat Commun 9:1–8

Trubl G, Solonenko N, Chittick L, Solonenko SA, Rich VI, Sullivan MB (2016) Optimization of viral resuspension methods for carbon-rich soils along a permafrost thaw gradient. PeerJ 4: e1999

Tu Q, Yu H, He Z, Deng Y, Wu L, Van Nostrand JD, Zhou A, Voordeckers J, Lee YJ, Qin Y (2014) GeoChip 4: a functional gene-array-based high-throughput environmental technology for microbial community analysis. Mol Ecol Resour 14:914–928

Urich T, Lanzén A, Qi J, Huson DH, Schleper C, Schuster SC (2008) Simultaneous assessment of soil microbial community structure and function through analysis of the meta-transcriptome. PLoS One 3:e2527

Valverde A, Makhalanyane TP, Seely M, Cowan DA (2015) Cyanobacteria drive community composition and functionality in rock–soil interface communities. Mol Ecol 24:812–821

Vaneechoutte M, Rossau R, De Vos P, Gillis M, Janssens D, Paepe N, De Rouck A, Fiers T, Claeys G, Kersters K (1992) Rapid identification of bacteria of the Comamonadaceae with amplified ribosomal DNA-restriction analysis (ARDRA). FEMS Microbiol Lett 93:227–233

Van Goethem MW, Makhalanyane TP, Cowan DA, Valverde A (2017) Cyanobacteria and alphaproteobacteria may facilitate cooperative interactions in niche communities. Front Microbiol 8:2099

Van Goethem MW, Swenson TL, Trubl G, Roux S, Northen TR (2019) Characteristics of wetting-induced bacteriophage blooms in biological soil crust. mBio 10(6)

Vidal A, Hirte J, Bender SF, Mayer J, Gattinger A, Höschen C, Schädler S, Iqbal TM, Mueller CW (2018) Linking 3D soil structure and plant-microbe-soil carbon transfer in the rhizosphere. Front Environ Sci 6:9

Warren CR (2016) Do microbial osmolytes or extracellular depolymerisation products accumulate as soil dries? Soil Biol Biochem 98:54–63

Wiegand S, Jogler M, Boedeker C, Pinto D, Vollmers J, Rivas-Marín E, Kohn T, Peeters SH, Heuer A, Rast P (2020) Cultivation and functional characterization of 79 Planctomycetes uncovers their unique biology. Nat Microbiol 5:126–140

Wilmes P, Heintz-Buschart A, Bond PL (2015) A decade of metaproteomics: where we stand and what the future holds. Proteomics 15:3409–3417

Winderl C, Penning H, Von Netzer F, Meckenstock RU, Lueders T (2010) DNA-SIP identifies sulfate-reducing Clostridia as important toluene degraders in tar-oil-contaminated aquifer sediment. ISME J 4:1314

Wu Y-W, Simmons BA, Singer SW (2015) MaxBin 2.0: an automated binning algorithm to recover genomes from multiple metagenomic datasets. Bioinformatics 32:605–607

Yamaguchi T, Kawakami S, Hatamoto M, Imachi H, Takahashi M, Araki N, Yamaguchi T, Kubota K (2015) In situ DNA-hybridization chain reaction (HCR): a facilitated in situ HCR system for the detection of environmental microorganisms. Environ Microbiol 17:2532–2541

Zablocki O, van Zyl L, Adriaenssens EM, Rubagotti E, Tuffin M, Cary SC, Cowan D (2014) High-level diversity of tailed phages, eukaryote-associated viruses, and virophage-like elements in the metaviromes of Antarctic soils. Appl Environ Microbiol 80:6888–6897

Zablocki O, Adriaenssens EM, Cowan D (2016) Diversity and ecology of viruses in hyperarid desert soils. Appl Environ Microbiol 82:770–777

Zhalnina K, Louie KB, Hao Z, Mansoori N, da Rocha UN, Shi S, Cho H, Karaoz U, Loqué D, Bowen BP (2018) Dynamic root exudate chemistry and microbial substrate preferences drive patterns in rhizosphere microbial community assembly. Nat Microbiol 3:470

Ziels RM, Sousa DZ, Stensel HD, Beck DA (2018) DNA-SIP based genome-centric metagenomics identifies key long-chain fatty acid-degrading populations in anaerobic digesters with different feeding frequencies. ISME J 12:112–123

Phototrophic Mats of the Desert: The Bacteria of the Biological Soil Crust Community

3

Blaire Steven

Abstract

Biological soil crusts (biocrusts) are surface soil communities that can be made up of cyanobacteria, lichens, heterotrophic bacteria, archaea, algae, mosses, liverworts, and fungi. These communities play important ecological roles, especially in global drylands, where they can colonize patches of soil between the sparsely distributed vegetation. In the early stages of development, bacteria play a central role in the colonization and development of biocrusts. Thus, this chapter focuses on those keystone bacterial populations that establish and nourish biocrusts. This chapter is not meant to be a comprehensive review of biocrusts, as this would require substantially more space than allotted to a single book chapter. Instead, the goal of this chapter is to introduce biocrusts to the non-specialist, in hopes that their importance and roles in desert ecology will be recognized, particularly in hot deserts where they have been critically understudied.

Keywords

Biological soil crusts · Bacteria · Drylands · Cyanobacteria · Desert microbiology

3.1 Introduction: What Constitutes a Biocrust?

Various crusts can form on the surface of soils. Broadly these crusts can be categorized into two distinct types: physical crusts and biological crusts. Physical crusts are created from various abiotic influences including compression from

B. Steven (✉)
Department of Environmental Sciences, Connecticut Agricultural Experiment Station, New Haven, CT, USA
e-mail: blaire.steven@ct.gov

J.-B. Ramond, D. A. Cowan (eds.), *Microbiology of Hot Deserts*, Ecological Studies 244, https://doi.org/10.1007/978-3-030-98415-1_3

Fig. 3.1 Differing biocrust structures. (**a**) Pinnacled cyanobacteria-dominated biocrusts on the Colorado Plateau, Utah, USA. Photo credit Bowen Dou. (**b**) Rugose biocrust with some lichenization, Utah, USA. (**c**) Lichenized biocrust Utah, USA. Lichen appear as colored patches on the biocrust surface. Photo credit **b**, **c** Cheryl Kuske. (**d**) A smooth biocrust from a hot desert in Australia. Photo credit David Elliott

rainfall, animal trampling, evaporative processes, or a combination of any or all of the above (Belnap 2003a; Williams et al. 2018). In contrast, biological crusts consist of a community of organisms that colonize surface soils. The diversity in appearance, structure, and function of biocrusts is underpinned by the number of terms that have collectively been employed to describe these communities. They have variously been referred to as microbiotic, microphytic, cryptogamic, or cryptobiotic crusts (Kleiner and Harper 1972; West 1990; Belnap 1993). In this chapter, the term biological soil crusts (shortened to biocrusts) is employed to refer to the essential biological nature of these crusts. The organisms that make up biocrusts include cyanobacteria, algae, lichens, mosses, liverworts, and associated heterotrophic fungi, bacteria, and archaea (Belnap 2003b; Martínez et al. 2006; Steven et al. 2015; Fig. 3.1). The biomass and distribution of biocrusts are largely controlled by the local climate, with the lowest diversity and abundance of biocrust organisms generally occurring in the hottest driest deserts (Bowker et al. 2016). In these environments, a biocrust may appear as not much more than consolidated sand (Fig. 3.1d). Thus, in the most inhospitable environments it can be a challenge to distinguish biocrusts from physical crusts at a macroscopic scale.

3.2 Distribution of Biocrusts

Biocrusts have been found on every continent. They are present at the most northern latitudes of the arctic to the limited patches of soil present in Antarctica (e.g., Cary et al. 2010; Steven et al. 2013b; Colesie et al. 2014; Williams et al. 2017). Biocrusts generally inhabit places where vascular plants are rare or absent. Because the limiting factor for plant productivity is often related to water availability, biocrusts are primarily associated with arid environments. However, aridity may not be the only predictor of biocrust presence. For example, biocrusts occur in temperate regions, sand dunes, high altitude alpine environments, glacier forefields, and reclaimed mining sites that exclude vascular plants, but may not be classified as arid zones (Yoshitake et al. 2010; Büdel et al. 2014; Gypser et al. 2016a, b; Corbin and Thiet 2020).

Biocrusts can occur as distinct small patches, only colonizing the hospitable soil niches suitable to sustain microbial life. For instance, in hyper-arid desert biocrusts may occur on the underside of translucent rocks, most commonly quartz, which allows for the transmission of sufficient light to allow for the development of photosynthetic communities (Cowan et al. 2011; Makhalanyane et al. 2013). These small desert oases under rocks have been referred to as hypoliths, but can also be considered isolated biocrust patches (Chan et al. 2012; Pointing 2016). At the other extreme, biocrusts can cover a significant proportion of the surface in some ecosystems (*ca.* 20–70%; Briggs and Morgan 2008; Viles 2012; Bowker et al. 2014; Fig. 3.1a).

Recent advances in remote sensing and modeling have increased our ability to monitor and predict the distribution of biocrusts. A recent study employing meta-analysis and environmental niche modeling identified 24% of the global land surface to be suitable for biocrust colonization (Rodriguez-Caballero et al. 2018; Fig. 3.2). Yet, studies on biocrusts have largely been conducted in just four regions, the USA, Europe, Israel, and Australia (Bowker et al. 2018). Smaller numbers of studies have been conducted in South America (e.g., Szyja et al. 2019; Jung et al. 2020), Africa (Ullmann and Büdel 2001; Büdel et al. 2009), and Antarctica (Colesie et al. 2014). For instance, there is limited or no information on biocrusts from some of the hottest deserts including the Lut Desert in Iran and the deserts of the Pakistani/Indian subcontinent, central Asia, and much of Africa. In this regard, there is a critical research need for the biocrust of these regions to be investigated in order to gain a truly global inventory of biocrust microorganisms.

3.3 Ecosystem Services of Biocrusts

Biocrusts often represent one of the largest collections of biomass in an arid ecosystem. As such, they provide many services to the surrounding landscape. Below are outlined some of the specific functions that biocrusts play in drylands.

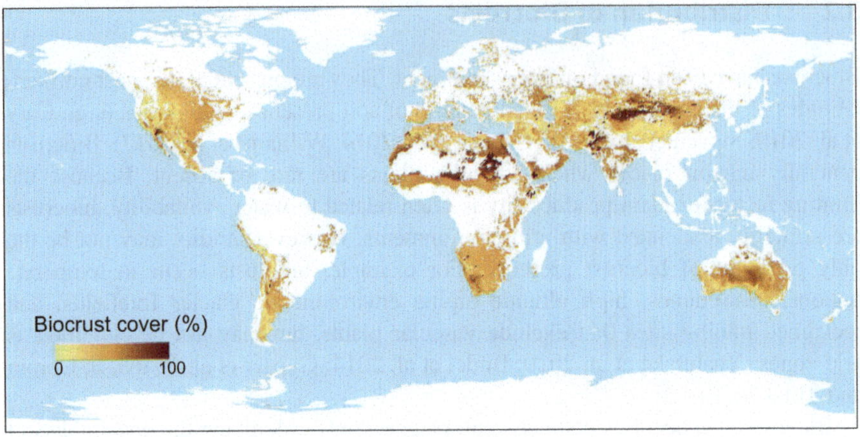

Fig. 3.2 Estimated global distribution of habitat for biological soil crusts. Up to 24% of the global surface is potentially colonizable by biocrusts. These data do not include polar regions. Figure reproduced with permission from Rodriguez-Caballero et al. (2018)

3.3.1 Soil Stabilization

Arid ecosystems are susceptible to wind erosion, as the lack of vegetative cover leaves surface soils exposed. Additionally, many arid soils are a mixture of sand and fine particles that are particularly prone to erosion (Belnap and Büdel 2016). Wind erosion increases soil nutrient heterogeneity by scouring and depositing nutrients across the landscape (Li et al. 2008; Ravi et al. 2010; Sankey et al. 2010). In dryland regions, soil loss can greatly exceed soil formation, as the genesis of arid soils is extremely slow (Pillans 1997). Millions of tons of soil are lost from just North American drylands annually (Duniway et al. 2019). This loss of topsoil reduces minerals and organic matter as well as coarsening the soil, which can lead to further soil degradation (Neff et al. 2005; McTainsh and Strong 2007).

Biocrusts stabilize soils through physical binding of soil particles with the cyanobacteria filaments and polysaccharides that act as a soil "glue" (Eldridge and Leys 2003; Zhang et al. 2006). Soils with a 5 mm thick biocrust have ~400% higher stability than uncrusted soils (Pi and Sharratt 2019). In wind tunnel experiments, biocrusts stabilized soils at wind speeds of up to 40 m s^{-1} (>100 km h^{-1}; Marticorena et al. 1997; Belnap and Gillette 1998)). Furthermore, because biocrusts are present year-round and retain their coverage in drought years, they often provide more stability than vegetation (Belnap and Gillette 1998). The ability of biocrusts to stabilize soils is related to their species composition and level of development (McKenna Neuman et al. 1996; McKenna Neuman and Maxwell 1999; Hu et al. 2002). For instance, there is a threshold biomass at which biocrusts begin to stabilize the soils. A recent study found that the most important factor in predicting the resistance of biocrusts to wind erosion was the content of cyanobacterial chlorophyll *a* (a metric of cyanobacteria biomass), which needed to exceed >0.014 mg g^{-1} to provide a significant erosion resistance (Belnap et al. 2014). As mosses and lichens colonize biocrusts they also tend to better resist wind erosion (Zhao et al. 2014; Belnap and Büdel 2016). In this regard, mature well-established biocrusts appear to increase the soil stabilizing process. Thus, predicting and modeling biocrust effects on soil stability will require ecosystem-specific measurements performed in the field.

3.3.2 Water Infiltration

The majority of research on biocrusts has been performed in arid environments. As such, much of the focus of this research has been on the interaction between biocrusts and hydrology and, specifically, how surface water enters the soil, known as water infiltration. Biocrusts have a multitude of effects on their surrounding environment, which can influence the movement of water. Pinnacled biocrusts that occur in cold deserts absorb raindrop energy and inhibit runoff, thereby increasing water infiltration (Barger et al. 2006). In contrast, in hot sandy deserts with poorly developed smooth biocrusts, the crusts can increase runoff resulting in lowered infiltration (Yair 2001; Belnap 2006). Biocrusts can also capture and store aerosolic dust (Reynolds et al. 2001). Silt and clay particles increase the absorptivity of the soil, but small dust particles can block soil pores reducing infiltration (Verrecchia et al. 1995; Felde et al. 2014). In fact, the main variable found to predict water infiltration of biocrusts of the Negev Desert of Israel was the content of fine soil particles (Felde et al. 2018). Finally, the dark color of some biocrusts increases albedo that can increase evaporative water loss from biocrusts (Zhang et al. 2014). The biology of biocrusts also interacts with how water is moved and stored in soils. The filaments that make up the biocrust absorb water and swell, which can fill soil pores. Once the biocrust is water-saturated, there can be substantial runoff and lower water infiltration in comparison to uncrusted soils (Verrecchia et al. 1995; Eldridge et al. 2000; Yair 2001). The polysaccharides produced by biocrust organisms can also interact with water infiltration. Extracellular polymers produced by microorganisms that make up the biocrust can absorb water and increase water retention in the soil (Adessi et al. 2018). In this manner, even if the total amount of water that moves into the biocrust is reduced, it is present for a longer period of time. Furthermore, biocrust organisms also differ in their ability to absorb and store water. Cyanobacteria can uptake ten times their total dry weight as water (de Winder et al. 1989; Wynn-Williams 2002). In contrast, mosses trap water in specialized structures and can store significantly more water than their cyanobacterial counterparts (Green et al. 2011; Tao and Zhang 2012; Coe et al. 2014).

These observations all point to the difficulty in predicting the influence of biocrusts on water infiltration. There is no simple maxim such as biocrusts will increase water infiltration. Instead, water infiltration is a metric that has to be determined in the field, taking into account soil edaphic factors, the biological communities that make up the biocrusts, and their level of development. A consequential outcome of the multifaceted influences of biocrusts on water infiltration is that biocrusts may act to increase the spatial heterogeneity of the landscape. Bare sandy soils have a high water infiltration capacity and will experience a relatively uniform depth of water retention. In contrast, biocrusts can redistribute water with differing levels of runoff between biocrust patches, thereby increasing the diversity of soil conditions in their vicinity (Maestre et al. 2005; Bowker et al. 2013; Concostrina-Zubiri et al. 2013). In this regard, biocrusts and their relationship with water is one of the mechanisms through which biocrusts act to shape the environment at a landscape scale. This is further discussed in Chap. 1.

3.3.3 Nitrogen Fixation

The low productivity of drylands results in ecosystem nitrogen limitations. Biocrusts are nitrogen hotspots in deserts, often containing 4–10 times more nitrogen content than the surrounding uncrusted soils (Johnson et al. 2007; Housman et al. 2007; Drahorad et al. 2013). Biological nitrogen fixation is likely the dominant path for nitrogen accumulation in biocrusts (Barger et al. 2016). Estimates for global biocrust nitrogen fixation are on the order of 49 Tg of nitrogen per year, almost half of the total estimate for total terrestrial nitrogen fixation (Elbert et al. 2012). Yet, as metabolic activity of biocrusts is generally limited to periods with soil moisture, nitrogen accumulation occurs for only a fraction of the year.

Biocrusts harbor a variety of nitrogen-fixing microorganisms. The predominant nitrogen-fixing organisms are free-living cyanobacteria, although heterotrophic nitrogen-fixing bacteria are also present and contribute to nitrogen fixation (Yeager et al. 2004, 2007; Pepe-Ranney et al. 2016). The nitrogen-fixing organisms in biocrusts show high excretion of nitrogenous compounds, with measured rates of 5–70% of fixed nitrogen being released from the cells to the community (Silvester et al. 1996). In this manner, biocrusts enrich the soil and surrounding community with fixed nitrogen. This nitrogen release may not even be limited to nitrogen-fixing organisms. The cyanobacterium *Microcoleus vaginatus* (non-nitrogen fixing) continuously releases nitrogen-containing metabolites, a feature not common to other cyanobacteria (Baran et al. 2015). In fact, *M. vaginatus* may recruit heterotrophic nitrogen-fixing bacteria to a region termed the "cyanosphere," the zone directly adjacent to cyanobacterial filaments, and exploit their ability to fix nitrogen in exchange for carbon (Couradeau et al. 2019). Thus, individual bacteria and bacterial cross-feeding all contribute to the enrichment of nitrogen in biocrusts.

A question remains as to how much of the biocrust fixed nitrogen is transported to underlying soils or outside of the immediate vicinity of the biocrusts. The exopolysaccharide produced by biocrusts readily sorbs compounds, effectively trapping nitrogen (Swenson et al. 2018). Yet, subsurface nitrogen was enriched under biocrusts, suggesting a vertical transfer of nitrogen (Guo et al. 2008). Furthermore, stable isotope analyses suggest that soils under biocrusts are nitrogen-enriched, consistent with the biocrusts contributing to the fertility of subsurface soils (Evans and Lange 2001). At larger scales, water runoff and wind erosion are potential pathways of transferring nitrogen from biocrusts at the ecosystem scale (Ludwig et al. 1996; Ravi et al. 2010). Studies have found nitrogen transfers from biocrusts to downslope plants and soils (Eldridge et al. 2000, 2002; Maestre et al. 2002). Taken together, these observations all point to the importance of biocrusts in the nitrogen cycle of drylands. Nonetheless, considerable uncertainty remains in determining the nitrogen balance in biocrusts. There is significantly less information on pathways of nitrogen transformation and loss from biocrusts, such as nitrification and denitrification (Giguere et al. 2015; Weber et al. 2015; Meusel et al. 2018; Abed et al. 2013). There is also little information on how processes such as nitrogen deposition or climate change will affect nitrogen-fixing populations in biocrusts (Steven et al. 2014, 2018; Mueller et al. 2015). Thus, the nitrogen cycle in biocrusts remains a fertile ground for future studies.

3.3.4 Other Ecosystem Services of Biocrusts

Biocrusts potentially play a multitude of other roles that may be less well documented than those above or cannot be generalized to biocrusts as a whole. For instance, an increasing number of studies are identifying higher levels of phosphorus in biocrusts, suggesting they may play a role in the phosphorus cycle (Reynolds et al. 2001; Pointing and Belnap 2012; Wu et al. 2013; Baumann et al. 2019). Biocrusts may also influence local vegetation through altering hydrology or nutrient cycles. Nutrient and water levels of plants in the vicinity of biocrusts are both elevated compared to in uncrusted sites (Harper and Belnap 2001; Serpe et al. 2006). Additionally, biocrusts can trap plant seeds along with dust. Thus, biocrusts may increase the local seed bank (Kinloch and Friedel 2005; Li et al. 2005; Zhang et al. 2016a, b). However, some negative effects on seed retention have been documented (Boeken et al. 2004). These differences appear to be mainly due to biocrust morphology, with rough pinnacled biocrusts of cold deserts trapping more seeds than the smooth biocrust of hot deserts (Su et al. 2007; Bertiller and Ares 2011). There are also complex interactions between biocrusts and seed germination with studies identifying both positive and negative effects (Rivera-Aguilar et al. 2005; Deines et al. 2007; Su et al. 2009; Gilbert and Corbin 2019).

In summary, biocrusts exert a multitude of functions in drylands. However, the effects of biocrusts are complex and often dependent on the level of biocrust development, the species composition, and local edaphic and environmental factors. Including biocrusts in process models requires ecosystem-specific measurements and an understanding of the local climate, geography, and land use. In this regard, the diversity in form and structure of biocrusts is matched by the variety of roles they play in supporting dryland ecosystems.

3.4 The Bacteria of Biocrusts

Biocrusts in their initial stages of development are predominantly composed of bacterial biomass (Yoshitake et al. 2010; Liu et al. 2013; Lan et al. 2013). Pioneering species of cyanobacteria colonize the soils, stabilizing the soil particles and creating niches for other organisms to inhabit. Later in development, other organisms such as bryophytes and lichens are common or even dominant biocrust members (Bowker 2007; Lalley and Viles 2008; Dojani et al. 2011; Muggia et al. 2013). Additionally, microfauna such as nematodes and protists have also been identified in biocrusts (Darby et al. 2007, 2010; Bamforth 2008; Dumack et al. 2016), but are not discussed in this chapter.

3.4.1 Bacterial Diversity

Biocrusts are a diverse assemblage of organisms made up of multiple species that span diverse phyla and even kingdoms. Many studies have attempted to catalogue the microbial diversity of biocrusts. Early studies employed culturing (Gundlapally and Garcia-Pichel 2006; Yeager et al. 2007), or low-resolution culture-independent methods such as clone libraries (Steven et al. 2014), denaturing gradient gel electrophoresis (DGGE; Garcia-Pichel et al. 2003; Smith et al. 2004; Gundlapally and Garcia-Pichel 2006; Zaady et al. 2010), automated rRNA intergenic spacer analysis (ARISA; Abed et al. 2012), or terminal restriction fragment length polymorphism (T-RFLP; Redfield et al. 2002; Yeager et al. 2004; Kuske et al. 2012) to characterize the microbial populations in biocrusts. The advent of high-throughput sequence analysis has allowed for the interrogation of biocrust diversity at much more comprehensive levels, and with greater replication to gain a deeper insight into these microbial communities (Steven et al. 2012a). Table 3.1 presents a review of recent studies that have employed next-generation sequencing to describe biocrust microbial diversity.

As can be seen from Table 3.1, the number of bacterial OTUs in biocrusts ranges from *ca.* 200 to >10,000. The number of recovered bacterial OTUs is typically in the thousands, suggesting that this is a likely estimate for the species diversity in biocrusts. The highest bacterial diversity was recovered from the Succulent Karoo of South Africa, with OTU counts of 5000–11,000. The Succulent Karoo is renowned for its biodiversity, often considered the most diverse semi-arid ecosystem on Earth, a characteristic that seems to also extend to its soil habitats. One pattern that becomes apparent when comparing the diversity of biocrusts across studies is that biocrusts collected from polar regions are not significantly less diverse than those from more temperate regions (Table 3.1). Thus, biocrust species do not seem to follow the well-established latitudinal gradient as observed for many macrofauna (Pereira 2016). As such, biocrusts represent a model system in which to test the limits of microbial dispersal, diversity, community assembly, and how these characteristics may differ from vegetation or animal populations.

Table 3.1 Presents a review of recent studies that have employed next-generation sequencing to describe biocrust microbial diversity

Study site	Biocrust type	OTUs (% sequence identity)	Shannon's diversity	Reference
Bacteria				
Colorado Plateau, USA	Cyanobacteria dominated	367–408 (97%)	6.4–6.8	Steven et al. (2013a)
High Arctic, Canada	Cyanobacteria dominated	1801–2264 (97%)	6.4–7.2	Steven et al. (2013b)
Negev Desert, Israel	Microcosms	200–400 (97%)	NA	Angel and Conrad (2013)
Region, China	Lichenized	1915–2403 (97%)	6.8–7.3	Wang et al. (2015)
Gurbantunggut Desert, China	Successional stages	2572–3157 (97%)	5.5–6.4	Zhang et al. (2016a)
Colorado Plateau, USA	Successional stages	835–1229 (97%)	6.0–8.0	Lee et al. (2016)
Colorado Plateau, USA	Successional stages	1099–1871 (97%)	NA	Couradeau et al. (2016)
Shapotou Desert, China	Stabilized dunes	1197–2307 (97%)	3.1–6.2	Liu et al. (2017b)
Shapotou Desert, China	Successional stages	2907–4327 (97%)	6.4–7.3	Liu et al. (2017a)
Succulent Karoo, South Africa	Successional stages	5588–11,669 (97%)	8.0–11.2	Maier et al. (2018)
Chihuahuan Desert, USA	Grassland	3620–6488 (97%)	6.5–8.4	Fernandes et al. (2018)
Svalbard, High Arctic and Livingston Island, Antarctica	Polar deserts	984–2032 (97%)	2.0–5.0	Rippin et al. (2018)
Sultanate of Oman	Cyanobacteria/ lichen	181–4855 (97%)	1.7–7.7	Abed et al. (2019)
Colorado Plateau, USA	Microcosms	1074–1871 (97%)	NA	Couradeau et al. (2019)
New South Wales, Australia	Successional stages	2134–2720 (97%)	NA	Chilton et al. (2018)
Cyanobacteria specific				
Nevada, USA	Cyanobacteria	15–85 (95%)	0.9–2.6	Steven et al. (2012b)
Svalbard, High Arctic	Successional stages	32–71 (97.5%)	3.5–5.0	Pushkareva et al. (2015)
Chihuahuan Desert, USA	Grassland	59–144 (97%)	0.5–1.7	Fernandes et al. (2018)
Sør Rondane Mountains, Antarctica	Polar desert	5–10 (97.5%)	NA	Pushkareva et al. (2018)

3.4.2 Bacterial Taxonomy and Function in Biocrusts

Biocrust microbes can be separated into two broad functional categories. Phototrophs perform photosynthesis and fix carbon into the biocrusts and heterotrophs that utilize and recycle that fixed carbon. A more detailed description of the functional potentials of hot desert BSCs is given in Chap. 7. The bacterial phototrophs of biocrusts are predominantly cyanobacteria, although other phototrophic organisms may also contribute to the net primary productivity. The heterotrophic microflora are more diverse consisting of bacteria, archaea, and fungi (Porras-Alfaro et al. 2011; Steven et al. 2015; Marusenko et al. 2015; Zhao et al. 2020). In general, heterotrophic bacterial populations are predominant and are the focus of this chapter.

3.4.2.1 Cyanobacteria

Cyanobacteria are thought to be among the most common, conserved, and abundant organisms in biocrusts. Data from cyanobacteria-specific surveys show that cyanobacteria populations in biocrusts are diverse. A single biocrust sample may harbor 30–100 cyanobacteria OTUs (Table 3.1). A recent review of cyanobacteria diversity in biocrusts puts the total number of described cyanobacteria from biocrusts at 320 species in over 70 genera (Büdel et al. 2016). As proposed by Büdel et al. (2016), biocrust cyanobacteria can be broadly categorized into three groups: (a) structural filamentous cyanobacteria, the "ecosystem engineers" of biocrusts. These organisms form the structure of the biocrust by gluing soil particles together. (b) Cosmopolitan biocrust cyanobacteria. These cyanobacteria are not directly associated with the structural integrity of the biocrust but perform other essential ecosystem services, such as nitrogen fixation. These cyanobacteria are consistently found in biocrusts around the globe. (c) Sporadic cyanobacteria. These cyanobacteria are only occasionally found in biocrusts and may arise by deposition from the atmosphere or surrounding environments. The ecological role of these bacteria is less clear, and they are only occasionally detected in biocrusts or are limited in geographic distribution. In the following sections, each of these three groups is discussed in brief.

3.4.2.2 Structural Cyanobacteria

Many different cyanobacteria can play a role in soil stabilization and the structural basis of biocrust formation. These include cyanobacteria within the genera *Nostoc, Coleofasciculus*, and *Trichocoleus* (e.g., Hagemann et al. 2015; Schulz et al. 2016; Čapková et al. 2016). However, a single organism *Microcoleus vaginatus* has been found in biocrusts on every continent (Dvořák et al. 2012; Cano-Díaz et al. 2019). In this respect, *M. vaginatus* has arisen as a model organism for studying filamentous bacteria and their role in biocrust formation. In fact, inoculation of *M. vaginatus* onto degraded soils has been pursued as a method to induce biocrust formation and restoration, highlighting its importance as a pioneering species in biocrust formation (Zhang et al. 2013; Ayuso et al. 2017; Bethany et al. 2019)

M. vaginatus produces an exopolysaccharide matrix (EPM), which is the glue that binds soil particles together and forms the structural basis of the biocrust (Rossi and De Philippis 2015; Tiwari et al. 2019). The EPM absorbs water and expands under wetting, which may act to delay desiccation as well as affecting material transfer through soil pores (Belnap and Gardner 1993). Thus, the EPM may affect various ecosystem level dynamics such as water infiltration and carbon flux (Gypser et al. 2016a, b; Rossi et al. 2018; Xiao et al. 2019; Kheirfam 2020). Furthermore, pigments produced by *M. vaginatus* such as scytonemin are embedded in the EPM, which protect *M. vaginatus,* and potentially other biocrust organisms, from UV damage (Chen et al. 2009; Pathak et al. 2019). This pigment is what gives mature biocrusts their dark color, increasing albedo and acting as a sunscreen (Fig. 3.1; Belnap et al. 2007). *M. vaginatus* is also a prodigious producer of extracellular small carbon compounds. Gas chromatography of *M. vaginatus* cultures identified more than 60 alkanes, fatty acids, and other hydrocarbons in spent culture media (Dembitsky et al. 2001). Many of these metabolites can be employed as a growth substrate by heterotrophic bacteria isolated from biocrusts, suggesting that *M. vaginatus* may be able to enrich specific heterotrophic partners (Baran et al. 2015). Thus, by stabilizing soils and creating a substrate suitable for biocrust formation while also recruiting heterotrophic bacteria further facilitating biocrust development, *M. vaginatus* truly lives up to its designation as an ecosystem engineer (Belnap and Gardner 1993; Starkenburg et al. 2011).

While *M. vaginatus* appears to be a globally distributed bacterium, there are limits to its colonization and dominance. High temperature denatures Photosystem II reaction centers for *M. vaginatus* (Lan et al. 2014). *M. vaginatus* appears to have a maximal growth of approximately 40 °C and consistent temperatures around this limit could induce ecological shifts favoring other filamentous cyanobacteria. This has been experimentally demonstrated along a latitudinal gradient, spanning the Sonoran, Mojave, and Chihuahuan deserts, as well as the Great Basin, and the Colorado Plateau. *Microcoleus steenstrupii* outcompeted *M. vaginatus* in the hottest sites (Garcia-Pichel et al. 2013). This relationship could also be recapitulated in the laboratory, suggesting that elevated temperatures favor *M. steenstrupii* dominance (Garcia-Pichel et al. 2013). Given the lack of studies on biocrusts from hot deserts, there is a critical research need for biocrusts that may present a different ecological state compared to the temperate biocrusts that have been predominantly studied. This will better inform processes such as ecological succession and what may be in store for some biocrusts in the face of climate change.

3.4.2.3 Cosmopolitan Cyanobacteria

A variety of other cyanobacteria are commonly identified in biocrusts. These include filamentous cyanobacteria in genera such as *Scytonema, Stigonema,* and *Calothrix.* In general, these cyanobacteria do not appear to produce EPM at the same levels as *M. vaginatus* so they seem to play less of a role in biocrust formation and stability (Garcia-Pichel et al. 2001; Steven et al. 2012b). Unicellular forms of cyanobacteria are also identified, such as *Chroococcidiopsis* (Yeager et al. 2004; Büdel et al. 2016). Many of these cyanobacteria are capable of fixing heterocystous nitrogen, pointing to their potential role in biocrusts (Yeager et al. 2007; Pietrasiak et al. 2013). Yet, there remains a significant knowledge gap concerning what supports the diversity of cyanobacteria in biocrusts and untangling their contribution to biocrust function.

3.4.2.4 Sporadic Cyanobacteria

A multitude of cyanobacteria have been identified in only a small number of biocrusts. These "sporadic" cyanobacteria often arise from culture-independent molecular characterizations of microbial diversity. These sequences are often most closely related to aquatic cyanobacteria and could arise from transport from outside environments. Alternatively, due to the poor representation of terrestrial cyanobacteria in publicly available sequence datasets, there is a high misclassification rate of biocrust cyanobacteria (Steven et al. 2012a). The misclassification is often for aquatic species of cyanobacteria as these organisms tend to dominate the available sequence datasets. Thus, a goal of the biocrust research community should be to increase the diversity of cyanobacteria genomes in sequence repositories to improve sequence classification.

3.4.3 Other Phototrophs

Cyanobacteria are not the only phototrophs in biocrusts. Another population that uses light to facilitate carbon cycling is the aerobic anoxygenic phototrophs (AAP; Fuchs et al. 2007). These organisms are especially well adapted to oligotrophic environments where their photosynthetic capabilities facilitate carbon cycling in the absence of available organic carbon (Cho et al. 2007). AAP isolates can make up to 6% of culturable cells recovered from biocrusts and belong to the genera *Methylobacterium*, *Belnapia*, *Muricoccus*, and *Sphingomonas* (Csotonyi et al. 2010). Stimulation of the AAP populations in biocrusts in the laboratory increased biocrust thickness and organic carbon content, suggesting that these organisms positively influence biocrust development (Tang et al. 2018). Other photosynthetic bacterial taxa such as the Chloroflexi have been identified in biocrusts. For instance, Chloroflexi were dominant in red-pigmented hypoliths in the hyper-arid Atacama Desert (Lacap et al. 2011). Yet, defining the role that these organisms play in biocrust function and development remains elusive.

3.4.4 Heterotrophic Bacteria

The heterotrophic bacterial populations in biocrusts are diverse. Common phyla recovered from biocrusts include Proteobacteria, Actinobacteria, and Bacteroidetes (Garcia-Pichel et al. 2003; Steven et al. 2013a). Yet, unlike for the dominant cyanobacteria, there does not appear to be highly conserved heterotrophs that are present between biocrusts among different environments or ecoregions. However, there does seem to be some environmental filtering that selects for certain taxa. The same bacterial taxa are recovered from biocrusts and other uncrusted patches, such

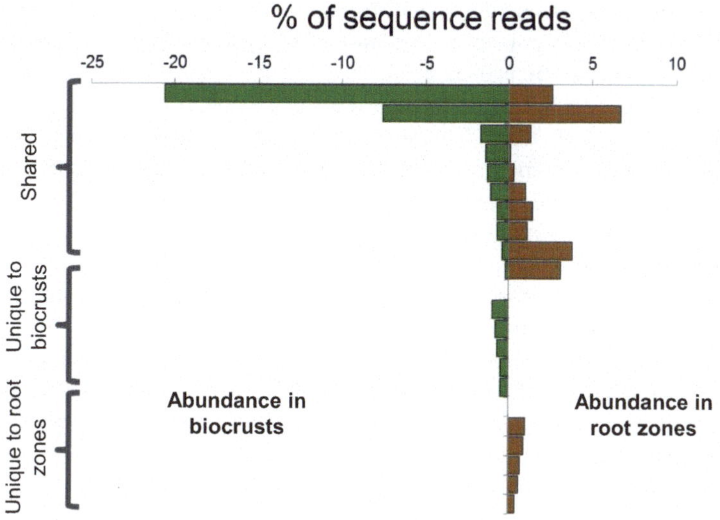

Fig. 3.3 Conservation of bacterial OTUs between biocrusts and uncrusted soil patches in the root zones of local shrubs. Each bar represents an OTU detected in the datasets, and they could be grouped into three categories, OTUs unique to the root zones, those unique to the biocrusts, and those shared between the two habitats. The most abundant OTUs were those shared between the two patches, although the abundance differed widely between patches. These observations suggest that the biocrusts and root zone soils are largely composed of bacteria selected from the same species pool. Data reproduced from Steven et al. (2014); Steven (2017)

as those occurring in the shrub root zones of the local vegetation, although in different proportions (Fig. 3.3). These observations suggest that there is a level of environmental filtering in drylands and that biocrusts enrich specific bacterial taxa out of the local species pool.

The role of heterotrophic bacteria in the establishment of biocrust is largely under-recognized. There is a "chicken or egg" paradox in the initial stages of biocrust colonization. Arid soils are generally devoid of available nitrogen (Evans and Ehleringer 1993; Hooper and Johnson 1999; Plaza et al. 2018). As discussed above, the cyanobacterium *M. vaginatus* is considered to be a pioneering species of biocrusts, yet is unable to perform nitrogen fixation (Starkenburg et al. 2011). In this respect, nitrogen fixation for *M. vaginatus* needs to be outsourced to other community members, including the abundant heterotrophic nitrogen-fixing bacteria (Pepe-Ranney et al. 2016). So the question arises, which comes first, the cyanobacteria or heterotrophic bacteria that lay the groundwork for colonization by providing a source of fixed nitrogen? There is some evidence that *M. vaginatus* growth is facilitated by pre-colonization by heterotrophic bacteria (Roeselers et al. 2007) particularly by bacteria within the phylum Proteobacteria (Chan et al. 2003). Thus, an oft-stated truism of biocrust establishment that biocrusts are initiated by the colonization by filamentous cyanobacteria may require further examination. A solution to this apparent paradox may be explained by the recent description of a critical biocrust zone termed the "cyanospshere" (Couradeau et al. 2019). This is the zone of soil directly in contact with filaments of cyanobacteria, analogous to the rhizosphere of plants (Berendsen et al. 2012). This cyanosphere is colonized by diverse microbial populations and enriched for members encoding nitrogen fixation genes (Couradeau et al. 2019). This suggests that *M. vaginatus* may predominantly exist as a small metacommunity. As *M. vaginatus* colonizes new soil patches, either by lateral growth or wind dispersal, it likely carries its heterotrophic associates with it. In this regard, *M. vaginatus* may be more rightly be viewed as biocrust propagule rather than an axenic species.

Taken together these observations all point to the fact that we have only begun to scratch the surface of the role heterotrophic bacteria play in biocrusts. The methods to fully enumerate and describe their diversity have only recently been developed. The next challenge will be to document the multitude of functional roles these diverse populations are likely to play in the founding, organization, development, and maintenance of biocrusts.

3.5 Conclusion

Biocrusts are phototrophic mats that colonize arid soils worldwide. These are not just microbial oddities, but provide ecosystem services such as increasing soil stability and altering hydrology at a landscape scale. Yet, there remain critical research needs concerning the establishment, rehabilitation, management, and conservation of biocrusts, particularly for hot deserts. This is especially important as these biocrusts often occur in degraded or extreme environments that limit or preclude other biology, such as vegetation. As such, biocrusts are often the last line of defense to maintain the productivity of marginal ecosystems. The preservation and recruitment of biocrusts will only become more important as we face a warmer future with the potential of desertification and increasing aridity.

Acknowledgments This work was supported by the USDA National Institute of Food and Agriculture, Hatch project 1006211. The author would like to thank Dr. Josephine Hyde for critical reading of the manuscript.

References

Abed RMM, Ramette A, Hübner V et al (2012) Microbial diversity of eolian dust sources from saline lake sediments and biological soil crusts in arid Southern Australia. FEMS Microbiol Ecol 80:294–304. https://doi.org/10.1111/j.1574-6941.2011.01289.x

Abed RMM, Lam P, De Beer D, Stief P (2013) High rates of denitrification and nitrous oxide emission in arid biological soil crusts from the Sultanate of Oman. ISME J 7(9):1862–1875

Abed RMM, Tamm A, Hassenrück C et al (2019) Habitat-dependent composition of bacterial and fungal communities in biological soil crusts from Oman. Sci Rep 9:1–16. https://doi.org/10.1038/s41598-019-42911-6

Adessi A, Cruz de Carvalho R, De Philippis R et al (2018) Microbial extracellular polymeric substances improve water retention in dryland biological soil crusts. Soil Biol Biochem 116:67–69. https://doi.org/10.1016/j.soilbio.2017.10.002

Angel R, Conrad R (2013) Elucidating the microbial resuscitation cascade in biological soil crusts following a simulated rain event. Environ Microbiol 15:2799–2815. https://doi.org/10.1111/1462-2920.12140

Ayuso SV, Silva AG, Nelson C et al (2017) Microbial nursery production of high-quality biological soil crust biomass for restoration of degraded dryland soils. Appl Environ Microbiol 83. https://doi.org/10.1128/AEM.02179-16

Bamforth SS (2008) Protozoa of biological soil crusts of a cool desert in Utah. J Arid Environ 72: 722–729. https://doi.org/10.1016/j.jaridenv.2007.08.007

Baran R, Brodie EL, Mayberry-Lewis J et al (2015) Exometabolite niche partitioning among sympatric soil bacteria. Nat Commun 6:8289. https://doi.org/10.1038/ncomms9289

Barger NN, Herrick JE, Van Zee J, Belnap J (2006) Impacts of biological soil crust disturbance and composition on C and N Loss from water erosion. Biogeochemistry 77:247–263. https://doi.org/10.1007/s10533-005-1424-7

Barger NN, Weber B, Garcia-Pichel F et al (2016) Patterns and controls on nitrogen cycling of biological soil crusts. In: Weber B, Büdel B, Belnap J (eds) Biological soil crusts: an organizing principle in drylands. Springer, Cham, pp 257–285

Baumann K, Siebers M, Kruse J et al (2019) Biological soil crusts as key player in biogeochemical P cycling during pedogenesis of sandy substrate. Geoderma 338:145–158. https://doi.org/10.1016/j.geoderma.2018.11.034

Belnap J (1993) Recovery rates of cryptobiotic crusts: inoculant use and assessment methods. Great Basin Nat 53:89–95

Belnap J (2003a) Comparative structure of physical and biological soil crusts. In: Belnap J, Lange OL (eds) Biological soil crusts: structure, function, and management. Springer, Berlin, pp 177–191

Belnap J (2003b) The world at your feet: desert biological soil crusts. Front Ecol Environ 1:181–189

Belnap J (2006) The potential roles of biological soil crusts in dryland hydrologic cycles. Hydrol Process 20:3159–3178. https://doi.org/10.1002/hyp.6325

Belnap J, Büdel B (2016) Biological soil crusts as soil stabilizers. In: Weber B, Büdel B, Belnap J (eds) Biological soil crusts: an organizing principle in drylands. Springer, Cham, pp 305–320

Belnap J, Gardner J (1993) Soil microstructure in soils of the Colorado Plateau: the role of the cyanobacterium *Microcoleus vaginatus*. Great Basin Nat 53:40–47

Belnap J, Gillette DA (1998) Vulnerability of desert biological soil crusts to wind erosion: the influences of crust development, soil texture, and disturbance. J Arid Environ 39:133–142

Belnap J, Phillips SL, Smith SD (2007) Dynamics of cover, UV-protective pigments, and quantum yield in biological soil crust communities of an undisturbed Mojave Desert shrubland. Flora Morphol Distrib Funct Ecol Plant 202:674–686. https://doi.org/10.1016/j.flora.2007.05.007

Belnap J, Walker BJ, Munson SM, Gill RA (2014) Controls on sediment production in two U.S. deserts. Aeolian Res 14:15–24. https://doi.org/10.1016/j.aeolia.2014.03.007

Berendsen RL, Pieterse CMJ, Bakker PAHM (2012) The rhizosphere microbiome and plant health. Trends Plant Sci 17:478–486. https://doi.org/10.1016/j.tplants.2012.04.001

Bertiller MB, Ares JO (2011) Does sheep selectivity along grazing paths negatively affect biological crusts and soil seed banks in arid shrublands? A case study in the Patagonian Monte, Argentina. J Environ Manage 92:2091–2096. https://doi.org/10.1016/j.jenvman.2011.03.027

Bethany J, Giraldo-Silva A, Nelson C et al (2019) Optimizing the production of nursery-based biological soil crusts for restoration of arid land soils. Appl Environ Microbiol 85:e00735-19. https://doi.org/10.1128/AEM.00735-19

Boeken B, Ariza C, Gutterman Y, Zaady E (2004) Environmental factors affecting dispersal, germination and distribution of *Stipa capensis* in the Negev Desert, Israel: *Stipa* germination among desert patches. Ecol Res 19:533–540. https://doi.org/10.1111/j.1440-1703.2004.00666.x

Bowker MA (2007) Biological soil crust rehabilitation in theory and practice: an underexploited opportunity. Restor Ecol 15:13–23. https://doi.org/10.1111/j.1526-100X.2006.00185.x

Bowker MA, Eldridge DJ, Val J, Soliveres S (2013) Hydrology in a patterned landscape is co-engineered by soil-disturbing animals and biological crusts. Soil Biol Biochem 61:14–22. https://doi.org/10.1016/j.soilbio.2013.02.002

Bowker MA, Maestre FT, Eldridge D et al (2014) Biological soil crusts (biocrusts) as a model system in community, landscape and ecosystem ecology. Biodivers Conserv 23:1619–1637. https://doi.org/10.1007/s10531-014-0658-x

Bowker MA, Belnap J, Büdel B et al (2016) Controls on distribution patterns of biological soil crusts at micro- to global scales. In: Weber B, Büdel B, Belnap J (eds) Biological soil crusts: an organizing principle in drylands. Springer, Cham, pp 173–197

Bowker MA, Reed SC, Maestre FT, Eldridge DJ (2018) Biocrusts: the living skin of the earth. Plant and Soil 429:1–7. https://doi.org/10.1007/s11104-018-3735-1

Briggs A, Morgan JW (2008) Morphological diversity and abundance of biological soil crusts differ in relation to landscape setting and vegetation type. Aust J Bot 56:246–253. https://doi.org/10.1071/BT07194

Büdel B, Darienko T, Deutschewitz K, Dojani S, Friedl T, Mohr KI, Salisch M, Reisser W, Weber B (2009) Southern African biological soil crusts are ubiquitous and highly diverse in drylands, being restricted by rainfall frequency. Microb Ecol 57(2):229–247

Büdel B, Colesie C, Green TG, Grube M, Lázaro Suau R, Loewen-Schneider K, Stefanie M, Thomas P, Ana P, José R, Ulrike R, Leopoldo S, Schroeter B, Roman T, Bettina W, Mats W, Martin W, Laura B-W, Zheng L (2014) Improved appreciation of the functioning and importance of biological soil crusts in Europe: the Soil Crust International Project (SCIN). Biodivers Conserv 23(7):1639–1658

Büdel B, Dulić T, Darienko T et al (2016) Cyanobacteria and algae of biological soil crusts. In: Weber B, Büdel B, Belnap J (eds) Biological soil crusts: an organizing principle in drylands. Springer, Cham, pp 55–80

Cano-Díaz C, Maestre FT, Eldridge DJ et al (2019) Ecological niche differentiation in soil cyanobacterial communities across the globe. bioRxiv. https://doi.org/10.1101/531145

Čapková K, Hauer T, Řeháková K, Doležal J (2016) Some like it high! Phylogenetic diversity of high-elevation cyanobacterial community from biological soil crusts of Western Himalaya. Microb Ecol 71:113–123. https://doi.org/10.1007/s00248-015-0694-4

Cary SC, McDonald IR, Barrett JE, Cowan DA (2010) On the rocks: the microbiology of Antarctic Dry Valley soils. Nat Rev Microbiol 8:129–138. https://doi.org/10.1038/nrmicro2281

Chan BK, Chan WK, Walker G (2003) Patterns of biofilm succession on a sheltered rocky shore in Hong Kong. Biofouling 19:371–380. https://doi.org/10.1080/08927010310001645229

Chan Y, Lacap DC, Lau MCY et al (2012) Hypolithic microbial communities: between a rock and a hard place: hypolithic microbial communities. Environ Microbiol 14:2272–2282. https://doi.org/10.1111/j.1462-2920.2012.02821.x

Chen L-Z, Wang G-H, Hong S et al (2009) UV-B-induced Oxidative damage and protective role of exopolysaccharides in desert cyanobacterium Microcoleus vaginatus. J Integr Plant Biol 51:194–200. https://doi.org/10.1111/j.1744-7909.2008.00784.x

Chilton AM, Neilan BA, Eldridge DJ (2018) Biocrust morphology is linked to marked differences in microbial community composition. Plant Soil 429:65–75. https://doi.org/10.1007/s11104-017-3442-3

Cho J-C, Stapels MD, Morris RM et al (2007) Polyphyletic photosynthetic reaction centre genes in oligotrophic marine Gammaproteobacteria. Environ Microbiol 9:1456–1463. https://doi.org/10.1111/j.1462-2920.2007.01264.x

Coe KK, Sparks JP, Belnap J (2014) Physiological ecology of dryland biocrust mosses. In: Hanson DT, Rice SK (eds) Photosynthesis in bryophytes and early land plants. Springer, Dordrecht, pp 291–308

Colesie C, Gommeaux M, Green TGA, Büdel B (2014) Biological soil crusts in continental Antarctica: Garwood Valley, southern Victoria Land, and Diamond Hill, Darwin Mountains region. Antarct Sci 26:115–123. https://doi.org/10.1017/S0954102013000291

Concostrina-Zubiri L, Huber-Sannwald E, Martínez I et al (2013) Biological soil crusts greatly contribute to small-scale soil heterogeneity along a grazing gradient. Soil Biol Biochem 64:28–36. https://doi.org/10.1016/j.soilbio.2013.03.029

Corbin JD, Thiet RK (2020) Temperate biocrusts: mesic counterparts to their better-known dryland cousins. Front Ecol Environ 18(8):456–464

Couradeau E, Karaoz U, Lim HC et al (2016) Bacteria increase arid-land soil surface temperature through the production of sunscreens. Nat Commun 7:10373. https://doi.org/10.1038/ncomms10373

Couradeau E, Giraldo-Silva A, De Martini F, Garcia-Pichel F (2019) Spatial segregation of the biological soil crust microbiome around its foundational cyanobacterium, Microcoleus vaginatus, and the formation of a nitrogen-fixing cyanosphere. Microbiome 7:55. https://doi.org/10.1186/s40168-019-0661-2

Cowan DA, Sohm JA, Makhalanyane TP et al (2011) Hypolithic communities: important nitrogen sources in Antarctic desert soils. Environ Microbiol Rep 3:581–586. https://doi.org/10.1111/j.1758-2229.2011.00266.x

Csotonyi JT, Swiderski J, Stackebrandt E, Yurkov V (2010) A new environment for aerobic anoxygenic phototrophic bacteria: biological soil crusts. Environ Microbiol Rep 2:651–656. https://doi.org/10.1111/j.1758-2229.2010.00151.x

Darby BJ, Neher DA, Belnap J (2007) Soil nematode communities are ecologically more mature beneath late- than early-successional stage biological soil crusts. Appl Soil Ecol 35:203–212. https://doi.org/10.1016/j.apsoil.2006.04.006

Darby BJ, Neher DA, Belnap J (2010) Impact of biological soil crusts and desert plants on soil microfaunal community composition. Plant Soil 328:421–431. https://doi.org/10.1007/s11104-009-0122-y

Deines L, Rosentreter R, Eldridge DJ, Serpe MD (2007) Germination and seedling establishment of two annual grasses on lichen-dominated biological soil crusts. Plant and Soil 295:23–35. https://doi.org/10.1007/s11104-007-9256-y

Dembitsky VM, Dor I, Shkrob I, Aki M (2001) Branched alkanes and other apolar compounds produced by the cyanobacterium Microcoleus vaginatus from the Negev Desert. Russ J Bioorg Chem 27:110–119. https://doi.org/10.1023/A:1011385220331

de Winder B, Matthijs HCP, Mur LR (1989) The role of water retaining substrata on the photosynthetic response of three drought tolerant phototrophic micro-organisms isolated from a terrestrial habitat. Arch Microbiol 152:458–462. https://doi.org/10.1007/BF00446929

Dojani S, Büdel B, Deutschewitz K, Weber B (2011) Rapid succession of Biological Soil Crusts after experimental disturbance in the Succulent Karoo, South Africa. Appl Soil Ecol 48:263–269. https://doi.org/10.1016/j.apsoil.2011.04.013

Drahorad S, Felix-Henningsen P, Eckhardt K-U, Leinweber P (2013) Spatial carbon and nitrogen distribution and organic matter characteristics of biological soil crusts in the Negev desert (Israel) along a rainfall gradient. J Arid Environ 94:18–26. https://doi.org/10.1016/j.jaridenv.2013.02.006

Dumack K, Koller R et al (2016) Estimated abundance and diversity of heterotrophic protists in South African biocrusts. S Afr J Sci 112. https://doi.org/10.17159/sajs.2016/20150302

Duniway MC, Pfennigwerth AA, Fick SE et al (2019) Wind erosion and dust from US drylands: a review of causes, consequences, and solutions in a changing world. Ecosphere 10:e02650. https://doi.org/10.1002/ecs2.2650

Dvořák P, Hašler P, Poulíčková A (2012) Phylogeography of the Microcoleus vaginatus (Cyanobacteria) from three continents – a spatial and temporal characterization. PLoS One 7:e40153. https://doi.org/10.1371/journal.pone.0040153

Elbert W, Weber B, Burrows S et al (2012) Contribution of cryptogamic covers to the global cycles of carbon and nitrogen. Nat Geosci 5:459–462. https://doi.org/10.1038/ngeo1486

Eldridge DJ, Leys JF (2003) Exploring some relationships between biological soil crusts, soil aggregation and wind erosion. J Arid Environ 53:457–466. https://doi.org/10.1006/jare.2002.1068

Eldridge D, Zaady E, Shachak M (2000) Infiltration through three contrasting biological soil crusts in patterned landscapes in the Negev, Israel. Catena 40:323–326

Eldridge DJ, Zaady E, Shachak M (2002) Microphytic crusts, shrub patches and water harvesting in the Negev Desert: the Shikim system. Landsc Ecol 17:587–597. https://doi.org/10.1023/A:1021575503284

Evans RD, Ehleringer JR (1993) A break in the nitrogen cycle in aridlands? Evidence from δp15N of soils. Oecologia 94:314–317

Evans RD, Lange OL (2001) Biological soil crusts and ecosystem nitrogen and carbon dynamics. In: Belnap J, Lange OL (eds) Biological soil crusts: structure, function, and management. Springer, Berlin, pp 263–279

Felde VJMNL, Peth S, Uteau-Puschmann D et al (2014) Soil microstructure as an under-explored feature of biological soil crust hydrological properties: case study from the NW Negev Desert. Biodivers Conserv 23:1687–1708. https://doi.org/10.1007/s10531-014-0693-7

Felde VJMNL, Chamizo S, Felix-Henningsen P, Drahorad SL (2018) What stabilizes biological soil crusts in the Negev Desert? Plant Soil 429:9–18. https://doi.org/10.1007/s11104-017-3459-7

Fernandes VMC, de Lima NMM, Roush D et al (2018) Exposure to predicted precipitation patterns decreases population size and alters community structure of cyanobacteria in biological soil crusts from the Chihuahuan Desert. Environ Microbiol 20:259–269. https://doi.org/10.1111/1462-2920.13983

Fuchs BM, Spring S, Teeling H et al (2007) Characterization of a marine gammaproteobacterium capable of aerobic anoxygenic photosynthesis. Proc Natl Acad Sci U S A 104:2891–2896. https://doi.org/10.1073/pnas.0608046104

Garcia-Pichel F, López-Cortés A, Nübel U (2001) Phylogenetic and morphological diversity of Cyanobacteria in soil desert crusts from the Colorado Plateau. Appl Environ Microbiol 67: 1902–1910. https://doi.org/10.1128/AEM.67.4.1902-1910.2001

Garcia-Pichel F, Johnson SL, Youngkin D, Belnap J (2003) Small-scale vertical distribution of bacterial biomass and diversity in biological soil crusts from arid lands in the Colorado Plateau. Microb Ecol 46:312–321. https://doi.org/10.1007/s00248-003-1004-0

Garcia-Pichel F, Loza V, Marusenko Y et al (2013) Temperature drives the continental-scale distribution of key microbes in topsoil communities. Science 340:1574–1577. https://doi.org/10.1126/science.1236404

Giguere AT, Taylor AE, Myrold DD, Bottomley PJ (2015) Nitrification responses of soil ammonia-oxidizing archaea and bacteria to ammonium concentrations. Soil Sci Soc Am J 79:1366–1374. https://doi.org/10.2136/sssaj2015.03.0107

Gilbert JA, Corbin JD (2019) Biological soil crusts inhibit seed germination in a temperate pine barren ecosystem. PLoS One 14(2):e0212466

Green TGA, Sancho LG, Pintado A (2011) Ecophysiology of desiccation/rehydration cycles in mosses and lichens. In: Lüttge U, Beck E, Bartels D (eds) Plant desiccation tolerance. Springer, Berlin, pp 89–120

Gundlapally SR, Garcia-Pichel F (2006) The community and phylogenetic diversity of biological soil crusts in the Colorado plateau studied by molecular fingerprinting and intensive cultivation. Microb Ecol 52:345–357. https://doi.org/10.1007/s00248-006-9011-6

Guo Y, Zhao H, Zuo X et al (2008) Biological soil crust development and its topsoil properties in the process of dune stabilization, Inner Mongolia, China. Environ Geol 54:653–662. https://doi.org/10.1007/s00254-007-1130-y

Gypser S, Herppich WB, Fischer T et al (2016a) Photosynthetic characteristics and their spatial variance on biological soil crusts covering initial soils of post-mining sites in Lower Lusatia, NE Germany. Flora Morphol Distrib Funct Ecol Plant 220:103–116. https://doi.org/10.1016/j.flora.2016.02.012

Gypser S, Veste M, Fischer T, Lange P (2016b) Infiltration and water retention of biological soil crusts on reclaimed soils of former open-cast lignite mining sites in Brandenburg, north-east Germany. J Hydrol Hydromech 64:1–11. https://doi.org/10.1515/johh-2016-0009

Hagemann M, Henneberg M, Felde VJMNL et al (2015) Cyanobacterial diversity in biological soil crusts along a precipitation gradient, Northwest Negev Desert, Israel. Microb Ecol 70:219–230. https://doi.org/10.1007/s00248-014-0533-z

Harper KT, Belnap J (2001) The influence of biological soil crusts on mineral uptake by associated vascular plants. J Arid Environ 47:347–357. https://doi.org/10.1006/jare.2000.0713

Hooper DU, Johnson L (1999) Nitrogen limitation in dryland ecosystems: responses to geographical and temporal variation in precipitation. Biogeochemistry 46:247–293. https://doi.org/10.1007/BF01007582

Housman DC, Yeager CM, Darby BJ et al (2007) Heterogeneity of soil nutrients and subsurface biota in a dryland ecosystem. Soil Biol Biochem 39:2138–2149. https://doi.org/10.1016/j.soilbio.2007.03.015

Hu C, Liu Y, Song L, Zhang D (2002) Effect of desert soil algae on the stabilization of fine sands. J Appl Phycol 14:281–292. https://doi.org/10.1023/A:1021128530086

Johnson SL, Neuer S, Garcia-Pichel F (2007) Export of nitrogenous compounds due to incomplete cycling within biological soil crusts of arid lands. Environ Microbiol 9:680–689. https://doi.org/10.1111/j.1462-2920.2006.01187.x

Jung P, Baumann K, Lehnert LW, Samolov E, Achilles S, Schermer M, Wraase LM et al (2020) Desert breath—How fog promotes a novel type of soil biocenosis, forming the coastal Atacama Desert's living skin. Geobiology 18(1):113–124

Kheirfam H (2020) Increasing soil potential for carbon sequestration using microbes from biological soil crusts. J Arid Environ 172:104022. https://doi.org/10.1016/j.jaridenv.2019.104022

Kinloch JE, Friedel MH (2005) Soil seed reserves in arid grazing lands of central Australia. Part 2: availability of 'safe sites'. J Arid Environ 60:163–185. https://doi.org/10.1016/j.jaridenv.2004.03.006

Kleiner EF, Harper KT (1972) Environment and community organization in grasslands of Canyonlands National Park. Ecology 53:299–309. https://doi.org/10.2307/1934086

Kuske CR, Yeager CM, Johnson S et al (2012) Response and resilience of soil biocrust bacterial communities to chronic physical disturbance in arid shrublands. ISME J 6:886–897. https://doi.org/10.1038/ismej.2011.153

Lacap DC, Warren-Rhodes KA, McKay CP, Pointing SB (2011) Cyanobacteria and chloroflexi-dominated hypolithic colonization of quartz at the hyper-arid core of the Atacama Desert, Chile. Extremophiles 15:31–38. https://doi.org/10.1007/s00792-010-0334-3

Lalley JS, Viles HA (2008) Recovery of lichen-dominated soil crusts in a hyper-arid desert. Biodivers Conserv 17:1–20. https://doi.org/10.1007/s10531-007-9153-y

Lan S, Wu L, Zhang D, Hu C (2013) Assessing level of development and successional stages in biological soil crusts with biological indicators. Microb Ecol 66:394–403. https://doi.org/10.1007/s00248-013-0191-6

Lan S, Wu L, Zhang D, Hu C (2014) Desiccation provides photosynthetic protection for crust cyanobacteria Microcoleus vaginatus from high temperature. Physiol Plant 152:345–354. https://doi.org/10.1111/ppl.12176

Lee KC, Archer SDJ, Boyle RH et al (2016) Niche filtering of bacteria in soil and rock habitats of the Colorado Plateau Desert, Utah, USA. Front Microbiol 7. https://doi.org/10.3389/fmicb.2016.01489

Li X-R, Jia X-H, Long L-Q, Zerbe S (2005) Effects of biological soil crusts on seed bank, germination and establishment of two annual plant species in the Tengger Desert (N China). Plant Soil 277:375–385. https://doi.org/10.1007/s11104-005-8162-4

Li J, Okin GS, Alvarez L, Epstein H (2008) Effects of wind erosion on the spatial heterogeneity of soil nutrients in two desert grassland communities. Biogeochemistry 88:73–88. https://doi.org/10.1007/s10533-008-9195-6

Liu Y, Li X, Xing Z et al (2013) Responses of soil microbial biomass and community composition to biological soil crusts in the revegetated areas of the Tengger Desert. Appl Soil Ecol 65:52–59. https://doi.org/10.1016/j.apsoil.2013.01.005

Liu L, Liu Y, Hui R, Xie M (2017a) Recovery of microbial community structure of biological soil crusts in successional stages of Shapotou desert revegetation, northwest China. Soil Biol Biochem 107:125–128. https://doi.org/10.1016/j.soilbio.2016.12.030

Liu Y-R, Delgado-Baquerizo M, Trivedi P et al (2017b) Identity of biocrust species and microbial communities drive the response of soil multifunctionality to simulated global change. Soil Biol Biochem 107:208–217. https://doi.org/10.1016/j.soilbio.2016.12.003

Ludwig J, Tongway D, Hodgkinson K et al (1996) Landscape ecology, function and management: principles from Australia's Rangelands. CSIRO, Melbourne

Maestre FT, Huesca M, Zaady E et al (2002) Infiltration, penetration resistance and microphytic crust composition in contrasted microsites within a Mediterranean semi-arid steppe. Soil Biol Biochem 34:895–898. https://doi.org/10.1016/S0038-0717(02)00021-4

Maestre FT, Escudero A, Martinez I et al (2005) Does spatial pattern matter to ecosystem functioning? Insights from biological soil crusts. Funct Ecol 19:566–573. https://doi.org/10.1111/j.1365-2435.2005.01000.x

Maier S, Tamm A, Wu D et al (2018) Photoautotrophic organisms control microbial abundance, diversity, and physiology in different types of biological soil crusts. ISME J 12:1032–1046. https://doi.org/10.1038/s41396-018-0062-8

Makhalanyane TP, Valverde A, Lacap DC et al (2013) Evidence of species recruitment and development of hot desert hypolithic communities. Environ Microbiol Rep 5:219–224. https://doi.org/10.1111/1758-2229.12003

Marticorena B, Bergametti G, Gillette D, Belnap J (1997) Factors controlling threshold friction velocity in semiarid and arid areas of the United States. J Geophys Res Atmos 102:23277–23287. https://doi.org/10.1029/97JD01303

Martínez I, Escudero A, Maestre FT et al (2006) Small-scale patterns of abundance of mosses and lichens forming biological soil crusts in two semi-arid gypsum environments. Aust J Bot 54:339. https://doi.org/10.1071/BT05078

Marusenko Y, Garcia-Pichel F, Hall SJ (2015) Ammonia-oxidizing archaea respond positively to inorganic nitrogen addition in desert soils. FEMS Microbiol Ecol 91:1–11. https://doi.org/10.1093/femsec/fiu023

McKenna Neuman C, Maxwell C (1999) A wind tunnel study of the resilience of three fungal crusts to particle abrasion during aeolian sediment transport. Catena 38:151–173. https://doi.org/10.1016/S0341-8162(99)00043-0

McKenna Neuman C, Maxwell CD, Boulton JW (1996) Wind transport of sand surfaces crusted with photoautotrophic microorganisms. Catena 27:229–247. https://doi.org/10.1016/0341-8162(96)00023-9

McTainsh G, Strong C (2007) The role of aeolian dust in ecosystems. Geomorphology 89:39–54. https://doi.org/10.1016/j.geomorph.2006.07.028

Meusel H, Tamm A, Kuhn U et al (2018) Emission of nitrous acid from soil and biological soil crusts represents an important source of HONO in the remote atmosphere in Cyprus. Atmos Chem Phys 18:799–813. https://doi.org/10.5194/acp-18-799-2018

Mueller RC, Belnap J, Kuske CR (2015) Soil bacterial and fungal community responses to nitrogen addition across soil depth and microhabitat in an arid shrubland. Front Microbiol 6. https://doi.org/10.3389/fmicb.2015.00891

Muggia L, Klug B, Berg G, Grube M (2013) Localization of bacteria in lichens from Alpine soil crusts by fluorescence in situ hybridization. Appl Soil Ecol 68:20–25. https://doi.org/10.1016/j.apsoil.2013.03.008

Neff JC, Reynolds RL, Belnap J, Lamothe P (2005) Multi-decadal impacts of grazing on soil physical and biogeochemical properties in Southeast Utah. Ecol Appl 15:87–95. https://doi.org/10.1890/04-0268

Pathak J, Pandey A, Maurya PK et al (2019) Cyanobacterial secondary metabolite scytonemin: a potential photoprotective and pharmaceutical compound. Proc Natl Acad Sci India Sect B Biol Sci. https://doi.org/10.1007/s40011-019-01134-5

Pepe-Ranney C, Koechli C, Potrafka R et al (2016) Non-cyanobacterial diazotrophs mediate dinitrogen fixation in biological soil crusts during early crust formation. ISME J 10:287–298. https://doi.org/10.1038/ismej.2015.106

Pereira HM (2016) A latitudinal gradient for genetic diversity. Science 353:1494–1495. https://doi.org/10.1126/science.aah6730

Pi H, Sharratt B (2019) Threshold friction velocity influenced by the crust cover of soils in the Columbia Plateau. Soil Sci Soc Am J 83:232. https://doi.org/10.2136/sssaj2018.06.0230

Pietrasiak N, Regus JU, Johansen JR et al (2013) Biological soil crust community types differ in key ecological functions. Soil Biol Biochem 65:168–171. https://doi.org/10.1016/j.soilbio.2013.05.011

Pillans B (1997) Soil development at snail's pace: evidence from a 6 Ma soil chronosequence on basalt in north Queensland, Australia. Geoderma 80:117–128. https://doi.org/10.1016/S0016-7061(97)00068-2

Plaza C, Zaccone C, Sawicka K et al (2018) Soil resources and element stocks in drylands to face global issues. Sci Rep 8:13788. https://doi.org/10.1038/s41598-018-32229-0

Pointing SB (2016) Hypolithic communities. In: Weber B, Büdel B, Belnap J (eds) Biological soil crusts: an organizing principle in drylands. Springer, Cham, pp 199–213

Pointing SB, Belnap J (2012) Microbial colonization and controls in dryland systems. Nat Rev Microbiol 10:551–562. https://doi.org/10.1038/nrmicro2831

Porras-Alfaro A, Herrera J, Natvig DO et al (2011) Diversity and distribution of soil fungal communities in a semiarid grassland. Mycologia 103:10–21. https://doi.org/10.3852/09-297

Pushkareva E, Pessi IS, Wilmotte A, Elster J (2015) Cyanobacterial community composition in Arctic soil crusts at different stages of development. FEMS Microbiol Ecol 91. https://doi.org/10.1093/femsec/fiv143

Pushkareva E, Pessi IS, Namsaraev Z et al (2018) Cyanobacteria inhabiting biological soil crusts of a polar desert: Sør Rondane Mountains, Antarctica. Syst Appl Microbiol 41:363–373. https://doi.org/10.1016/j.syapm.2018.01.006

Ravi S, Breshears DD, Huxman TE, D'Odorico P (2010) Land degradation in drylands: interactions among hydrologic–aeolian erosion and vegetation dynamics. Geomorphology 116:236–245. https://doi.org/10.1016/j.geomorph.2009.11.023

Redfield E, Barns SM, Belnap J et al (2002) Comparative diversity and composition of cyanobacteria in three predominant soil crusts of the Colorado Plateau. FEMS Microbiol Ecol 40:55–63. https://doi.org/10.1111/j.1574-6941.2002.tb00936.x

Reynolds R, Belnap J, Reheis M et al (2001) Aeolian dust in Colorado Plateau soils: nutrient inputs and recent change in source. PNAS 98:7123–7127. https://doi.org/10.1073/pnas.121094298

Rippin M, Lange S, Sausen N, Becker B (2018) Biodiversity of biological soil crusts from the Polar Regions revealed by metabarcoding. FEMS Microbiol Ecol 94. https://doi.org/10.1093/femsec/fiy036

Rivera-Aguilar V, Godínez-Alvarez H, Manuell-Cacheux I, Rodríguez-Zaragoza S (2005) Physical effects of biological soil crusts on seed germination of two desert plants under laboratory conditions. J Arid Environ 63:344–352. https://doi.org/10.1016/j.jaridenv.2005.03.012

Rodriguez-Caballero E, Belnap J, Büdel B et al (2018) Dryland photoautotrophic soil surface communities endangered by global change. Nat Geosci 11:185–189. https://doi.org/10.1038/s41561-018-0072-1

Roeselers G, van Loosdrecht MCM, Muyzer G (2007) Heterotrophic pioneers facilitate phototrophic biofilm development. Microb Ecol 54:578–585. https://doi.org/10.1007/s00248-007-9238-x

Rossi F, De Philippis R (2015) Role of cyanobacterial exopolysaccharides in phototrophic biofilms and in complex microbial mats. Life 5:1218–1238. https://doi.org/10.3390/life5021218

Rossi F, Mugnai G, De Philippis R (2018) Complex role of the polymeric matrix in biological soil crusts. Plant Soil 429:19–34. https://doi.org/10.1007/s11104-017-3441-4
Sankey JB, Glenn NF, Germino MJ et al (2010) Relationships of aeolian erosion and deposition with LiDAR-derived landscape surface roughness following wildfire. Geomorphology 119: 135–145. https://doi.org/10.1016/j.geomorph.2010.03.013
Schulz K, Mikhailyuk T, Dreßler M et al (2016) Biological soil crusts from coastal dunes at the Baltic sea: cyanobacterial and algal biodiversity and related soil properties. Microb Ecol 71: 178–193. https://doi.org/10.1007/s00248-015-0691-7
Serpe MD, Orm JM, Barkes T, Rosentreter R (2006) Germination and seed water status of four grasses on moss-dominated biological soil crusts from arid lands. Plant Ecol 185:163–178. https://doi.org/10.1007/s11258-005-9092-1
Silvester WB, Parsons R, Watt PW (1996) Direct measurement of release and assimilation of ammonia in the Gunnera–Nostoc symbiosis. New Phytol 132:617–625. https://doi.org/10.1111/j.1469-8137.1996.tb01880.x
Smith SM, Abed RMM, Gercia-Pichel F (2004) Biological soil crusts of sand dunes in Cape Cod National Seashore, Massachusetts, USA. Microb Ecol 48:200–208. https://doi.org/10.1007/s00248-004-0254-9
Starkenburg SR, Reitenga KG, Freitas T et al (2011) Genome of the cyanobacterium Microcoleus vaginatusFGP-2, a photosynthetic ecosystem engineer of arid land soil biocrusts worldwide. J Bacteriol 193:4569–4570. https://doi.org/10.1128/JB.05138-11
Steven B (2017) 1. An introduction to arid soils and their biology. In: Steven B (ed) The biology of arid soils. De Gruyter, Boston, pp 1–14
Steven B, Gallegos-Graves LV, Starkenburg SR et al (2012a) Targeted and shotgun metagenomic approaches provide different descriptions of dryland soil microbial communities in a manipulated field study. Environ Microbiol Rep 4:248–256. https://doi.org/10.1111/j.1758-2229.2012.00328.x
Steven B, Gallegos-Graves LV, Yeager CM et al (2012b) Dryland biological soil crust cyanobacteria show unexpected decreases in abundance under long-term elevated CO_2. Environ Microbiol 14:3247–3258. https://doi.org/10.1111/1462-2920.12011
Steven B, Gallegos-Graves LV, Belnap J, Kuske CR (2013a) Dryland soil microbial communities display spatial biogeographic patterns associated with soil depth and soil parent material. FEMS Microbiol Ecol 86:101–113. https://doi.org/10.1111/1574-6941.12143
Steven B, Lionard M, Kuske CR, Vincent WF (2013b) High bacterial diversity of biological soil crusts in water tracks over permafrost in the high Arctic polar desert. PLoS One 8:e71489. https://doi.org/10.1371/journal.pone.0071489
Steven B, Gallegos-Graves LV, Yeager CM et al (2014) Common and distinguishing features of the bacterial and fungal communities in biological soil crusts and shrub root zone soils. Soil Biol Biochem 69:302–312
Steven B, Hesse C, Gallegos-Graves LV et al (2015) Fungal diversity in biological soil crusts of the Colorado plateau. In: Proceedings of the 12th biennial conference of science and management of the Colorado plateau
Steven B, Belnap J, Kuske CR (2018) Chronic physical disturbance substantially alters the response of biological soil crusts to a wetting pulse, as characterized by metatranscriptomic sequencing. Front Microbiol 9. https://doi.org/10.3389/fmicb.2018.02382
Su Y-G, Li X-R, Cheng Y-W et al (2007) Effects of biological soil crusts on emergence of desert vascular plants in North China. Plant Ecol 191:11–19. https://doi.org/10.1007/s11258-006-9210-8
Su Y-G, Li XR, Zheng J-G, Huang G (2009) The effect of biological soil crusts of different successional stages and conditions on the germination of seeds of three desert plants. J Arid Environ 73:931–936. https://doi.org/10.1016/j.jaridenv.2009.04.010
Swenson TL, Couradeau E, Bowen BP et al (2018) A novel method to evaluate nutrient retention by biological soil crust exopolymeric matrix. Plant Soil 429:53–64. https://doi.org/10.1007/s11104-017-3537-x

Szyja M, de Souza Menezes AG, Oliveira FDA, Leal I, Tabarelli M, Büdel B, Wirth R (2019) Neglected but potent dry forest players: ecological role and ecosystem service provision of biological soil crusts in the human-modified Caatinga. Front Ecol Evol 7:482

Tang K, Jia L, Yuan B et al (2018) Aerobic anoxygenic phototrophic bacteria promote the development of biological soil crusts. Front Microbiol 9. https://doi.org/10.3389/fmicb.2018.02715

Tao Y, Zhang YM (2012) Effects of leaf hair points of a desert moss on water retention and dew formation: implications for desiccation tolerance. J Plant Res 125:351–360. https://doi.org/10.1007/s10265-011-0449-3

Tiwari ON, Bhunia B, Mondal A et al (2019) System metabolic engineering of exopolysaccharide-producing cyanobacteria in soil rehabilitation by inducing the formation of biological soil crusts: a review. J Clean Prod 211:70–82. https://doi.org/10.1016/j.jclepro.2018.11.188

Ullmann I, Büdel B (2001) Biological soil crusts of Africa. In: Biological soil crusts: structure, function, and management. Springer, Berlin, pp 107–118

Verrecchia E, Yair A, Kidron GJ, Verrecchia K (1995) Physical properties of the psammophile cryptogamic crust and their consequences to the water regime of sandy soils, north-western Negev Desert, Israel. J Arid Environ 29:427–437. https://doi.org/10.1016/S0140-1963(95)80015-8

Viles HA (2012) Microbial geomorphology: a neglected link between life and landscape. Geomorphology 157–158:6–16. https://doi.org/10.1016/j.geomorph.2011.03.021

Wang J, Bao J, Su J et al (2015) Impact of inorganic nitrogen additions on microbes in biological soil crusts. Soil Biol Biochem 88:303–313. https://doi.org/10.1016/j.soilbio.2015.06.004

Weber B, Wu D, Tamm A et al (2015) Biological soil crusts accelerate the nitrogen cycle through large NO and HONO emissions in drylands. PNAS 112:15384–15389. https://doi.org/10.1073/pnas.1515818112

West NE (1990) Structure and function of microphytic soil crusts in wildland ecosystems of arid to semi-arid regions. In: Begon M, Fitter AH, Macfadyen A (eds) Advances in ecological research. Academic, pp 179–223

Williams L, Borchhardt N, Colesie C et al (2017) Biological soil crusts of Arctic Svalbard and of Livingston Island, Antarctica. Polar Biol 40:399–411. https://doi.org/10.1007/s00300-016-1967-1

Williams AJ, Pagliai M, Stoops G (2018) Chapter 19—Physical and biological surface crusts and seals. In: Stoops G, Marcelino V, Mees F (eds) Interpretation of micromorphological features of soils and regoliths, 2nd edn. Elsevier, pp 539–574

Wu Y, Rao B, Wu P et al (2013) Development of artificially induced biological soil crusts in fields and their effects on top soil. Plant Soil 370:115–124. https://doi.org/10.1007/s11104-013-1611-6

Wynn-Williams DD (2002) Cyanobacteria in deserts—life at the limit? In: Whitton BA, Potts M (eds) The ecology of cyanobacteria. Kluwer Academic, Dordrecht, pp 341–366

Xiao B, Sun F, Hu K, Kidron GJ (2019) Biocrusts reduce surface soil infiltrability and impede soil water infiltration under tension and ponding conditions in dryland ecosystem. J Hydrol 568:792–802. https://doi.org/10.1016/j.jhydrol.2018.11.051

Yair A (2001) Effects of biological soil crusts on water redistribution...—Google Scholar. In: Biological soil crusts: structure, function, and management. Springer, Berlin, pp 303–314

Yeager CM, Kornosky JL, Housman DC et al (2004) Diazotrophic community structure and function in two successional stages of biological soil crusts from the Colorado Plateau and Chihuahuan Desert. Appl Environ Microbiol 70:973–983. https://doi.org/10.1128/AEM.70.2.973-983.2004

Yeager CM, Kornosky JL, Morgan RE et al (2007) Three distinct clades of cultured heterocystous cyanobacteria constitute the dominant N_2-fixing members of biological soil crusts of the Colorado Plateau, USA. FEMS Microbiol Ecol 60:85–97. https://doi.org/10.1111/j.1574-6941.2006.00265.x

Yoshitake S, Uchida M, Koizumi H et al (2010) Production of biological soil crusts in the early stage of primary succession on a High Arctic glacier foreland. New Phytol 186:451–460. https://doi.org/10.1111/j.1469-8137.2010.03180.x

Zaady E, Ben-David EA, Sher Y et al (2010) Inferring biological soil crust successional stage using combined PLFA, DGGE, physical and biophysiological analyses. Soil Biol Biochem 42:842–849. https://doi.org/10.1016/j.soilbio.2010.02.002

Zhang YM, Wang HL, Wang XQ et al (2006) The microstructure of microbiotic crust and its influence on wind erosion for a sandy soil surface in the Gurbantunggut Desert of Northwestern China. Geoderma 132:441–449. https://doi.org/10.1016/j.geoderma.2005.06.008

Zhang B, Zhang Y, Su Y et al (2013) Responses of microalgal-microbial biomass and enzyme activities of biological soil crusts to moisture and inoculated Microcoleus vaginatus gradients. Arid Land Res Manag 27:216–230. https://doi.org/10.1080/15324982.2012.754514

Zhang Y, Wang X, Hu R et al (2014) Variation of albedo to soil moisture for sand dunes and biological soil crusts in arid desert ecosystems. Environ Earth Sci 71:1281–1288. https://doi.org/10.1007/s12665-013-2532-7

Zhang B, Kong W, Wu N, Zhang Y (2016a) Bacterial diversity and community along the succession of biological soil crusts in the Gurbantunggut Desert, Northern China. J Basic Microbiol 56:670–679. https://doi.org/10.1002/jobm.201500751

Zhang Y, Aradottir AL, Serpe M, Boeken B (2016b) Interactions of biological soil crusts with vascular plants. In: Weber B, Büdel B, Belnap J (eds) Biological soil crusts: an organizing principle in drylands. Springer, Cham, pp 385–406

Zhao Y, Qin N, Weber B, Xu M (2014) Response of biological soil crusts to raindrop erosivity and underlying influences in the hilly Loess Plateau region, China. Biodivers Conserv 23:1669–1686. https://doi.org/10.1007/s10531-014-0680-z

Zhao L, Liu Y, Yuan S et al (2020) Development of archaeal communities in biological soil crusts along a revegetation chronosequence in the Tengger Desert, north central China. Soil Tillage Res 196:104443. https://doi.org/10.1016/j.still.2019.104443

Microbial Ecology of Hot Desert Soils

4

Jean-Baptiste Ramond and Don A. Cowan

Abstract

Deserts are one of the major terrestrial biomes on Earth and, with the impacts of global climate change, are expanding. Given the fact that over two billion humans currently live in dryland areas, understanding how these major ecosystems function is particularly important. The poly-extreme environmental conditions of deserts, particularly (hyper)aridity, (hyper)oligotrophy, and high temperatures, limit the distribution and density of higher organisms, particularly plants. As a result, indigenous microbial communities in hot deserts are the dominant sources of primary productivity and other critical ecosystem services.

Counterintuitively, hot desert surface soils, which are exposed to the harshest of abiotic conditions, exhibit high phylogenetic and functional microbial diversity. Independent of geographical localization, these soil communities are dominated by members of the bacterial Actinobacteria and Proteobacteria phyla. Furthermore, community assembly is primarily driven by deterministic niche-partitioning and habitat-filtration processes, with evidence that biotic interactions also play a role. The functionality of desert soil communities is highly water dependent. However, despite the observations that these communities are strongly activated by water input, there is evidence that desert soil microbiomes retain functionality during dry periods, particularly driven

J.-B. Ramond (✉)
Departamento Genética Molecular y Microbiología, Pontificia Universidad Católica de Chile, Santiago, Chile

Centre for Microbial Ecology and Genomics, Department of Biochemistry, Genetics and Microbiology, University of Pretoria, Pretoria, South Africa
e-mail: jbramond@bio.puc.cl

D. A. Cowan
Centre for Microbial Ecology and Genomics, Department of Biochemistry, Genetics and Microbiology, University of Pretoria, Pretoria, South Africa

J.-B. Ramond, D. A. Cowan (eds.), *Microbiology of Hot Deserts*, Ecological Studies 244, https://doi.org/10.1007/978-3-030-98415-1_4

toward nutrient and energy acquisition. Together, these studies clearly confirm that hot desert soil microbial communities are well adapted to water scarcity, oligotrophy, and resource patchiness.

Keywords

Soil microbiology · Dryland ecology · Desertification · Microbial interactions · Microbiome

4.1 Introduction

Hot arid lands cover a substantial portion of the earth's surface and are present on every continent except Antarctica. The poly-extreme conditions in these environments tend to limit the presence of higher organisms, particularly plants. In consequence, edaphic microbial communities are regarded as the dominant players in desert biogeochemical cycling (Pointing and Belnap 2012; Makhalanyane et al. 2015) and, for much of any annual cycle, are the basis of desert primary production webs (Ramond et al. 2018).

The combined effects of land usage practices and climate change are expected to dramatically expand the total global surface area of hot deserts in a relatively near-term future (Huang et al. 2016; UNCCD 2017). Consequently, a deeper understanding of the function of these important and microbe-dominated biomes is critical to better understand the contribution of these vast systems to global processes such as carbon sequestration.

A large volume of research on desert biological soil crust (BSC) and lithic communities already exists (Chaps. 3, 5, and 9, and references therein), probably stemming from the fact that these communities are both *clearly* visible and constitute microbial biodiversity and functional hotspots (Pointing and Belnap 2012; Elbert et al. 2012; Ramond et al. 2018). However, the fact that edaphic communities seed cryptic niches in hot deserts, such as hypoliths (Makhalanyane et al. 2013), and specialized plant-associated environments such as rhizosheaths (Marasco et al. 2018 and Chap. 8), makes them fundamental to the global functioning of desert ecosystems. This is strengthened by recent experimental evidence suggesting that (hyper)arid hot desert soil microbial communities, even when desiccated, are active and functional (e.g., Angel et al. 2012; Gunnigle et al. 2014, 2017; Schulze-Makuch et al. 2018; León-Sobrino et al. 2019; Jordaan et al. 2020).

In this chapter, we review the body of literature relating to the microbial ecology of the dominant desert soil niche: open soils. We describe the taxonomic composition and functional capacities of hot desert edaphic microbial communities, as well as the key drivers for both composition and function.

4.2 Hot Desert Edaphic Microbial Community Diversity and Assembly

Environmental microbial communities are composed of myriads of interacting microorganisms, ranging from viruses to microeukaryotes but usually dominated by bacteria and fungi (Cordero and Datta 2016; Fierer 2017; Jansson and Hofmockel 2019). Hot desert soils generally have low microbial biomass levels (Fuhrmann et al. 2017) and harbor microbial communities with significantly different functional and taxonomical diversities from non-desert edaphic environments (Fierer et al. 2012). Furthermore, within drylands, an inverse relationship between aridity levels and soil bacterial and fungal abundances has also been observed (Maestre et al. 2015). For example, in the hyperarid region of the Atacama Desert, bacterial cell numbers have been estimated to range from 2×10^5 to 7×10^6 bacterial cell equivalents per gram of soil (estimated from PFLA analyses; Connon et al. 2007) or from 3.1 to 511×10^3 cells per gram of soil (via DAPI staining; Crits-Christoph et al. 2013). In the less arid Mojave Desert, estimates range from 10^4 to 10^7 cells per gram of soil (via soil fluorescence; Smith et al. 2012). In contrast, more productive soils harbor generally between 10^8 and 10^{10} bacterial cells per gram of soils (Fuhrmann et al. 2017).

4.2.1 Taxonomic Composition and Assembly Processes of Hot Desert Edaphic Communities

4.2.1.1 Prokaryotic Communities

DNA-based high-throughput sequencing studies show that bacterial communities from hot desert soils around the world generally comprise between 20 and 30 bacterial phyla (see references in Table 4.1) and are largely dominated by members of the Actinobacteria and Proteobacteria phyla, with Bacteroidetes, Acidobacteria, Cyanobacteria, Chloroflexi, and Firmicutes each generally representing $> 1\%$ of the communities (Table 4.1; Crits-Christoph et al. 2013; Makhalanyane et al. 2015; Bull et al. 2016; Cowan et al. 2020; Knief et al. 2020). The same phyla also dominate in soils in other biomes, but the high prevalence of Actinobacteria and Cyanobacteria is characteristic of hot desert soils. Meta-transcriptomics analyses have also confirmed the overall dominance of Actinobacteria and Proteobacteria in the active fraction of hot desert microbial communities (Gunnigle et al. 2017; León-Sobrino et al. 2019). The importance of these phyla in desert soils is further supported by the fact that the isolation in mixed cultures of dwarf bacteria (<0.45 μm; which is indicative of nutritionally stressed organisms or organisms that are adapted to highly oligotrophic conditions; Kieft 2000) from a semiarid Chihuahuan desert soil yielded taxa almost exclusively associated with the actinobacterial phylum and the α- and β-proteobacterial classes (Rutz and Kieft 2004). Atacama Desert edaphic bacterial communities are particularly rich in Actinobacteria, typically representing over 70% of all bacteria identified (Crits-Christoph et al. 2013; Knief et al. 2020). The overall dominance of Actinobacteria in hot desert soils is most probably linked to their numerous UV repair mechanisms (enabling them to survive on the soil surface), their

Table 4.1 Dominant microbial phyla detected in hot desert soil communities using metabarcoding

Region	Desert	Biotope	Most abundant phyla	References
America	Sonoran	Sandy soil	Actinobacteria (~17%)/ Proteobacteria (~15%)/ Planctomycetes (~10%)/ Firmicutes (~6%)/Chloroflexi (~5%)	Andrew et al. (2012)
	Atacama	Aridity transect	Actinobacteria (72–88%)/ Acidobacteria (3.8–6.6%)/ Proteobacteria (2.2–9.2%)/ Unknown Bacteria/ Gemmatimonadetes	Crits-Christoph et al. (2013)
		Elevation, temperature, pH, and aridity gradient	Actinobacteria (18–83%)/ Proteobacteria (5–41%)/ Acidobacteria (3–34%)/ Bacteroidetes (0.8–4.3%)/ Chloroflexi (1.1–2.8%)	Mandakovic et al. (2018)
		Aridity and elevational transect	Actinobacteria (26–95%)/ Proteobacteria (0.3–52%)/ Chloroflexi (1.4–37%)/ Gemmatimonadetes/ Bacteroidetes	Knief et al. (2020)
		Bulk	Actinobacteria (35–45%)/ Proteobacteria (18–22%)/ Chloroflexi (8–14%)/Firmicutes (1–12%)/Acidobacteria (2–12%)	Araya et al. (2019)
			Ascomycota (60–85%)/ Basidiomycota (1–12%)/ Unidentified (2–10%)	
Africa	Namib	Gravel plain	Proteobacteria (23%)/ Actinobacteria (22%)/ Firmicutes (4%)/Acidobacteria (4%)/Chloroflexi (4%)	Armstrong et al. (2016)
			Actinobacteria (36%)/ Crenarchaeota (12–18%)/ Proteobacteria (22–26%)/ Acidobacteria (10–22%)/ Bacteroidetes (2–4%)	Van der Walt et al. (2016)
			Dothideomycetes (27–43%)/ Unassigned (35–47%)/ Agaricostilbomycetes (6.5–12%)/Unidentified Ascomycota (5–7%)/Tremellomycetes (1–1.5%)	
		Dune	Proteobacteria (34.9–49.7%)/ Actinobacteria (12.9–50.5%)/ Bacteroidetes (5.8–31.3%)/ Firmicutes (0.5–9.4%)/ Chloroflexi (0.2–10.7%)	Ronca et al. (2015)

(continued)

Table 4.1 (continued)

Region	Desert	Biotope	Most abundant phyla	References
			Actinobacteria (28–40%)/ Crenarchaeota (16–34%)/ Proteobacteria (18–28%)/ Bacteroidetes (4–10%)/ Acidobacteria (4%)	Van der Walt et al. (2016)
			Unidentified Ascomycota (15–42%)/Unassigned (28–35%)/ Agaricostilbomycetes (2–22%)/ Dothideomycetes (7–10%)/ Sordariomycetes (2–10%)	
			Proteobacteria (19–57.5%)/ Actinobacteria (17–42.7%)/ Chloroflexi (8.8–55.3%)/ Acidobacteria (0.5–7.4%)/ Planctomycetes (0.2–2%)	Marasco et al. (2018)
			Saccharomycetes (0–98%)/ Sordariomycetes (0–97%)/ Dothideomycetes (0.1–60%)/ Agaricomycetes (0.1–26%)/ Pezizomycetes (0–21%)	
	Sahara	Tunisian Oasis	Proteobacteria (25.7–49.7%)/ Actinobacteria (15.8–43.9%)/ Chloroflexi (4.6–20.7%)/ Bacteroidetes (2.2–15.5%)/ Acidobacteria (1.3–12.2%)	Mosqueira et al. (2019)
Middle East	Negev	Rocky open soils	Proteobacteria (~40%)/ Actinobacteria (~20%)/ Unclassified Bacteria (~20%)/ Bacteroidetes (~5%)/ Cyanobacteria (~5%)	Wasserstrom et al. (2017)
			Dothideomycetes (~40%)/ Unclassified Fungi (~15%)/ Unclassified Ascomycota (~8%)/Sordariomycetes (~4%)/ Tremellomycetes (~3%)	
	Kuwait	Sand dune	Actinobacteria > Unclassified Bacteria > Proteobacteria > Planctomycetes > Acidobacteria	Quoreshi et al. (2019)
	Saoudi Arabia	Dryland soils	Ascomycota (74.3–99.6%)/ Basidiomycota (0.4–25.7%)/ Chytridiomycota (0–0.04%)/ Glomeromycota (0–0.03%)	Moussa et al. (2017)
		Elevation gradient	Proteobacteria (7–35%)/ Actinobacteria (4.9–47.1%)/ Acidobacteria (1.5–18.3%)/ Chloroflexi (1.7–10.3%)/ Firmicutes (0.3–32.9%)	Yasir et al. (2015)

(continued)

Table 4.1 (continued)

Region	Desert	Biotope	Most abundant phyla	References
Asia	Gobi	Sand dune	Firmicutes (62–82%)/ Proteobacteria (~12%)/ Bacteroidetes (2–18%)/ Actinobacteria (1–2%)	An et al. (2013)
	Thar	dryland soils	Actinobacteria (39%)/ Proteobacteria (33%)/ Acidobacteria (9.2%)/ Firmicutes (3.5%)/ Bacteroidetes (3.5%)/ Chloroflexi (3.5%)	Sivakala et al. (2018)
	Taklamakan	Sand dune	Firmicutes (~2–60%)/ Proteobacteria (10–47%)/ Actinobacteria (~3–16%)/ Bacteroidetes (1–39%)/ Unclassified Bacteria (~4–6%)	An et al. (2013)
Australia	Australian Deserts	Dryland soils	Actinobacteria (24%)/ Proteobacteria (21.8%)/ Acidobacteria (13.2%) / Planctomycetes (12.9%)/ Chloroflexi (11.1%)	Eldridge et al. (2018)
			Ascomycota (58%)/ Basidiomycota (30%)/ Zygomycota (5%)/ Chytridiomycota (2.5%)	
		Dune shrubland	Actinobacteria > Proteobacteria > Acidobacteria > Cyanobacteria	Eldridge et al. (2015)
			Lecanoromycetes > Archaeosporomycetes > M-Ascomycota > Leotiomycetes > Eurotiomycetes	

metabolic plasticity, their capacity to sporulate, and their ability to produce a wide range of secondary metabolites (Makhalanyane et al. 2015), which gives them a competitive edge in the nutrient-poor desert soils. There is now strong evidence that hot desert soils represent "hotspots" for so-called Actinobacterial Dark Matter, i.e., isolates from rare actinobacterial genera which may, or may not, possess genes and pathways of potential biotechnological value (Takahashi et al. 1996; Goodfellow et al. 2018)

Hot desert edaphic archaeal communities are generally dominated by Thaumarchaeota/Crenarchaeota and generally constitute 1 to ~7% of the total prokaryotic phylotypes (Bates et al. 2011; Andrew et al. 2012; Fierer et al. 2012; Armstrong et al. 2016). It should, however, be noted that edaphic communities from the dunes and gravel plains of the hyperarid Namib Desert harbored from 10 to 25% of Crenarchaeota/Thaumarchaeota (Table 4.1; Van der Walt et al. 2016), making

Fig. 4.1 Photographs of drylands clearly showing the increasing vegetation cover with decreasing aridity (photo courtesy J-B Ramond)

these some of the most archaeal-rich soils on the planet. It is noted that a deep soil community from the Atacama Desert was composed of >50% of Euryarchaeota (Schulze-Makuch et al. 2018), although other soils from the hyperarid Atacama showed only low relative abundances of archaeal phyla, with on average 0.03% Euryarchaeota and 0.01% Thaumarchaeota (Araya et al. 2019). At present, the ecological functions of hot desert archaea remain poorly understood, although they probably play important roles in N cycling.

Deterministic niche partitioning and habitat filtration have been found to be the dominant drivers of prokaryotic community assembly in hot desert soils (Angel et al. 2010; Andrew et al. 2012; Elliott et al. 2014; Johnson et al. 2017; Scola et al. 2018). Recent studies also suggest that biotic interactions among prokaryotic community members are important drivers of community assembly in these soils (e.g., Gunnigle et al. 2017). Aridity status (from semi- to hyperaridity; Fig. 4.1) and precipitation patterns (e.g., fog vs rain or light rain vs heavy rain) also impact significantly on the structuring and composition of arid land prokaryotic communities (Angel et al. 2010; Maestre et al. 2015; Scola et al. 2018; Bay et al. 2021). A recent global dryland soil survey suggested that increases in aridity were associated with higher relative abundances of Chloroflexi and α-Proteobacteria, and reduced abundances of

Acidobacteria and Verrucomicrobia (Maestre et al. 2015). This effect has also been observed at lower taxonomic ranks, with extremely hyperarid Atacama soils being dominated by actinobacterial *Friedmaniella*, *Nocardioides*, *Sanguibacter*, *Sporichthya*, and *Streptomyces* species, and hyperarid soils by *Aciditerramonas* and *Corynebacterium* spp. (Idris et al. 2017). Aridity has also been shown to differentially influence prokaryotic functional guilds in Negev Desert soils. The structure of ammonia oxidizing archaeal (AOA) communities differed between arid and semiarid soils, while the effect was not observed for ammonia oxidizing bacteria (AOB) (Sher et al. 2013). Similarly, in Namib Deserts, proteobacterial taxa displayed distinct population structures depending on the local soil microenvironment, with dune tops dominated by β-Proteobacteria (>70%) and interdunes dominated by α-Proteobacteria (>85%) (Ronca et al. 2015).

Different environmental parameters may influence the bacterial and archaeal components of hot desert prokaryotic edaphic communities (Angel et al. 2010; Johnson et al. 2017). In Namib Desert gravel plain soils, climate factors segregated bacterial communities between the arid and hyperarid zones, while archaeal communities showed no segregation. Furthermore, the parameters most significantly influencing the bacterial communities in the arid zone were soil cation exchange capacity (CEC) and P content, while soil Ca^{2+}, Mg^{2+}, and K^{+} contents were the dominant variables in the hyperarid zone. In both zones, soil CEC and NH_4^+ content significantly impacted archaeal community structures (Johnson et al. 2017). In a similar study across an aridity gradient in Israel, both archaeal and bacterial communities were influenced by soil water, $CaCO_3$, and organic carbon content, while bacterial (but not archaeal) communities were impacted by soil Mg^{2+} and NO_3-N content (Angel et al. 2010). In Sonoran Desert soils, Acidobacteria were principally influenced by soil pH, and the Desulfurococcaceae family (Crenarchaeota archaeal phylum) are affected by soil percent carbon (Andrew et al. 2012).

Hot desert soil prokaryote communities appear to sustain some activity under arid conditions (e.g., León-Sobrino et al. 2019), but rapidly adapt to water input before returning to their initial state (Cowan et al. 2020). These observations can be interpreted as evidence for the strong resistance/adaptation and resilience capabilities of the edaphic microbial communities (Armstrong et al. 2016; Štovícek et al. 2017; Chap. 10). A single rain event in the Namib Desert gravel plains resulted in the detection of 137 additional prokaryotic taxa (using 16S rRNA gene metabarcoding: Armstrong et al. 2016) and, in the central Negev Desert, in an enrichment of members of the Enterobacterales, Clostridiales, Lactobacillales, and Bacteroidales classes (Štovícek et al. 2017). However, there is some evidence, from RNA-based metabarcoding, that even in arid conditions edaphic microbial community compositions fluctuate over diel cycles (Gunnigle et al. 2017). This highlights the possibility that different conclusions may be drawn from the use of different phylogenetic and/or functional markers, and the general use of multi-omics approaches is strongly advised (Jansson and Hofmockel 2018). This may be particularly relevant for very dry environments, such as hot desert soils, where *relic DNA* may persist for very long periods and skew DNA-based studies (Carini et al. 2016).

4.2.1.2 Microeukaryotic Communities

Among the different soil microeukaryotes, fungal communities have been the most intensively studied in hot desert soils, while other groups of microeukaryotes have been largely ignored. This is perhaps unfortunate since (1) microalgae represent a substantial portion of biological soil crusts and are critical for their primary production (Büdel et al. 2016) (see Chap. 3), (2) protists are the most diverse eukaryotes in many soils and actively participate in food webs (from primary production to predation; Geisen et al. 2017, 2018), (3) amoeba-infecting mimiviruses have been detected in many hot desert soils (see the next subsection), and (4) diatoms were present in all Namib Desert soils investigated (Cowan et al. 2020).

A recent global protist diversity survey of 180 soils demonstrated that arid soils harbored higher relative abundances of phototrophic protistans (up to 40%) when compared to less arid environments, suggesting that protists may play an important role in hot desert primary productivity (Oliverio et al. 2020). Furthermore, mean annual precipitation (MAP) was found to be the best predictor for protistan community structures (Oliverio et al. 2020). This further suggests that protistan communities in hot deserts may be highly specific to an aridity zone and/or water regime history, even at the local desert scale. However, as most of the diatom taxa detected in hyperarid Namib Desert soils are also usually found in less extreme environments, it is possible that diatom communities are essentially composed of eurytopic species (Cowan et al. 2020). Namib Desert soils were found to be dominated by the diatom *Achnanthidium exiguum* (Cowan et al. 2020), a species also recently identified in Atacama Desert salars (Rivera and Cruces 2018).

Until recently, research on desert soil fungal diversity was largely culture-based (Makhalanyane et al. 2015). An extensive list of fungal isolates from diverse soil biotopes from the hyperarid Atacama and Namib Deserts can be found in Santiago et al. (2018) and Cowan et al. (2020), respectively. Despite the ubiquity of modern phylogenetics methods in microbial biodiversity studies, the marker sequences typically used to determine fungal diversity (the internal transcribed spacer (ITS) region and/or the 18S rRNA gene) often lack the phylogenetic resolution to characterize fungal diversity beyond the class level (e.g., Tedersoo et al. 2014; Van der Walt et al. 2016). In addition, the primers commonly used to amplify fungal ITS2 (Ihrmark et al. 2012) may under-represent certain taxonomic groups: for example, this primer set only partly targets taxa within the Glomeromycota, which encompasses the arbuscular mycorrhiza that infect many species of grasses and herbs. It has been reported that while many *Glomus* species (Glomeromycota phylum) have been isolated from Namib Desert soils (Stutz et al. 2000; Uhlmann et al. 2006), members of this phylum were barely detectable using NGS technologies (Cowan et al. 2020). This emphasizes the fact that the use of both culture-independent and -dependent approaches would be essential to fully evaluate desert soil fungal diversity.

As in other biomes, hot desert soil fungal communities are typically dominated by ascomycetes (Table 4.1; Grishkan and Nevo 2010; Sterflinger et al. 2012; Maestre et al. 2015; León-Sobrino et al. 2019). Metabarcoding has shown that the Sordariomycetes, Saccharomycetes, and Dothideomycetes Ascomycota classes

(Marasco et al. 2018) and the *Chaetomium*, *Pleosporales*, and *Stachybotrys* genera (Van der Walt et al. 2016) were abundant in Namib Desert soils. A global fungal metagenomics survey (on 365 soils) demonstrated that mean annual precipitation (MAP) was an important environmental factor positively influencing soil fungal diversity and richness (Tedersoo et al. 2014). While it should be noted that this "global" survey did not include any hot desert soil samples, these results nevertheless suggest that arid environments may harbor less diverse edaphic fungal communities than non-arid soils. However, while some fungal endemism was detectable on a global scale (Tedersoo et al. 2014), most taxa identified showed a cosmopolitan distribution. There is some localized evidence that hot desert soils, even if less diverse than moist soils, harbor very high levels of unique fungal taxa (Van der Walt et al. 2016). For example, ~50% of the fungal OTUs detected in Namib Desert soils could not be assigned to representative taxa (Van der Walt et al. 2016).

The composition of edaphic fungal communities is influenced by local physicochemical, and maybe mineralogical, properties (Van der Walt et al. 2016; Johnson et al. 2017). A comparison of the fungal communities of Namib Desert dune sands and gravel plain soils noted that the latter were dominated by members of the Dothideomycetes class (27.5–46.2%) [Ascomycota phylum], while the former was dominated by members of the Agaricostilbomycetes class [Basidiomycota phylum] and Chytridiomycota phylum (Van der Walt et al. 2016).

4.2.1.3 The Hot Desert Soil Virosphere

Hot desert soil viral communities are thought to be the least rich of all edaphic meta-viromes, with abundances ranging from 10^3 to 10^7 viral particles (VLP) per gram of dry soil (VLP gdw^{-1}) c.f. 10^7–10^{10} VLP gdw^{-1} for other soils (Fuhrmann et al. 2017). In terrestrial biomes, epifluorescence microscopy clearly showed that viruses outnumbered prokaryotes, with virus-to-bacteria ratios ranging from 5:1 to > 5000:1 (Srinivasiah et al. 2008). However, in a study comparing viral and bacterial abundances in dryland top soils originating from the Middle East (Saudi Arabia, Kuwait, and the United Arabic Emirates), Africa (Morocco), Asia (Afghanistan), and the North America (USA—Texas, California, and Arizona), it was, perhaps surprisingly, observed that bacterial counts were often higher than viral counts (Gonzalez-Martin et al. 2013).

Studies characterizing viral communities in hot deserts soils have essentially aimed at cataloguing diversity, while the roles of viruses and phages in the functioning of these ecosystems remain to be investigated in depth. While the positive correlation between bacterial and viral abundance suggests that viruses may control bacterial host abundances in (desert) soils (Fuhrmann et al. 2017), their genetic diversities and functional characteristics in these soils remain understudied and poorly understood. This is highlighted by the fact that a large majority of viral sequences in soil meta-viromes/genomes are typically unassigned (up to 97%; Fuhrmann et al. 2017). Consequently, more extensive and in-depth studies of hot desert edaphic *viral functional and phylogenetic dark matter* are undoubtedly necessary (Zablocki et al. 2016; Fuhrmann et al. 2017).

Hot desert soil viral communities, like most soils and independently of the technology used, have been found to be dominated by members of the dsDNA Caudovirales order (Prestel et al. 2012; Makhalanyane et al. 2015; Zablocki et al. 2016; Schulze-Makuch et al. 2018; Scola et al. 2018). Sahara Desert soil viral communities were particularly enriched in tailed viruses from the Myoviridae family (Prestel et al. 2012), while Namib Desert soils were in addition dominated by members of the Siphoviridae and Podoviridae families (Zablocki et al. 2017; Scola et al. 2018). Giant viruses from the Mimiviridae family have also been detected in soils from the Kutch (India), the Namib, the Chihuahuan, and the Mojave Deserts (Kerepesi and Grolmusz 2016, 2017; Scola et al. 2018). In Namib Desert soils, sequences assigned to the large Phycodnaviridae family have also been observed (Scola et al. 2018). Globally, viral sequence abundances have been found to mirror those of their respective hosts (Zablocki et al. 2016; Scola et al. 2018; Schulze-Makuch et al. 2018). Interestingly, a recent study detected, for the first time, transcriptionally active viruses in desiccated hot desert soils (León-Sobrino et al. 2019), challenging the general perception that viruses in such environments exist largely in a lysogenic state (Zablocki et al. 2016). For a more detailed review of desert *viromics* from soils and also other desert habitats, readers are referred to Chap. 6.

4.2.2 Desert Soil Microbial Interactomes: Lessons from Co-occurrence Networks

Co-occurrence networks have become popular means of inferring potential biotic interactions in environmental microbial communities, and for identifying potential linkages between community compositions and environmental parameters (Faust and Raes 2012; Berry and Widder 2014).

For example, in Namib Desert dune soils, metagenomic DNA-based co-occurrence networks, which combined fungal and prokaryotic single variants (SVs), confirmed the importance of the Actinobacteria phylum in desert soil microbial communities (Marasco et al. 2018). The two dominant network hubs, which were assigned to the Microbacteriaceae family and the *Microvirga* genus, and the only network keystone species (*Microvirga* sp.) belonged to this phylum. The centrality of *Microvirga* sp. in this microbial network may be related to the N-fixation and nutrient mineralization capacities of this taxon (Bailey et al. 2014). Similarly, a bacterial community network from Sahara Desert soil was dominated by OTUs from the Actinobacteria, α- and γ-Proteobacteria, and Chloroflexi, but with OTUs from the Verrucomicrobia, Bacteroidetes, and Actinobacteria phyla showing high numbers of interactions (Mosqueira et al. 2019). Furthermore, of the seven network hubs, the three actinobacterial ones were also network keystone species and belonged to the Acidimicrobiales order.

Edaphic bacterial community networks from two contrasting "regions" of the Atacama Desert, with significant differences in soil pH and nutrient status, showed significantly different topologies while confirming the importance of Actinobacteria and Proteobacteria in hot hyperarid desert communities (Mandakovic et al. 2018). The network from the less oligotrophic and more basic (pH 7.7–8.6) soils was dominated by Proteobacteria (34%) and Actinobacteria (26.9%), while the opposite was observed in the network from the more oligotrophic and acidic (pH 5.7–6.9) soils (i.e., 42.2% Actinobacteria vs 27.7% Proteobacteria). In both networks, the keystone species were low abundance Actinobacteria, Proteobacteria, and Acidobacteria OTUs. Overall, this strongly suggested that desert soil community stability is related to the presence of low abundant OTUs with specific (i.e., non-redundant) functions and that their disappearance (due for example to climate change) may have critical and detrimental consequences in their global functioning.

Perhaps not surprisingly, microbial co-occurrence networks in plant-created niches, i.e., rhizosheath/root systems (Marasco et al. 2018) and rhizospheric zones (Mosqueira et al. 2019), have been found to be more complex that those of bulk soils in hot deserts. For example, Namib Desert dune bulk sand co-occurrence network had fewer nodes (or SVs) and fewer bacteria–bacteria, bacteria–fungi, and fungi–fungi interactions than networks from speargrass rhizosheath microbial communities (Marasco et al. 2018). The implication from this result is that hot desert edaphic microbial communities are isolated, more disturbed, and less interactive than those of more structured and nutrient-rich niches (rhizosheaths). This result is entirely consistent with the concept (Zablocki et al. 2016; Cowan et al. 2020) that microbial "communities" in desert soils are, under desiccating conditions, spatially isolated, i.e., as multispecies biofilms associated with individual soil and mineral particles, with limited physical contact and little "communication" with neighboring particles.

Meta-transcriptome networks of the functional community members in dry desert soils were also largely composed of Actinobacterial and Proteobacterial OTUs (Gunnigle et al. 2017). However, while environmental parameters could only explain 10.5% of variations over a diel timescale in the active community, the temporal dynamics in bacterial interactions exhibited different inter- and intra-phylum co-occurrence patterns throughout the day. In particular, the "midday" community (highest temperature, lowest %RH) was more "connected" than the morning (06:00 h) and evening (18:00 h) networks. Interestingly, the midday co-occurrence network was the only example where a Cyanobacterial OTU (order Oscillatoriales) was positioned as a module hub.

Furthermore, the third most abundant OTU in the complete dataset, assigned to Proteobacteria phylum, showed 11 positive and 1 negative co-occurrence links with other OTUs in the "morning" network, with ratios of 5:3 at "midday" and 3:5 in the "evening" sample. The implication of this observation is that this specific taxon, through diel cycles, switches from *mutualism* to *competition* (Faust and Raes 2012). It is surprising that a single community member can, over such a short time frame and under conditions when many organisms are inactive (León-Sobrino et al. 2019), switch from positive and negative community interactions. Nevertheless, this reinforces the concept that biotic interactions play a critical role in the function of desert soil edaphic communities.

4.3 Functional Properties of Desert Edaphic Communities

A shotgun metagenomic survey demonstrated that hot desert microbial communities contained larger numbers of genetic signatures assigned to stress resistance categories (e.g., genes related to osmoregulation, dormancy, and sporulation) and significantly less implicated in nutrient cycling and antibiotic resistance than those of more productive ecosystems such as forests, prairies, and tundra soils (Fierer et al. 2012). In contrast, shotgun meta-transcriptome analyses suggested that the active taxa in hyperarid soils were driven by nutrient acquisition processes and not stress resistance (León-Sobrino et al. 2019). These contrasting results demonstrate the need to perform more detailed comparative studies, meta-transcriptomics and meta-proteomics studies, in order to clarify the functional status of desert soil microbial communities.

4.3.1 Desert Edaphic Communities Are Active During Dry Spells

Deserts, by definition, are water-deficient environments, where water availability for cellular processes is assumed to be the dominant limiting factor for long periods (months to years). It is generally believed that desert edaphic microbial communities are essentially inactive/dormant during such periods and are only activated during occasional rain events (Noy-Meir 1973; Belnap et al. 2005). However, a recent series of studies using meta-transcriptomics, meta-proteomics, and metabolomics has clearly demonstrated that microbial communities in (hyper)arid soil environments retain a substantial level of activity during periods of desiccation (e.g., Angel et al. 2013: Gunnigle et al. 2014; Bastida et al. 2016, 2018; Schulze-Makuch et al. 2018; León-Sobrino et al. 2019). It has notably been observed that even in the most arid section of the world's driest Atacama Desert intracellular ATP was detected and therefore that microbial activity was retained (Schulze-Makuch et al. 2018).

Extracellular enzyme activities, particularly fluorescein diacetate hydrolysis, β-glucosidase, β-N-acetylglucosaminidase, leucine aminopeptidase, and alkaline phosphatase, are used as proxies for microbial community biogeochemical cycling capacities. In the Namib Desert, differences in titers of extracellular enzyme activities in different rainfall zones (Scola et al. 2018) and in different soil zones (Frossard et al. 2015; Ronca et al. 2015) suggest that community functionality is driven by multiple abiotic factors, not just water activity. Intracellular ATP titers, as a measure of the functional (i.e., non-dormant) fraction of the microbial community, and relative metabolite abundances decreased with increasing soil aridity (Schulze-Makuch et al. 2018). Conversely, ammonia oxidation potential (determined as the nitrite production rate) increased in soils across a semiarid to arid transition (Sher et al. 2013).

A recent study of desiccated soils using shotgun meta-transcriptomics clearly showed that nitrate reduction (*nar* genes) and nitrite reduction (*nir* genes) were the dominant processes for N acquisition (León-Sobrino et al. 2019), and that *nif* gene expression (and, by implication diazotrophy) was almost undetectable in dry soils.

This is consistent with suggestions that dinitrogen fixation processes are highly sensitive to desiccation (Liu et al. 2018), and is perhaps not surprising, given that nitrate is the dominant form of fixed N in many hot desert soils (Prentice 2008). In contrast, it should be noted that cyanobacterial-rich BSCs and hypoliths represent important atmospheric N fixation hubs (e.g., Su et al. 2011; Ramond et al. 2018), providing surrounding bulk soils with bioavailable N. Interestingly, the metabolic capacity for nitrate reduction showed almost no redundancy, as the *nar* genes identified were exclusively transcribed by members of the Nitrospiraceae family (Nitrospirae phylum). Many different actinobacterial taxa were apparently involved in nitrite reduction (from *nir* gene expression). The very narrow range of soil taxa involved in nitrate reduction, as the entry point for inorganic nitrogen, suggests that desert soils may possess very low functional resilience (Allison and Martiny 2008), at least with respect to N acquisition under desiccating conditions.

In dry soils, photosynthetic pathway genes (RUBISCOs) were not transcribed at high levels, while much higher levels of transcription for chemoautotrophic ("dark") carbon fixation and CO_2 re-assimilation genes were observed (León-Sobrino et al. 2019). This strongly suggested that chemoautotrophy, which particularly involved members of the actinobacterial Rubrobacteraceae family, is favored over photosynthesis in desiccated soils. Photoautotrophic processes are well known to be particularly sensitive to desiccation (Potts 2001).

Phosphorus cycling under dry conditions was also dominated by the Actinobacteria and Proteobacteria phyla, with potential synergetic interactions between the Geodermatophilaceae family (Actinobacteria) and Rhizobiales order (α-Proteobacteria), while the acquisition of sulfur particularly involved members of the β-proteobacterial Burkholderiales order (León-Sobrino et al. 2019).

The desiccated soil meta-transcriptomics dataset was also used to generate microbial meta-transcriptional networks. Despite a significant transcription of "resistance genes," these were, surprisingly, not central to the transcriptional network, which was structured around hub genes related to nutrient acquisition: *nir*A and *gud*B (N assimilation), *cys*D and *cys*N/C (S assimilation), *ugp*B (P acquisition), and DLD and *sco*B (carbohydrate metabolism) (León-Sobrino et al. 2019). The overall conclusion was that some microbial taxa in hot desert microbial communities remain fully functional under desiccating conditions and that the principal "driver" is nutrient acquisition rather than stress responses (León-Sobrino et al. 2019).

Recent metagenomic studies suggest that atmospheric gas (H_2, CO, CH_4) oxidation is a critical energy harvesting process performed by edaphic microbial communities in (hyper)oligotrophic and (hyper)arid soils (e.g., Bay et al. 2018, 2021; Leung et al. 2020; Ortiz et al. 2021). The wider ecological relevance of these novel energy and carbon capture processes is yet to be determined. However, it is noteworthy that the various primary production strategies evolved by desert soil microbial communities is particularly well adapted to hot desert oligotrophy and resource patchiness (see Chaps. 7 and 10 for more details).

4.3.2 The Impact of Water on Arid Soil Microbial Communities

Not surprisingly, the addition of water to dry soils triggers a major pulse of activity in desert soil microbial communities (Xu et al. 2004; Chap. 11). This effect has been described in mechanistic, spatial, and temporal terms, within the trigger–transfer–reserve–pulse (TTRP) framework (Belnap et al. 2005). As described in this framework, water is the *trigger* which *transfers* nutrients (and aqueous solvent) to a receiving area, the *reserve*, where a functional *pulse* occurs.

Microbially mediated processes activated by water (e.g., Cable et al. 2008), such as respiration, almost immediately deplete dryland soils of readily accessible nutrients/resources, and the *pulse* occurs through the utilization rather than the establishment of a *reserve*. An alternative model has been presented to address this apparent anomaly: namely the pulse-dynamic framework, which more accurately described the processes triggered by water addition to arid soils, in the absence of a *reserve* production (Collins et al. 2014). The pulse-dynamic framework takes into account different spatio-temporal scales in order to explain the effect(s) of a single or multiple water pulses (Collins et al. 2014). In summary, over a limited timescale (from a single water pulse to multiple occurring in a year), (a)biotic nutrient and water *exchanges* occur at the individual, the population, and community levels with associated shifts in community structures and functions. However, over longer periods (i.e., decade to centuries) water pulses, while still triggering *exchanges*, will have different effects depending on the scale. Within an ecosystem, pulses can affect species *interactions* and ultimately lead to *transitions* within landscapes (e.g., from a shrubland to a grassland biome) and *transfers* of material from terrestrial to aquatic systems. This framework, while particularly adapted to vegetation patterns and even biological soil crust communities, could be developed in the context of barren desert soils and their indigenous microbial communities.

Globally, hydration of open soils in hot deserts leads to the activation of numerous microbial processes, including carbon fixation, respiration, hydrogen oxidation, decomposition, and N mineralization (Jacobson and Jacobson 1998; Armstrong et al. 2016; Jordaan et al. 2020; Chap. 11). For example, 3 days after a substantial (38 mm) rain event in the Namib Desert gravel plains, soil respiration rates increased 4-fold and both soil chlorophyll a concentration (a proxy for photosynthesis) and fluorescein diacetate hydrolysis rates increased 15-fold (Armstrong et al. 2016). Similarly, a 30 mm rainfall event almost doubled soil extracellular enzyme activities (alkaline phosphatase, β-glucosidase, leucine aminopeptidase, and phenol oxidase) in the less arid Chihuahuan Desert (Ladwig et al. 2015). In the Sonoran Desert, in situ watering experiments showed that a minimum of 7 mm rain was necessary to trigger a soil respiration response, although the size of the response was dependent on soil texture (coarse > fine grain; Cable et al. 2008). Rapid changes in microbial community structures were also observed after a watering in the Negev (Štovícek et al. 2017) and Namib Deserts (Armstrong et al. 2016). However, both the community structures and measurable functions returned to pre-rainfall levels and composition rapidly (Armstrong et al. 2016; Štovícek et al. 2017; Ladwig et al. 2015). The responses of desert edaphic communities to hydration have also been shown to be influenced by water regime history and rainfall intensity, rather than frequency (Vargas et al. 2012; Frossard et al. 2015). It should, however, be noted that an intense water event in deserts can lead to the death, through osmolysis, of up to half of the edaphic microbial biomass, releasing substantial concentrations of nutrients (Belnap et al. 2005).

4.4 Conclusions and Perspectives: Hot Desert Soil Microorganisms in a Changing World

Independent of their aridity status, hot deserts harbor phylogenetically diverse and adapted edaphic microbial communities that retain some functions under desiccated conditions and react dramatically to episodic water inputs (Chap. 11). Contributors to these communities include niche habitats (such as hypoliths), soil surface assemblages (BSCs), and the microbial populations disbursed throughout open soils. Members of the Actinobacterial and Proteobacterial phyla often dominate the phylogenetic compositions of these dryland soil microbial communities and are the principal drivers of soil biological processes during periods of water insufficiency.

Despite the dominance of drylands across the planet, and the appreciation that in such soils most ecosystem services are provided by microbial communities, many dryland regions have received little or no attention from researchers (e.g., the Arabic Peninsula, the Australian Deserts, and most of the largest hot desert region, the Sahara; Table 4.1).

One of the clear predictions of the effects of climate change is a global increase in dryland surface area and many drylands will experience increased aridity (Huang et al. 2016; UNCCD 2017). However, given that drylands are widely dispersed on Earth, it is difficult to predict how any given dryland region will be impacted by climate change effects (IPCC 2001; Safriel et al. 2006; Jansson and Hofmockel 2019). Models generally predict, on a global scale, that drylands will experience changes in the hydrological cycles, with longer periods of drought and rarer but more intense precipitation events, in conjunction with significantly higher temperatures (Tsonis et al. 2005). In consequence, environmental conditions in dryland regions will tend to become more extreme, with likely increases in stress effect for desert organisms and communities (including microorganisms).

Many of the informative studies on the impact of warming on dryland soil communities have been performed in semiarid ecosystems (see review by Romero-Olivares et al. 2017). A 4-year warming experiment (+~1.6 °C) in a semiarid grassland suggested that warming (and associated changes in water regimes, derived from increased evapotranspiration) can influence carbon usage and microbial growth rates, but may not grossly affect community structures (Zhang et al. 2013b). The implication from this study is that dryland soil microbial communities are already adapted to climate perturbations. This suggestion was perhaps confirmed by a 5-year warming experiment (+1.8 °C), which showed no significant change in the abundances of bacterial genes involved in N cycling (i.e.,

N_2 fixation [*nif*H], nitrification [*amo*AB], mineralization [*chi*A], and denitrification [*nir*S, *nir*K, and *nos*Z]) (Zhang et al. 2013a). However, under the most optimistic representative concentration pathway (RCP) scenario, RCP2.6, global temperatures are expected to increase from 2.6 to 4.8 °C while under the least optimistic, RCP8.5, temperatures may rise from over 10 °C in some regions (IPCC 2014). It is currently not known how microbial communities in semiarid soils will adapt to such dramatic changes and whether they will retain their biogeochemical capacities and ecosystem services?

Such questions can be addressed by (1) generating quantitative process data from drylands to improve actual climate models and (2) performing short- and long-term in situ and controlled experiments. We argue that there is scope, if not an immediate need, to better characterize the microbial communities of global dryland soils, in terms of their compositions, functions and processes, and responses to the effects of climate change.

References

Allison SD, Martiny JBH (2008) Resistance, resilience, and redundancy in microbial communities. Proc Natl Acad Sci U S A 105:11512–11519

An S, Couteau C, Luo F, Neveu J, DuBow MS (2013) Bacterial diversity of surface sand samples from the Gobi and Taklamaken deserts. Microb Ecol 66:850–860

Andrew DR, Fitak RR, Munguia-Vega A, Racolta A, Martinson VG, Dontsova K (2012) Abiotic factors shape microbial diversity in Sonoran Desert soils. Appl Environ Microbiol 78:7527–7537

Angel R, Soares MIM, Ungar ED, Gillor O (2010) Biogeography of soil archaea and bacteria along a steep precipitation gradient. ISME J 4:553–563

Angel R, Claus P, Conrad R (2012) Methanogenic archaea are globally ubiquitous in aerated soils and become active under wet anoxic conditions. ISME J 6:847–862

Angel R, Pasternak Z, Soares MIM, Conrad R, Gillor O (2013) Active and total prokaryotic communities in dryland soils. FEMS Microbiol Ecol 86:130–138

Araya JP, Gonzalez M, Cardinale M, Schnell S, Stoll A (2019) Microbiome dynamics associated with the Atacama Flowering Desert. Front Microbiol 10:3160

Armstrong A, Valverde A, Ramond JB, Makhalanyane TP, Jansson JK, Hopkins DW, Aspray TJ, Seely M, Trindade MI, Cowan DA (2016) Temporal dynamics of hot desert microbial communities reveal structural and functional responses to water input. Sci Rep 6:34434

Bailey AC, Kellom M, Poret-Peterson AT, Noonan K, Hartnett HE, Raymond J (2014) Draft genome sequence of *Microvirga* sp. strain BSC39, isolated from biological soil crust of Moab, Utah. Genome Announc 2:e01198

Bastida F, Torres IF, Moreno JL, Baldrian P, Ondoño S, Ruiz-Navarro A, Hernández T, Richnow HH, Starke R, García C, Jehmlich N (2016) The active microbial diversity drives ecosystem multifunctionality and is physiologically related to carbon availability in Mediterranean semi-arid soils. Mol Ecol 25:4660–4673

Bastida F, Jehmlich N, Torres IF, García C (2018) The extracellular metaproteome of soils under semiarid climate: a methodological comparison of extraction buffers. Sci Total Environ 619: 707–711

Bates ST, Berg-Lyons D, Caporaso JG, Walters WA, Knight R, Fierer N (2011) Examining the global distribution of dominant archaeal populations in soil. ISME J 5:908–917

Bay S, Ferrari B, Greening C (2018) Life without water: how do bacteria generate biomass in desert ecosystems. Microbiol Aust 39:28–32

Bay SK, Waite DW, Dong X, Gillor O, Chown SL, Hugenholtz P, Greening C (2021) Chemosynthetic and photosynthetic bacteria contribute differentially to primary production across a steep desert aridity gradient. ISME J 1–18

Belnap J, Welter JR, Grimm NB, Barger N, Ludwig JA (2005) Linkages between microbial and hydrologic processes in arid and semiarid watersheds. Ecology 86:298–307

Berry D, Widder S (2014) Deciphering microbial interactions and detecting keystone species with co-occurrence networks. Front Microbiol 5:219

Büdel B, Dulić T, Darienko T, Rybalka N, Friedl T (2016) Cyanobacteria and algae of biological soil crusts. In: Biological soil crusts: an organizing principle in drylands. Springer, Cham, pp 55–80

Bull AT, Asenjo JA, Goodfellow M, Gomez-Silva B (2016) The Atacama Desert: technical resources and the growing importance of novel microbial diversity. Annu Rev Microbiol 70: 215–234

Cable JM, Ogle K, Williams DG, Weltzin JF, Huxman TE (2008) Soil texture drives responses of soil respiration to precipitation pulses in the Sonoran Desert: implications for climate change. Ecosystems 11:961–979

Carini P, Marsden PJ, Leff JW, Morgan EE, Strickland MS, Fierer N (2016) Relic DNA is abundant in soil and obscures estimates of soil microbial diversity. Nat Microbiol 2:1–6

Collins SL, Belnap J, Grimm NB, Rudgers JA, Dahm CN, D'Odorico P, Litvak M, Natvig DO, Peters DC, Pockman WT, Sinsabaugh RL (2014) A multiscale, hierarchical model of pulse dynamics in arid-land ecosystems. Annu Rev Ecol Evol Syst 45:397–419

Connon SA, Lester ED, Shafaat HS, Obenhuber DC, Ponce A (2007) Bacterial diversity in hyperarid Atacama Desert soils. J Geophys Res 112(G4)

Cordero OX, Datta MS (2016) Microbial interactions and community assembly at microscales. Curr Opin Microbiol 31:227–234

Cowan DA, Hopkins DW, Jones BE, Maggs-Kölling G, Majewska R, Ramond JB (2020) Microbiomics of Namib Desert habitats. Extremophiles 24:17–29

Crits-Christoph A, Robinson CK, Barnum T, Fricke WF, Davila AF, Jedynak B, McKay CP, DiRuggiero J (2013) Colonization patterns of soil microbial communities in the Atacama Desert. Microbiome 1:28

Elbert W, Weber B, Burrows S, Steinkamp J, Büdel B, Andreae MO, Pöschl U (2012) Contribution of cryptogamic covers to the global cycles of carbon and nitrogen. Nat Geosci 5:459–462

Eldridge DJ, Woodhouse JN, Curlevski NJ, Hayward M, Brown MV, Neilan BA (2015) Soil-foraging animals alter the composition and co-occurrence of microbial communities in a desert shrubland. ISME J 9:2671–2681

Eldridge DJ, Maestre FT, Koen TB, Delgado-Baquerizo M (2018) Australian dryland soils are acidic and nutrient-depleted, and have unique microbial communities compared with other drylands. J Biogeogr 45:2803–2814

Elliott DR, Thomas AD, Hoon SR, Sen R (2014) Niche partitioning of bacterial communities in biological crusts and soils under grasses, shrubs and trees in the Kalahari. Biodivers Conserv 23: 1709–1733

Faust K, Raes J (2012) Microbial interactions: from networks to models. Nat Rev Microbiol 10: 538–550

Fierer N (2017) Embracing the unknown: disentangling the complexities of the soil microbiome. Nat Rev Microbiol 15:579

Fierer N, Lauber CL, Ramirez KS, Zaneveld J, Bradford MA, Knight R (2012) Comparative metagenomic, phylogenetic and physiological analyses of soil microbial communities across nitrogen gradients. ISME J 6:1007–1017

Frossard A, Ramond JB, Seely M, Cowan DA (2015) Water regime history drives responses of soil Namib Desert microbial communities to wetting events. Sci Rep 5:12263
Geisen S, Mitchell EA, Wilkinson DM, Adl S, Bonkowski M, Brown MW, Fiore-Donno AM, Heger TJ, Jassey VE, Krashevska V, Lahr DJ (2017) Soil protistology rebooted: 30 fundamental questions to start with. Soil Biol Biochem 111:94–103
Geisen S, Mitchell EA, Adl S, Bonkowski M, Dunthorn M, Ekelund F, Fernández LD, Jousset A, Krashevska V, Singer D, Spiegel FW (2018) Soil protists: a fertile frontier in soil biology research. FEMS Microbiol Rev 42:293–323
Gonzalez-Martin C, Teigell-Perez N, Lyles M, Valladares B, Griffin DW (2013) Epifluorescent direct counts of bacteria and viruses from topsoil of various desert dust storm regions. Res Microbiol 164:17–21
Goodfellow M, Nouioui I, Sanderson R, Xie F, Bull AT (2018) Rare taxa and dark microbial matter: novel bioactive actinobacteria abound in Atacama Desert soils. Antonie Van Leeuwenhoek 111: 1315–1332
Grishkan I, Nevo E (2010) Spatiotemporal distribution of soil microfungi in the Makhtesh Ramon area, central Negev desert, Israel. Fungal Ecol 3(4):326–337
Gunnigle E, Ramond JB, Frossard A, Seely M, Cowan D (2014) A sequential co-extraction method for DNA, RNA and protein recovery from soil for future system-based approaches. J Microbiol Methods 103:118–123
Gunnigle E, Frossard A, Ramond JB, Guerrero L, Seely M, Cowan DA (2017) Diel-scale temporal dynamics recorded for bacterial groups in Namib Desert soil. Sci Rep 7:1–12
Huang J, Yu H, Guan X, Wang G, Guo R (2016) Accelerated dryland expansion under climate change. Nat Clim Change 6:166–171
Idris H, Goodfellow M, Sanderson R, Asenjo JA, Bull AT (2017) Actinobacterial rare biospheres and dark matter revealed in habitats of the Chilean Atacama Desert. Sci Rep 7:1–11
Ihrmark K, Bödeker I, Cruz-Martinez K, Friberg H, Kubartova A, Schenck J, Strid Y, Stenlid J, Brandström-Durling M, Clemmensen KE, Lindahl BD (2012) New primers to amplify the fungal ITS2 region–evaluation by 454-sequencing of artificial and natural communities. FEMS Microbiol Ecol 82:666–677
IPCC (2001) IPCC third assessment report. World Meteorological Organisation and UNEP
IPCC (2014) Mitigation of climate change. Contribution of Working Group III to the Fifth Assessment Report of the Intergovernmental Panel on climate Change
Jacobson KM, Jacobson PJ (1998) Rainfall regulates decomposition of buried cellulose in the Namib Desert. J Arid Environ 38:571–583
Jansson JK, Hofmockel KS (2018) The soil microbiome—from metagenomics to metaphenomics. Curr Opin Microbiol 43:162–168
Jansson JK, Hofmockel KS (2019) Soil microbiomes and climate change. Nat Rev Microbiol 1–12
Johnson RM, Ramond JB, Gunnigle E, Seely M, Cowan DA (2017) Namib Desert edaphic bacterial, fungal and archaeal communities assemble through deterministic processes but are influenced by different abiotic parameters. Extremophiles 21:381–392
Jordaan K, Lappan R, Dong X, Aitkenhead IJ, Bay SK, Chiri E, Wieler N, Meredith LK, Cowan DA, Chown SL, Greening C (2020) Hydrogen-oxidizing bacteria are abundant in desert soils and strongly stimulated by hydration. Msystems 5:e01131–e01120
Kerepesi C, Grolmusz V (2016) Giant viruses of the Kutch desert. Arch Virol 161:721–724
Kerepesi C, Grolmusz V (2017) The “Giant Virus Finder” discovers an abundance of giant viruses in the Antarctic dry valleys. Arch Virol 162:1671–1676
Kieft TL (2000) Size matters: dwarf cells in soil and subsurface terrestrial environments. In: Colwell RR, Grimes DJ (eds) Nonculturable microorganisms in the environment. ASM Press, Washington, DC, pp 19–46
Knief C, Bol R, Amelung W, Kusch S, Frindte K, Eckmeier E, Jaeschke A, Dunai T, Fuentes B, Mörchen R, Schütte T (2020) Tracing elevational changes in microbial life and organic carbon sources in soils of the Atacama Desert. Global Planet Change 184:103078
Ladwig LM, Sinsabaugh RL, Collins SL, Thomey ML (2015) Soil enzyme responses to varying rainfall regimes in Chihuahuan Desert soils. Ecosphere 6:1–10

León-Sobrino C, Ramond JB, Maggs-Kölling G, Cowan DA (2019) Nutrient acquisition, rather than stress response over diel cycles, drives microbial transcription in a hyper-arid Namib Desert soil. Front Microbiol 10:1054

Leung PM, Bay SK, Meier DV, Chiri E, Cowan DA, Gillor O, Woebken D, Greening C (2020) Energetic basis of microbial growth and persistence in desert ecosystems. Msystems 5. https://doi.org/10.1128/mSystems.00495-19

Liu S, Liu W, Shi X, Li S, Hu T, Song L, Wu C (2018) Dry-hot stress significantly reduced the nitrogenase activity of epiphytic cyanolichen. Sci Total Environ 619:630–637

Maestre FT, Delgado-Baquerizo M, Jeffries TC, Eldridge DJ, Ochoa V, Gozalo B, Quero JL, Garcia-Gomez M, Gallardo A, Ulrich W et al (2015) Increasing aridity reduces soil microbial diversity and abundance in global drylands. Proc Natl Acad Sci U S A 112:15684–15689

Makhalanyane TP, Valverde A, Lacap DC, Pointing SB, Tuffin MI, Cowan DA (2013) Evidence of species recruitment and development of hot desert hypolithic communities. Environ Microbiol Rep 5:219–224

Makhalanyane TP, Valverde A, Gunnigle E, Frossard A, Ramond JB, Cowan DA (2015) Microbial ecology of hot desert edaphic systems. FEMS Microbiol Rev 39:203–221

Mandakovic D, Rojas C, Maldonado J, Latorre M, Travisany D, Delage E, Bihouée A, Jean G, Díaz FP, Fernández-Gómez B, Cabrera P (2018) Structure and co-occurrence patterns in microbial communities under acute environmental stress reveal ecological factors fostering resilience. Sci Rep 8:5875

Marasco R, Mosqueira MJ, Fusi M, Ramond JB, Merlino G, Booth JM, Maggs-Kölling G, Cowan DA, Daffonchio D (2018) Rhizosheath microbial community assembly of sympatric desert speargrasses is independent of the plant host. Microbiome 6:215

Mosqueira MJ, Marasco R, Fusi M, Michoud G, Merlino G, Cherif A, Daffonchio D (2019) Consistent bacterial selection by date palm root system across heterogeneous desert oasis agroecosystems. Sci Rep 9:1–12

Moussa TA, Al-Zahrani HS, Almaghrabi OA, Abdelmoneim TS, Fuller MP (2017) Comparative metagenomics approaches to characterize the soil fungal communities of western coastal region, Saudi Arabia. PLoS One 12(9)

Noy-Meir I (1973) Desert ecosystems: environment and producers. Ann Rev Ecol Syst 4:25–51

Oliverio AM, Geisen S, Delgado-Baquerizo M, Maestre FT, Turner BL, Fierer N (2020) The global-scale distributions of soil protists and their contributions to belowground systems. Sci Adv 6:eaax8787

Ortiz M, Leung PM, Shelley G, Jirapanjawat T, Nauer PA, Van Goethem MW, Bay SK, Islam ZF, Jordaan K, Vikram S, Chown SL, Hogg ID, Makhalanyane TP, Grinter R, Cowan DA, Greening C (2021) Multiple energy sources and metabolic strategies sustain microbial diversity in Antarctic desert soils. PNAS 118(45)

Pointing SB, Belnap J (2012) Microbial colonization and controls in dryland systems. Nat Rev Microbiol 10:551–562

Potts M (2001) Desiccation tolerance: a simple process? Trends Microbiol 9:553–559

Prentice IC (2008) Terrestrial nitrogen cycle simulation with a dynamic global vegetation model. Glob Chang Biol 14:1745–1764

Prestel E, Regeard C, Andrews J, Oger P, DuBow MS (2012) A novel bacteriophage morphotype with a ribbon-like structure at the tail extremity. Res J Microbiol 7:75–81

Quoreshi AM, Suleiman MK, Kumar V, Manuvel AJ, Sivadasan MT, Islam MA, Khasa DP (2019) Untangling the bacterial community composition and structure in selected Kuwait desert soils. App Soil Ecol 138:1–9

Ramond JB, Woodborne S, Hall G, Seely M, Cowan DA (2018) Namib desert primary productivity is driven by cryptic microbial community N-fixation. Sci Rep 8:1–9

Rivera P, Cruces F (2018) *Achnanthidium exiguum* (Bacillariophyta): nuevas citas para localidades andinas del norte de Chile. Gayana Botánica 75:646–649

Romero-Olivares AL, Allison SD, Treseder KK (2017) Soil microbes and their response to experimental warming over time: a meta-analysis of field studies. Soil Biol Biochem 107:32–40

Ronca S, Ramond JB, Jones BE, Seely M, Cowan DA (2015) Namib desert dune/interdune transects exhibit habitat-specific edaphic bacterial communities. Front Microbiol 6:845
Rutz BA, Kieft TL (2004) Phylogenetic characterization of dwarf archaea and bacteria from a semiarid soil. Soil Biol Biochem 36:825–833
Safriel U, Ezcurra E, Tegen I, Schlesinger WH, Nellemann C, Batjes NH, Dent D, Groner E, Morrison S, Resenfeld D, Avner U (2006) Deserts and the planet-linkages between deserts and non-deserts. In: Global deserts outlook. UNEP, pp 50–72
Santiago IF, Gonçalves VN, Gómez-Silva B, Galetovic A, Rosa LH (2018) Fungal diversity in the Atacama Desert. Antonie Van Leeuwenhoek 111:1345–1360
Schulze-Makuch D, Wagner D, Kounaves SP, Mangelsdorf K, Devine KG, de Vera JP, Schmitt-Kopplin P, Grossart HP, Parro V, Kaupenjohann M, Galy A (2018) Transitory microbial habitat in the hyperarid Atacama Desert. Proc Natl Acad Sci U S A 115:2670–2675
Scola V, Ramond JB, Frossard A, Zablocki O, Adriaenssens EM, Johnson RM, Seely M, Cowan DA (2018) Namib desert soil microbial community diversity, assembly, and function along a natural xeric gradient. Microb Ecol 75:193–203
Sher Y, Zaady E, Nejidat A (2013) Spatial and temporal diversity and abundance of ammonia oxidizers in semi-arid and arid soils: indications for a differential seasonal effect on archaeal and bacterial ammonia oxidizers. FEMS Microbiol Ecol 86:544–556
Sivakala KK, Jose PA, Anandham R, Thinesh T, Jebakumar SRD, Samaddar S, Chatterjee P, Sivakumar N, Sa T (2018) Spatial physicochemical and metagenomic analysis of desert environment. J Microbiol Biotechnol 28:1517–1526
Smith HD, Duncan AG, Neary PL, Lloyd CR, Anderson AJ, Sims RC, McKay CP (2012) In situ microbial detection in Mojave Desert soil using native fluorescence. Astrobiology 12:247–257
Srinivasiah S, Bhavsar J, Thapar K, Liles M, Schoenfeld T, Wommack KE (2008) Phages across the biosphere: contrasts of viruses in soil and aquatic environments. Res Microbiol 159:349–357
Sterflinger K, Tesei D, Zakharova K (2012) Fungi in hot and cold deserts with particular reference to microcolonial fungi. Fungal Ecol 5(4):453–462
Štovícek A, Kim M, Or D, Gillor O (2017) Microbial community response to hydration-desiccation cycles in desert soil. Sci Rep 7:45735
Stutz JC, Copeman R, Martin CA, Morton JB (2000) Patterns of species composition and distribution of arbuscular mycorrhizal fungi in arid regions of southwestern North America and Namibia, Africa. Can J Bot 78:237–245
Su YG, Zhao X, Li AX, Li XR, Huang G (2011) Nitrogen fixation in biological soil crusts from the Tengger desert, northern China. Eur J Soil Biol 47:182–187
Takahashi Y, Matsumoto A, Seino A, Iwai Y, Omura S (1996) Rare actinomycetes isolated from desert soils. Actinomycetologica 10:91–97
Tedersoo L, Bahram M, Põlme S, Kõljalg U, Yorou NS, Wijesundera R, Ruiz LV, Vasco-Palacios AM, Thu PQ, Suija A, Smith ME (2014) Global diversity and geography of soil fungi. Science 346:1256688
Tsonis AA, Elsner JB, Hunt AG, Jagger TH (2005) Unfolding the relation between global temperature and ENSO. Geophys Res Lett 32(9)
Uhlmann E, Görke C, Petersen A, Oberwinkler F (2006) Arbuscular mycorrhizae from arid parts of Namibia. J Arid Environ 64:221–237
UNCCD (2017) Chapter 12: Drylands. In: The global land outlook, pp 246–269
Van der Walt AJ, Johnson RM, Cowan DA, Seely M, Ramond JB (2016) Unique microbial phylotypes in Namib Desert dune and gravel plain Fairy Circle soils. Appl Environ Microbiol 82:4592–4601
Vargas R, Collins SL, Thomey ML, Johnson JE, Brown RF, Natvig DO, Friggens MT (2012) Precipitation variability and fire influence the temporal dynamics of soil CO_2 efflux in an arid grassland. Glob Chang Biol 18:1401–1411
Wasserstrom H, Ben-Ezra VE, Sherman C, Unc A, Steinberger Y (2017) The influence of flint stones on a soil microbial community in the northern Negev Desert. AIMS Microbiol 3:580

Williamson KE, Fuhrmann JJ, Wommack KE, Radosevich M (2017) Viruses in soil ecosystems: an unknown quantity within an unexplored territory. Annu Rev Virol 4:201–219

Xu LK, Baldocchi DD, Tang J (2004) How soil moisture, rain pulses, and growth alter the response of ecosystem respiration to temperature. Global Biogeochem Cycles 18:GB4002

Yasir M, Azhar EI, Khan I, Bibi F, Baabdullah R, Al-Zahrani IA, Al-Ghamdi AK (2015) Composition of soil microbiome along elevation gradients in southwestern highlands of Saudi Arabia. BMC Microbiol 15:65

Zablocki O, Adriaenssens EM, Cowan D (2016) Diversity and ecology of viruses in hyperarid desert soils. Appl Environ Microbiol 82:770–777

Zablocki O, Adriaenssens EM, Frossard A, Seely M, Ramond JB, Cowan D (2017) Metaviromes of extracellular soil viruses along a Namib Desert aridity gradient. Genome Announc 5:e01470–e01416

Zhang X, Liu W, Schloter M, Zhang G, Chen Q, Huang J, Li L, Elser JJ, Han X (2013a) Response of the abundance of key soil microbial nitrogen-cycling genes to multi-factorial global changes. PLoS One 8(10)

Zhang N, Liu W, Yang H, Yu X, Gutknecht JL, Zhang Z, Wan S, Ma K (2013b) Soil microbial responses to warming and increased precipitation and their implications for ecosystem C cycling. Oecologia 173:1125–1142

Biology of Desert Endolithic Habitats

5

Evan B. Qu, Emine Ertekin, and Jocelyne DiRuggiero

Abstract

In extreme deserts around the world, endolithic microbial communities find refuge in the interior of semi-translucent rocks as a survival strategy. The pores and fissures of the rock matrix provide a space for colonization, protection from high solar irradiance, and promote the retention of scarce liquid water. Endolithic communities are rather complex ecosystems, spanning multiple domains of life and several trophic levels. These communities are based on the primary production of cyanobacteria, sometime algae, and constituted of an assemblage of heterotrophic bacteria and/or archaea, and viruses. Water availability, in the form of precipitation or relative humidity, is the primary limit on primary production, and thus overall activity in hot and arid deserts. However, other factors such as substrate properties, biotic interactions, and stochastic assembly processes may also influence endolithic community assembly and organization. Metagenomics studies have uncovered that these communities encode for diverse secondary metabolites, including many antimicrobial compounds, suggesting more complex community dynamics than previously thought. Longitudinal studies have revealed the resilience and modes of recovery of endolithic communities to severe climate perturbations but many more studies are needed; arid and hyper-arid deserts, at the dry limit for life, are fragile ecosystems, and their inhabitants will be on the front line of the major changes in climate ahead of us.

Evan B. Qu and Emine Ertekin contributed equally with all other contributors.

E. B. Qu · E. Ertekin · J. DiRuggiero (✉)
Department of Biology, Johns Hopkins University, Baltimore, MD, USA
e-mail: jdiruggiero@jhu.edu

J.-B. Ramond, D. A. Cowan (eds.), *Microbiology of Hot Deserts*, Ecological Studies 244, https://doi.org/10.1007/978-3-030-98415-1_5

Keywords

Microbial communities · Endolithic · Metagenomics · Extreme deserts · Cyanobacteria

5.1 Introduction

Endolithic (*endon* = within, *lithos* = stone) microbial communities inhabit the interior of semi-translucent rocks, colonizing pores and fissures within the rock matrix. While endolithic communities can be found in a diversity of environments including temperate montane zones (Matthes-Sears et al. 1997) and tropic marine systems (Campion-Alsumard et al. 1995), these communities play a particular importance in deserts as the last refuge for life in the most extreme hyper-arid conditions (Warren-Rhodes et al. 2007; Pointing and Belnap 2012). In deserts, the rock habitat protects microorganisms from high UV radiation and drastic temperature fluctuations and promotes the retention of scarce liquid water within the rock matrix (Walker and Pace 2007; Meslier and DiRuggiero 2019). Endolithic microbial communities are relatively simple, dominated by a few core photoautotrophic taxa along with a diversity of heterotrophic taxa. These communities have been described in nearly every desert on Earth and can colonize a diversity of rock substrates including sandstone (Friedmann et al. 1967; Friedmann and Ocampo 1976; Walker and Pace 2007), limestone (Wong et al. 2010), halite evaporites (Wierzchos et al. 2006; de los Ríos et al. 2010), ignimbrite (Wierzchos et al. 2013; Meslier et al. 2018), granite (de los Ríos et al. 2005) calcite and gypsum (DiRuggiero et al. 2013; Wierzchos et al. 2011; Wierzchos et al. 2015; Meslier et al. 2018), highlighting the ubiquity of this habitat in desert ecosystems.

The study of desert endolithic life originated with Imre Friedmann and colleagues' discovery of endolithic communities in sandstones from the Negev Desert, Israel (Friedmann et al. 1967) and the McMurdo Dry Valleys of Antarctica (Friedmann and Ocampo 1976). These findings immediately garnered interest in the fields of extremophile biology and astrobiology, as they were islands of life in otherwise barren environments. Initial research used in situ microscopy and physiological studies to characterize the communities and understand how endolithic microorganisms were adapted to their extreme environment. Early findings showed that highly tolerant Cyanobacteria dominated the endolithic community, were capable of surviving severe desiccation, and could quickly activate under milder conditions (Friedmann 1980; Billi et al. 2011). On the other hand, limitations of microscopic and culture-based methods meant that much of the remaining community went unstudied. Another important finding from early research was that conditions in hot deserts were more extreme than those in the polar deserts (Friedmann 1980), in contrast with what was initially assumed from climate data. It is now established that the combination of desiccation, drastic temperature fluctuations, and high solar radiation in hot, hyper-arid deserts constitutes one of the most extreme terrestrial environments on Earth (Wierzchos et al. 2018). Special attention has been focused on endolithic biology in the Atacama Desert, Chile, described as the "closest Earthly analog to Mars," where it was estimated that conditions are only suitable for photosynthesis nine days out of the year in its driest part (Davila et al. 2008). The Atacama Desert is also the only place where endolithic communities inhabiting halite evaporitic rocks are found, providing a unique environment for halophilic bacteria and Archaea (Wierzchos et al. 2006).

With the advent of next-generation sequencing technologies (NGS) and the use of geologic (mineralogy, soluble ions) and microscopic methods (3-D computed tomography, scanning electron microscopy with backscattered electrons), new aspects of the endolithic system have come into appreciation (Casero et al. 2020). Encased in rocks, endolithic communities have minimal biomass exchange with the outside environment and limited nutrient input beyond atmospheric water and gases. Endolithic communities are also orders of magnitude less diverse than most other microbial communities. Current estimates of terrestrial microbial diversity range from tens of thousands to millions of taxa per gram of soil (Torsvik and Øvreås 2002; Allison and Martiny 2009; Maron et al. 2018), while endolithic communities harbor only hundreds of taxa at most, ranging down to tens of taxa at the least diverse sites (Meslier et al. 2018; Qu et al. 2020). Put together, the endolithic system makes a highly tractable system for understanding basic principles of microbial ecology, including spatial and temporal patterning (Uritskiy et al. 2019, 2020b; Qu et al. 2020; Uritskiy et al. 2020b) and community response to environmental changes (Uritskiy et al. 2020a). More recently, metagenomics studies have uncovered diverse secondary metabolites, including many antimicrobial compounds, suggesting that these communities harbor more complex community dynamics than previously thought (Crits-Christoph et al. 2016a; Uritskiy et al. 2020a). Putative interactions have also been uncovered at the mineral–microbe interface and several chemolithotrophic pathways have been reported in endolithic communities from the Atacama Desert (Ertekin et al 2021; Huang et al 2020); however, these studies are still in their infancy in endolithic systems.

In this chapter, we introduce the members of the endolithic community in Sect. 5.2, describing the properties of this unique environment and its inhabitants. In Sect. 5.3, we discuss the climate and macroscopic factors that drive the generation and maintenance of biological diversity in endolithic systems, while in Sect. 5.4, we address the rock properties that drive that diversity. We finally discuss in Sect. 5.5 the adaptive strategies and the metabolic pathways that endolithic microbial communities have evolved not only to survive but also thrive in their rocky habitat.

5.2 The Endolithic Community

The endolithic community occupies a pigmented band, termed the colonization zone, directly below the surface of the rock substrate. These bands are often colored green from chlorophyll pigments, but may also be orange, brown, or black owing to

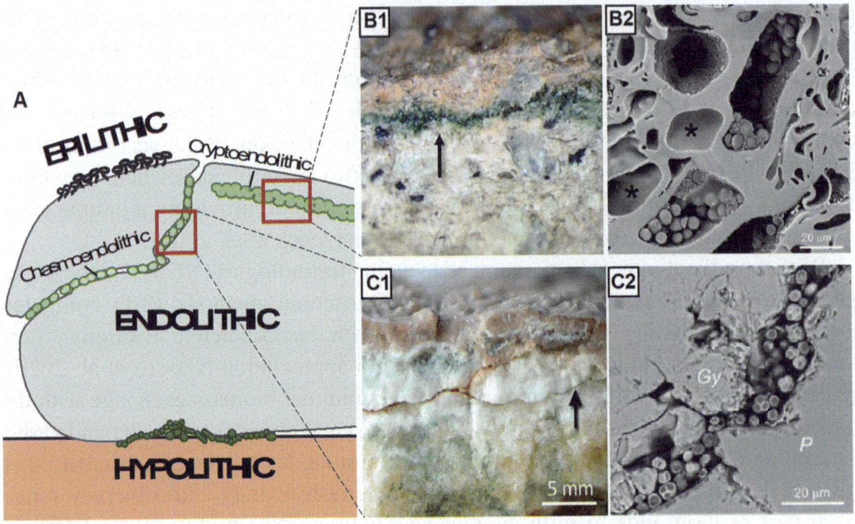

Fig. 5.1 The endolithic habitat. (**a**) Diagram of the possible types of lithobiontic colonization and their location in the rock substrate. (**b1**) Cryptoendolithic colonization (black arrow) in ignimbrite rock, (**b2**) scanning electron microscopy (SEM) image of cryptoendolithic colonization in ignimbrite rock, (**c1**) chasmoendolithic colonization (black arrow) in a crevice of calcite rock, (**c2**) SEM image of chasmoendolithic colonization in gypsum-covered rhyolite, all from the Atacama Desert (adapted from DiRuggiero et al. 2013; Wierzchos et al. 2015; Crits-Christoph et al. 2016a)

various carotenoids, melanins, and scytonemins (Wierzchos et al. 2015). Endolithic communities are categorized according to their specific colonization strategy (Fig. 5.1). Cryptoendolithic communities inhabit small pores and cavities in the rock matrix and are often found in highly porous rocks. Chasmoendolithic communities inhabit cracks and fissure space in the rocks and are more commonly found in highly sedimented rocks. (Golubic et al. 1981).

The hot desert endolithic community is dominated by Cyanobacteria, which constitute between 40 and 80% of the bacterial community (Walker and Pace, 2007; Crits-Christoph et al. 2016a; Meslier et al. 2018; Qu et al. 2020) (Fig. 5.2a). Cyanobacteria predominantly belong to the coccoidal genus *Chroococcidiopsis* (Friedmann 1982), which is known for its superlative desiccation and radiation resistance (Friedmann et al. 1967; Billi et al. 2000, 2011). Occasionally in hot deserts, *Chroococcidiopsis* spp. are accompanied by additional cyanobacterial taxa including *Chroococcales, Synechococcales,* and members from filamentous orders such as *Oscillatoriales* and *Nostocales*, although additional cyanobacterial taxa become more prevalent in colder or wetter climate regimes (Pointing et al. 2009; Goordial et al. 2016; Khomutovska et al. 2017). Diverse heterotrophic bacteria have also been reported, with the phyla *Actinobacteria*, *Proteobacteria*, *Chloroflexi*, *Bacteroidetes*, and *Deinococcus-Thermus* particularly represented (Meslier et al. 2018; Qu et al. 2020) (Fig. 5.2a). Archaea are minimally present in endolithic communities except for unique hyper-saline communities inhabiting halite nodules

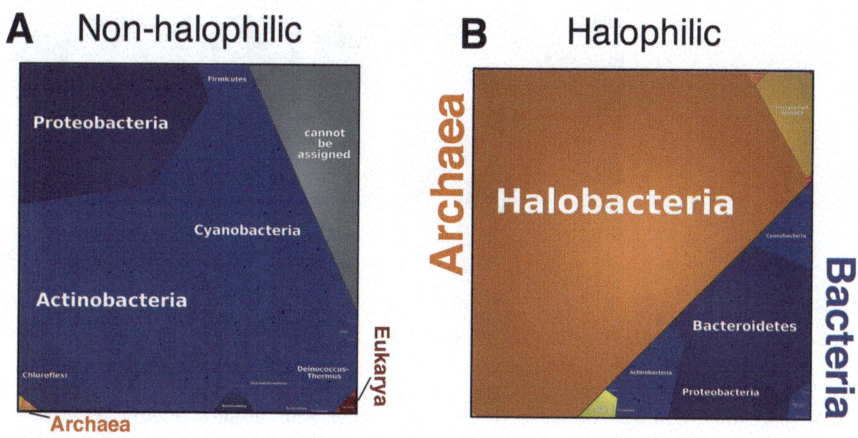

Fig. 5.2 Taxonomic composition of endolithic communities. Voronoi treemap showing relative abundance of taxa within (**a**) a gypsum endolithic community and (**b**) a halite endolithic community from the Atacama Desert (shotgun sequencing of environmental DNA data obtained from Uritskiy et al. 2019, Ertekin et al. 2021)

in the Atacama Desert; in these communities, highly salt-adapted *Halobacteria* (Euryarchaeota) comprise over 75% of the community (Robinson et al. 2015) (Fig. 5.2b). Bacteria make up the remainder of the halite community and are dominated by the phyla *Bacteroidetes*, *Proteobacteria*, *Actinobacteria*, and *Cyanobacteria* (Robinson et al. 2015) (Fig. 5.2b).

Very few eukaryotic taxa have been reported in hot desert endolithic communities, and it is believed that eukaryotes are generally excluded from hot deserts due to harsher environmental conditions (Friedmann 1980; Qu et al. 2020). This is in contrast with endolithic communities from cold and polar deserts, which are known to contain high abundances of symbiotic lichens (Friedmann 1982; Pointing et al. 2009; Wong et al. 2010; Goordial et al. 2016; Archer et al. 2017). However, microalgae and fungi have been reported in low relative abundances (<5%) in gypsum rocks and halite nodules from the Atacama Desert, with the algae specifically belonging to the genera *Trebouxia (Trebouxiophyceae)* and *Dolichomastix* (*Mamiellaceae*), respectively (Crits-Christoph et al. 2016b) (Fig. 5.3a). Despite their low relative abundance in hot deserts, recent findings suggest that algae still make substantial contributions to overall primary productivity. In halite nodules from the Atacama Desert, algae make up ~2% of the genomic abundance of the community, but metatranscriptomic analysis shows that they are the most active members and express more photosynthetic genes than the relatively more abundant cyanobacteria (Uritskiy et al. 2020a).

Endolithic microorganisms are often found in association with large amounts of extracellular polymeric substances (EPSs) in the rock matrix (Fig. 5.3b and c). EPSs are secreted by photosynthetic organisms within the endolithic community and thought to serve a variety of ecosystem functions, most important of which is trapping scarce water for the community due to their hygroscopic nature (Lebre

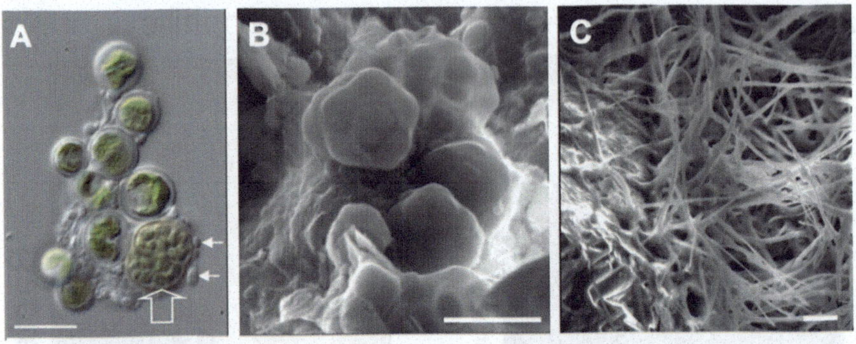

Fig. 5.3 Micrographs of algae and *Cyanobacteria* from endolithic substrates. (**a**) Differential interference contrast image of algae (large round cells) with distinct chloroplasts (green), Cyanobacteria (open arrow), and heterotrophic bacteria (small white arrows) from halite nodules from the Atacama Desert; scale bar = 10 μm (from Robinson et al. 2015). (**b**) SEM of coccoid (embedded in a sheath) and (**c**) filamentous cyanobacteria from sandstone from the Negev and Namib Deserts, respectively; scale bars = 5 μm

et al. 2017). The amount and arrangement of EPSs vary greatly between communities; for example, halite nodules from the Atacama Desert harbor cyanobacterial cells embedded in concentric sheets of EPS and surrounded by heterotrophic bacteria (Wierzchos et al. 2006). *Chroococcidiopsis* spp. from sandstone rocks in the Negev Desert were reported to surround themselves in envelopes of EPS, which may be a possible stress response to desiccation (Lebre et al. 2017). However, it remains unknown what factors determine the organization of EPSs in different communities.

5.3 Climate and Macroscopic Drivers of Diversity

The endolithic system is relatively simple and, as such, processes that drive the diversity and organization of endolithic microbial communities have been relatively well-studied. Endolithic communities are thought to be seeded from a relatively small metacommunity that is highly adapted to the extreme environment, evidenced by the remarkably similar community compositions across deserts and continents (Meslier et al. 2018; Qu et al. 2020). Recruitment sources for these communities are less clear, but evidence exists for local recruitment from nearby biological soil crusts (Pointing and Belnap 2012), as well as long-range airborne transport (Walker and Pace 2007).

Water availability, in the form of precipitation or relative humidity, is the primary limit on primary production, and thus overall activity in hot and arid deserts, and explains much of the variability in endolithic community composition. In the most hyper-arid environments, taxonomic diversity is low, with fewer than 100 unique taxa reported in endolithic communities from the most hyper-arid regions of the Atacama and Namib Deserts (Robinson et al. 2015; Qu et al. 2020). Endolithic communities in the most extreme conditions support a very high ratio of Cyanobacteria, reaching up to 80% of the total community (Crits-Christoph et al. 2016a; Meslier et al. 2018). It is thought that the relative abundance of photosynthesizers increases with aridity, as a higher ratio of these primary producers is needed to support a community under lower-productivity conditions (Lacap-Bugler et al. 2017). As available moisture increases, community diversity has been found to increase at both global and regional scales, with even relatively small increases in water availability leading to increases in diversity (DiRuggiero et al. 2013; Robinson et al. 2015; Qu et al. 2020). Higher water availability is also associated with the emergence of eukaryotic and mesophilic taxa in endolithic communities. A study of a precipitation gradient in the Namib Desert found the emergence of several filamentous cyanobacterial taxa accompanying *Chroococcidiopsis*-like taxa at higher moisture regimes (Qu et al. 2020). In the Atacama Desert, halite endolithic communities from a less arid region exposed to coastal fogs were found to harbor a unique green alga not found in other parts of the desert (Robinson et al. 2015). At the highest moisture regimes, such as those in the semi-arid Colorado Plateau region in the United States, endolithic communities have been found to support robust algal and fungal populations and many non-desert adapted taxa, which are otherwise entirely excluded from desert endolithic habitats (Walker and Pace, 2007; Qu et al. 2020).

Temperature is also thought to be a significant driver of endolithic community assembly, and a significant difference in composition has been observed between hot and cold desert conditions. While in hot deserts the endolithic community is mostly bacterial with *Chroococcidiopsis* as the dominant phototroph, cold deserts, defined as having mean annual temperatures below 18 °C, harbor significant eukaryotic populations, with lichens often replacing cyanobacteria as the dominant primary producer in the community (Friedmann 1980; Goordial et al. 2016). While cyanobacteria are still found in cold deserts, the cyanobacterial population is often dominated by non-*Chroococcidiopsis* genera, including *Phormidium* and *Cyanothece* (Qu et al. 2020). This change in the producer profile has been attributed to differences in evapotranspiration potential, and thus water availability, between hot and cold deserts. In hot deserts, high daytime evapotranspiration causes any available moisture to quickly evaporate; however, this is not the case with cold deserts, where high saturations of water vapor can be maintained in the rock interior even at low air relative humidity values (Friedmann 1980). Palmer and Friedmann (1990) found that endolithic lichens can photosynthesize at lower relative humidity values (minimum 70% RH) than cyanobacteria (minimum 90% RH), which may contribute to their success in cold deserts.

Nevertheless, climate on its own is not sufficient to explain all variabilities in community composition. In the Canadian Arctic, differences in endolithic community composition were observed between sites despite no difference in moisture regime (Qu et al. 2020), suggesting that we still do not have a comprehensive understanding of endolithic community assembly mechanisms. Studies of other lithic systems suggest that biotic interactions (Valverde et al. 2015) and stochastic assembly processes (Caruso et al. 2011) may also influence endolithic community organization, but these are poorly studied in the endolithic system. Recent evidence also indicates that the drivers of endolithic community assembly differ between producers and consumers, with producers being more sensitive to climate-based selection than consumers (Qu et al. 2020). Furthermore, it is poorly understood how the endolithic system responds temporally to changes in environmental conditions. Investigation into halite community response after a heavy rain event in the hyper-arid core of the Atacama Desert did not report an increase in diversity, but rather a significant die-off of highly specialized members of the community, followed by stochastic recolonization and a gradual adjustment of community composition over time (Uritskiy et al. 2019).

5.4 The Rock Habitat

The rock habitat plays an essential role in supporting and shaping the endolithic community. The bioreceptivity of a substrate is the product of the physico-chemical properties that mediate light transmittance, water retention, nutrient diffusion rates, and provide spaces for microbial colonization (Walker and Pace, 2007; Wierzchos et al. 2015).

While UV is almost completely attenuated within 1–2 mm by the rock substrate (Hughes and Lawley, 2003; Cockell et al. 2008), the decline in visible light transmission occurs at a much lower rate, providing sufficient photosynthetically active radiation (PAR) for phototrophs within the colonization zone (Amaral et al. 2007; Meslier et al. 2018; McKay 2012; Horath et al. 2006; Matthes-Sears et al. 2001). Meslier et al. (2018) measured the light transmittance in several substrates from the Atacama Desert and found a direct relationship between the spectral properties of the substrate and the depth of the colonization zone; higher light-transmitting substrates (calcite, gypsum, and granite) showed microbial colonization located deeper in the substrate, while communities from ignimbrite, a darker substrate, were located closer to the surface (Meslier et al. 2018). Light attenuation from 1 to 0.02% of the incident PAR has been reported in the colonization zone of endoliths (Horath et al. 2006; Matthes-Sears et al. 2001; Meslier et al. 2018; McKay 2012; Robinson et al. 2015) suggesting that the Cyanobacteria inhabiting these substrates might have low-light photoacclimation photosystems such as those described in cyanobacteria ecotype of microbial mats (Ho et al. 2017). Despite the low incident light reported within halite nodules, Davila et al. (2015) observed a 20–40% decline in the quantum yield of photosystem II (PSII) from in situ pulse amplitude modulated (PAM) measurements of halite nodules between 12 p.m. and 4 p.m. suggesting transient photoinhibition during the brightest part of the day.

Molecular adaptations of phototrophs in response to the high solar irradiance found in deserts (Cordero et al. 2014) include the production of UV-absorbing and/or screening pigments such as scytonemin in Cyanobacteria and carotenoids in both Cyanobacteria and algae (Wierzchos et al. 2015; Vítek et al. 2016; Casero et al. 2020). Light-dependent changes in carotenoid conjugation to mediate non-photochemical quenching was also reported in Cyanobacteria (Vítek et al.

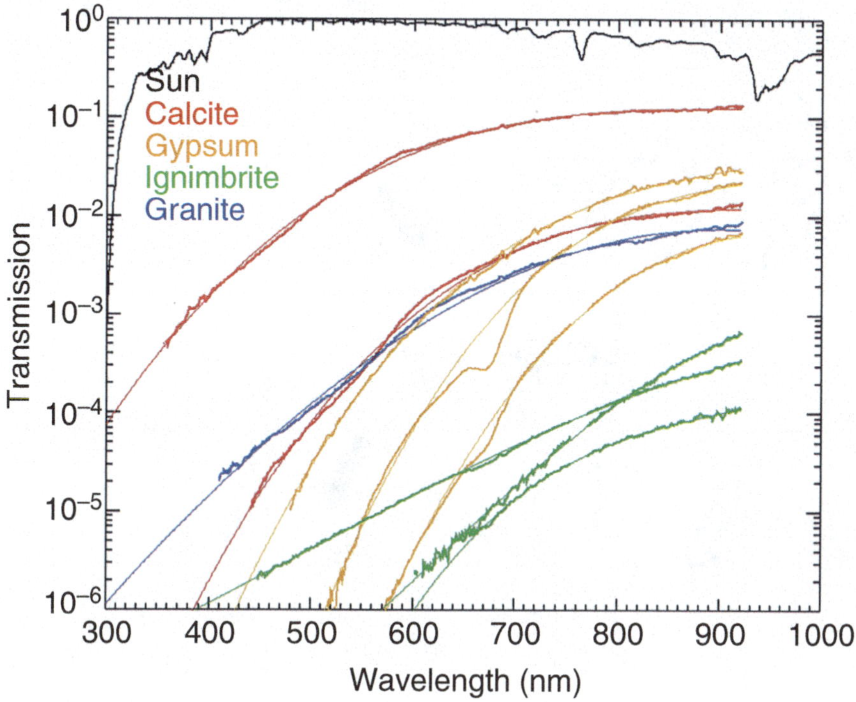

Fig. 5.4 Light transmission spectra at the nominal depth of 2 mm from the surface of the rocks for calcite, granite, gypsum, and ignimbrite. Thick colored lines are data and thin colored lines are polynomial fits to the data. Black, solar spectrum with its peak normalized to 1 on the y-axis; red, calcite; blue, granite; orange, gypsum; green, ignimbrite. The telluric molecular oxygen band at 760 nm was masked (from Meslier et al. 2018)

2017). Additionally, in gypsum from the Atacama Desert and sandstone from polar deserts, protective cell-layering has been observed as a light and UV protection strategy (Wierzchos et al. 2015; Friedmann 1982). In gypsum, algae located in the most upper part of the substrate were shown to accumulate large amounts of carotenoids, suggesting that they may act as a light protective barrier for the community (Vítek et al. 2017; Wierzchos et al. 2015).

Surprisingly, when measuring the light transmission spectra through several endolithic substrates, Meslier et al. (2018) found that the solar spectrum was significantly shifted toward far-red light (FRL; Fig. 5.4). When characterizing several cyanobacteria isolated from endolithic substrates from the Atacama and Negev Deserts, Murray and DiRuggiero (pers. comm.) found that they all produced chlorophyll *f* (chl *f*), a hallmark of far-red light photoacclimation (FaRLiP; Ho et al. 2016). Twenty-one genes encoding for a Chl *f* synthase, transcriptional activators, and paralogous copies of pigment-binding core subunits of PSI, PSII, and phycobilisome (PBS) constitute the FaRLiP gene cluster (Ho et al. 2017). Activation of the FaRLiP cluster by FRL results in structural remodeling of the core of the

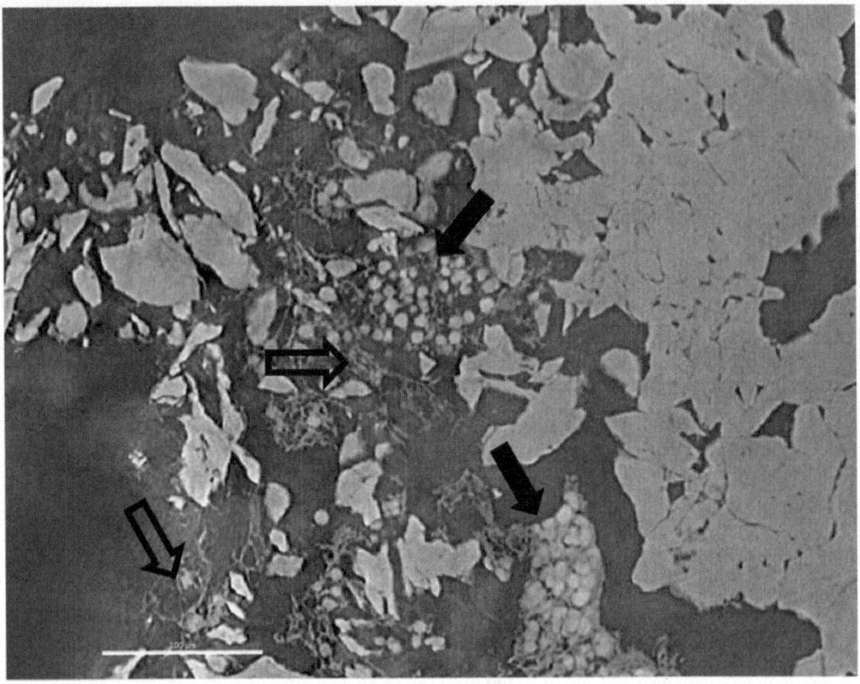

Fig. 5.5 2D slice of gypsum rock from the Atacama Desert at 0.55 μm/voxel resolution, obtained by micro-computed tomography, showing the association of algal cells and fungal hyphae within the pores of the rock; close arrows point to algal cells and open arrows point to fungal hyphae. Scale bars indicate 100 μm

photosynthesis apparatus and of the PBS and confers the ability to absorb light between 700 and 800 nm (Ho et al. 2017). Genome sequences from gypsum and halite cyanobacteria isolates revealed that the FaRLiP gene cluster was indeed present in these organisms (Murray and DiRuggiero, pers. comm.).

Endolithic microorganisms rely on scare rain events and/or the condensation of atmospheric moisture into their substrate for their water supply. Recent studies have demonstrated that the rock architecture—defined as the space available for colonization and characterized by the cracks, fissures, pores and their connection to the surface—is tightly linked to substrate water retention capacities and a main driver of community structure and diversity (Wierzchos et al. 2015; Crits-Christoph et al. 2016a; Meslier et al. 2018) (Fig. 5.5). Within that context, the high potential for water retention of calcite and gypsum substrates from the Atacama Desert was linked to high community diversity, and vice versa with granite and ignimbrite substrates with low potential for water retention and low community diversity (Meslier et al. 2018). The presence of minerals such as sepiolite, a magnesium silicate clay with high efficiency for water absorption and retention (Caturla et al. 1999), was shown to significantly enhance the moisture content within gypsum rocks (Wierzchos et al. 2015; Vitek et al. 2017). The water budget of an endolithic community can also be augmented by the high thermal conductivity properties of a substrate, such as calcite, promoting dew formation on its surface (DiRuggiero et al. 2013). A recent study, using a combination of microscopy and spectroscopy, has also shown that endolithic microorganisms can extract crystalline water from gypsum rock ($CaSO_4 \cdot 2H_2O$), resulting in phase transformation to anhydrite ($CaSO_4$) (Huang et al. 2020).

A unique process of water absorption and retention which has been found in halite nodules of the Atacama Desert is that of mineral deliquescence (Davila et al. 2008). When the air relative humidity reaches ~75%, water condenses inside the halite pores into concentrated brine, providing liquid water to the community. In places such as the Yungay area of the Atacama Desert, where water from rain, fog, or dew is exceedingly rare, only the hygroscopic properties of salt provide enough water for endolithic colonization (Davila et al. 2013).

5.5 Adaptive Strategies and Metabolism

In their rocky habitat, endolithic microorganisms not only have to withstand desiccation stress but also have to compete for nutrients and colonization space (Pointing and Belnap 2012; Lebre et al. 2017; Meslier and DiRuggiero 2019). The genetic makeup (i.e., the "functional potential") of the endolithic community has therefore been selected to encode and regulate adaptive mechanisms for meeting these challenges at the utmost efficiency. Gaining insights into this functional potential is essential to further our understanding of the molecular adaptations and metabolic strategies used by endolithic communities in hyper-arid deserts. To that extent, recent advances in metagenomics have provided us with unprecedented access to the genetic landscape of microbial communities. Here, we summarize the recent findings on molecular adaptations and metabolic strategies used by endolithic communities in hot deserts.

Desiccation leads to the accumulation of reactive oxygen species (ROS), which damage nucleic acids, proteins, and lipids, hence resulting in oxidative stress throughout the cell (Potts 1994). Alleviating the harmful effects of ROS is therefore essential in adapting to an endolithic lifestyle (Fig. 5.6a). Studies have shown that general oxidative stress response functions were more abundant in metagenomes from desert lithic habitats compared to soils (Goordial et al. 2016), and that hot desert lithic metagenomes harbored a significantly higher number of oxidative stress response-related genes compared to cold deserts (Le et al. 2016). To date, the best documented oxidative stress response mechanism in hot desert endolithic communities is the production of ROS scavenging antioxidants (Crits-Christoph et al. 2016a). For example, gene clusters encoding for two classes of antioxidants, phenylpropanoids, and phytoalexins were identified in ignimbrite and calcite community metagenomes (Crits-Christoph et al. 2016a). Carotenoids are another group of antioxidants that are of particular interest for the endolithic habitat because of their role in protecting the photosynthetic machinery of cyanobacteria (Vítek et al. 2017).

Withstanding osmotic pressure imposed by low intracellular water content or by high ion concentration outside the cell is one of the main adaptations in adjusting to

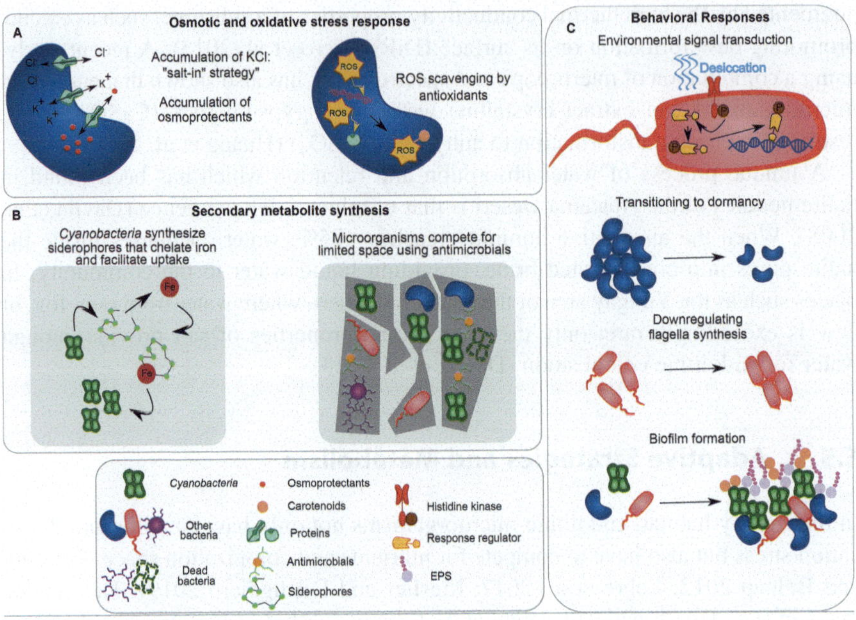

Fig. 5.6 Adaptations of endolithic microorganisms to the environmental conditions of their habitats. (**a**) Osmoprotectants such as glycine, proline, and trehalose are accumulated in the cell to prevent removal of water. Highly specialized microorganisms inhabiting halite nodules use the "salt-in" strategy that is the intracellular accumulation of potassium to maintain osmotic homeostasis. Antioxidant molecules are used for scavenging reactive oxygen species (ROS) resulting from desiccation-driven cellular damages. (**b**) In the endolithic habitat, siderophores are used to enhance iron uptake and polyketides and non-ribosomal peptides with antimicrobial properties are used in competition for space. (**c**) Rapid physiological response upon desiccation is ensured by environmental signal transduction systems such as the two-component system, mediated by histidine kinases, and response regulators. Common behavioral responses to desiccation are transitioning to dormancy/cyst formation, suppressing flagella synthesis and forming biofilms

an endolithic lifestyle in hyper-arid deserts (Potts 1994; Pointing and Belnap 2012). A well-documented strategy to alleviate osmotic stress is the accumulation of compatible solutes (i.e., osmoprotectants) to prevent removal of water from the cell (Potts 1994; Lebre et al. 2017; Omelon et al. 2016). Examples of osmoprotectants are glutamate, glutamine, proline, quaternary amines (i.e. glycine betaine), glucosyl glycerol, and sugars such as trehalose (Potts 1994, 2001; Lebre et al. 2017). These are either synthesized by the cell in high amounts or taken up from the environment. Crits-Christoph et al. (2016a) showed that molecular pathways related to trehalose biosynthesis and uptake, betaine and ectoine biosynthesis, and synthesis of glucans were abundant in the metagenomes of ignimbrite and calcite endolithic communities from the Atacama Desert. Genes encoding for ABC transporters that facilitate the uptake of polyols (glycerol and inositol), choline, betaine, and glycine betaine were also detected, together with genes encoding

Table 5.1 Example secondary metabolites identified in endolithic communities (Crits-Christoph et al. 2016a)

Compound	Class	Function	Organism
Penicillin and cephalosporin precursor	NRPs	Antimicrobial	*Cyanobacteria*
Coronatine	NRPs-T1PKs	Antimicrobial	*Cyanobacteria*
Largamide	NRPs-T1PKs	Protease inhibitor	*Cyanobacteria*
Pyoverdine	NRPs	Siderophore	*Cyanobacteria*

transporters mediating flux of water molecules, such as the aquaporin Z water channel protein (Crits-Christoph et al. 2016a).

Among endolithic substrates, halite nodules represent the most challenging osmotic conditions for microbial communities (Wierzchos et al. 2012; Artieda et al. 2015). Under these conditions, the primary mechanism for maintaining osmotic homeostasis is the "salt-in" strategy where cells accumulate potassium chloride to balance the high external salt concentrations (Oren 2008). The salt-in strategy is thought to be more energetically favorable for life at the highest salt concentrations and is mostly found among haloarchaea and a few bacterial taxa, including *Salinibacter* (Oren 2013). To remain functional, the proteins of "salt-in" strategists have an increased number of acidic amino acid residues on their surface, resulting in a proteome with a low isoelectric point (Oren 2013). Potassium uptake systems such as those encoded in the metagenome-assembled genome (MAG) of *Nanopetramus* SG9, a nanohaloarchaeon found in a halite community of the Atacama Desert, play essential roles for "salt-in" strategists (Crits-Christoph et al. 2016b), underscoring the role of intracellular K^+ concentrations for maintaining homeostasis in a highly specialized endolithic community.

Recent metagenomic sequencing uncovered a diversity of secondary metabolite clusters in endolithic communities (Table 5.1), suggesting that these compounds may have a significant ecological role for adaptation to the endolithic habitat (Crits-Christoph et al. 2016a) (Fig. 5.6b). For example, the metagenome of a community colonizing ignimbrite, a volcanic rock characterized by small pores, was enriched in secondary metabolite clusters when compared to that of a community inhabiting calcite, which had large colonized cracks and fissures (Crits-Christoph et al. 2016a). These secondary metabolite clusters mainly encoded non-ribosomal peptides (NRPs) and polyketides (PKs), specifically involved in the synthesis of siderophores and antimicrobial compounds. These findings lead to the hypothesis that increasingly unfavorable conditions and limited colonization space in the endolithic habitat resulted in fierce competition for space and nutrients.

In response to hyper-aridity, microorganisms have been found to regulate certain behavioral responses such as dormancy, cell motility, and biofilm formation, ensuring that energy-intensive processes, such as cell growth or flagella synthesis, only occur under favorable conditions (Lebre et al. 2017; see also Chap. 10) (Fig. 5.6c). For example, Louati et al. (2019) recently reported that genes involved in the regulation and synthesis of flagella-dependent cell motility and sporulation functions were enriched in stone dwelling communities from hot deserts compared to those from Mediterranean/semi-arid locations in North Africa. Increasing evidence indicates that the modulation of these behaviors, in response to fluctuating conditions during desiccation cycles, might be regulated by environmental signal transduction pathways (He and Bauer 2014). Indeed, the discovery of an extensive repertoire of genes involved in a two-component system in the genome of *Ramlibacter tataouinensis* TTB310, isolated from a meteorite in Southern Tunisia, suggests that this organism might anticipate dew formation periods for proliferated and might form desiccation resistant heterocysts under non-favorable conditions (de Luca et al. 2011).

5.5.1 Metabolic Pathways

While endolithic communities from the McMurdo Dry Valleys in Antarctica have some of the slowest reported growth of any microorganism on Earth (Bonani et al. 1988; Johnston and Vestal 1991), isotope composition analyses from endolithic communities in the Atacama Desert find carbon cycling times on the order of years to decades, indicating that hot desert endolithic communities metabolize more frequently and actively than those of cold environments (Ziolkowski et al. 2013). Limited water availability in hot deserts often restricts the window for microbial activity to short periods, requiring microorganisms to be highly efficient at turning on/off their metabolism (see Chap. 11). While the rock micro-environment, characterized by light transmission, water adsorption capacity, and available minerals, defines the initial constraints for microbial metabolism (Wierzchos et al. 2015; Meslier and DiRuggiero 2019), microorganisms can also modify the surrounding rock, impacting the biogeochemistry of their environment (Viles 1995, 2011; Buedel et al. 2004; Mergelov et al. 2018). Furthermore, the increased availability of metagenomic data has revealed that endolithic communities meet their carbon and energy requirements in more versatile ways than once assumed (Davila et al. 2015), providing insights into microbial dynamics and the fate of biogeochemical cycles at extreme aridity.

In endolithic communities, the main energy source for carbon fixation is oxygenic photosynthesis and it is carried out by cyanobacteria and algae (Davila et al. 2013; Davila et al. 2015; Meslier and DiRuggiero 2019) (Fig. 5.7a). The efficiency by which these primary producers carry out oxygenic photosynthesis is directly dependent on water availability (Davila et al. 2013; DiRuggiero et al. 2013). Chemolithotrophic carbon fixation has also been suggested for primary production in endolithic systems; for example, in halite communities from the Atacama Desert, dark carbon fixation was enhanced in the presence of ammonia (Davila et al. 2015), indicating potential chemolithoautotrophic ammonia oxidation (Davila et al. 2015). In gypsum communities from the Atacama Desert, genes for RubisCO type I were identified in MAGs of *Actinobacteria*, *Proteobacteria*, and *Chloroflexi* (Ertekin et al. 2021). These results suggest that, in carbon starved ecosystems such as hot deserts, chemolithoautotrophy can potentially provide additional fixed carbon to the lithobiontic community (Omelon 2008; Jordaan et al. 2020; Bay et al. 2021; Jones

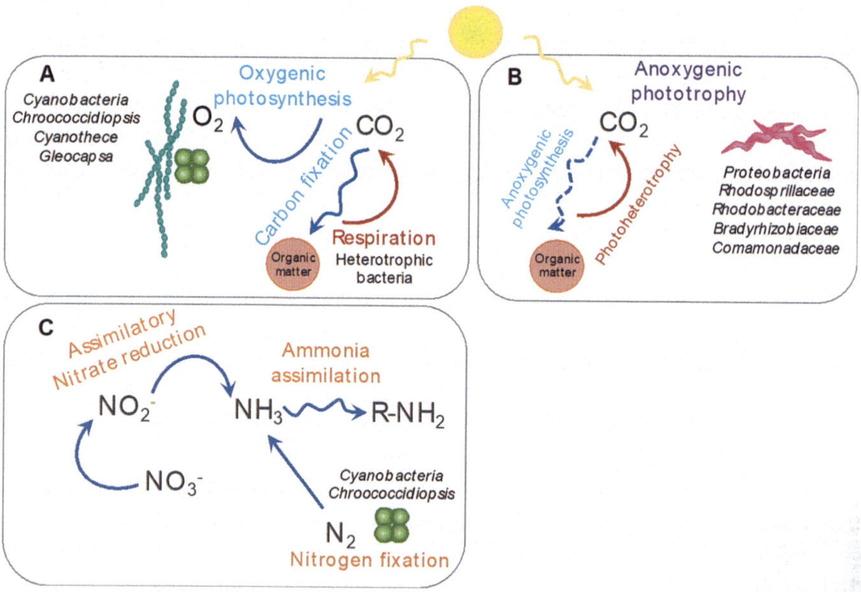

Fig. 5.7 Metabolic pathways in endolithic microorganisms. (**a**) Oxygenic photosynthesis carried out by Cyanobacteria is the main source of organic carbon in the communities. Various heterotrophic bacteria respire on this organic carbon and produce CO_2. (**b**) Anoxygenic phototrophy, the use of energy from light with organic carbon or reduced inorganic compounds as electron donors, is another form of metabolism that uses light but does not produce oxygen. (**c**) Nitrogen gas, fixed by Cyanobacteria, and nitrate are the main nitrogen sources in deserts. Dotted lines are potential pathways

and Bennett 2014; 2017). While metagenome analysis may provide predictions for chemolithotrophic pathways in endoliths, in situ measurements and laboratory experiments are needed to test these predictions and obtain a comprehensive view of carbon flux in these communities.

Anoxygenic photosynthesis has also been suggested as a means of energy acquisition in endolithic systems (Fig. 5.7b). Several families with known anoxygenic phototrophic capability have been reported in sandstone communities from multiple deserts, including Rhodospirillaceae, Rhodobacteraceae, Bradyrhizobiaceae, and Comamonadaceae (Qu et al. 2020). Additionally, several genes encoding bacteriochlorophylls (*bchl*) were detected in the metagenomes of gypsum endolithic community from the Atacama Desert (Ertekin et al. 2021) and in that of a hypolithic community from the Namib Desert (Vikram et al. 2016). The analysis of MAGs recovered from the former revealed that most organisms capable of anoxygenic photosynthesis also encoded enzymes for carbon fixation. These findings further highlight the importance of anoxygenic photosynthesis for carbon flux in hot desert lithic habitats (Finke et al. 2013; Overmann and Garcia-Pitchel 2013). Another mean of harvesting energy from light is that of bacteriorhodopsin, a light-driven proton pump found in halophilic archaea (and xanthorhodopsin in

Salinibacter), which play important roles in enhancing the energy budget of these microorganisms (Oren 2013). Once CO_2 is fixed into organic carbon in the endolithic community, it can be utilized by a diversity of heterotrophic bacteria and, in hypersaline environments, heterotrophic archaea (Crits-Christoph et al. 2016a, b). Consequently, the endolithic communities harbor a diverse array of genes involved in heterotrophic metabolic pathways such as glycolysis and gluconeogenesis, central carbohydrate metabolism, lactate fermentation, and photoheterotrophy (Crits-Christoph et al. 2016a, b; Ertekin et al. 2021).

Hot deserts are nitrogen-limited, thus nitrogen acquisition is a serious constraint for growth and primary production in endolithic habitats (Pointing and Belnap 2012) (Fig. 5.7c). While nitrogen fixation has been reported in sandstone communities from the Dry Valleys of Antarctica (Goordial et al. 2016), no *nif* gene was detected in the metagenomes of endolithic communities from the Atacama Desert (Finstad et al. 2017; Crits-Christoph et al. 2016a, b; Uritskiy et al. 2019; Ertekin et al. 2021). Because of the abundance of atmospherically deposited nitrate in the Atacama Desert, nitrogen assimilation, a much less energy-intensive pathway than nitrogen fixation, has been suggested as the main source of nitrogen for these communities (Finstad et al. 2017; Uritskiy et al. 2019).

Iron, sulfur, and phosphorus are all essential nutrients for microorganisms; iron is a cofactor in many enzymatic reactions and sulfur may stabilize proteins with disulfide bonds under desiccating conditions (Potts 1994). Atmospheric deposition from weathered bedrocks and allochthonous sediments has been suggested as the main source for these nutrients in deserts (Eckardt and Spiro 1999; Michalski et al. 2004; Johnston and Vestal 1989; Omelon et al. 2006) and their uptake is, therefore, essential for the functioning of endolithic communities. Pathways for sulfate assimilation and phosphorus uptake have been detected in the metagenomes of endolithic communities from the Atacama Desert and the Dry Valleys of Antarctica (Crits-Christoph et al. 2016a; Finstad et al. 2016; Goordial et al. 2016) and in hypolithic communities from hot deserts (Vikram et al. 2016). Transporters for both ferric and ferrous iron were present in the metagenomes of ignimbrite and calcite endolithic communities from the Atacama Desert, suggesting that both oxidation states of iron could be utilized by the microorganisms (Crits-Christoph et al. 2016a). Competition for iron also plays an important role in the survival of these communities, and multiple copies of the ferrichrome-iron receptor gene, the iron(III) dicitrate transport system *fecB* gene, and of siderophores were found in the draft genomes of Cyanobacteria from calcite and ignimbrite rocks (Crits-Christoph et al. 2016a).

5.6 Conclusion and Future Directions

Endolithic microbial communities from hot deserts are constituted of a core group of desiccation-adapted phototrophs and heterotrophs. Recent high-throughput molecular analyses have demonstrated that there is a fine scale diversification of the microbial reservoir driven by substrate properties, the so-called substrate architecture (Meslier et al. 2018; Qu et al. 2020). Novel metagenomic and metatranscriptomic studies have provided insights into the metabolic potential of endolithic communities, possible interactions between hosts and viruses, and highlighted the most active members for these communities (Uritskiy et al. 2019; Uritskiy et al. 2020a; Ertekin et al. 2021). However, these studies have been limited to a small number of substrates because of the challenges in obtaining enough samples and biomass (Uriskiy and DiRuggiero 2019). The next step is now to test these putative functions and interactions derived from metagenomic studies with laboratory experiments using enrichment cultures or reconstituted communities more amenable to genetic manipulations.

Many knowledge gaps remain. In particular, little is known of the interactions between the biotic and abiotic components of these unique ecosystems. The use of sophisticated methods traditionally applied to material sciences, such as microscopic (atomic force microscopy) and spectroscopic (Fourier transform infrared spectroscopic, Raman spectroscopy, X-ray photoelectron spectroscopy) analyses to interrogate surface textures and chemistries of various rocky substrates, is needed to elucidate mineral–microbe interactions within the rock environments and how these interactions impact the assembly, adaptations, and activities of these communities. Interactions between functional groups in the community and the role of viruses in shaping endolithic communities also remain to be investigated and it may require starting with simple, reconstituted communities and/or enrichment cultures.

Finally, we need long-term longitudinal studies to give us insights into the functioning of these communities at the ecosystem level, their response to environmental stresses, and help us predict how resistant and resilient these communities are to climate change. Because arid and hyper-arid deserts, at the dry limit for life, are fragile ecosystems, their inhabitants will be on the front line of the major changes in climate ahead of us.

References

Allison SD, Martiny JBH (2008) PNAS 105:11512–11519. https://doi.org/10.1073/pnas.0801925105

Allison SD, Martiny JBH (2009) Resistance, resilience, and redundancy in microbial communities. Light Evol 2:149–166. https://doi.org/10.17226/12501

Amaral G, Martinez-Frias J, Vázquez L (2007) UV shielding properties of jarosite vs. gypsum: astrobiological implications for Mars. World Appl Sci J 2:112–116

Archer SDJ et al (2017) Endolithic microbial diversity in sandstone and granite from the McMurdo Dry Valleys. Antarct Polar Biol 40:997–1006

Artieda O, Davila A, Wierzchos J et al (2015) Surface evolution of salt-encrusted playas under extreme and continued dryness. Earth Surf Process Landforms 40:1939–1950. https://doi.org/10.1002/esp.3771

Bay SK, Waite DW, Dong X, Gillor O, Chown SL et al (2021) Chemosynthetic and photosynthetic bacteria contribute differentially to primary production across a steep desert aridity gradient. ISME J 15:3339–3356

Billi D, Friedmann EI, Hofer KG et al (2000) Ionizing-radiation resistance in the desiccation-tolerant cyanobacterium *Chroococcidiopsis*. Appl Environ Microbiol 66:1489–1492. https://doi.org/10.1128/AEM.66.4.1489-1492.2000

Billi D, Viaggiu E, Cockell CS et al (2011) Damage escape and repair in dried *Chroococcidiopsis* spp. from hot and cold deserts exposed to simulated space and Martian conditions. Astrobiology 11:65–73. https://doi.org/10.1089/ast.2009.0430

Bonani G, Friedmann EI, Ocampo-Friedmann R et al (1988) Preliminary report on radiocarbon dating of cryptoendolithic microorganisms. Polarforschung 58:199–200

Buedel B et al (2004) Reshaping of sandstone surfaces by cryptoendolithic cyanobacteria: bioalkalization causes chemical weathering in arid landscapes. Geobiology 2:261–268. https://doi.org/10.1111/j.1472-4677.2004.00040.x

Campion-Alsumard TL, Golubic S, Hutchings P (1995) Microbial endoliths in skeletons of live and dead corals: Porites lobata (Moorea, French Polynesia). Mar Ecol Prog Ser 117:149–158. https://doi.org/10.3354/meps117149

Caruso T, Chan Y, Lacap DC et al (2011) Stochastic and deterministic processes interact in the assembly of desert microbial communities on a global scale. ISME J 5:1406–1413. https://doi.org/10.1038/ismej.2011.21

Casero C, Meslier V, Wierzchos J et al (2020) Preandean Atacama Desert endolithic microbiology. In: Farias ME (ed) Extremophile microbial ecosystems in central Andes extreme environments. Springer, Berlin. In Press

Caturla F, Molina-Sabio M, Rodriguez-Reinoso F (1999) Adsorption- desorption of water vapor by natural and heat-treated sepiolite in ambient air. App Clay Sci 15:367–380. https://doi.org/10.1016/S0169-1317(99)00030-7

Cockell CS, McKay CP, Warren-Rhodes K et al (2008) Ultraviolet radiation-induced limitation to epilithic microbial growth in arid deserts - Dosimetric experiments in the hyperarid core of the Atacama Desert. J Photochem Photobiol 90:79–87. https://doi.org/10.1016/j.jphotobiol.2007.11.009

Cordero RR, Seckmeyer G, Damiani A et al (2014) The world's highest levels of surface UV. Photochem Photobiol Sci 13:70–81. https://doi.org/10.1039/c3pp50221j

Crits-Christoph A, Robinson CK, Ma B et al (2016a) Phylogenetic and functional substrate specificity for endolithic microbial communities in hyper-arid environments. Front Microbiol 7:1–15. https://doi.org/10.3389/fmicb.2016.00301

Crits-Christoph A, Gelsinger DR, Ma B et al (2016b) Functional interactions of archaea, bacteria and viruses in a hypersaline endolithic community. Environ Microbiol 18:2064–2077. https://doi.org/10.1111/1462-2920.13259

Davila AF, Gómez-Silva B, de los Rios A et al (2008) Facilitation of endolithic microbial survival in the hyperarid core of the Atacama Desert by mineral deliquescence. J Geophys Res Biogeosciences 113:1–9. https://doi.org/10.1029/2007JG000561

Davila AF, Hawes I, Ascaso C, Wierzchos J (2013) Salt deliquescence drives photosynthesis in the hyperarid Atacama Desert. Environ Microbiol Rep 5:583–587. https://doi.org/10.1111/1758-2229.12050

Davila AF, Hawes I, Araya JG et al (2015) In situ metabolism in halite endolithic microbial communities of the hyperarid Atacama Desert. Front Microbiol 6:1–8. https://doi.org/10.3389/fmicb.2015.01035

de los Ríos A, Sancho L, Grube M et al (2005) Endolithic growth of two Lecidea lichens in granite from continental Antarctica detected by molecular and microscopy techniques. New Phytol 165:181–190. https://doi.org/10.1111/j.1469-8137.2004.01199.x

de los Ríos A, Valea S, Ascaso C et al (2010) Comparative analysis of the microbial communities inhabiting halite evaporites of the Atacama Desert. Int Microbiol 13:79–89. https://doi.org/10.2436/20.1501.01.113

de Luca G, Barakat M, Ortet P et al (2011) The cyst-dividing bacterium ramlibacter tataouinensis TTB310 genome reveals a well-stocked toolbox for adaptation to a desert environment. PLoS One 6:e23784. https://doi.org/10.1371/journal.pone.0023784

DiRuggiero J, Wierzchos J, Robinson CK et al (2013) Microbial colonisation of chasmoendolithic habitats in the hyper-arid zone of the Atacama Desert. Biogeosciences 10:2439–2450. https://doi.org/10.5194/bg-10-2439-2013

Eckardt FD, Spiro B (1999) The origin of Sulphur in gypsum and dissolved sulphate in the central Namib Desert, Namibia. Sediment Geol 123:255–273. https://doi.org/10.1016/S0037-0738(98)00137-7

Ertekin E, Meslier V, Browning A, Treadgold J, DiRuggiero J (2021) Functional and taxonomic diversity is driven by substrate architecture in endolithic communities from extreme environments. Env Microbiol 23:3937–3956. PubMed PMID: 33078515

Finke N, Hoehler TM, Polerecky L et al (2013) Competition for inorganic carbon between oxygenic and anoxygenic phototrophs in a hypersaline microbial mat, Guerrero Negro, Mexico. Environ Microbiol 15:1532–1550. https://doi.org/10.1111/1462-2920.12032

Finstad K, Pfeiffer M, McNicol G et al (2016) Rates and geochemical processes of soil and salt crust formation in Salars of the Atacama Desert, Chile. Geoderma 284:57–72. https://doi.org/10.1016/j.geoderma.2016.08.020

Finstad KM, Banfield JF, Thomas BC et al (2017) Microbial community structure and the persistence of cyanobacterial populations in salt crusts of the Hyperarid Atacama Desert from genome-resolved metagenomics. Front Microbiol 8:1–10. https://doi.org/10.3389/fmicb.2017.01435

Friedmann EI (1982) Endolithic microorganisms in the Antarctic Cold Desert. Science 215:1045–1053

Friedmann EI, Ocampo R (1976) Endolithic blue-green algae in the dry valleys: primary producers in the Antarctic Desert ecosystem published by: American Association for the Advancement of Science stable URL: https://www.jstor.org/stable/1742400 linked references are available on JS. Science 193:1247–1249

Friedmann I (1980) Endolithic microbial life in hot and cold deserts. Orig Life 10:223–235

Friedmann I, Lipkin Y, Ocampo-Paus R (1967) Desert algae of the Negev (Israel). Phycologia 6: 185–200. https://doi.org/10.2216/i0031-8884-6-4-185.1

Golubic S, Friedmann I, Schneider J (1981) The lithobiontic ecological niche, with special reference to microorganisms. J Sediment Res 51:475–478

Goordial J, Davila A, Greer CW et al (2016) Comparative activity and functional ecology of permafrost soils and lithic niches in a hyper-arid polar desert. Environ Microbiol 19:443–458. https://doi.org/10.1111/1462-2920.13353

He K, Bauer CE (2014) Chemosensory signaling systems that control bacterial survival. Trends Microbiol 22:389–398. https://doi.org/10.1016/j.tim.2014.04.004

Huang W, Ertekin E, Wang T et al (2020) Mechanism of water extraction from gypsum rock by desert colonizing microorganisms. Proc Natl Acad Sci U S A 117:10681–10687. https://doi.org/10.1073/pnas.2001613117

Ho MY, Shen G, Canniffe DP et al (2016) Light-dependent chlorophyll f synthase is a highly divergent paralog of PsbA of photosystem II. Science 26:353. https://doi.org/10.1126/science.aaf9178

Ho MY, Soulier NT, Canniffe DP et al (2017) Light regulation of pigment and photosystem biosynthesis in cyanobacteria. Curr Op Plant Biol 37:24–33. https://doi.org/10.1016/j.pbi.2017.03.006

Horath T, Neu T, Bachofen R (2006) An endolithic microbial community in dolomite rock in Central Switzerland: characterization by reflection spectroscopy, pigment analyses, scanning electron microscopy, and laser scanning microscopy. Microb Ecol 51:353–364. https://doi.org/10.1007/s00248-006-9051-y

Hughes KA, Lawley B (2003) A novel Antarctic microbial endolithic community within gypsum crusts. Environ Microbiol 7:555–565. https://doi.org/10.1046/j.1462-2920.2003.00439.x

Johnston CG, Vestal JR (1989) Distribution of inorganic species in two Antarctic cryptoendolithic communities. Geomicrobiol J 7:137–153. https://doi.org/10.1080/01490458909377858

Johnston CG, Vestal JR (1991) Photosynthetic carbon incorporation and turnover in Antarctic cryptoendolithic microbial communities: are they the slowest-growing communities on earth? Appl Environ Microbiol 57:2308–2311. https://doi.org/10.1128/aem.57.8.2308-2311.1991

Jones AA, Bennett PC (2014) Mineral microniches control the diversity of subsurface microbial populations. Geomicrobiol J 31(3):246–261

Jones AA, Bennett PC (2017) Mineral ecology: surface specific colonization and geochemical drivers of biofilm accumulation, composition, and phylogeny. Front Microbiol 8:1–14

Jordaan K, Lappan R, Dong X, Aitkenhead IJ, Bay SK et al (2020) Hydrogen oxidizing bacteria are abundant in desert soils and strongly stimulated by hydration. mSystems 5:e01131–e01120

Khomutovska N, Jerzak M, Kostrzewska-Szlakowska I et al (2017) Life in extreme habitats: diversity of endolithic microorganisms from Cold Desert ecosystems of eastern Pamir. Polish J Ecol 65:303–319. https://doi.org/10.3161/15052249PJE2017.65.4.001

Lacap-Bugler DC, Lee KK, Archer S et al (2017) Global diversity of desert hypolithic cyanobacteria. Front Microbiol 8:867. https://doi.org/10.3389/fmicb.2017.00867

Le PT, Makhalanyane TP, Guerrero LD et al (2016) Comparative metagenomic analysis reveals mechanisms for stress response in hypoliths from extreme hyperarid deserts. Genome Biol Evol 8:2737–2747. https://doi.org/10.1093/gbe/evw189

Lebre PH, De Maayer P, Cowan DA (2017) Xerotolerant bacteria : surviving through a dry spell. Nat Rev Microbiol 15:285–296. https://doi.org/10.1038/nrmicro.2017.16

Louati M, Ennis NJ, Ghodhbane-Gtari F et al (2019) Elucidating the ecological networks in stone-dwelling microbiomes. Environ Microbiol 00:1–14. https://doi.org/10.1111/1462-2920.14700

Maron P-A, Sarr A, Kaisermann A et al (2018) High microbial diversity promotes soil ecosystem functioning. Appl Environ Microbiol 84:1–13. https://doi.org/10.1128/AEM.02738-17

Matthes-Sears U, Gerrath JA, Larson DW (1997) Abundance, biomass, and productivity of endolithic and Epilithic lower plants on the temperate-zone cliffs of the Niagara escarpment. Int J Plant Sci 158:451–460

Matthes-Sears U, Turner SJ, Larson DW (2001) Light attenuation by limestone rock and its constraint on the depth distribution of endolithic algae and cyanobacteria. Int J Plant Sci 162: 263–270. https://doi.org/10.1086/319570

McKay CP (2012) Full solar spectrum measurements of absorption of light in a sample of the Beacon sandstone containing the Antarctic cryptoendolithic microbial community. Antarct Sci 24:243–248. https://doi.org/10.1017/S0954102011000915

Mergelov N, Mueller CW, Prater I et al (2018) Alteration of rocks by endolithic organisms is one of the pathways for the beginning of soils on earth. Sci Rep 8:1–15. https://doi.org/10.1038/s41598-018-21682-6

Meslier V, Casero MC, Dailey M et al (2018) Fundamental drivers for endolithic microbial community assemblies in the hyperarid Atacama Desert. Environ Microbiol 20:1765–1781. https://doi.org/10.1111/1462-2920.14106

Meslier V, DiRuggiero J (2019) Endolithic microbial communities as model systems for ecology and astrobiology. In: Seckbach J, Rampelotto PH (eds) Model ecosystems in extreme environments. Elsevier, Amsterdam, pp 145–168

Michalski G, Böhlke JK, Thiemens M (2004) Long term atmospheric deposition as the source of nitrate and other salts in the Atacama Desert, Chile: new evidence from mass-independent oxygen isotopic compositions. Geochim Cosmochim Acta 68:4023–4038. https://doi.org/10.1016/j.gca.2004.04.009

Omelon CR (2008) Endolithic microbial communities in polar desert habitats. Geomicrobiol J 25: 404–414. https://doi.org/10.1080/01490450802403057

Omelon CR, Pollard WH, Ferris FG (2006) Environmental controls on microbial colonization of high Arctic cryptoendolithic habitats. Polar Biol 30:19–29. https://doi.org/10.1007/s00300-006-0155-0

Omelon CR (2016) Endolithic microorganisms and their habitats. In: Hurst CJ (ed) Their world: a diversity of microbial environments. Springer, pp 171–201

Oren A (2008) Microbial life at high salt concentrations: phylogenetic and metabolic diversity. Saline Systems 4:1–13. https://doi.org/10.1186/1746-1448-4-2

Oren A (2013) Life at high salt concentrations, intracellular KCl concentrations, and acidic proteomes. Front Microbiol 4:315. https://doi.org/10.3389/fmicb.2013.00315

Overmann J, Garcia-Pichel F (2013) The phototrophic way of life. In: Rosenberg E, DeLong EF, Lory S, Stackebrandt E, Thompson F (eds) The prokaryotes: prokaryotic communities and ecophysiology. Springer, Berlin, pp 203–257

Palmer RJ, Friedmann EI (1990) Water relations and photosynthesis in the Cryptoendolithic microbial habitat of hot and cold deserts. Microb Ecol 19:111–118

Pointing SB, Belnap J (2012) Microbial colonization and controls in dryland systems. Nat Rev Microbiol 10:551–562. https://doi.org/10.1038/nrmicro2831

Pointing SB, Chan Y, Lacap DC et al (2009) Highly specialized microbial diversity in hyper-arid polar desert. PNAS 107:19964–19969. https://doi.org/10.1073/pnas.0913882107

Potts M (1994) Desiccation tolerance of prokaryotes. Microbiol Rev 58:755–805

Potts M (2001) Desiccation tolerance: a simple process? Trends Microbiol 9(11):553–559

Qu E, Omelon C, Oren O, Meslier V, Cowan DA, Maggs-Kölling G, DiRuggiero J (2020) Trophic selective pressures organize the composition of endolithic microbial communities from global deserts. Front Microbiol 10:2952. PubMed PMID: 31969867

Robinson CK, Wierzchos J, Black C et al (2015) Microbial diversity and the presence of algae in halite endolithic communities are correlated to atmospheric moisture in the hyper-arid zone of the Atacama Desert. Environ Microbiol 17:299–315. https://doi.org/10.1111/1462-2920.12364

Torsvik V, Øvreås L (2002) Microbial diversity and function in soil: from genes to ecosystems. Curr Opin Microbiol 3:240–245. https://doi.org/10.1016/j.jenvp.2016.12.005

Uriskiy G, DiRuggiero J (2019) Applying genome-resolved metagenomics to de-convolute the halophilic microbiome. Gene 10:220. https://doi.org/10.3390/genes10030220

Uritskiy G, Getsin S, Munn A et al (2019) Halophilic microbial community compositional shift after a rare rainfall in the Atacama Desert. ISME J 13:2737–2749. https://doi.org/10.1038/s41396-019-0468-y

Uritskiy G, Tisza MJ, Gelsinger DR, Munn A, Taylor J, DiRuggiero J (2020a) Cellular life from the three domains and viruses are transcriptionally active in a hypersaline desert community. Environ Microbiol 23(7). https://doi.org/10.1111/1462-2920.15023

Uritskiy G, Munn A, Dailey M, Gelsinger DR, Getsin S, Davila A, McCullough PR, Taylor J, DiRuggiero J (2020b) Environmental factors driving spatial heterogeneity in desert halophile microbial communities. Front Microbiol 11:578660

Valverde A, Makhalanyane TP, Seely M, Cowan DA (2015) Cyanobacteria drive community composition and functionality in rock-soil interface communities. Mol Ecol 24:812–821. https://doi.org/10.1111/mec.13068

Vikram S, Guerrero LD, Makhalanyane TP et al (2016) Metagenomic analysis provides insights into functional capacity in a hyperarid desert soil niche community. Environ Microbiol 18: 1875–1888. https://doi.org/10.1111/1462-2920.13088

Viles H (1995) Ecological perspectives on rock surface weathering: towards a conceptual model. Geomorphology 13:21–35. https://doi.org/10.1016/0169-555X(95)00024-Y

Viles H (2011) In: Thomas DSG (ed) Arid zone geomorphology: process, form and change in drylands. Wiley-Blackwell, Chichester, pp 131–180

Vítek P, Ascaso C, Artieda O et al (2016) Raman imaging in geomicrobiology: endolithic phototrophic microorganisms in gypsum from the extreme sun irradiation area in the Atacama Desert. Anal Bioanal Chem 408(15):4083–4092. https://doi.org/10.1007/s00216-016-9497-9

Vítek P, Ascaso C, Artieda O et al (2017) Discovery of carotenoid red-shift in endolithic cyanobacteria from the Atacama Desert. Sci Rep 7(1):11116. https://doi.org/10.1038/s41598-017-11581-7

Walker JJ, Pace NR (2007) Endolithic microbial ecosystems. Annu Rev Microbiol 61:331–347. https://doi.org/10.1146/annurev.micro.61.080706.093302

Warren-Rhodes KA, Dungan JL, Piatek J et al (2007) Ecology and spatial pattern of cyanobacterial community island patches in the Atacama Desert, Chile. J Geophys Res Biogeosciences 112:1–14. https://doi.org/10.1029/2006JG000305

Wierzchos J, Ascaso C, McKay CP (2006) Endolithic cyanobacteria in halite rocks from the Hyperarid Core of the Atacama Desert. Astrobiology 6:415–422. https://doi.org/10.1089/ast.2006.6.415

Wierzchos J, Cámara B, De los Ríos A et al (2011) Microbial colonization of ca-sulfate crusts in the hyperarid core of the Atacama Desert: implications for the search for life on Mars. Geobiology 9:44–60. https://doi.org/10.1111/j.1472-4669.2010.00254.x

Wierzchos J, Davila AF, Sánchez-Almazo IM et al (2012) Novel water source for endolithic life in the hyperarid core of the Atacama Desert. Biogeosciences 9:2275–2286. https://doi.org/10.5194/bg-9-2275-2012

Wierzchos J, Davila AF, Artieda O et al (2013) Ignimbrite as a substrate for endolithic life in the hyper-arid Atacama Desert: implications for the search for life on Mars. Icarus 224:334–346. https://doi.org/10.1016/j.icarus.2012.06.009

Wierzchos J, DiRuggiero J, Vítek P et al (2015) Adaptation strategies of endolithic chlorophototrophs to survive the hyperarid and extreme solar radiation environment of the Atacama Desert. Front Microbiol 6:1–17. https://doi.org/10.3389/fmicb.2015.00934

Wierzchos J, Casero MC, Artieda O, Ascaso C (2018) Endolithic microbial habitats as refuges for life in polyextreme environment of the Atacama Desert. Curr Opin Microbiol 43:124–131. https://doi.org/10.1016/j.mib.2018.01.003

Wong FKY, Lau MCY, Lacap DC et al (2010) Endolithic microbial colonization of limestone in a high-altitude arid environment. Microb Ecol 59:689–699. https://doi.org/10.1007/s00248-009-9607-8

Ziolkowski LA, Wierzchos J, Davila AF, Slater GF (2013) Radiocarbon evidence of active endolithic microbial communities in the hyperarid core of the Atacama Desert. Astrobiology 13:607–616. https://doi.org/10.1089/ast.2012.0854

6 Journey of a Thousand Miles: The Evolution of Our Understanding of Viruses in Hot Deserts

Leonardo Joaquim van Zyl, Laura Martinez Alvarez, and Marla Trindade

Abstract

Our understanding of soil virus ecology is in its infancy with the first study using modern techniques only published in 2007. Since then, many new tools have been developed and we now have a greater understanding of aquatic and especially marine viral communities, however soil viral communities and in particular those from hot Deserts remain poorly characterized. In this chapter, we explore the evolution of studies describing the diversity and distribution of viruses in hot Deserts both in soils and the water bodies found here and relay the impact researchers think they have on these communities. Although traits common to all Deserts studied are difficult to define, the main finding is that hot Desert soils, as in other soils, are dominated by lysogenic bacteriophages adapted to take advantage of bursts in host metabolic activity following stochastic inputs such as rainfall. In some instances, the physicochemical makeup of these environments appears to play a dominant role in shaping the host and viral communities. The virus–host communities of the few aquatic environments found in hot Deserts reflect the dynamics of better studied aquatic environments; however, they feature an abundance of unique viruses. The hot Desert also has a few microbiological mysteries that need solution such as the absence of cyanobacterial viruses in samples from communities dominated by these bacteria. Far more work is needed to address the nuances of the virus–host interactions in this environment and to understand the drivers of community establishment, development, maturation, and adaptation.

L. J. van Zyl (✉) · M. Trindade
Institute for Microbial Biotechnology and Metagenomics, University of the Western Cape, Bellville, South Africa
e-mail: lvanzyl@uwc.ac.za

L. M. Alvarez
Center for Microbial Ecology and Genomics, University of Pretoria, Pretoria, South Africa

J.-B. Ramond, D. A. Cowan (eds.), *Microbiology of Hot Deserts*, Ecological Studies 244, https://doi.org/10.1007/978-3-030-98415-1_6

Keywords

Desert · Phage · Virus · Lysogen · Adaptation · Namib

6.1 Introduction

Soil viral diversity and its possible impact on the soil biome is greatly underappreciated. This is chiefly due to a lack of studies investigating soil viral diversity (Kimura et al. 2008; Williamson et al. 2017). Although soils are home to a range of plant and animal species all of which could be hosts for soil viruses, it has become accepted that, when investigating viral diversity, most environments are dominated by bacteriophages (viruses that specifically infect bacteria) (Kimura et al. 2008). Therefore, studies investigating soil viral diversity have mainly focused on these (Motlagh et al. 2017; Yang et al. 2019). Soil matrices are thought to harbor 90% of the earth's bacteria with estimates ranging between 590 and 100,000 species per gram of soil (Curtis et al. 2002; Prigent et al. 2005; Schloss and Handelsman 2006). In marine environments, 1–4% of bacteria are visibly infected with phage and the estimates are ten virus particles to each bacterial cell (Wommack and Colwell 2000). These observations cannot be directly translated to soil systems; however, given these estimates, the number of characterized phages described thus far indicates that we have a long way to go toward a complete understanding of the microbial virus–host interactions in soil (Kimura et al. 2008). It is also accepted that microorganisms hold sway over global biogeochemical cycles (Anderson 2018; Liang et al. 2019) and, although considered non-living, bacteriophages play an important role in modifying microbial biodiversity, species distribution, metabolic capacity, and evolution (Brüssow et al. 2004; Zhang et al. 2017; Emerson et al. 2018; Trubl et al. 2018). It is commonly thought that they affect this through "predation" (lytic effect of bacteriophages on bacteria) or through lysogenic conversion of their host and the ability to exchange DNA with their hosts (Clokie et al. 2011; Peduzzi et al. 2014). The separation of the virus lifecycle into these two prominently recognized stages paints a very monotone picture of the virus–host interaction which, in reality, is far more complex and includes aspects of mutual inhibition, competition, amensalism, and commensalism, which themselves can occur over a range of scales. Naturally, the phage and host involved dictate what their interaction would be, however the changes that occur in their immediate environment (moisture availability, pH, nutrient availability, elemental composition) all play a role in modulating their interaction. They can either affect how the bacterial host grows which may promote infection and the lysis–lysogeny decision; how well the phage persists as a free particle; modulate the efficiency of infection or enable physical contact between virus and host (mobilization of phage particles in response to water input) as well as lead to irreversible binding to soil particles, which both limits their ability to infect and how efficiently they can be recovered for study.

Given the abundance of viral hosts and their importance in soil ecosystems and the influence viruses have on these populations (Rodriguez-Caballero et al. 2018), it is surprising that the viral component of soils in general has been so poorly studied, unlike aquatic and human-related microbiomes (Breitbart et al. 2003; Breitbart and Rohwer 2005; Williamson et al. 2017; Schulz et al. 2018; Emerson et al. 2018; Graham et al. 2019; Chap. 4; Fig. 4.1). In fact, the first study to use sequence-based analysis to study soil viral communities was only published in 2007 (Fierer et al. 2007). Of the soil ecosystems that have been studied, desert ecosystems and in particular hot deserts are the most recent and are the environments we know least about (Zablocki et al. 2016). It is debatable whether the water bodies such as saline springs (playa's) or perennial ponds (gueltas/wadi's) present in desert environments qualify as "desert" environments. They certainly qualify as aquatic, moderately extreme (often have high salinity), and are poorly studied. These water bodies have significantly different bacterial composition and production rates (and therefore viral compositions) as compared with the sand and soil of the surrounding Desert, mimicking better studied aquatic environments. Apart from a few single isolate studies (Prigent et al. 2005; Prestel et al. 2008; Meiring et al. 2012; Swanson et al. 2012), the era of metagenomics to investigate viral diversity in both hot (and cold) desert ecosystems only started roughly a decade ago.

From a viral ecology perspective, it would be of interest to answer questions such as: Do dryland viral communities or individual viruses evolve at the same pace and along the same paths as those in aquatic or wetter terrestrial climates? Do they reflect their host populations to the same degree as has been found in other environments, or are there greater portions of host species exempt from infection? What are the specific adaptations of these viral particles to persist (do they persist?) in this environment and are they different to the adaptations of viruses from other environments? Are the viruses found here unique (niche partitioning) and novel or are they a mixture of autochthonous communities and those seeded from elsewhere through stochastic processes? Are there unique virus–host interactions taking place (new lifestyles) or do they share mechanisms common to most bacteriophages (lytic, lysogenic, pseudo-lysogeny)? How do viral communities change (respond) over time or through the influence of stochastic events such as rainfall or along aridity and other gradients? What is the crosstalk between some of the unique environments that exist within hot deserts (wadi's/gueltas/saline springs) and the areas surrounding them, do these viruses jump hosts? What factors, physical and biological (intra-microbial interactions), influence phage and host communities?

Many questions about soil microbial ecology remain to be answered. One of these is the scale at which ecological effects play out in soil (Wilpiszeski et al. 2019). In their review, Wilpiszeski and coauthors suggest that soil microaggregates (<250 μm) should be considered the unit at which these questions need to be asked and answered as chemoorganic and physical gradients (soil moisture, temperature, ionic strength, pH, dissolved gas concentrations, nutrient concentrations) exist between and within these aggregates. Changes have also been noted at other scales (meters, kilometers; metacommunity vs local community) and will depend on which environment is under investigation (rainforest or deserts) (Scola et al. 2018) and distance-decay functions may describe many of the features observed for these communities. Aquatic systems can be seen as homogenous and soil matrices as non-homogenous, at least during dry periods (Zablocki et al. 2016). The soil matrix therefore stands in stark contrast to aquatic systems where free flow/exchange of particles over great distance is possible, whereas the flow of particles may only be permitted in soils that are partially wetted or waterlogged. In the 2008 study by Srinivasiah and colleagues they describe the contrast between the findings of early viromic studies of soils and marine environments stating that:

> As might be expected, cross-assemblies of the three viral metagenome libraries from desert, rainforest, and prairie soils showed that soil viral assemblages are locally unique, with almost no overlap in viral genotypes. In contrast, a similar analysis of viral metagenome libraries from four oceanic biomes indicated that any pair of samples could share a majority of viral genotypes, albeit with differing positions along the rank abundance distribution

Soil matrices are non-homogenous, therefore microbial dynamics are likely very different in soil systems compared with aquatic systems. Soil viral assemblages have been shown to be locally unique with almost no overlap in viral genotypes (Srinivasiah et al. 2008). Therefore, the soil type and the resolution at which soil microbial ecology questions are asked may have an important impact on the interpretation of data generated and the dynamics described (Williamson et al. 2007). Of importance to this chapter are the gradients that exist in xeric (very dry) soils of hot deserts. It was demonstrated for the Namib desert that, as might be expected, surface soil moisture was higher than in deeper soil layers following rain events and that different sites, depending on soil type, retain more or less water over time (Li et al. 2016). Thus, even in one desert system, there can exist thousands of microenvironments with their own mixtures of nutrients and physicochemical factors influencing the virus–host relationship. As far as we are aware, no studies have yet been designed to test the dynamics of soil microbial systems at these levels and most soil virome studies have been performed on what can be considered low-resolution bulk soil samples. Here we summarize the results from studies which have been performed on hot desert systems, including waterbodies (Wadi's, gueltas, and springs) found in these deserts. We draw on soil virus studies performed on more temperate environments and those performed on cold deserts, to "better understand" the influence and evolution of viruses in these settings and synthesize the best current understanding of viral ecology in hot deserts (Table 6.1).

6.2 Terrestrial Desert Phage Communities: Soil, Sand, and Rock

Prestel, du Bow and colleagues were the first to study the viruses of hot deserts (Sahara, the Namib, and Mojave). The studies conducted in the Sahara and Namib were mainly based on electron microscopy and sequence analysis of a small number of clones to gauge the diversity of bacterial viruses in these biomes. As with many environments they found that the bulk of the observed morphotypes were that of tailed viruses belonging to the order Caudovirales and DNA sequencing affirmed

Table 6.1 Soil-based virome studies since 2005

Authors, year of publication	Soil type(s)	Location(s) of sample collection	Method(s) used
Prigent et al. (2005)	Hot desert surface sand	Sahara Desert in Morocco and Tunisia	EM, PFGE, lytic induction
Williamson et al. (2005)	Agricultural, forest	Delaware, USA	Epifluorescence microscopy, EM
Fierer et al. (2007)	Hot arid desert, tallgrass prairie, tropical rainforest	USA, Peru	Sanger sequencing of random viral metagenomics clones
Williamson et al. (2007)	Loamy and sandy soils, agricultural, forested wetlands	Antarctica (Tom and Obelisk Pond); Delaware, USA	Induction assays, epifluorescence counting
Prestel et al. (2008)	Surface sand	Namib Desert	EM, PFGE, Sanger sequencing of cloned DNA fragments (LASL)
Swanson et al. (2009)	Dystric-fluvic Cambisol soil	Dundee, Scotland	EM, epifluorescence counting
Meiring et al. (2012)	Soil under hypoliths	Miers Valley, Antarctica	Lytic induction, EM, phage isolation from culture
Pearce et al. (2012)	Surface soil	Alexander Island, Antarctica	Shotgun metagenome pyrosequencing
Swanson et al. (2012)	Surface soil (Antarctica)	Antarctica	EM, lytic induction, phage isolation
Prestel et al. (2013)	Dune surface sand	Mojave Desert, USA	EM, random amplification for viral DNA (Sanger)
Srinivasiah et al. (2013)	Surface soil (Antarctica); silt loamy soil (USA)	Antarctica (Tom and Obelisk pond); Delaware, USA	RAPD viral community fingerprinting
Zablocki et al. (2014)	Antarctic surface soil and hypoliths	Miers Valley, Antarctica	Shotgun viral metagenome sequencing (Illumina)
Adriaenssens et al. (2015)	Soil-associated rocks (hypoliths)	Namib Desert	Shotgun viral metagenome sequencing (Illumina)
Reavy et al. (2015)	Machair soil	Western Scotland and Ireland	Shotgun viral metagenome sequencing (Roche 454)
Srinivasiah et al. (2015)	Silt loamy soil	Delaware, USA	Microcosms, RAPD viral community fingerprinting, epifluorescence counting
Segobola et al. (2018)	Fynbos soil	Kogelberg Biosphere Reserve in the Cape Floral Kingdom in South Africa	Shotgun viral metagenome sequencing (Illumina)
Emerson et al. (2018)	Permafrost soil	Stordalen mire (Sweden)	Shotgun viral metagenome sequencing (Illumina) and lytic induction
Graham et al. (2019)	Global soil metagenome comparison	Global soil metagenome comparison	Global soil metagenome comparison

(continued)

Table 6.1 (continued)

Authors, year of publication	Soil type(s)	Location(s) of sample collection	Method(s) used
Van Goethem et al. (2019)	Biological soil crust (biocrust) from the Moab Desert	Southwestern United States	Shotgun metagenome sequencing (Illumina)
Göller et al. (2020)	Agricultural soil	Zurich Switzerland	Shotgun metagenome sequencing (Illumina)
Roy et al. (2020)	Agricultural soil	Hickory Corners, Michigan, United States	Epifluorescence microscopy, RAPD-PCR

Updated from Zablocki et al. (2016)

that these were mostly bacteriophages (Prestel et al. 2008). Particle density was so low that these investigators had to enrich for them using mitomycin C induction to study the lysogenic viruses in the bacterial hosts from these samples (Prigent et al. 2005; Prestel et al. 2008, 2013). Given the lack of culturability of bacteria in lab settings (Staley and Konopka 1985), enrichment likely biased this early analysis, and may have led to a misinterpretation of the most abundant hosts and phages in these environments as infecting Gram-positive bacteria (*Bacillus* species and other Firmicutes) and perhaps highlighted the need for meta-study. A number of *Bacillus* isolates and *Bacillus*-like phages were also found in Death Valley samples (Mojave Desert). The argument against Firmicute hosts and phages dominating these environments are the bacterial community studies that have been performed, often on the same samples (Valverde et al. 2015). Studies looking at the community profile at the rock/soil interface for hypoliths in the Namib desert as well as open soil showed that these communities are dominated by photoautotrophic cyanobacteria (34%), alphaproteobacteria (22%), and actinobacteria (17% rock/soil interface/44% open soil). A similar trend of actinobacterial dominance has been seen in other desert soils (Liu et al. 2009; Goswami et al. 2013; Makhalanyane et al. 2013). This seems to confirm that the early studies cultivated and enriched for the Firmicutes that likely form spores leading to the interpretation that these hosts and their phages dominate.

The metagenomic analysis of the viral diversity of the Namib hypoliths also showed the presence of many *Bacillus*/*Geobacillus*-type phages. Here, however, the DNA was extracted directly from the original sample without enrichment (Adriaenssens et al. 2015). It was very surprising that there was the absence of, or the inability to identify, phages that infect cyanobacteria, specifically *Chroococcidiopsis* species, which are visibly and molecularly identifiable in hypolithons. This is in contrast to the halite endolithic communities from the Atacama where the presence of phages related to haloviruses, but more importantly, phages that showed identity to cyanobacterial viruses, was observed (Crits-Christoph et al. 2016). Several hypotheses were put forward to explain this discrepancy. It could be that the phages that do infect *Chroococcidiopsis* species have not been characterized and added to current databases used to classify phages in metagenomic datasets. It is therefore possible that the *Chroococcidiopsis*-infecting phages are present in the ~40% of sequence that could not be annotated. If this is the case, it suggests that the genomes of these phages are very different to cyanophages from aquatic environments. Another explanation for this observation is that the seemingly *Bacillus/Geobacillus*-related viruses identified do infect cyanobacteria, however due to poor database representation, *Bacillus* and *Geobacillus*-related viruses are the closest relatives in current databases. Yet another interpretation could be that the Firmicute-like hosts are the dominant lysogens in this environment with few of the other species harboring prophages.

One interesting observation from these early studies was the presence of a Siphovirus with a peculiar spiral tail fiber at the distal end of its tail. This morphotype is unusual and was observed in the Namib and Mojave deserts (Prestel et al. 2008, 2013). The presence of such an unusual morphotype in biogeographically distant environments suggested the presence of related, if not identical, bacteriophages in these biomes. Without sequence data to compare the viral composition of these environments with each other it was not possible to say if this is the case, showing the limitations of studies based on electron microscopic investigation of viral morphotypes. Metagenomic studies would however give researchers the opportunity to do just that.

Deserts are extreme environments subject to desiccation, high solar irradiation, and extremes of temperature. An open question is how the phages associated with desert microbial communities resist degradation in these extreme environments? Phages can either be present in the environment as free viral particles or inside bacterial cells (prophage), and each of these will offer different opportunities for the virus to evolve to persist. Even prior to 1950, there was a large body of information on the persistence of pathogenic microorganisms in soil and later on the persistence of artificially introduced pathogenic viruses in both aquatic and terrestrial systems (Duboise et al. 1979). However, a large-scale study of naturally occurring microbial communities and their associated viruses in soil has only come about following the genomics revolution. More recently Jończyk reviewed studies looking at the effect of certain physicochemical properties (pH, temperature, humidity, type of ions as well as ionic strength) on bacteriophages in particular. This indicated that several physical factors could play a role in the stability and therefore persistence of particles in a given environment, which do not affect viral types equally (Jończyk et al. 2011). In the Namib Desert, the combination of xeric zone and soil chemistry explained 37.5% of the variation in the assembly of bacterial communities along a transect (Scola et al. 2018). Thus, soil chemistry definitely has a role to play in shaping soil microbial communities in hot desert ecosystem and therefore should have an effect on prevalence of certain viral types. In the Mojave, researchers found that soil properties were heavily influenced by climate and that climate was likely the best predictor of community variability, with mean annual precipitation being the strongest influence on soil pH, for example (McHugh et al. 2017). Soil pH is one of the major factors in stickiness of phage particles to the soil matrix (Kinoshita et al. 1993; Kimura et al. 2008). Goyal and Gerba demonstrated that for a range of soil types, soils with a pH less than 5 generally tended to adsorb phages better (Goyal and Gerba 1979), however even among the five phages tested there was marked variability in their ability to adsorb to the different soil types. Given the effect of precipitation on soil pH described by McHugh for Desert soils, which usually have a neutral to alkaline pH (McHugh et al. 2017; Scola et al. 2018), it should be expected that precipitation plays a role in mobilization of phage particles following moisture input not only as carrier liquid, but perhaps due to modification of the pH. What the direct effect of soil chemistry is on free viral particles in these environments has not been investigated.

The abundance of firmicutes in initial culturing studies (Prigent et al. 2005; Prestel et al. 2008, 2013; Schulze-Makuch et al. 2018) and the presence of sequences related to phages infecting these in the Namib hypolith data (Adriaenssens et al. 2015), the presence of actinobacteria in desert soils (Fierer et al. 2012; Makhalanyane et al. 2013; Ronca et al. 2015) and cyanobacteria in both hypolithon communities and those responsible for dryland soil crusts, suggest that these microorganisms have each adapted to persisting in this environment either through dormancy or by developing ways to overcome the main obstacle to life in deserts, low water availability (accumulation of compatible solutes). This immediately suggests a mechanism by which viruses that infect these bacteria can persist: using host dormancy instead of producing physically and chemically stable free viral particles. For each of these phyla, sequences related to phages that infect them have been identified in hot desert viromes (Fierer et al. 2007; Adriaenssens et al. 2015; Zablocki et al. 2017; Scola et al. 2018).

The presence of viral particles in any environment is the result of the production and removal rate. The production rate is determined by the metabolic rate and abundance of their host bacterium. When assessing viral lifestyles (lytic versus lysogenic) two approaches have been described: Direct counts of bacterial and viral particles including the use of lytic induction to determine the inducible fraction as well as genomic studies using the presence or absence of certain genes (integrase) or training datasets based on the gene composition of known lysogenic and lytic phage (PHACTS) to infer lysogeny. However, virus to bacterial ratios (VBR) derived from induction by mitomycin C do not support a model of low virus to microbe ratios, because of low microbial counts in a given environment (Knowles et al. 2016). Studies looking at VBRs from a range of soil types (including the ability to induce phages from these) suggest that lysogeny may be a common feature of soil bacteria (Williamson et al. 2007; Ghosh et al. 2008; Meiring et al. 2012). Although no formal particle counts have been published from hot deserts, a low free particle concentration is expected for open soil samples given that early studies could not see phage particles by TEM in the samples taken, however in later metagenomic studies particles could be visualized following concentration (Prigent et al. 2005; Prestel et al. 2008, 2013; Hesse et al. 2017). For more productive settings such as hypolithons, viral particles could be observed without concentration to the same degree, suggesting higher particle production rates and/or lower removal rates, likely due to higher metabolic activity of host bacteria (Adriaenssens et al. 2015). Therefore, the absence of large numbers of viral particles from the Sahara and Mojave Desert samples speaks either to the metabolic activity of their hosts or higher particle turnover. The ability to induce phages from samples collected from several hot

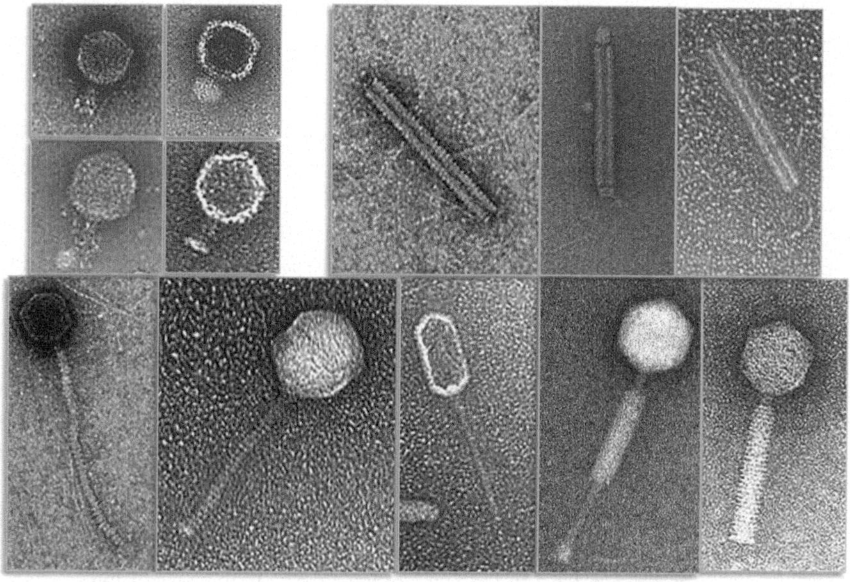

Fig. 6.1 Transmission electron microscope images of a selection of phage morphotypes observed from an open soil sample collected in the Namib Desert (S 23°28.900′; E 15°19.198′), following TFF of suspended soil particles

Desert environments together with estimates from metagenomic data suggests that the majority of phages in these biomes are maintained as lysogens (84% estimate for the Mojave) (Prigent et al. 2005; Prestel et al. 2008, 2013; Zablocki et al. 2016; Van Goethem et al. 2019). Studies of Antarctic open soils and hypoliths indicated that the bacteriophages were also predominantly lysogenic (Zablocki et al. 2014). Perhaps phages can, through lysogenizing their respective hosts, take advantage of the same mechanisms employed by their host to survive for extended periods. For cyanobacteria, production of extracellular polymeric substances (EPS) reduces water loss while firmicutes and actinobacteria are both sporulating microorganisms (Schimel 2018).

Several studies have noted the presence of free viral particles in hot and cold desert soils (Williamson et al. 2007; Adriaenssens et al. 2015; Hesse et al. 2017), which at the very least shows the presence of "intact" phage particles, suggesting that they may still be infective (Fig. 6.1). Studies of Antarctic cold desert systems suggest that although the particle production rate there may be low (due to low microbial metabolic activity), viral particles persist for longer periods, perhaps due to the cold climate (slow removal rate) which yielded high virus to bacterial ratios (Williamson et al. 2007). Unlike aquatic systems where this ratio can range from 1:1 to 70:1 VLPs/bacterium, soil ecosystems appear to show much higher variability and appear to be strongly influenced by ecosystem factors (Srinivasiah et al. 2008). As no quantitative studies have been conducted for hot desert ecosystems it is not possible to say if the same holds true here and as no studies describing the isolation of phage through infection assays from desert soils have been published it is difficult to say how well the particles withstand desiccation, variations in soil chemical composition, and exposure to UV.

Both hot and cold deserts are considered arid/hyper-arid (Pointing et al. 2009) which has implications for bacterial growth and therefore viral abundance and diversity. A Namib Desert transect study showed higher viral diversity (higher numbers of viral OTUs) in the low rainfall region samples compared with drier sites toward the coast and although TRFLP analysis to assess host diversity along the transect did not indicate significantly increased viral richness correlated with increased water availability (α-diversity of 9.6 ± 2.8 vs. 13.4 ± 2.7), enzymatic data did show an increase in population metabolic activity with increased water availability (Scola et al. 2018). Both iREP analysis of bacterial metagenomic data collected from the Atacama Desert and wetting experiments on biological soil crust from the Moab Desert indicated bursts of microbial activity following a rain event in 2015 or wetting events in the lab with a subsequent bloom in phage sequence (Schulze-Makuch et al. 2018; Van Goethem et al. 2019). These three studies demonstrate just how dependent microbial metabolism and therefore viral production rates are on water availability in these environments. In the case of the study by Van Goethem, they demonstrate that questions about virus–host interactions can be brought into the lab through the use of microcosms which appear to faithfully reproduce the events seen by Schulze-Makuch and colleagues in situ. Although the distance-decay pattern observed in all domains of life might be thought to describe the diversity distribution of bacteria in desert systems, McHugh and colleagues demonstrated that the local microbial community composition in the Great Basin, Mojave and Sonoran deserts were predominantly shaped by climate and soil properties (McHugh et al. 2017). The study by Scola and coworkers' contrasts that of McHugh in that they find that precipitation does not appear to be an important driving factor of community diversity in the Namib whereas it appears to be in the southwestern United States. Both studies however indicate that soil chemistry plays a determinant role. Thus, local weather patterns together with soil physicochemical properties should play a determining role in host diversity and by inference virus diversity, abundance and therefore virus–host dynamics.

Furthermore, it was shown that the microbial diversity in hot deserts displays low endemism with 70% of OTUs shared across the three deserts studied (McHugh et al. 2017). This does not seem to agree with the observation by Srinivasiah and colleagues regarding the uniqueness of viral communities in soils. Their observations were based on the data from the 2007 study by Fierer comparing Desert, prairie, and rainforest soils. Scola and coworkers showed little variation in bacterial alpha diversity along a xeric gradient in the Namib desert which supports the findings in the McHugh study. This supports a view that highly similar environments (Deserts, Rainforests, or Marine) select for largely similar bacterial populations and therefore viral populations. As hot deserts cover ~20% of Earth's surface, this means that there may be large areas with highly similar viral composition. Three hot desert virome studies have noted what appears to be significant inputs to the viral diversity from "external" sources, that being the presence of marine-associated viruses found deep inland in the Namib desert (Adriaenssens et al. 2015; Hesse et al. 2017; Scola et al. 2018). The reasons advanced by McHugh and colleagues for the even distribution of bacteria in those hot Deserts (dispersal of bacteria on dust particles and rain clouds), may explain why these signatures are seen in regions they are not expected in. In a 2017 study by Johnson et al., it was demonstrated that when looking at radically different soil types in the Namib (dune, gravel plain, salt spring soil) they found radically different bacterial communities unlike the Scola study would suggest (Johnson et al. 2017). This re-iterates how important it is to delineate the scale at which the questions of microbial diversity are asked at.

It has been demonstrated in several studies that microbial diversity decreases with decreased soil moisture and extremes of temperature compared with more temperate environments (Fierer et al. 2012; McHugh et al. 2017; Schulze-Makuch et al. 2018). Whether viral diversity mirrors host diversity in hot Deserts has not been determined. In other words, are there more or less viral OTU's per bacterium in hot Deserts than in temperate environments? In the 2007 study by Fierer and coworkers they found increased viral diversity in rainforest soil compared with Desert and prairie soils, therefore it seems as though higher microbial diversity in soils with increased moisture content supports higher viral diversity (Fierer et al. 2007). The study by Scola et al. seems to counter this finding with higher viral diversity with increasing soil moisture content along a xeric gradient, but no increase in bacterial alpha diversity. Again, the increased enzymatic activity they observed may explain this phenomenon, whereby the low bacterial diversity present may become more metabolically active when water is available, therefore producing a greater diversity of free viral particles which the study assessed.

As indicated earlier, virus abundance appears to fluctuate in hot Desert systems following stochastic inputs such as rainfall. It may be expected that this is a consequence of host lysis events indicating an initial increase in host metabolic activity and numbers followed by lysis which leads to a decrease in host numbers and release of nutrients. Similar events were seen in soil microcosms supplemented with a nutrient source (yeast extract) suggesting that soil microbial communities can rapidly respond to these inputs and further suggested rapid removal/decay of viruses in that system (Srinivasiah et al. 2008). Whether this rapid removal happens in hot Desert environments remains to be tested and it may be that the removal rate stratifies along soil depth with the particles exposed to harsh conditions at the surface breaking down first.

It would therefore appear that in hyper-arid deserts, phage production and the effect they have on shaping the microbial community is a consequence of their evolution to take advantage of increased microbial metabolic activity as a result of stochastic inputs such as rainfall or fog events analogous to what has been observed in single virus–host studies. Indeed, the recent study by Van Goethem seems to shed light on the microbial dynamics at play in biological soil crusts in relation to such inputs (Van Goethem et al. 2019). These authors found that following a controlled wetting event of a biological soil crust microcosm the initial dominance by cyanobacteria rapidly tapered off giving way to firmicutes and their viruses. The reduction in cyanobacterial numbers occurred just 3 min following the wetting event, suggesting that osmotic swelling led to the lysis of these cells. This also means that, even if present, no cyanobacterial viruses had the opportunity to show themselves during this experiment. This may partly explain why several studies have found low or no indication of cyanobacterial viruses even though the hosts are present and abundant in desert soils and especially hypolith samples. This paints a picture where, in soil crusts, cyanobacteria may be responsible for primary productivity during periods of low water input while firmicutes with their ability to survive extreme conditions as spores, rapid response to water input together with lytic action of their viruses are dominant nutrient cyclers. Although this was clearly the case for *Bacillus* species, other genera such as *Paenibacillus* and *Brevibacillus* continued to increase in numbers following the wetting event, indicating that it might only be a limited number of genera which take part in extreme nutrient cycling. It would seem that starting with Prestel and Du Bow who isolated firmicute infecting viruses from Saharan desert soil samples right through to using our most modern approaches to dissecting which bacteria and viruses are key drivers in these environments (Van Goethem et al. 2019) firmicute viruses have cemented their place.

Lysogenic conversion describes the process whereby bacteriophages impart genes with special functions to bacterial cells without these functions (Waldor and Mekalanos 1996). These genes, which are presumed to assist in rate-limiting steps in host metabolism, are termed auxiliary metabolic genes (AMG) and have been classified into Class I, those genes that encode a protein with a central metabolic function and Class II, those that encode proteins that have undefined metabolic roles (Warwick-Dugdale et al. 2019). They include enzymes involved in nucleotide metabolism (Nrd—ribonucleotide reductase related), photosynthesis (PsbA, PsbB), phosphate uptake (PstS), carbon turnover (TalC), polysaccharide binding, polysaccharide degradation (xylosidases, amylases, galactosidases, arabinases, fucosidases), central carbon metabolism and sporulation (Trubl et al. 2018; Emerson et al. 2018).

As large portions of virome data cannot be annotated, we do not yet have a sense of what portion of the viral community carry such genes. Even so, as it has been found that most viruses are habitat specific (Paez-Espino et al. 2016), and the presence of AMGs whose role can easily be linked to the habitat these viruses and hosts find themselves have been described and in particular for soil systems (Trubl et al. 2018; Emerson et al. 2018), it should be expected that this phenomenon assists in the adaptation of bacterial species to hot Deserts as in other environments (Crummett et al. 2016). In the Namib hypolith study it was found that ribonulcleotide reductases (RNR) were significantly more abundant in this dataset compared with the then available viromes. These enzymes mediate the synthesis of deoxyribonucleotides from ribonucleotides through conversion of the four ribonucleotide triphosphates (NTPs) into their corresponding dNTPs through the reduction of the C2′-OH, essential for bacterial life (Torrents 2014). As they are classified into three families that show various levels of oxygen dependence or intolerance, the relative abundance of the three families can be used as a tool to infer host metabolic dependence on oxygen. In this virome, class II RNR's (oxygen independent) were dominant, with class I RNR's (oxygen dependent) the second most abundant. Class III was present but at much lower levels than class I and II suggesting a dominantly aerobic community with the presence of some anaerobic members.

In the case of firmicute infecting viruses in biological soil crusts and permafrost, several phage genomes were found to encode sporulation associated genes (Trubl et al. 2018; Van Goethem et al. 2019), a phenomenon which has been recorded for many *Bacillus* infecting phages (Grose et al. 2014; Asare et al. 2015). Although several roles other than sporulation have been proposed for these genes, it is possible that their presence ensures induction of sporulation in infected hosts as soon as conditions become unfavorable, thereby ensuring their continued persistence. It would be of interest to determine if *Bacillus* infecting phage genomes from more temperate environments show higher or lower prevalence of sporulation genes. If lower, it could perhaps show the effect environmental factors such as stochastic water inputs have in shaping phage genomes by selecting for those which have incorporated these genes as part of their genomic repertoire.

The absence of photosynthesis related genes in these viral genomes may again point to these phages not infecting cyanobacteria, or that unlike what it found in aquatic systems, that provision of extra copies of these genes or improved versions by the phage to the host, do not benefit this viral community to the same degree. In the study by Crits-Christoph of Atacama halite communities they identified several genes related to photosystem II which may be virus encoded leaving the question as to whether or not, at least for cyanobacteria, these terrestrial viruses mimic their aquatic counterparts (Crits-Christoph et al. 2016). A similar lack of AMG's was observed for viromes from a permafrost mire and other soil viromes, suggesting that there may be a common lack of AMGs in soil virus genomes (Trubl et al. 2018). There is too little data now, as well as a lack of appropriate tools (with specific reference to genome identification from metagenomic data) to accurately model the way in which the AMG's carried by phage may influence a portion of the microbial population or the population as a whole.

6.3 Aquatic Desert Phage Communities: Wadi's, Gueltas, and Springs

Comparison of the datasets generated from two Namibian saline springs with those from several other environments showed that they had more phage species in common between them than with samples from other biomes, supported by the Lake Pavin and Bourget study (Adriaenssens et al. 2016). This suggests that niche specialization plays a role in shaping the associated communities and that the environment selects for the microorganism's present. The phenomenon also appears to be limited by distance-decay as illustrated by the two hypolith viral communities studied thus far and the suggestion from a meta-study showing distance-decay for roughly ten percent of known viral protein clusters (similar environments share fewer sequences the farther they are from each other geographically) (Fierer et al. 2007; Adriaenssens et al. 2016; Graham et al. 2019). One feature of the Namib

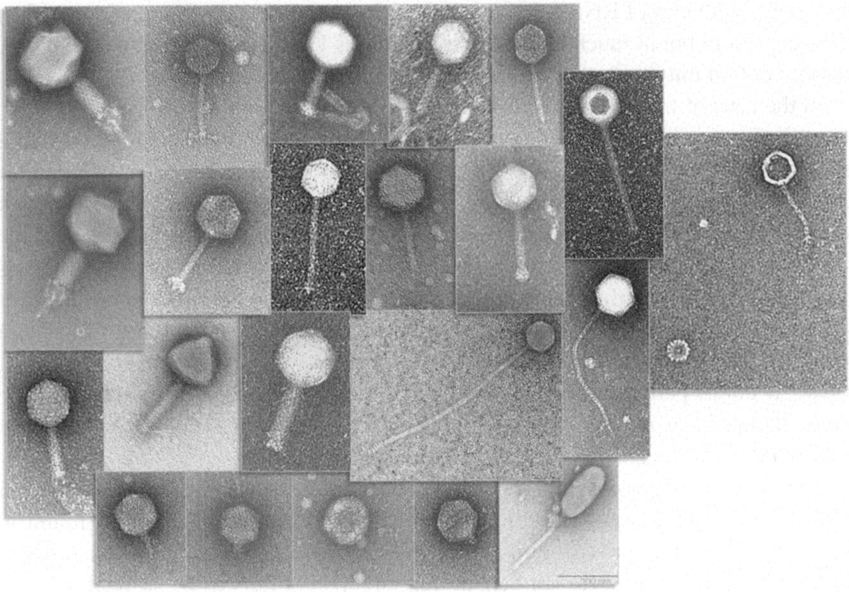

Fig. 6.2 Transmission electron microscope images of a selection of phage morphotypes observed in samples from the Eisfeld and Hosabes saline springs collected in the Namib Desert following TFF concentration of water collected at these sites

spring dataset was the presence of a high diversity of single stranded DNA viruses suggesting that these viruses and their hosts were better suited to aquatic environments. The dominant morphotypes observed under TEM in the Namibian salt playas (Fig. 6.2) were that of phages belonging to the order Caudovirales (dsDNA genomes), and it therefore might be expected that the metagenomic sequence data would be dominated by genomes of tailed viruses, however the data was instead dominated by ssDNA viral genomes. The discrepancy is likely due to a methodological bias (phi29 amplification) and shows how direct observation of a sample using TEM can to some level assist in corroborating the viral composition. The dominance of ssDNA viruses in this dataset did however allow for a deeper analysis of this portion of the viral community in these unique samples. For the Swakopmund saline site (Hosabes), most of these belonged to the family *Circoviridae*, specifically viruses that infect avian species. This demonstrates that viruses may be seeded from "outside" the system through animal movement. Although visits to the site by birds that alight there, may be seen as transitory, we would suggest seeing them as part of these systems. One might expect that deposition and persistence of viral pathogens in these environments means that these springs could serve as disease reservoirs and this is supported by comparison of *Salinivibrio* phages isolated from this environment (unpublished data).

When compared with marine-derived ssDNA datasets and other saline viromes no significant overlap was seen in the viral communities. This indicates that the playa communities are composed of highly unique viruses. Unpublished data of 16S rRNA sequence from these microbial communities indicate that they are dominated by members of the cyanobacteria and plantomycetes, and the virome data supported the presence of phages related to those that infect cyanobacteria, which could be expected. Many of the ssDNA viruses appear to be related to Gokushovirinae, a subfamily specialized to infect intracellular parasites (Zheng et al. 2018). The relatively recent discovery of highly abundant Gokushovirinae-related viruses associated with Gammaproteobacteria in freshwater settings might suggest that the Namib spring viruses infect non-pathogenic bacteria or non-human intracellular pathogens (Roux et al. 2012). Many of the sequences found here may represent new bacteriophage subfamilies and genera in the family *Microviridae*, infecting cyanobacteria or planctomycetes (Zheng et al. 2018). The discovery of phages infecting planctomycetes would be highly novel as only two phages have been described thus far (Ward et al. 2006).

Although the data from these springs did show some evidence of external viral input (avian viruses in a saline spring), there was no evidence of viruses related to those seen in the transect or hypolith studies in the saline spring datasets (Adriaenssens et al. 2016; Scola et al. 2018). In these more metabolically productive environments, it could be that the viruses present in the local community dominate over the small fraction input from neighboring sources.

Fancello and colleagues found that many of the viruses in the Mauritanian gueltas likely infect photosynthetic bacteria such as *Prochlorococcus* and *Synechococcus* (Fancello et al. 2013), and that the virus communities in these rocky pools were dominated by Myoviruses. Although the authors also employed phi29 amplification of the extracted DNA, they did not see the same bias toward ssDNA viruses observed for the Namibian samples. In many ways the viral composition of these sites reflects the composition of other aquatic systems, particularly marine virus community composition (Sepulveda et al. 2016). The Shannon diversity estimates for the gueltas ($H = 2.21$–4.83) are well below some reports for diversity in other environments (≤ 13) (Williamson et al. 2012; Hannigan et al. 2015), suggesting greater constraints on the presence of a range of viruses (hosts are not supported in these environments), or constraints on their evolution in these settings.

To highlight the novelty of the viruses found in these environments, Adriaenssens and coworkers detailed the relationship of three haloviral-like contigs found in the Hosabes playa to that of the archaeal salterprovirus His1 (Adriaenssens et al. 2016). In their 2018 review of archaeal viruses, Krupovic et al. compared the genomes of several pleolipovirus with that of the only known salterprovirus and a contig from the Namib saline spring study (Adriaenssens et al. 2016) highlighting domain swapping events among these viruses (Krupovic et al. 2018). In this instance, this domain swapping may have generated viruses with radically different morphologies, which may share similar replication machinery and vice versa further supporting a genome-based phylogeny of viruses. The study of viruses in these extreme environments is therefore serving to help establish links between previously unrelated viral families whose classification perhaps depended solely on morphology and limited genomic analysis by revealing how successful domain swapping can lead to new viruses that seem to straddle these families.

Again, the question can be asked as to which environmental factors influence the virus–host association in these settings. Stochastic inputs such as rainfall should be expected to have an influence on these communities as this changes the chemical composition of these streams through dilution, even if just transiently (Eckardt and Drake 2011). How the diurnal and seasonal temperature and light conditions may affect these communities have yet to be investigated. The lack of temporal studies together with little overlap in the genomes of phages found in these environments with those elsewhere makes it difficult at this point in time to discuss the drivers of evolution of the viruses present as well as the influence they may have on adaptation of their hosts to these environments.

A criticism of all meta studies conducted on hot Desert viromes thus far is that most have looked exclusively at VLP's as opposed to sequencing and analyzing full metagenomes or a combination. The argument for and against assessing just viral particles as opposed to the complete metagenome was recently made by Trubl and colleagues (Trubl et al. 2018) concluding that:

> These findings suggest that viromes (which greatly enrich for viral particles) and bulk-soil metagenomes (which are less methodologically intensive and provide simultaneous information on both viruses and microbes) offer complementary views of viral communities in soils, and if only one method can be applied, its selection will depend on the goal of the study.

Therefore, this may require that some of these studies to be revisited using the complimentary approach of whole metagenome sequencing and analysis, especially to better establish phage–host links.

6.4 Eukaryotic Viruses

Viral ecology studies in desert environments have preferentially focused on bacteriophages; however, Eukaryotic viruses are sure to play a huge role in modifying the abundance and diversity of smaller Eukaryotes in desert environments.

Since their initial discovery in 2003, several species of Nucleocytoplasmic Large DNA Viruses (NCLDVs) have been described and found to infect six different supergroups of eukaryotic protists (Claverie and Abergel 2018). These viruses not only show great genetic and host diversity, they are also distributed worldwide, found in aquatic and terrestrial habitats (Halary et al. 2016; Gallot-Lavallée and Blanc 2017; Moniruzzaman et al. 2017; Schulz et al. 2018), but also extreme environments such as Siberian permafrost and Yellowstone lake (Legendre et al. 2015; Zhang et al. 2015).

Due to the recent discovery of these "giant" viruses, few studies have addressed their presence in hot desert ecosystems, although several lines of evidence suggest that they could be represented. First of all, amoeba and algae, two of the protists infected by NCLDVs, are present in desert soils, including the hyper-arid Mojave Desert (Whitford 2002; Rodriguez-Zaragoza et al. 2005; Bamforth 2008; Cardon et al. 2008; Fierer et al. 2012; Bates et al. 2013). These protozoans are often found together with other microbial species in cryptic and refuge niches such as biological soil crusts and hypoliths, or in association with the scarce vegetation, which presumably helps them to withstand desiccation and the extremely oligotrophic conditions of arid ecosystems (Belnap and Weber 2013; Makhalanyane et al. 2015). Protists are also abundant in other desert aquatic niches such as saltern ponds, riverbeds, or ephimeral lakes. Thus, giant viruses are potentially widespread in hot desert ecosystems.

Additionally, members of the NCLDV families *Mimiviridae* and *Phycodnaviridae*, which infect predominantly amoeba and unicellular algae, have been detected in the metagenomic datasets for the ponds and salterns in the Sahara and the Namib Deserts (Fancello et al. 2013; Adriaenssens et al. 2016), the Namib Desert hypoliths (Adriaenssens et al. 2015) and soils from the Mojave, the Chihuahuan and the Namib Deserts (Kerepesi and Grolmusz 2017; Scola et al. 2018). The detection of NCLDV sequences in the viromes is surprising because the filtration step used to remove cellular microorganisms and enrich viruses should remove these giant viruses as well, considering they all have sizes greater than the 0.2–0.45 μm pore of the membrane used during filtration. It is speculated that filtration may allow the crossing of partial virion particles or viral DNA present in the sample. Nevertheless, the presence of NCLDV sequences in the desert metagenomic cellular fraction of samples collected worldwide strongly supports the presence of these viruses in hot and arid ecosystems. The development of better enrichment techniques for isolation of NCLDV particles detected in viromes and the application of bioinformatic pipelines that identify NCLDV signature genes (Verneau et al. 2016; Kerepesi and Grolmusz 2017, Roux et al. 2017; Zhao et al. 2017) should help our understanding of the diversity of eukaryotic large nucleocytoplasmic DNA viruses in hyper-arid soils.

6.5 Conclusion and Outlook

In their recent review of soil virus studies Kuzyakov and Mason-Jones explored the five concepts (Viral shunt; Forever young concept; Exocellular metabolism (EXOMET); Microscale divergence of C/N/P stoichiometry; Carbon stabilization and sequestration in soil as microbial necromass) proposed to explain the role played by phages in shaping soil bacterial communities, in depth (Kuzyakov and Mason-Jones 2018). It is clear that, depending on the environment, elements of these mechanisms will be present and affect bacterial community change and nutrient cycling to various degrees. In hot Deserts, "forever young" and EXOMET may only be expected to be dominant mechanisms when the viral shunt is in operation after rainfall events, followed by relatively long periods of stasis, compared with wetter climates and more productive soils.

In the Namib (and elsewhere) as much as 62% of the bacterial community variability remains unexplained, and it is tempting to suggest that phages, which are rarely factored in, account for these shifts. Although a radically different setting, Zhang and coworkers found that phage activity described 40.6% of community diversity versus 14.5% attributed to abiotic factors in anaerobic digesters. This suggests that viruses have likely had a much larger influence on regulating the microbial diversity reported in microbial ecology assessments, than is appreciated (Zhang et al. 2017). This is supported by Kuzyakov and Mason-Jones citing high infection rates in soil bacteria, short times to lysis, and large burst sizes being indicators of high rates of viral predation on these communities.

The current picture that has developed for desert environments is that they are dominated by tailed viruses whose diversity reflects that of their hosts and most of which are lysogenic. Not much can be said about their abundance as no quantitative studies have been performed and many are skewed by biases introduced during sample processing. As eluded to earlier, anecdotal evidence would suggest that, as in other environments, those with higher metabolic activity (hypolithons) yield higher numbers of free phage, whereas the opposite is true for niches (topsoil) with low host metabolic activity. Slow metabolic host metabolic activity in turn is a consequence of oligotrophic habitats and perhaps more importantly low water activity. As yet, there appears to be no hot Desert specific adaptations in the viral genomes studied thus far and their persistence and evolution appear to be a mixture of features already described for other virus–host pairs in a range of ecological settings.

There is mounting evidence of how lysogenic conversion of the host is a beneficial trait, often aiding in the adaptation of their hosts to their environment and with the mountain of sequence data that has become available, it is clear that phages contribute significantly to the accessory metabolic genes harbored by bacteria (Trubl et al. 2018). Given the interdependence of phages and their hosts the current data for hot Deserts supports the notion that at the extremes of life selection favors a relationship between viruses and their hosts that sees them act more like advanced programmed cell death routines rather than purely bacterial predators. Together with their role in adding to the nutrient pool on lysing their host following rainfall or other moisture inputs, it suggests that phages are a crucial component to maintaining the health of these environments.

One of the central tenets of viral ecology is that the viral population reflects the diversity and metabolic activity of their host(s) (Thurber 2009). Such host dependence lends itself well to sequence-based investigations of the microbial diversity in hot desert ecosystems to infer viral lifestyles, evolution, host interactions, and ecological influence. Therefore, a common opening statement made in sequence-based studies is: "To better understand the role of viruses in these environments....". While analysis of viral genomic sequence in many of the studies referred to in this chapter has served to confirm the presence of viruses in these extreme environments, most often, apart from cataloging new and existing viruses and trying to deduce viral lifestyles and host interactions, no genuine new knowledge is gained about actual interactions between specific phages and their hosts, how individual or communities of viruses evolve in these environments, nor how the

Table 6.2 Percentage of unknown (unclassified sequence) in viral metagenomes

Study	Percentage unknown contigs / reads	Reference
Namib Gobabeb Saline	80/95	Adriaenssens et al. (2016)
Namib Swakopmund Saline	79/92	Adriaenssens et al. (2016)
Gueltas from Sahara	34–93	Fancello et al. (2013)
Namib desert transect	81–91	Zablocki et al. (2017)
Namib hypolith	40	Adriaenssens et al. (2015)

viruses in the corresponding environment shape the microbial community. Furthermore, the high percentage of sequence that cannot be annotated or taxonomically assigned (Table 6.2) in virome datasets suggests that the conclusions reached in most of the studies are drawn from a small fraction of the viral sequence, likely leading to incorrect interpretation of the microbial dynamics. Where distant relatives were assigned, it was not immediately obvious that they infect the same hosts or share similar lifestyles with these relatives. For example, deletion or mutation of the integrase can render a formerly lysogenic phage lytic (Bruttin et al. 1997), making it difficult at this point in time to accurately model the interactions between environmental viruses and hosts and estimate their impact on these environments. This is especially true for soil environments for which very little data is available. This puts great emphasis on the software and databases used for analyzing (annotating and describing relationships between the identified genomes) the large volume of data generated. Over the past 10 years much progress has been made in developing software that allows us to first identify which contigs are likely truly of viral origin, followed by annotation and comparison (Table 6.3). Additionally, investigators have proposed standards for describing viral genomes from metagenomic datasets, which, in time should lead to greater understanding of what drives these communities through analysis of the associated metadata (Roux et al. 2018). However, what is sorely needed is for databases to be populated with properly curated sequences and characterization data for virus–host pairs. More attention should be focused on developing methods for bringing viral hosts into culture and to describe novel phage–host pairs and their interactions in controlled settings to truly begin understanding the role and function of microorganisms in these ecological systems.

Our ability to model these interactions is pivotal to our fundamental understanding of ecology and biology, but may also become important in a future where every advantage is needed to enable the most efficient use of arable land, understanding disease reservoirs, assist in desert reclamation, rebalancing of biogeochemical cycles, or monitoring of climate change. The development of new approaches, such as investigating the lysis–lysogeny switch at a single cell level, has led to a paradigm shift in what was thought to be a well understood system. Recently, Shao and coworkers demonstrated that instead of just choosing lysis or lysogeny, phage λ can exist in a "lyso-lysis" state within its host where phage DNA is integrated into host genomic DNA even when the phage enters the lytic cycle (Shao et al. 2016). It was observed that the first phage genomes to replicate would physically separate from one another inside the cell, possibly allowing each phage to make an

Table 6.3 Programs and databases used for viral metagenome analysis

Category	Program	Function	Reference
Databases	IMG/VR v.2.0	Integrated data management and analysis system for cultivated and environmental viral genomes	Paez-Espino et al. (2019)
	GenBank	A comprehensive database that contains publicly available nucleotide sequences	Clark et al. (2016)
Prophage identification	Prophage Finder	Tool for prophage prediction in prokaryotic genomes	Bose and Barber (2006)
	Phage Finder	Tool for prophage prediction in prokaryotic genomes	Fouts (2006)
	Prophinder	Tool for prophage prediction in prokaryotic genomes	Lima-Mendez et al. (2008)
	PhiSpy	A novel algorithm for finding prophages in bacterial genomes	Akhter et al. (2012)
	PHASTER	Tool for prophage prediction in prokaryotic genomes	Arndt et al. (2016)
Virus identification from metagenome data	PHACTS	Computational approach to classifying the lifestyle of phages	McNair et al. (2012)
	VirSorter	Tool that uses a reference database of known viral genomes/proteins to identify viral contigs in metagenomic data	Roux et al. (2015)
	MG-Digger	An automated pipeline to search for giant virus-related sequences in metagenomes	Verneau et al. (2016)
	MetaPhinder	Tool for extracting phage contigs from previously assembled metagenomic contigs to classify a contig as of phage origin or non-phage	Jurtz et al. (2016)
	VirusSeeker	Computational pipeline for virus discovery and virome composition analysis	Zhao et al. (2017)
	vConTact 2.0	Tool to classify double-stranded DNA viruses	Jang et al. (2019)
	DeepvirFinder	k-mer based tool for identifying phage contigs from metagenomic data	Ren et al. (2020)
	FastViromeExplorer	Pipeline for virus and phage identification and abundance profiling in metagenomics data	Tithi et al. (2018)

(continued)

Table 6.3 (continued)

Category	Program	Function	Reference
	MARVEL	Tool for prediction of double-stranded DNA bacteriophage sequences in metagenomic bins	Amgarten et al. (2018)
	virMine	Tool for the identification of viral genomes within metagenomic data sets from raw sequence, read quality control through assembly and annotation	Garretto et al. (2019)
	VirMiner	Tool that employs the random forest model to identify phage contigs from metagenomic data	Zheng et al. (2019)
	VIBRANT	Automated recovery, annotation and curation of microbial viruses, and evaluation of viral community function from genomic sequences	Kieft et al. (2020)
Visualization	Phage Eco-Locator (Phantome)	Tool for visualization and analysis of phage genomes in metagenomic data sets	Aziz et al. (2011)
	Easyfig	Application for creating linear comparison figures of multiple genomic loci	Sullivan et al. (2011)
	Phamerator	A bioinformatic tool for comparative bacteriophage genomics	Cresawn et al. (2011)
Viral ecology	PHACCS	Estimating the structure and diversity of uncultured viral communities	Angly et al. (2005)
	Circonspect	Tool for calculating contig spectra	Angly et al. (2006)
Virus-Host linkage	CRISPRFinder	Web service offering fundamental tools for CRISPR detection	Grissa et al. (2007)
	WISH	Predicts prokaryotic hosts of phages from their genomic sequences	Galiez et al. (2017)

independent decision whether to enter the lytic cycle, or not. This seemed to support their earlier work which suggested that each phage which infects a host makes an independent decision and the fate of the cell is dependent on the consensus "vote" of all phages that infect. They then demonstrated that the decision to enter either cycle *is* made at the level of each individual phage genome (Trinh et al. 2017). If even a well-studied system can teach us something new today, it serves to show just how far we have yet to go toward a working understanding of the dynamics between phages and hosts in the wide range of ecological settings they occupy.

As we conclude, it should also be kept in mind that most of these studies highlight the easily observable phenomena, and that the nuanced interactions of thousands of phage–host pairs as well as host–host interactions perhaps across the domains of life are likely to be missed. However, the techniques employed in studying these microbial communities from sample prep to data analysis are constantly being improved and should lead us to deeper insight (Trubl et al. 2020; Göller et al. 2020). The question of whether or not these viruses evolve differently to those from more temperate or aquatic environments can only be answered with more study. As more virome datasets are generated from environments such as hot deserts and more temperate environments for comparison, the data will become available to better assess the questions posed at the beginning of this chapter.

References

Adriaenssens EM, Van Zyl L, De Maayer P, Rubagotti E, Rybicki E, Tuffin M, Cowan DA (2015) Metagenomic analysis of the viral community in Namib Desert hypoliths. Environ Microbiol 17:480–495

Adriaenssens EM, van Zyl LJ, Cowan DA, Trindade MI (2016) Metaviromics of Namib Desert salt pans: a novel lineage of Haloarchaeal salterproviruses and a rich source of ssDNA viruses. Viruses 8:14. https://doi.org/10.3390/v8010014

Akhter S, Aziz RK, Edwards RA (2012) PhiSpy: a novel algorithm for finding prophages in bacterial genomes that combines similarity- and composition-based strategies. Nucleic Acids Res 40:e126

Amgarten D, Braga LPP, da Silva AM, Setubal JC (2018) MARVEL, a tool for prediction of bacteriophage sequences in metagenomic bins. Front Genet 9:304. https://doi.org/10.3389/fgene.2018.00304

Anderson OR (2018) Evidence for coupling of the carbon and phosphorus biogeochemical cycles in freshwater microbial communities. Front Mar Sci 5:20. https://doi.org/10.3389/fmars.2018.00020

Angly F, Rodriguez-Brito B, Bangor D, McNairnie P, Breitbart M, Salamon P, Felts B, Nulton J, Mahaffy J, Rohwer F (2005) PHACCS, an online tool for estimating the structure and diversity of uncultured viral communities using metagenomic information. BMC Bioinformatics 6:41

Angly FE, Felts B, Breitbart M, Salamon P, Edwards RA, Carlson C, Chan AM, Haynes M, Kelley S, Liu H, Mahaffy JM, Mueller JE, Nulton J, Olson R, Parsons R, Rayhawk S, Suttle CA, Rohwer F (2006) The marine viromes of four oceanic regions. PLoS Biol 4:e368

Arndt D, Grant JR, Marcu A, Sajed T, Pon A, Liang Y, Wishart DS (2016) PHASTER: a better, faster version of the PHAST phage search tool. Nucleic Acids Res 44:W16–W21

Asare PT, Jeong T, Ryu S et al (2015) Putative type 1 thymidylate synthase and dihydrofolate reductase as signature genes of a novel bastille-like group of phages in the subfamily Spounavirinae. BMC Genomics 16:582

Aziz RK, Dwivedi B, Breitbart M et al (2011) Phage Eco-Locator: a web tool for visualization and analysis of phage genomes in metagenomic data sets. BMC Bioinformatics 12:A9

Bamforth SS (2008) Protozoa of biological soil crusts of a cool desert in Utah. J Arid Environ 72: 722–729

Bates ST, Clemente JC, Flores GE, Walters WA, Parfrey LW, Knight R, Fierer N (2013) Global biogeography of highly diverse protistan communities in soil. ISME J 7:652–659

Belnap J, Weber B (2013) Biological soil crusts as an integral component of desert environments. Ecol Process 2:11

Bose M, Barber RD (2006) Prophage Finder: a prophage loci prediction tool for prokaryotic genome sequences. In Silico Biol 6:223–227

Breitbart M, Rohwer F (2005) Here a virus, there a virus, everywhere the same virus? Trends Microbiol 13:278–284

Breitbart M, Hewson I, Felts B, Mahaffy JM, Nulton J, Salamon P, Rohwer F (2003) Metagenomic analyses of an uncultured viral community from human feces. J Bacteriol 85:6220–6223

Brüssow H, Canchaya C, Hardt W-D (2004) Phages and the evolution of bacterial pathogens: from genomic rearrangements to lysogenic conversion. Microbiol Mol Biol Rev 68:560–602

Bruttin A, Desiere F, Lucchini S, Foley S, Brüssow H (1997) Characterization of the lysogeny DNA module from the temperate Streptococcus thermophilus bacteriophage phi Sfi21. Virology 233: 136–148

Cardon ZG, Gray DW, Lewis LA (2008) The green algal underground: evolutionary secrets of desert cells. BioScience 58:114–122

Clark K, Karsch-Mizrachi I, Lipman DJ, Ostell J, Sayers EW (2016) GenBank. Nucleic Acids Res 44:D67–D72. https://doi.org/10.1093/nar/gkv1276

Claverie J-M, Abergel C (2018) Mimiviridae: an expanding family of highly diverse large dsDNA viruses infecting a wide phylogenetic range of aquatic eukaryotes. Viruses 10:E506. https://doi.org/10.3390/v10090506

Clokie MRJ, Millard AD, Letarov AV, Heaphy S (2011) Phages in nature. Bacteriophage 1:31–45

Cresawn SG, Bogel M, Day N, Jacobs-Sera D, Hendrix RW, Hatfull GF (2011) Phamerator: a bioinformatic tool for comparative bacteriophage genomics BMC. Bioinformatics 12:395

Crits-Christoph A, Gelsinger DR, Ma B, Wierzchos J, Ravel J, Davila A, Casero MC, DiRuggiero J (2016) Functional interactions of archaea, bacteria and viruses in a hypersaline endolithic community. Environ Microbiol 18:2064–2077

Crummett LT, Puxty RJ, Weihe C, Marston MF, Martiny JBH (2016) The genomic content and context of auxiliary metabolic genes in marine cyanomyoviruses. Virology 499:219–229

Curtis TP, Sloan WT, Scannell JW (2002) Estimating prokaryotic diversity and its limits. Proc Natl Acad Sci U S A 99:10494–10499

Duboise SM, Moore BE, Sorber CA, Sagik BP (1979) Viruses in soil systems. CRC Crit Rev Microbiol 7:245–285

Eckardt FD, Drake N (2011) Introducing the Namib Desert playas. In: Öztürk M, Böer B, Barth H-J, Clüsener-Godt M, Khan MA, Breckle S-W (eds) Sabkha ecosystems, vol 46. Springer, Dordrecht, pp 19–25

Emerson JB, Roux S, Brum JR et al (2018) Host-linked soil viral ecology along a permafrost thaw gradient. Nat Microbiol 3:870–880. https://doi.org/10.1038/s41564-018-0190-y

Fancello L, Trape S, Robert C, Boyer M, Popgeorgiev N, Raoult D, Desnues C (2013) Viruses in the desert: a metagenomic survey of viral communities in four perennial ponds of the Mauritanian Sahara. ISME J 7:359–369

Fierer N, Breitbart M, Nulton J, Salamon P, Lozupone C, Jones R, Robeson M, Edwards RA, Felts B, Rayhawk S, Knight R, Rohwer F, Jackson RB (2007) Metagenomic and small-subunit rRNA analyses reveal the genetic. Diversity of bacteria, archaea, fungi, and viruses in soil. Appl Environ Microbiol 73:7059–7066

Fierer N, Leff JW, Adams BJ, Nielsen UN, Bates ST, Lauber CL, Owens S, Gilbert JA, Wall DH, Caporaso JG (2012) Cross-biome metagenomic analyses of soil microbial communities and their functional attributes. Proc Natl Acad Sci U S A 109:21390–21395

Fouts DE (2006) *Phage_Finder*: automated identification and classification of prophage regions in complete bacterial genome sequences. Nucleic Acids Res 34:5839–5851

Galiez C, Siebert M, Enault F, Vincent J, Söding J (2017) WIsH: who is the host? Predicting prokaryotic hosts from metagenomic phage contigs. Bioinformatics 33:3113–3114

Gallot-Lavallée L, Blanc G (2017) A Glimpse of nucleo-cytoplasmic large DNA virus biodiversity through the eukaryotic genomics window. Viruses 9:17

Garretto A, Hatzopoulos T, Putonti C (2019) virMine: automated detection of viral sequences from complex metagenomic samples. PeerJ 7:e6695

Ghosh D, Roy K, Williamson KE, White DC, Wommack KE, Sublette KL, Radosevich M (2008) Prevalence of lysogeny among soil bacteria and presence of 16S rRNA and *trzN* genes in viral-community DNA. Appl Environ Microbiol 74:495–502

Göller PC, Haro-Moreno JM, Rodriguez-Valera F et al (2020) Uncovering a hidden diversity: optimized protocols for the extraction of dsDNA bacteriophages from soil. Microbiome 8:17

Goswami D, Pithwa S, Dhandhukia P et al (2013) Delineating *Kocuria turfanensis* 2M4 as a credible PGPR: a novel IAA producing bacteria isolated from saline desert. J Plant Interact 9: 56676

Goyal SM, Gerba CP (1979) Comparative Adsorption of Human enteroviruses, simian rotavirus, and selected bacteriophages to soils. Appl Environ Microbiol 38:241–247

Graham EB, Paez-Espino D, Brislawn C, Neches RY, Hofmockel KS, Wu R, Kyrpides NC, Jansson JK, McDermott JE (2019) Untapped viral diversity in global soil metagenomes. BioRxiv. https://doi.org/10.1101/583997

Grissa I, Vergnaud G, Pourcel C (2007) CRISPRFinder: a web tool to identify clustered regularly interspaced short palindromic repeats. Nucleic Acids Res 35:W52–W57

Grose JH, Jensen GL, Burnett SH et al (2014) Genomic comparison of 93 *Bacillus* phages reveals 12 clusters, 14 singletons and remarkable diversity. BMC Genomics 15:855

Halary S, Temmam S, Raoult D, Desnues C (2016) Viral metagenomics: are we missing the giants? Curr Opin Microbiol 31:34–43

Hannigan GD, Meisel JS, Tyldsley AS, Zheng Q, Hodkinson BP, SanMiguel AJ, Minot S, Bushman FD, Grice EA (2015) The human skin double-stranded DNA virome: topographical and temporal diversity, genetic enrichment, and dynamic associations with the host microbiome. mBio 6:e01578–e01515. https://doi.org/10.1128/mBio.01578-15

Hesse U, van Heusden P, Kirby BM, Olonade I, van Zyl LJ, Trindade M (2017) Virome assembly and annotation: a surprise in the Namib Desert. Front Microbiol 8:13

Jang HB, Bolduc B, Zablocki O, Kuhn JH, Roux S, Adriaenssens EM, Brister JR, Kropinski AM, Krupovic M, Turner D, Sullivan MB (2019) Gene sharing networks to automate genome-based prokaryotic viral taxonomy. BioRxiv. https://doi.org/10.1101/533240

Johnson RM, Ramond J-B, Gunnigle E, Seely M, Cowan DA (2017) Namib Desert edaphic bacterial, fungal and archaeal communities assemble through deterministic processes but are influenced by different abiotic parameters. Extremophiles 21:381–392

Jończyk E, Kłak M, Międzybrodzki R, Górski A (2011) The influence of external factors on bacteriophages-review. Folia Microbiol (Praha) 56:191–200

Jurtz VI, Villarroel J, Lund O, Voldby Larsen M, Nielsen M (2016) MetaPhinder-identifying bacteriophage sequences in metagenomic data sets. PLoS One 11:e0163111

Kerepesi C, Grolmusz V (2017) The "Giant Virus Finder" discovers an abundance of giant viruses in the Antarctic dry valleys. Arch Virol 162:1671–1676

Kieft K, Zhou Z, Anantharaman K (2020) VIBRANT: automated recovery, annotation and curation of microbial viruses, and evaluation of viral community function from genomic sequences. Microbiome 8:90. https://doi.org/10.1186/s40168-020-00867-0

Kimura M, Jia Z-J, Nakayama N, Asakawa S (2008) Ecology of viruses in soils: past, present and future perspectives. Soil Sci Plant Nutr 54:1–32

Kinoshita T, Bales RC, Maguire KM, Gerba CP (1993) Effect of pH on bacteriophage transport through sandy soils. J Contam Hydrol 14:55–70

Knowles B, Silveira CB, Bailey BA et al (2016) Lytic to temperate switching of viral communities. Nature 531:466–470

Krupovic M, Cvirkaite-Krupovic V, Iranzo J, Prangishvili D, Koonin EV (2018) Viruses of archaea: structural, functional, environmental and evolutionary genomics. Virus Res 244:181–193

Kuzyakov Y, Mason-Jones K (2018) Viruses in soil: nano-scale undead drivers of microbial life, biogeochemical turnover and ecosystem functions. Soil Biol Biochem 127:305–317

Legendre M, Lartigue A, Bertaux L, Jeudy S, Bartoli J, Lescot M, Alempic J-M, Ramus C, Bruley C, Labadie K, Shmakova L, Rivkina E, Couté Y, Abergel C, Claverie J-M (2015)

In-depth study of *Mollivirus sibericum*, a new 30,000-y-old giant virus infecting Acanthamoeba. Proc Natl Acad Sci U S A 112:E5327–E5335
Li B, Wang L, Kaseke KF, Li L, Seely MK (2016) The impact of rainfall on soil moisture dynamics in a foggy desert. PLoS One 11(10):e0164982
Liang X, Wagner RE, Zhuang J, DeBruyn JM, Wilhelm SW, Liu F, Yang L, Staton ME, Sherfy AC, Radosevich M (2019) Viral abundance and diversity vary with depth in a southeastern United States agricultural ultisol. Soil Biol Biochem 137:107546
Lima-Mendez G, Van Helden J, Toussaint A, Leplae R (2008) Prophinder: a computational tool for prophage prediction in prokaryotic genomes. Bioinformatics 24:863–865
Liu M, Peng F, Wang Y et al (2009) *Kineococcus xinjiangensis* sp. nov., isolated from desert sand. Int J Syst Evol Microbiol 59:10903
Makhalanyane TP, Valverde A, Lacap DC, Pointing SB, Tuffin MI, Cowan DA (2013) Evidence of species recruitment and development of hot desert hypolithic communities. Environ Microbiol Rep 5:219–224
Makhalanyane TP, Valverde A, Gunnigle E, Frossard A, Ramond J-B, Cowan DA (2015) Microbial ecology of hot desert edaphic systems. FEMS Microbiol Rev 39:203–221
McHugh TA, Compson Z, van Gestel N, Hayer M, Ballard L, Haverty M, Hines J, Irvine N, Krassner D, Lyons T, Musta EJ, Schiff M, Zint P, Schwartz E (2017) Climate controls prokaryotic community composition in desert soils of the southwestern United States. FEMS Microbiol Ecol 93. https://doi.org/10.1093/femsec/fix116
McNair K, Bailey BA, Edwards RA (2012) PHACTS, a computational approach to classifying the lifestyle of phages. Bioinformatics 28:614–618
Meiring TL, Tuffin IM, Cary C, Cowan DA (2012) Genome sequence of temperate bacteriophage Psymv2 from Antarctic Dry Valley soil isolate *Psychrobacter* sp. MV2. Extremophiles 16:715–726
Moniruzzaman M, Wurch LL, Alexander H, Dyhrman ST, Gobler CJ, Wilhelm SW (2017) Virus-host relationships of marine single-celled eukaryotes resolved from metatranscriptomics. Nat Commun 8:16054
Motlagh AM, Bhattacharjee AS, Coutinho FH, Dutilh BE, Casjens SR, Goel RK (2017) Insights of phage-host interaction in hypersaline ecosystem through metagenomics analyses. Front Microbiol 8:352. https://doi.org/10.3389/fmicb.2017.00352
Paez-Espino D, Eloe-Fadrosh EA, Pavlopoulos GA, Thomas AD, Huntemann M, Mikhailova N, Rubin E, Ivanova NN, Kyrpides NC (2016) Uncovering earth's virome. Nature 536:425–430
Paez-Espino D, Roux S, Chen I-MA, Palaniappan K, Ratner A, Chu K, Huntemann M, Reddy TBK, Pons TC, Llabrés M, Eloe-Fadrosh EA, Ivanova NN, Kyrpides NC (2019) IMG/VR v.2.0: an integrated data management and analysis system for cultivated and environmental viral genomes. Nucleic Acids Res 47:D678–D686
Pearce DA, Newsham KK, Thorne MAS, Calvo-Bado L, Krsek M, Laskaris P, Hodson A, Wellington EM (2012) Metagenomic analysis of a southern maritime Antarctic soil. Front Microbiol 3:403
Peduzzi P, Gruber M, Gruber M, Schagerl M (2014) The virus's tooth: cyanophages affect an African flamingo population in a bottom-up cascade. ISME J 8:1346–1351
Pointing SB, Chan Y, Lacap DC, Lau MCY, Jurgens JA, Farrell RL (2009) Highly specialized microbial diversity in hyper-arid polar desert. Proc Natl Acad Sci U S A 106:19964–19969
Prestel E, Salamitou S, DuBow MS (2008) An examination of the bacteriophages and bacteria of the Namib Desert. J Microbiol 46:364–372
Prestel E, Regeard C, Salamitou S, Neveu J, DuBow MS (2013) The bacteria and bacteriophages from a Mesquite Flats site of the Death Valley desert. Antonie Van Leeuwenhoek 103:1329–1341
Prigent M, Leroy M, Confalonieri F, Dutertre M, DuBow MS (2005) A diversity of bacteriophage forms and genomes can be isolated from the surface sands of the Sahara Desert. Extremophiles 9:289–296

Reavy B, Swanson MM, Cock PJA, Dawson L, Freitag TE, Singh BK, Torrance L, Mushegian AR, Talianskya M (2015) Distinct circular single-stranded DNA viruses exist in different soil types. Appl Environ Microbiol 81:3934–3945

Ren J, Song K, Deng C, Ahlgren NA, Fuhrman JA, Li Y, Xie X, Poplin R, Sun F (2020) Identifying viruses from metagenomic data using deep learning. Quant Biol 8:64–77

Rodriguez-Caballero E, Belnap J, Büdel B, Crutzen PJ, Andreae MO, Pöschl U, Weber B (2018) Dryland photoautotrophic soil surface communities endangered by global change. Nat Geosci 11:185–189

Rodriguez-Zaragoza S, Mayzlish E, Steinberger Y (2005) Vertical distribution of the free-living amoeba population in soil under desert shrubs in the Negev desert, Israel. Appl Environ Microbiol 71:2053–2060

Ronca S, Ramond J-B, Jones BE, Seely M, Cowan DA (2015) Namib Desert dune/interdune transects exhibit habitat-specific edaphic bacterial communities. Front Microbiol 6:845

Roux S, Enault F, Robin A, Ravet V, Personnic S, Theil S et al (2012) Assessing the diversity and specificity of two freshwater viral communities through metagenomics. PLoS One 7:e33641. https://doi.org/10.1371/journal.pone.0033641

Roux S, Enault F, Hurwitz BL, Sullivan MB (2015) VirSorter: mining viral signal from microbial genomic data. PeerJ 3:e985. https://doi.org/10.7717/peerj.985

Roux S, Chan L-K, Egan R, Malmstrom RR, McMahon KD, Sullivan MB (2017) Ecogenomics of virophages and their giant virus hosts assessed through time series metagenomics. Nat Commun 8:1–12

Roux S, Adriaenssens EM, Dutilh BE, Koonin EV, Kropinski AM, Krupovic M, Kuhn JH, Lavigne R, Brister JR, Varsani A et al (2018) Minimum information about an uncultivated virus genome (MIUViG). Nat Biotechnol 37:29–37

Roy K, Ghosh D, DeBruyn JM, Dasgupta T, Wommack KE, Liang X, Wagner RE, Radosevich M (2020) Temporal dynamics of soil virus and bacterial populations in agricultural and early plant successional soils. Front Microbiol 11:1494. https://doi.org/10.3389/fmicb.2020.01494

Schimel JP (2018) Life in dry soils: effects of drought on soil microbial communities and processes. Annu Rev Ecol Evol Syst 49:409–432

Schloss PD, Handelsman J (2006) Toward a census of bacteria in soil. PLoS Comput Biol 2:e92. https://doi.org/10.1371/journal.pcbi.0020092

Schulz F, Alteio L, Goudeau D, Ryan EM, Yu FB, Malmstrom RR, Blanchard J, Woyke T (2018) Hidden diversity of soil giant viruses. Nat Commun 9:1–9

Schulze-Makuch D, Wagner D, Kounaves SP et al (2018) Transitory microbial habitat in the hyperarid Atacama Desert. Proc Natl Acad Sci U S A 115:2670–2675

Scola V, Ramond J-B, Frossard A, Zablocki O, Adriaenssens EM, Johnson RM, Seely M, Cowan DA (2018) Namib desert soil microbial community diversity, assembly, and function along a natural xeric gradient. Microb Ecol 75:193–203. https://doi.org/10.1007/s00248-017-1009-8

Segobola J, Adriaenssens E, Tsekoa T, Rashamuse K, Cowan D (2018) Exploring viral diversity in a unique South African soil habitat. Sci Rep 8:111

Sepulveda BP, Redgwell T, Rihtman B, Pitt F, Scanlan DJ, Millard A (2016) Marine phage genomics: the tip of the iceberg. FEMS Microbiol Lett 363:fnw158

Shao Q, Trinh JT, McIntosh CS, Christenson B, Balázsi G, Zeng L (2016) Lysis-lysogeny coexistence: prophage integration during lytic development. Microbiol Open 6:e00395

Srinivasiah S, Bhavsar J, Thapar K, Liles M, Schoenfeld T, Wommack KE (2008) Phages across the biosphere: contrasts of viruses in soil and aquatic environments. Res Microbiol 159:349–357

Srinivasiah S, Lovett J, Polson S, Bhavsar J, Ghosh D, Roy K, Fuhrmann JJ, Radosevich M, Wommack KE (2013) Direct assessment of viral diversity in soils by random PCR amplification of polymorphic DNA. Appl Environ Microbiol 79:5450–5457

Srinivasiah S, Lovett J, Ghosh D, Roy K, Fuhrmann JJ, Radosevich M, Wommack KE (2015) Dynamics of autochthonous soil viral communities parallels dynamics of host communities under nutrient stimulation. FEMS Microbiol Ecol 91:fiv063

Staley JT, Konopka A (1985) Measurement of in situ activities of nonphotosynthetic microorganisms in aquatic and terrestrial habitats. Annu Rev Microbiol 39:321–346
Sullivan MJ, Petty NK, Beatson SA (2011) Easyfig: a genome comparison visualizer. Bioinformatics 27:1009–1010
Swanson MM, Fraser G, Daniell TJ, Torrance L, Gregory PJ, Taliansky M (2009) Viruses in soils: morphological diversity and abundance in the rhizosphere. Ann Appl Biol 155:51–60
Swanson MM, Reavy B, Makarova KS, Cock PJ, Hopkins DW, Torrance L, Koonin EV, Taliansky M (2012) Novel bacteriophages containing a genome of another bacteriophage within their genomes. PLoS One 7:e40683
Thurber RV (2009) Current insights into phage biodiversity and biogeography. Curr Opin Microbiol 12:582–587
Tithi SS, Aylward FO, Jensen RV, Zhang L (2018) FastViromeExplorer: a pipeline for virus and phage identification and abundance profiling in metagenomics data. PeerJ 6:e4227
Torrents E (2014) Ribonucleotide reductases: essential enzymes for bacterial life. Front Cell Infect Microbiol 4:52
Trinh JT, Székely T, Shao Q, Balázsi G, Zeng L (2017) Cell fate decisions emerge as phages cooperate or compete inside their host. Nat Commun 8:14341
Trubl G, Jang HB, Roux S, Emerson JB, Solonenko N, Vik DR, Solden L, Ellenbogen J, Runyon AT, Bolduc B, Woodcroft BJ, Saleska SR, Tyson GW, Wrighton KC, Sullivan MB, Rich VI (2018) Soil viruses are underexplored players in ecosystem carbon processing. mSystems 3: e00076-18
Trubl G, Hyman P, Roux S, Abedon ST (2020) Coming-of-age characterization of soil viruses: a user's guide to virus isolation, detection within metagenomes, and viromics. Soil Syst 4:23
Valverde A, Makhalanyane TP, Seely M, Cowan DA (2015) Cyanobacteria drive community composition and functionality in rock-soil interface communities. Mol Ecol 24:812–821
Van Goethem MW, Swenson TL, Trubl G, Roux S, Northen TR (2019) Characteristics of wetting-induced bacteriophage blooms in biological soil crust. MBio 10:e02287–e02219
Verneau J, Levasseur A, Raoult D, La Scola B, Colson P (2016) MG-digger: an automated pipeline to search for giant virus-related sequences in metagenomes. Front Microbiol 7:428
Waldor MK, Mekalanos JJ (1996) Lysogenic conversion by a filamentous phage encoding cholera toxin. Science 272:1910–1914
Ward N, Staley JT, Fuerst JA et al (2006) The order Planctomycetales, including the genera *Planctomyces*, *Pirellula*, *Gemmata* and *Isosphaera* and the candidatus genera *Brocadia*, *Kuenenia* and *Scalindua*. In: Dworkin M (ed) The prokaryotes, vol 7. Springer, New York, pp 757–793
Warwick-Dugdale J, Buchholz HH, Allen MJ, Temperton B (2019) Host-hijacking and planktonic piracy: how phages command the microbial high seas. Virology 16:15
Whitford WG (2002) Ecology of desert systems, 1st edn. Academic, San Diego
Williamson KE, Radosevich M, Wommack KE (2005) Abundance and diversity of viruses in six Delaware soils. Appl Environ Microbiol 71:3119–3125
Williamson KE, Mark Radosevich M, Smith DW, Wommack KE (2007) Incidence of lysogeny within temperate and extreme soil environments. Environ Microbiol 9:2563–2574
Williamson SJ, Allen LZ, Lorenzi HA, Fadrosh DW, Brami D et al (2012) Metagenomic exploration of viruses throughout the Indian Ocean. PLoS One 7:e42047
Williamson KE, Fuhrmann JJ, Wommack KE, Radosevich M (2017) Viruses in soil ecosystems: an unknown quantity within an unexplored territory. Annu Rev Virol 4:201–219
Wilpiszeski RL, Aufrecht JA, Retterer ST, Sullivan MB, Graham DE, Pierce EM, Zablocki OD, Palumbo AV, Elias DA (2019) Soil aggregate microbial communities: towards understanding microbiome interactions at biologically relevant scales. Appl Environ Microbiol 85:e00324-19
Wommack KE, Colwell RR (2000) Virioplankton: viruses in aquatic ecosystems. Microbiol Mol Biol Rev 64:69–114

Yang Q, Gao C, Jiang Y, Wang M, Zhou X, Shao H, Gong Z, McMinn A (2019) Metagenomic characterization of the viral community of the South Scotia Ridge. Viruses 11:95. https://doi.org/10.3390/v11020095

Zablocki O, van Zyl L, Adriaenssens EM, Rubagotti E, Tuffin M, Cary SC, Cowan D (2014) High-level diversity of tailed phages, eukaryote-associated viruses, and virophage-like elements in the metaviromes of Antarctic soils. Appl Environ Microbiol 80:6888–6897

Zablocki O, Adriaenssens EM, Cowan D (2016) Diversity and ecology of viruses in hyperarid desert soils. Appl Environ Microbiol 82:770–777

Zablocki O, Adriaenssens EM, Frossard A, Seely M, Ramond J-B, Cowan D (2017) Metaviromes of extracellular soil viruses along a Namib Desert aridity gradient. Genome Announc 5:e01470–e01416. https://doi.org/10.1128/genomeA.01470-16

Zhang W, Zhou J, Liu T, Yu Y, Pan Y, Yan S, Wang Y (2015) Four novel algal virus genomes discovered from Yellowstone Lake metagenomes. Sci Rep 5:15131

Zhang J, Gao Q, Zhang Q, Wang T, Yue H, Wu L, Shi J, Qin Z, Zhou J, Zuo J, Yang Y (2017) Bacteriophage–prokaryote dynamics and interaction within anaerobic digestion processes across time and space. Microbiome 5:57

Zhao G, Wu G, Lim ES, Droit L, Krishnamurthy S, Barouch DH, Virgin HW, Wang D (2017) VirusSeeker, a computational pipeline for virus discovery and virome composition analysis. Virology 503:21–30

Zheng Q, Chen Q, Xu Y, Suttle CA, Jiao N (2018) A virus infecting marine photoheterotrophic alphaproteobacteria (*Citromicrobium* spp.) defines a new lineage of ssDNA viruses. Front Microbiol 9:1418. https://doi.org/10.3389/fmicb.2018.01418

Zheng T, Li J, Ni Y, Kang K, Misiakou M-A, Imamovic L, Chow BKC, Rode AA, Bytzer P, Sommer M, Panagiotou G (2019) Mining, analyzing, and integrating viral signals from metagenomic data. Microbiome 7:42

C, N, and P Nutrient Cycling in Drylands

7

Karen Jordaan, Karina Stucken, and Beatriz Díez

Abstract

Drylands play a significant role in the global biogeochemical cycling of nutrients (carbon, nitrogen, and phosphorus) through abiotic (geological, atmospheric, and hydrological) and biotic (animals, insects, plants, and microorganisms) pathways. They act as important carbon reservoirs and are estimated to store over 30% of the global soil organic carbon reserve. However, nitrogen and phosphorus availability are major limiting factors for biological activity in these oligotrophic environments, affecting community structure, species diversity, and other ecosystem functions (e.g., nutrient cycling and productivity). Nutrient cycling in desert soils is primarily achieved by plant and microbial communities, in particular soil microbial communities, biological soil crusts, hypoliths, and endoliths. Drylands are highly sensitive and prone to disturbance and land degradation resulting from desertification. Changes induced by climate (e.g., precipitation and temperature), structural and temporal variability (nutrient accumulation and distribution of minerals, seasonal variation, and differences in turnover rates), and human activity often alter nutrient cycles that negatively affect the structure and function of these ecosystems (e.g., decreasing carbon storage capacity, increasing NOx emissions, and reducing phosphorus cycling).

K. Jordaan
Department of Molecular Genetics and Microbiology, Biological Sciences Faculty, Pontifical Catholic University of Chile, Santiago, Chile

K. Stucken
Food Engineering Department, University of La Serena, La Serena, Chile

B. Díez (✉)
Department of Molecular Genetics and Microbiology, Biological Sciences Faculty, Pontifical Catholic University of Chile, Santiago, Chile

Center for Climate and Resilience Research (CR)2, Santiago, Chile
e-mail: bdiez@bio.puc.cl

J.-B. Ramond, D. A. Cowan (eds.), *Microbiology of Hot Deserts*, Ecological Studies 244, https://doi.org/10.1007/978-3-030-98415-1_7

Comprehending the extent, nature, magnitude, and reversibility of such changes is urgent, given the global importance of drylands in terms of carbon sequestration, greenhouse gas emissions, ecology and biodiversity, and human habitation.

Keywords

Drylands · Carbon · Nitrogen · Phosphorus · Nutrient cycling

The focus of this chapter is motivated by the high sensitivity of drylands to perturbations affecting the past and current nutrient cycling of most of these ecosystems around the world. It shows how changes in the abiotic and mostly biotic components and their interactions of each particular dryland's microclime ecosystems can affect the dynamics of their nutrient cycling both in time and space. Thus, in this chapter, we present a general discussion of the dynamics of the soil C, N, and P cycles in drylands with special attention to the role of Biological Soil Crusts, the influence of hydration-desiccation pulses, and the impact of Climate Change on nutrient cycling.

7.1 Carbon in Drylands Soils

Studies have indicated that drylands currently act as a large carbon (C) sink and play a more significant role in the terrestrial C balance than previously expected (Lal 2001; Wohlfahrt et al. 2008; Rotenberg and Yakir 2010). Dryland ecosystems are an important component of the global C cycle as they store ~32% (~470 Pg; Plaza et al. 2018a) of the global soil organic carbon (SOC), account for about one-third of the global vegetation C storage (Allen-Diaz et al. 1996), and represent 30–35% of terrestrial net primary production (NPP; Field et al. 1998). Drylands, therefore, dominate the positive global land CO_2 sink, although large fractions of the interannual variability (IAV) of net CO_2 flux have been mostly associated with variations in gross primary production (GPP; Beer et al. 2010; Yao et al. 2020). Model projects showed that the majority of dryland GPP variability is attributed to precipitation and air temperature (Ahlström et al. 2015; Zhang et al. 2016a; Yao et al. 2020), although other environmental variables such as changes in ecosystem types also play relatively small roles in regulating GPP variability (Yao et al. 2020). With the increasing frequency of extreme climatic events, the IAV of GPP is also projected to increase and will likely cause significant impacts on the global C cycle (Zscheischler et al. 2014). Although the physical-chemical properties of dryland soils limit their potential to store C (Ewing et al. 2006; Serrano-Ortiz et al. 2012; Weil and Brady 2017; Plaza et al. 2018a), their large coverage of the global land area, and the fact that many of these soils have been degraded, means that drylands have the greatest potential to sequester C (Scurlock and Hall 1998; Rosenberg et al. 1999). In fact, the amount of SOC present in drylands is ~42 times more than what is added into the atmosphere through anthropogenic activities, estimated at 11.3 Pg C year^{-1} in 2017

(Lal 2004a; Le Quéré et al. 2018). Studies have indicated that drylands currently act as a large C sink and play a more significant role in the terrestrial C balance than previously expected (Lal 2001; Wohlfahrt et al. 2008; Rotenberg and Yakir 2010).

In this chapter, we present a general discussion of the dynamics of the soil C cycle in drylands with special attention to C sequestration and climate change in arid environments. Although greenhouse gases methane (CH_4) and nitrous oxide (N_2O) are important components in the C cycle, carbon dioxide (CO_2) is the most prevalent greenhouse gas in the atmosphere and will be the primary gas considered when discussing C sequestration.

Soil C pools in arid environments exist as three main components: (1) soil organic C (SOC); (2) soil inorganic C (SIC); and (3) biomass organic C (Serrano-Ortiz et al. 2012). Important features and current stocks will be briefly discussed for each pool in the following sections.

7.1.1 Soil Organic Carbon in Drylands Soils

SOC is a strong determinant of soil quality, particularly in arid environments, as it influences the physical, chemical, and biological properties of soil (Gaitán et al. 2019) and is critical for improving soil fertility (Zhang and Shao 2014). As the SOC pool is strongly affected by precipitation and temperature (Lal 2002), SOC storage tends to decrease exponentially with an increase in temperature, but increases with increments in soil water content (Lal 2002; Follett et al. 2012; Yan et al. 2017; Chatterjee et al. 2019). Consequently, dryland soils contain small amounts of SOC, often less than 0.5% of the soil mass resulting in typical densities of 0–15 kg m^{-2} (Lal 2002, 2004a). However, due to the vast extension of dryland ecosystems soils are estimated to contain ~ 646 ± 9 Pg of SOC to a 2-m depth, representing about 32% of the world's total SOC pool (Plaza et al. 2018a; Lal 2019).

Arid and hyper-arid soils contain ~113 Pg organic C, whereas semiarid and sub-humid areas store ~ 318 Pg SOC (Plaza et al. 2018b). Studies have demonstrated that the SOC density is lower for bare soils (1–3 kg C m^{-2}; Rasmussen 2006; Woomer et al. 2004) compared to shrublands (2–6 kg C m^{-2}; Chen et al. 2007; Wiesmeier et al. 2011) and grasslands (5 ± 2 kg C m^{-2}; Chen et al. 2007; He et al. 2008).

7.1.2 Soil Inorganic C in Drylands Soils

SIC is also a critical component of arid regions and plays an important role in carbon sequestration and climate alleviation (Lal 2004b). Dryland SIC pools are estimated to be higher (1237 ± 15 Pg; Plaza et al. 2018a) than SOC pools and may even exceed SOC by a factor of 10 in some arid areas (Schlesinger 1985, 2006; Scharpenseel et al. 2000). The SIC stocks are generally highest for arid (487 ± 7 Pg) and semiarid (456 ± 7 Pg) environments (Plaza et al. 2018a) where the SIC pool is mainly present as lithogenic inorganic carbonates (LIC) that originates as detritus from parent materials (i.e., limestone), and pedogenic inorganic carbonates (PIC) formed by dissolution and re-precipitation of LIC or by the dissolution of CO_2 into bicarbonate (HCO_3^-) followed by precipitation with Ca^{2+} and/or Mg^{2+} originating from non-LIC minerals (e.g., silicate weathering, dust, and fertilizers) (Marion 1989; Schlesinger 1985, 2002; Tan et al. 2014; Wang et al. 2015; Sahrawat 2003; An et al. 2019).

These carbonates tend to precipitate at relatively shallow depths as a result of low precipitation and poor leaching of soils (Gocke et al. 2011). Consequently, about 80% of the global SIC pool is found in drylands (Eswaran et al. 2000; Plaza et al. 2018a). Reported SIC stocks can be modified significantly taking into account anthropogenic practices (Serrano-Ortiz et al. 2012). For example, the SIC pool can be affected by land management practices such as afforestation, irrigation, fertilization, and liming (Sanderman 2012; Gao et al. 2017). Such activities can cause elevated levels of CO_2 in soils, resulting in significant formation and precipitation of secondary carbonates, thereby contributing to soil C sequestration (Denef et al. 2008; Sanderman 2012; Gao et al. 2017).

7.1.3 Biomass Organic C in Dryland Soils

7.1.3.1 Vegetation

Dryland vegetation is highly variable and can range from barren or sparsely vegetated desert to grasslands, shrublands and savannahs, croplands, and dry woodlands (FAO 2004; Lal 2004a; Mander et al. 2017). Vegetation cover in drylands is influenced by various factors including drought stress, variations in annual temperatures, and precipitation intensity and frequency (FAO 2004; Lal 2004a; Mander et al. 2017). These factors strongly impact above-ground biomass productivity and therefore the level of SOC in soil.

Carbon storage for dryland vegetation is overall low (Eswaran et al. 2000), with the global average and maximum stock estimated at ~65 and 81 Pg C, respectively (Safriel et al. 2005; Serrano-Ortiz et al. 2012). However, these values can be modified significantly when considering anthropogenic activities, like grazing and desertification, which may reduce the biomass C pool by 10–20% of the given value (Serrano-Ortiz et al. 2012; An et al. 2019). Biomass C density can differ between arid regions, depending on vegetation coverage and land use. For example, hyper-arid and arid deserts have the capacity to store 0.04–0.40 kg C m^{-2} in biomass (Woomer et al. 2004; Fan et al. 2008; Perez-Quezada et al. 2011), while grasslands and sub-humid forested drylands have a biomass C density of ~1 (He et al. 2008) and 4–5 kg C m^{-2} (Glenday 2008), respectively.

7.1.3.2 Microbial Autotrophs

As most drylands represent sparse vegetation cover, autotrophic microbes (soil and BSC) act as important primary producers (Liu et al. 2018). Autotrophic CO_2 fixation processes are significant for C accumulation in dryland soils that may influence ecosystem succession processes (Agarwal et al. 2017; Liu et al. 2018). In fact, microbial autotrophy accounts for ~ 4% of total C sequestered by terrestrial ecosystems per year, of which drylands comprise a substantial fraction (Yuan et al. 2012). Microbial autotrophs are distributed amongst both bacteria and archaea with highly diverse phylogeny, metabolic activities, and ecological variants (Hügler and Sievert 2011). For example, studies have shown that autotrophic microbes in drylands belong to the taxa Actinobacteria, Proteobacteria, Chloroflexi, Acidobacteria, Gemmatimonadetes, Firmicutes, Thaumarchaeota, Nitrospirae, Planctomycetes, and Bacteroidetes (Yang et al. 2017; Liu et al. 2018).

Even though the Calvin cycle is the predominant pathway utilized by microbes in nutrient-rich conditions, many autotrophic microbes in drylands fix C through the energy-conserving reductive citrate cycle, reductive acetyl-CoA cycle and 3-hydroxypropionate bi-cycle pathway (Agarwal et al. 2017; Liu et al. 2018). Liu et al. (2018) were able to demonstrate the capacity of desert microbial autotrophs to directly incorporate $^{13}CO_2$ into SOC using the three above-mentioned pathways, which accounted for ~4% of the atmospheric CO_2 absorbed by desert soil. In addition, the study showed that the efficiency of autotrophic CO_2 fixation was impacted by soil properties, the autotrophic composition, and the abundance of genes associated with the CO_2 fixation pathways. Their results highlighted the underestimated importance of microbial autotrophy in the C cycle and storage in drylands.

7.1.4 Carbon Sequestration and Loss: The Impact of Abiotic and Biotic Factors

Carbon sequestration in drylands involves the transfer of atmospheric CO_2 into both SOC and SIC pools via management of vegetation, soil, and water resources that lead to a positive C budget (Lal 2009). Mechanisms involved in SOC and SIC sequestration, and the rate of each process are discussed by Lal (2009). In general, primary producers/plants convert atmospheric CO_2 to complex organic molecules through photosynthesis; these molecules enter the soil C cycle as decaying organic matter (OM; Batlle-Aguilar et al. 2011; Serrano-Ortiz et al. 2012). A significant portion of the OM is directly used to sustain energy for pedo-and microfauna metabolism and is released back to the atmosphere as a by-product of autotrophic (plant) and heterotrophic (e.g., microbial) respiration (Batlle-Aguilar et al. 2011; Thomey et al. 2014). Some of the soil C is assimilated by vegetation and finally transferred to the soil as plant litter, becoming part of SOM (Porporato et al. 2003). In addition to biological processes, there is growing evidence to suggest that abiotic processes contribute to CO_2 fluxes in soils and may even dominate the flux during dry seasons (Kowalski et al. 2008; Serrano-Ortiz et al. 2012; Cueva et al. 2019). The exact physical and biogeochemical mechanisms that promote CO_2 capture in soils are still uncertain and warrant future research to accurately understand and quantify the C balance, from local to global scales (Cueva et al. 2019).

Soil C stock in drylands is strongly impacted by biotic and abiotic processes, factors (e.g., climate, vegetation, soil properties), causes (e.g., urbanization, wildfires), and interactions among them (Lal 2019), that primarily control C fluxes between the atmosphere and both SOC and SIC pools via management of vegetation, soil and water resources that lead to a positive C budget (Lal 2019). The C stock is also affected by erosional processes, both aeolian and hydrologic, which influence transport, redistribution, and deposition of C over the landscape (Lal 2019).

Significant progress has been made to understand the controls on regional and global patterns of soil SOC in drylands. Correlative analyses and/or structural equation modeling (SEM) across natural environmental gradients indicate that SOC is largely controlled by climatic factors such as precipitation and temperature, as well as soil properties and plant productivity (White et al. 2009; He et al. 2014; Wang et al. 2014a–d; Mureva et al. 2018; Gaitán et al. 2019; Smith and Waring 2019; Zhu et al. 2019). For example, precipitation influences SOC storage by constraining primary productivity and decomposition (Wynn et al. 2006; Yang et al. 2007), whereas higher temperatures accelerate the microbial decomposition of SOM, thereby causing C loss (Giardina and Ryan 2000).

In addition to climate, soil properties, such as soil mineralogical characteristics (Smith and Waring 2019), pH (Min et al. 2014; Ou et al. 2017), bulk density (Feng et al. 2002), total phosphorus (Tian et al. 2017), and soil moisture content (Wang et al. 2014a–d), are strongly related to SOC. It should be noted that the main factors controlling SOC concentration may differ between regions and even within the same ecosystem (Cable et al. 2011), or can be scale dependent (Dai and Huang 2006; Liu et al. 2006; Evans et al. 2011; Wang et al. 2013; Qin et al. 2016). An important reason for such discrepancies is that soil CO_2 efflux is a combined result of biotic processes and abiotic factors (Ma et al. 2010, 2017; Gaitán et al. 2019), each of which exhibits its own flux behavior at various time scales and responds differently to the environment (Ryan and Law 2005; Li et al. 2010) to control SOC concentration and dynamics.

7.2 Nitrogen in Dryland Soils

Second to water, nitrogen (N) availability is the main limiting factor of biological activity in oligotrophic arid environments (Whitford 2002). Microorganisms in open soils, root systems, or cryptic niches (e.g., biological soil crusts, hypoliths, and endoliths) are the main providers of atmospheric fixed N but also act as the key players in other N-cycling processes such as nitrification and denitrification.

7.2.1 Biological Nitrogen Fixation (BNF) as N Input in Drylands

The main nitrogen input in many hyperarid environments comes from biological nitrogen fixation (BNF) (Johnson et al. 2005; Pointing and Belnap 2012). The BNF is an anaerobic process that is performed only by prokaryotes. The oxygen-sensitive nitrogenase complex catalyzes the reduction of atmospheric N_2 to bioavailable ammonia (NH_3). This is a highly energy-demanding process for cells (16 ATP molecules and 8 electrons per reduced N_2 molecule; Zehr et al. 2003), particularly in desert environments that are nutrient deprived. Acetylene reduction assays (ARA) are commonly used to measure the nitrogenase activity in oligotrophic environments *via* the reduction of acetylene to ethylene (Aranibar et al. 2003) and calculating the rates by which ^{15}N is incorporated into cells (Mayland and McIntosh 1966; Caputa et al. 2013; Alcamán-Arias et al. 2018).

The most widely used phylogenetic marker to study N-fixation is the *nifH* gene as its phylogeny is fairly concordant with that of the 16S rRNA gene (Zehr et al. 2003). Nitrogen-fixing bacteria (diazotrophs) belong to several phyla including the autotrophic Cyanobacteria and Chlorobi, as well as to heterotrophic Actinobacteria, Firmicutes, and Proteobacteria (α-, γ-, ε-classes; see Hartley et al. 2007); all of which have been detected in hot and cold deserts (Cowan et al. 2011; Ronca et al. 2015; Crits-Christoph et al. 2016; Lacap-Bugler et al. 2017; Meslier et al. 2018). Also, several archaeal taxa such as *Methanobacteria*, *Methanococci*, and *Methanomicrobia* have the potential to fix N (Chan et al. 2013; Wei et al. 2016) as recently demonstrated in wetland soils (Bae et al. 2018).

Biological soil crusts (BSCs) are abundant in cold and hot deserts (Belnap 2001a; Pointing and Belnap 2012; Makhalanyane et al. 2015; Chap. 3) and can contribute to ~30% of biologically fixed nitrogen in terrestrial ecosystems (Elbert et al. 2012). In cyanobacterial dominated crusts N is primarily fixed by filamentous heterocystous (e.g., *Anabaena, Nostoc, Scytonema*) and non-heterocystous (e.g., *Microcoleus, Chroococcidiopsis, Phormidium)* cyanobacteria (Belnap 2002; Wei et al. 2016; Lacap-Bugler et al. 2017; Mogul et al. 2017), although other organisms such as chlorophyte algae, heterotrophic bacteria, fungi, mosses and/or lichens are also present (Belnap 2003; Pointing and Belnap 2012). Studies have shown that N-fixation in cyanobacterial–algal crusts from the Tengger and the Gurbantunggut Deserts presented the highest nitrogenase activity (up to 16.6 mmol m^{-2} h^{-1}), followed by lichens (up to 6.9 mmol m^{-2} h^{-1}) and mosses (up to 2.6 mmol m^{-2} h^{-1}; Wu et al. 2009; Su et al. 2011). In contrast, lichen-dominated crusts in the Colorado Plateau were found to be the main N-fixers and the amount of N fixed was ten-fold higher than that of cyanobacteria-dominated crusts (Belnap 2002). Looking at ammonium oxidation (AO) in two types of BSCs from the Colorado Plateau, Johnson et al. (2005) found that N_2 fixation potential was appreciably higher (6.5–48 μmol C_2H_2 m^{-2} h^{-1}) in dark crusts, dominated by heterocystous cyanobacteria, than those in light crusts.

N-fixation and N-fixing activity in drylands are influenced by several environmental factors, of which water availability and temperature appear to be the most important parameters (Belnap 2001b; Hartley et al. 2007). Cyanobacteria and other heterotrophic bacteria are physiologically active only when water is available, thereby controlling photosynthesis and ultimately N-fixation (Belnap 2001b; Hartley et al. 2007). Moisture levels needed to initiate and optimize N-fixation vary widely with species, habitats, and pre-existing conditions (Belnap 2001b). N-fixation rates are also limited by temperature extremes. Most nitrogenase activity occurs at –5 to 30 °C, with the optimum at 20–28 °C for the majority of drylands (Belnap 2001b; Wu et al. 2009; Schwabedissen et al. 2017). Low temperatures can reduce photosynthetic rates and thus available ATP and reductant pools, creating a lag time before

N-fixation is initiated (Belnap 2001b). In contrast, at higher temperatures (i.e., above minimum temperature for species) N-fixation rates show a strong, positive response to increasing air temperature until an upper limit is reached, after which rates quickly decline (Belnap 2001b).

Other factors that affect N-fixation in drylands include pH, salinity, and nutrient content (Hartley et al. 2007). Nitrogenase activity in soil cyanobacteria and BSCs are greatest above pH 7 (Hartley et al. 2007; Schwabedissen et al. 2017); however, nitrogenase activity is reduced at high (8–10) and low pH (see Belnap 2001a for further explanations). Effects of salinity on nitrogenase activity in drylands are not well studied and show mixed results. For example, N-fixing cyanobacteria in Utah deserts preferred soils with high electrical conductivity (70 dS m^{-1}; Anderson et al. 1982). In contrast, experimental addition of NaCl to cyanobacterial lichen crusts in the Chihuahuan desert inhibited nitrogenase activity (Delwiche and Wijler 1956). Nutrient effects on nitrogenase activity vary in drylands and largely depend on the element. Phosphorus (P) and potassium (K) additions can stimulate cyanobacterial nitrogenase activity, likely through the stimulation of ATP synthesis (Dodds et al. 1995). Low amounts of zinc (Zn), cobalt (Co), molybdenum (Mo), and iron (Fe) can also stimulate cyanobacterial nitrogenase activity, whereas higher levels of the same elements have adverse effects (Granhall 1981; Dodds et al. 1995). Glucose (i.e., carbon) additions to soils and BSCs have been shown to increase heterotrophic nitrogenase activity, suggesting that C sources are essential in N dynamics (Hartley and Schlesinger 2002; Billings et al. 2003; Hartley et al. 2007).

Hypolithic microbial communities (biofilms on quartz and translucent rocks) represent major contributions to organic biomass in extreme deserts where communities are dominated by cyanobacterial photoautotrophs (Chan et al. 2012), although red Chloroflexi-dominated hypoliths have been encountered in the Atacama Desert (Lacap et al. 2011). In the Mojave Desert, hypolithic dinitrogen fixation potential was demonstrated using ARA measurements (Schlesinger et al. 2003), while a recent shotgun metagenome study revealed that Namib Desert hypolithic communities could mediate the full N-cycle (apart from ANAMMOX; Vikram et al. 2016). These results were later supported by stable N isotope ($\delta^{15}N$) measurements of Namib Desert hypolithic biomass, and surface and subsurface soils over a 3 year period across dune and gravel plain biotopes (Ramond et al. 2018). The authors demonstrated that photoautotrophic hypolithic communities are the main contributors to N_2 fixation for plant productivity events in the Namib Desert. In accordance with these findings, metatranscriptome profiling of Namib Desert soils revealed an active microbial community dominated by non-photosynthetic bacteria where nitrate appeared to be the source of bioavailable N (Leon-Sobrino et al. 2019).

7.2.2 Atmospheric N Deposition and N Discharges as N Inputs in Drylands

Although BNF is the most studied and described process to generate N in drylands, other inputs such as atmospheric depositions and accumulations, environmental runoff, discharge of N-rich untreated wastewater, and/or agriculture processes are also important N sources.

Drylands act as large reservoirs for nitrate-rich salts. N is delivered to desert landscapes by either wet or dry atmospheric deposition of particles produced by gas to particle conversion, and by biological N_2 fixation (Böhlke et al. 1997; Michalski et al. 2004; Graham et al. 2008). Wet deposition of N (e.g., precipitation, sea spray, and fog) has been implicated as sources of NH_4^+, NO_3^- and other salts in the Turpan-Hami area, northwestern China (Qin et al. 2012), and the Atacama (Michalski et al. 2004; Wang et al. 2014c, d), Mojave (Michalski et al. 2004), and Chihuahuan Deserts (Báez et al. 2007). Additionally, drylands receive external N inputs via the discharge of untreated, ammonia-rich wastewater (e.g., Negev and Tengger Deserts), or through agricultural processes where poultry manure is used for soil amendments (Abed et al. 2010; Posmanik et al. 2017; Zuo et al. 2018). Large N inputs have been deposited in areas surrounding the Gurbantunggut Desert following farming activities (Li et al. 2012a). All these practices induce nitrification and denitrification with the subsequent release of greenhouse gasses (N_2O).

7.2.3 Nitrogen Losses in Drylands

Biological fixed N does not accumulate over time and is lost to the atmosphere from drylands through several mechanisms including ammonia volatilization, wind erosion, leaching, nitrification, and denitrification (Peterjohn and Schlesinger 1990; Walvoord et al. 2003; McCalley and Sparks 2009; Wang et al. 2014a, b; Jin et al. 2015). Volatilization is considered to be the primary cause of N loss in deserts, with wind erosion contributing to 6.6% of total N loss (Peterjohn and Schlesinger 1990; McCalley and Sparks 2009). Volatile losses of N include nitric oxide (NO), nitrous oxide (N_2O), ammonia (NH_3), and dinitrogen (N_2) gas (Peterjohn and Schlesinger 1990). Of these compounds, NO, N_2O, and NH_3 are of particular consequence to the atmosphere affecting ozone levels and contributing to global warming (Logan et al. 1981; Peterjohn and Schlesinger 1990; van Amstel and Swart 1994). N loss from dryland soils through leaching was previously considered to be negligible, however, several studies challenged this assumption and demonstrated significant NH_3 leaching and accumulation in subsoil zones (Walvoord et al. 2003; Jin et al. 2015).

Nitrification and denitrification processes, especially in BSCs, are considered major contributors to N loss in dryland soils (Austin et al. 2004). BSCs have been shown to emit NO, N_2O, and nitrous acid (HONO) with amounts dependent on N_2 fixation rates, crust type, and season (Barger et al. 2005; Abed et al. 2013; Weber et al. 2015; Maier et al. 2018). Weber et al. (2015) showed that dark cyanobacteria-dominated crusts exhibit the highest NO and HONO emission fluxes and can even exceed those of neighboring uncrusted soils by 20 times. Using laboratory, field, and satellite measurement data, the authors estimated a global emission of reactive N from BSCs at ~1.7 Tg year^{-1} (1.1 Tg year^{-1} of NO-N and 0.6 Tg year^{-1} of HONO-N), corresponding to ~20% of global NO_x emissions from soils under natural vegetation. In the desert of northern Oman (Arabian Desert), N_2O gas was emitted from incomplete denitrification at very high potential rates of 387 ± 143 and 31 ± 6 μmol N m^{-2} h^{-1} from cyanobacterial and lichen crusts, respectively (Abed et al. 2013). This contributed up to 54–66% of the total produced gases during denitrification in both types of crusts. Although lower N_2O emissions have been detected from other drylands such as the Sonoran Desert (Guilbault and Matthias 1998), Great Basin (Mummey et al. 1994), and southeastern Utah (Belnap 2001b), their environmental implications should not be underestimated.

The potential for N loss from drylands is primarily affected by precipitation, substrate availability, and high soil-surface temperatures (McCalley and Sparks 2008, 2009; Yahdjian and Sala 2010). In the Mojave Desert, NH_3 dominated reactive N gas emissions were shown to be higher during the summer season, while a fall precipitation event yielded even larger N fluxes (McCalley and Sparks 2008). Further laboratory manipulations of environmental conditions on N gas emissions showed a large transient NH_3 pulse (~70–100 ng N m^{-2} s^{-1}) following water addition, presumably driven by an increase in soil NH_4 concentrations. As such, NO production was boosted with maximum NO flux rates of 34 ng N m^{-2} s^{-1}. Similar results have been observed for rainfall manipulated experiments in the Chihuahuan Desert (Hartley and Schlesinger 2000) and Patagonian steppe (Yahdjian and Sala 2010). N_2O flux in drylands is relatively constant over the dry seasons; however, N_2O emissions have been shown to considerably increase following precipitation and/or irrigation events (Hall et al. 2008; Chen et al. 2013; Liu et al. 2014; Yue et al. 2020). These brief "wetting-pulse" patterns of N_2O fluxes can account for the majority of annual N_2O emissions, making dryland N_2O patterns unique compared with other terrestrial ecosystems (Hu et al. 2017). Interestingly, studies have found that N gas fluxes are responsive to additions of labile C and elevated CO_2 (Schaeffer and Evans 2005; McCalley et al. 2011). Increased availability of C alters soil N dynamics by increasing biological (plant and microbial) activity and immobilization of N, thereby reducing N availability for nitrification and denitrification, and NH_3 volatilization (Schaeffer and Evans 2005; McCalley et al. 2011).

7.2.4 Nitrification and Denitrification in Desert Soils

Nitrification is a biological two-step process in which NH_3 is oxidized to nitrite (NO_2^-) followed by the oxidation of NO_2^- to NO_3^-. The transformation of ammonia to NO_2^- is usually the rate-limiting step of the process (Kowalchuk and Stephen 2001) performed by ammonia-oxidizing bacteria (AOB) and/or archaea (AOA) (reviewed by Offre et al. 2013; Stein 2019). Subsequent oxidation of NO_2^- to NO_3^- is performed by nitrite-oxidizing bacteria (NOB), mostly from the genera Nitrobacter and Nitrospira. Both steps produce energy coupled to ATP synthesis. Ammonia- and nitrite oxidation may also occur simultaneously in Comamox bacteria (Daims et al. 2015; van Kessel et al. 2015).

Nitrification potential for drylands has been demonstrated in both soils and BSCs across several deserts and drylands (e.g., Colorado Plateau, Great Basin, and Negev and Sonoran Deserts; Nejidat 2005; Strauss et al. 2012; Marusenko et al. 2013, 2015; Delgado-Baquerizo et al. 2016). Interestingly, results suggest that temperature and aridity are important modulators of AO communities and that AOA abundance dominates over that of AOB in bare soils at more extreme conditions (i.e., higher temperature and aridity; Marusenko et al. 2013; Delgado-Baquerizo et al. 2016). Conversely, vegetated dryland soils show a higher abundance and richness of AOB than AOA due to higher OM and available N contents (Delgado-Baquerizo et al. 2016). Collectively, results suggest that niche differentiation plays a key role between AOA and AOB communities (Marusenko et al. 2013, 2015; Zhou et al. 2016; Delgado-Baquerizo et al. 2016). Furthermore, AO estimates (mean of 110 μmol N m^{-2} h^{-1}) in BSCs from the Colorado Plateau and the Mojave, Sonoran, and Chihuahuan Deserts indicate that the magnitude of N cycling processes (e.g., nitrification) is subject to seasonal changes where soil temperature and moisture levels affect microbial activity (Strauss et al. 2012). Crust type and oxygen availability have also been implicated as important factors in ammonia oxidation fluxes (Johnson et al. 2005). For example, Johnson et al. (2005) found that AO rates (measured at depth of 2–3 mm) from light cyanobacterial crusts (53.4 ± 28.1 mmol N m^{-2} h^{-1} in light and 28.6 ± 13.0 mmol N m^{-2} h^{-1} in dark) were higher than for dark cyanobacterial crusts (42.0 ± 21.1 mmol N m^{-2} h^{-1} in light and 14.8 ± 10.2 mmol N m^{-2} h^{-1} in dark), irrespective of the incubation strategy (Johnson et al. 2005). It should be noted that potential rates were based on wet crusts and are likely to be overestimated.

Nitrate is removed from the immediate environment by processes such as denitrification, assimilation, or anaerobic ammonia oxidation (anammox). Denitrification, the biological process in which NO_2^-, NO, and N_2O are successively reduced to N_2 gas, is a facultative anaerobic process performed by heterotrophic bacteria (Williams et al. 1992). This process is highly dependent on NO_3^- and C source availability (Williams et al. 1992). In addition to bacteria, several fungi are able to perform denitrification during anaerobic respiration (Shoun et al. 2012). Drylands are not considered as denitrification hotspots, although denitrification potential and denitrifying enzyme activity have been detected in grasslands (Santa Barbara, CA; Homyak et al. 2016), semiarid ("El Romeral," Cajón del Maipo, Chile; Orlando et al. 2012) and hyperarid soils (Atacama Desert; Orlando et al. 2012), and BSCs (Abed et al. 2013). Studies suggest that semiarid soils have a higher denitrification potential (based on the presence of *nirK* and *nirS* genes) than hyperarid soils attributable to its higher soil moisture and nutrient content (Orlando et al. 2012). Denitrification rates in BSCs are likely to surpass those in uncrusted soils due to the formation of SOC in crusts (Barger et al. 2016; Brankatschk et al. 2013), however, contradictory results exist to support this hypothesis (see Barger et al. 2016 for full review on denitrification in BSCs).

7.3 Phosphorus in Dryland Soils

Phosphorus (P) is one of the less-abundant macronutrients in the lithosphere (0.1% of total) and then considered as a limiting nutrient (Bate et al. 2008), thus turnover of organic phosphorus (P_o) is critical for community structure, species diversity, and ecosystem processes (Runge-Metzger 1995; Sardans et al. 2012). P concentration is fundamental in the cycling of nutrients such as soil organic matter (SOM), nitrogen (N), and C (Stevenson and Cole 1999; Mackenzie et al. 2002; Vitousek et al. 2010; Griffiths et al. 2012; George et al. 2018), and vital to cellular organization (phospholipids) and genetic material (nucleic acids), intracellular signaling molecules, and for metabolism and energy transfer (ATP; Ruttenberg 2003; Butusov and Jernelöv 2013).

In mesic terrestrial environments, soil P is derived from three sources: (1) chemical and biological weathering of bedrock; (2) aeolian dust deposition; and (3) decomposition of biomass (Walker and Syers 1976; Lajtha and Schlesinger 1988; Okin et al. 2004; Turner et al. 2007; Yang and Post 2011; Porder and Ramachandran 2013). In most soils, P is the result of geochemical weathering of parent material (Lajtha and Schlesinger 1988), with relatively small inputs from aeolian deposition (Okin et al. 2004; Selmants and Hart 2010). Plants and microbes incorporate P into biomass and return it to the soil in organic forms, which can then be recycled by phosphatase enzymes to release inorganic phosphate for biological uptake (Walker and Syers 1976; Cross and Schlesinger 1995; Turner et al. 2007). Contrariwise, the input of P into dryland soils is primarily the result of atmospheric dust deposition (see Belnap 2011 for a comprehensive description), followed by biological weathering of parent materials (Reynolds et al. 2006; Belnap 2011), and then topsoil (0–15 cm) typically contain high concentrations of total P (Jobbagy and Jackson 2001; He et al. 2014).

7.3.1 P Stocks and Redistribution by Biological Processes in Drylands

Phosphorus stocks in dryland soils range from ~200 to 1200 mg P kg^{-1} (Tiessen et al. 1984; Turner et al. 2003a; Plaza et al. 2018a). More than 50% of the total P consists of labile inorganic and apatite P, that are mostly unavailable for plant or microbial uptake (Cross and Schlesinger 2001; Jones and Oburger 2011). Since P_o is mostly absent or at low concentrations, variation in P fractions can be closely related to the P_i content of parent materials (Neff et al. 2006; Buckingham et al. 2010). In addition, P_i is more likely to be found in mineral-associated forms such as oxides (e.g., aluminum and iron), clays, or carbonates (Sinsabaugh et al. 2008). In fact, calcium (Ca) minerals (e.g., calcium carbonate-and phosphate) are the predominant reservoir of P (Lajtha and Schlesinger 1988; Cross and Schlesinger 1995; Sims and Pierzynski 2005; Belnap 2011) in drylands due to reduced water inputs and high evapotranspiration.

The ability of P to change from one fraction to another is mainly controlled by mineralogy (Cross and Schlesinger 2001; Lajtha and Schlesinger 1988; Neff et al. 2006). Several different minerals play important roles in the stabilization and release of P (Buckingham et al. 2010) in the soil matrix. For example, studies have shown that carbonates sorbs P moderately due to its positive electrostatic charge, although the association can be reversed by changes in pH (Tiessen and Moir 1993), where goethite quickly sorbs P via ligand exchange until the sorption requirements of the mineral are met (Goldberg and Sposito 1985). Some minerals, such as aluminum (Al) and iron (Fe) oxides, may stabilize P through several mechanisms (Buckingham et al. 2010). These oxides can sorb P onto mineral surfaces that are reversible or may exchange ligands that sequester P into the physical structure of the oxides (Sollins et al. 1988). Amorphous minerals tend to sorb P more rapidly than crystalline materials (Ryan et al. 1985) and form tighter bonds with P because of their larger surface area (Carreira et al. 2006). Ultimately, the type and intensity of P-mineral bonds are important characteristics that determine its mobility and plant availability in soils (Buckingham et al. 2010).

In un-weathered or moderately weathered soils with neutral to alkaline pH, calcium-phosphates are the main supply of P_i. In contrast, Al and Fe phosphates and P_i bound and/or occluded by Al and Fe oxy(hydr)oxides predominate in acidic and more progressively weathered soils (Sims and Pierzynski 2005). In neutral and acidic soils, Al and Fe oxides and hydroxides exert a great impact on P availability given that various identified Fe and Al phosphates (e.g., wavellite, variscite, and strengite) are generally rare in occurrence (Harris 2002). In addition, the increased positive surface charge of the oxides forms strong covalent bonds (chemisorption) with the negatively charged P under acidic conditions, rendering it rather recalcitrant to exchange reactions (Jones and Oburger 2011). Nevertheless, changes in pH might directly or indirectly affect the oxides' surface potential and consequently P_i solubility. Also, LMW organic anions (e.g., gluconate, oxalate, etc.) released by P solubilizing organisms are capable of competing with P_i for sorption sites (Jones and Oburger 2011).

P_o influences P availability by contributing to the labile P_i pool, which is important to NPP as mentioned previously (Cross and Schlesinger 2001). Certain fractions of P_o are relatively labile and accessible for biological uptake (Haas et al. 1961; Wild and Oke 1966; Dalal 1977), while unavailable P_o is mineralized to labile P_i by simple autolysis or enzymatic phosphorylation (Cosgrove 1977). On average, P_o can comprise 30–65% of total P in mineral soils, while P_o approaches up to 90% in organic soils (Dalal 1977; Jones and Oburger 2011). The main P_o compounds in soil include inositol phosphates (>80%), phospholipids (0.5–7.0%), and nucleic acids (<3%; Dalal 1977; Quiquampoix and Mousain 2005; Turner et al. 2002; Jones and Oburger 2011). Other, less abundant forms of P_o include sugar-P (Anderson 1961), monophosphorylated carboxylic acids (Anderson and Malcolm 1974), and teichoic acids (Zhang et al. 1998).

The redistribution of P into various mineral and chemical fractions is impacted by mechanical and biological weathering, biological uptake and cycling, adsorption onto secondary minerals or soil constituents, or leaching (Walker and Syers 1976;

Tiessen et al. 1984; Ruttenberg 2003). These P pools are typically represented as soluble phosphate, P_o, and primary and secondary mineral fractions that vary in their potential mobility, relative quantities, and availability to biota (Hedley et al. 1982; Tiessen et al. 1984; Kuo 1996). The horizontal (e.g., wind, water, wildlife, livestock, human activities) and vertical (e.g., animals, insects, and plants) redistribution of soil P in dryland systems is imperative for plant productivity. Redistribution of P can determine the community, biomass, and distribution of a given plant community, which in turn affects animal distributions (Belnap 2011). For a comprehensive review on redistribution of soil P in drylands see Belnap (2011).

7.3.2 Abiotic and Biotic Control of P Cycling

Biogeochemical cycling of P in drylands is ultimately controlled by precipitation and temperature (Belnap 2011; Hou et al. 2018). The timing, intensity, and amount of water input directly affect the availability of P by influencing rates of geochemical reactions, ion diffusion, and biological activity (Belnap 2011). Because precipitation in arid environments is highly variable and inherently low, pulses of P-releasing activities vary on both spatial and temporal scales (Belnap 2011). While precipitation affects soil P leaching, temperature can directly enhance soil P sorption and desorption (Barrow 1983). Climate may also affect soil P availability through additional factors including: (1) different soil P forms; (2) soil properties such as soil particle size, pH, and SOM; and (3) plant and soil microbial activities (e.g., P uptake and return; Hou et al. 2018). For instance, high temperature facilitates the transformation of labile P and secondary mineral P to occluded P in soil (Barrow 1983; Siebers et al. 2017), whereas high precipitation may drive the loss of secondary mineral P or even occluded P in very humid environments (annual precipitation >2000 mm; Austin and Vitousek 1998). The interaction of precipitation and temperature can also affect the release of bound P (see Belnap 2011). Essentially, cool and wet conditions result in greater production of carbonic acid (H_2CO_3) that causes the soil pH to decrease, thus dissolving carbonates and increasing the transition of solid-phase P to solution-phase P (Lajtha and Schlesinger 1988; Magid and Nielsen 1992; Jungk and Claassen 1997; Miller et al. 2006a, b).

While the availability of P_i in dryland soils is mainly regulated by the dissolution properties of P-bearing minerals, the availability of P derived from P_o is largely governed by microbial activity (Jones and Oburger 2011). Soil microorganisms can solubilize P via three mechanisms including: (1) the active or passive release of protons, CO_2, and secondary organic metabolites (e.g., sugars, organic acid anions, amino acids, siderophores, enzymes, phenols); (2) the release of extracellular phosphatase enzymes (biochemical P_o mineralization); and (3) the release of P_o during substrate degradation (biological P_o mineralization) (Mcgill and Cole 1981; Jones and Oburger 2011). P incorporated into microbial biomass, to allow the use of organic C and root exudates for energy (Wu et al. 2007), may be temporarily immobilized but remains in a bioavailable form that can be released via microbial turnover (re-mineralization; Jones and Oburger 2011).

Approximately 1–50% of soil bacteria and 0.5–0.1% of soil fungi can be classified as P-solubilizing microorganisms (PSM; Kucey et al. 1989; Gyaneshwar et al. 2002); however, their population and activity vary with environment and depend upon various environmental factors (Chatli et al. 2008). P-solubilizing organisms are capable of hydrolyzing P_o and P_i compounds from insoluble P (Kalayu 2019). Among the PSM group, species from bacterial genera (*Bacillus, Pseudomonas, Rhizobium, Burkholderia, Achromobacter, Agrobacterium, Microccocus, Aereobacter, Flavobacterium*, and *Erwinia*), fungal general (*Penicillium, Aspergillus,* and *Trichoderma*), and actinomycetes (*Streptomyces* and *Streptoverticillium*) are significant in solubilizing phosphate (Rodriguez and Fraga 1999; Whitelaw 2000; Kumar et al. 2018; Kalayu 2019). Although the number of bacteria in soil classified as PSM generally outnumber those of fungi, the fungal isolates generally exhibit a greater P-solubilizing capacity (Banik and Dey 1982; Gyaneshwar et al. 2002). The mechanisms and processes involved in P mobilization by PSM in soil are discussed elsewhere (Jones and Oburger 2011; Richardson and Simpson 2011).

Microbial immobilization of P constitutes a significant component of the total soil P and is generally equivalent to, or exceeds that held in plant biomass (Richardson and Simpson 2011). Concentrations of microbial P in dryland soils typically account for 2–4% of total soil P (Lajtha and Schlesinger 1988; Xu et al. 2013; Perroni et al. 2014). Importantly, microbial P is a highly dynamic pool of soil P and release of P (i.e., orthophosphate and in organic forms) during the microbial biomass turnover is subject to significant change in response to environmental factors such as seasonal conditions, soil temperature and moisture, and C availability (Patra et al. 1990; He et al. 1997; Butterly et al. 2009; Richardson and Simpson 2011).

Carbon dynamics in soil have been closely linked to microbial biomass P (Achat et al. 2009) where immobilization of P is highly regulated by C availability (C:P ratio; Spohn and Kuzyakov 2013; Heuck et al. 2015). Microbial phosphatases and other enzymes are involved in organic C mineralization to promote the availability of P and C from organophosphorylated compounds (Spohn and Kuzyakov 2013; Heuck et al. 2015). Thus, P immobilization and transformation is inherently coupled, to a certain extent, with C mineralization (Luo et al. 2020). As soil microbial activity increases with the availability of C (i.e., SOM), immobilized P is released to increase the available P during microbial biomass turnover (Yang et al. 2010; Haripal and Sahoo 2014). When C sources become limited microbial biomass P will subsequently decrease (low C:P ratio; Yang et al. 2010; Haripal and Sahoo 2014). Given that the turnover of microbial P is largely driven by the C availability it is of particular importance in the rhizosphere (Richardson and Simpson 2011). As orthophosphate diffuses through the rhizosphere/mycorrhizosphere (Jakobsen et al. 2005) it is in direct competition for uptake and immobilization by microorganisms. Subsequently, the rate of release of P from microorganisms or microbial biomass turnover time within the rhizosphere will have major implications for P availability to plants (Richardson and Simpson 2011).

7.4 Role of Biological Soil Crust in Nutrient (C, N, and P) Cycles

Biological soil crusts (BSCs) or biocrust consist of photoautotrophic (green microalgae and cyanobacteria) and heterotrophic (bacteria and fungi) organisms, stable layers predominantly on bare soil surfaces in which inorganic particles are bound together by sticky extracellular polymeric substances (EPS; Belnap et al. 2001). The importance of biocrusts in the functioning of ecosystems in arid environments is well documented (Veste et al. 2001; Belnap et al. 2003; Maestre et al. 2011; Chap. 3).

It is estimated that BSCs are responsible for global C fixation of ~7% of the terrestrial vegetation, and for N fixation of ~50% of terrestrial biological N fixation (Elbert et al. 2012). Much of the C that is photosynthetically assimilated by biocrusts is released to the underlying soil shortly after fixation, thereby increasing the total amount of C and organic matter in soil (Pointing and Belnap 2012). Biocrusts also regulate the temporal dynamics of soil CO_2 efflux and net CO_2 uptake (Castillo-Monroy et al. 2011; Wilske et al. 2008, 2009). Studies have shown that temporal increase in soil moisture (i.e., higher and more intense precipitation frequencies) significantly increases C inputs in BSCs (Wilske et al. 2008; Huang et al. 2014), although the moisture levels at which organisms become photosynthetically active are highly variable (Belnap et al. 2004). Concurrently, higher moisture levels and temperature induce CO_2 emissions (Maestre et al. 2013) by several factors:

(i) Infiltration of water may physically displace the CO_2 accumulated in soil pore spaces during the dry season.
(ii) Rewetting triggers microbial activity in both biocrusts and underlying soil matrix that results in increased respiration rates.
(iii) Water addition may induce CO_2 release from microbial biomass pool (Li et al. 2018).

Biocrusts may also influence the activity of C-related soil enzymes, depending on the environmental conditions (e.g., β-glucosidase; Bowker et al. 2011; Yuan and Yue 2012; Miralles et al. 2013). When moisture and organic C content increases enough, sufficient resources are available to support higher microbial biomass and thus higher enzyme activity. High organic C content also allows existing enzymes to be stabilized by their absorption to SOM (Yuan and Yue 2012). Conversely, low moisture content and organic C result in relatively low microbial growth, enzyme stabilization, and activity (Yuan and Yue 2012). BSCs type plays an important role in C fixation at sites (Sancho et al. 2016; Maier et al. 2018). For example, Miralles et al. (2018) have demonstrated that late-successional biocrusts (i.e., lichens and mosses) had higher gross photosynthesis than early-successional biocrusts (cyanobacteria) in two semiarid ecosystems. Similar results were obtained by Housman et al. (2006) for the Colorado Plateau and Chihuahuan Desert where late-successional biocrusts (i.e., cyanobacteria [*Nostoc* and *Scytonema*] and lichens [*Placidium* and *Collema*]) had higher C fixation rates (1.2–1.3-fold) than early-successional biocrusts (cyanobacteria [*Microcoleus*]). Estimates for annual C inputs range from 0.4 to 2.3 g C m^{-2} $year^{-1}$ for early successional crusts to 12–37 g C m^{-2} $year^{-1}$ for late-successional crusts (Evans and Lange 2003; Li et al. 2012b; Yan-Gui et al. 2013), representing 1% of the NPP of terrestrial vegetation (~56 Pg $year^{-1}$; Zhao et al. 2005). In terms of dryland systems, biocrusts account for ~9% of the total NPP, corresponding to ~0.07 Pg $year^{-1}$ compared to the total NPP of ~0.8 Pg $year^{-1}$ (Zhao et al. 2005; Elbert et al. 2012; Sancho et al. 2016). In terms of C release, studies have found that 40–60% of soil respiration in dryland ecosystems were attributable to (algal) biocrust-dominated sites (Castillo-Monroy et al. 2011; Zhang et al. 2013). However, soil respiration is highly dependent on the type of crust, temperature, and precipitation (Tucker and Reed 2016; Zhang et al. 2016b; Guan et al. 2018; Li et al. 2018).

Given that most BSC components can fix C, the availability of P increases as organic matter accumulates in soil (Belnap 2011). The availability of P is further intensified through the secretion of extracellular phosphatases, organic acids (e.g., oxalic acid, citric acid, and malic acid), and metal chelators (e.g., siderochromes; Lange 1974; Belnap 2001a, 2011; Crain et al. 2018). Most BSC organisms (soil and hypolithic cyanobacteria, green algae, lichens, and mosses) contain phosphatases in their cell walls and mucilaginous sheaths, but they also release extracellular phosphatases into the surrounding soil (Belnap 2011). Phosphatases hydrolyze organic phosphates to subsequently release P (Turner et al. 2003b; Nannipieri et al. 2011; Baumann et al. 2017; Crain et al. 2018) which can then be immobilized by microbes, transferred to plant host roots, or stabilized by humic substances (Sinsabaugh 1994; Lindahl et al. 2005). As phosphatase activity is highly correlated with SOM, which occurs at low concentrations in dryland soils, reduced phosphatase activity is expected compared to more mesic environments (Sinsabaugh et al. 2008).

In addition to nutrient cycling, BSCs control the water balance (Warren 2003; Chamizo et al. 2016) by regulating the following factors (Chap. 3):

(i) Water infiltration and runoff (Belnap 2006; Chamizo et al. 2012, 2016; Zaady et al. 2013)
(ii) Increase soil moisture retention
(iii) Protect soil surface aggregates
(iv) Prevent soil erosion by water or wind
(v) Facilitate colonization by vascular plants (Belnap 2003)
(vi) Contribute to the stabilization of sand dunes (Eldridge and Greene 1994; Belnap 2002)

The presence of BSCs can also alter soil surface conditions in ways that affect the suitability of the habitat for other organisms (Bowker et al. 2005, 2006). However, their metabolic activity is strongly linked to moisture availability (Sancho et al. 2016). Due to low and highly variable precipitation pulses in drylands, biocrusts are sporadically active (Johnson et al. 2012; Yu et al. 2014; Sancho et al. 2016; Fernandes et al. 2018; Miralles et al. 2018). Nevertheless, their distinguishing characteristics highlight their overwhelming ecological importance in dryland environments, particularly in element cycling where the C, N, and P cycles are interlinked (Delgado-Baquerizo et al. 2013).

7.5 Influence of Hydration-Desiccation Pulses on Nutrient (C, N, and P) Cycles

Infrequent precipitation events occur in drylands; thus, soil surfaces are most often dry. During desiccation periods, nutrients accumulate on the soil surface due to dust deposition and the degradation and/or death of organisms from UV exposure (Belnap 2011). When precipitation occurs at higher temperatures, microbial diversity and function tend to rapidly increase thus affecting the rate of nutrient cycling in soils (Belnap 2011; Bell et al. 2014; Montiel-González et al. 2017; Chap. 11). Should precipitation occur at the proper time to stimulate annual plant activity for a few weeks to months, subsequent C inputs will stimulate soil microbial activity, including processes that release bound P (Whitford 1999). However, when a small amount of precipitation occurs (<5 mm rain) there is a shallow penetration of water into the soil, and substantial buildup of nutrients at the surface, that elicit more responses from soil surface organisms and their associated processes than vascular plants. The temporal dynamics of water use between microbial and plant communities can result in decoupling between their responses to precipitation and nutrients (Stursova et al. 2006; Belnap et al. 2004), but also in the accumulation of nutrients (hydrolyzable P_o and C) at the soil surface between large precipitation events (Whitford 2002; White et al. 2004).

In that sense, much attention has been given to hydration-desiccation cycles in drylands and the impact of rewetting on microbial communities and nutrient cycling (emphasis on C) in both laboratory conditions and field studies (Austin et al. 2004, Bell et al. 2014; Frossard et al. 2015; Montiel-González et al. 2017; Schimel 2018; Chap. 11). As soils dry, diffusion rates decline and semi-soluble materials precipitate (Schimel 2018) causing water stress. Microbes will increase their intracellular solute concentration to compensate for the extracellular concentration and counterbalance the increased osmotic pressure (Stark and Firestone 1995; Bell et al. 2008). This high concentration of solutes results in an inhibition of the enzymatic activity and therefore decreases cellular activity (Batlle-Aguilar et al. 2011). As a consequence, microbial activity and respiration is reduced (Stark and Firestone 1995; Austin et al. 2004; Schimel 2018), and a portion of microbial biomass is killed under such conditions (Austin et al. 2004). However, the magnitude of the respiration decrease may vary; in some cases, substantial reductions in soil moisture have limited effects on respiration (Lu et al. 2017).

Rewetting of soils after a drought period can result in large pulses of CO_2, also known as the "Birch effect" (Birch 1958). The CO_2 pulse can be many times greater than the basal respiration level (Schimel 2018), although the size of the pulse may largely depend on the labile C soil pools and the soil's history of physical disturbance (Austin et al. 2004; Schimel 2018). The exact mechanism that drives the Birch effect remains unclear and contrasting patterns emerge from rewetting experiments (see Schimel 2018 for a comprehensive overview). A study by Fierer et al. (2003) suggests that during each rewetting cycle, carbon is immediately lost from the (dead) microbial biomass and labile SOM is simultaneously mobilized from the soil matrix. The mobilized material may then be utilized by microbes to replace lost biomass C and to promote additional growth (Schimel 2018). Therefore, through multiple cycles, the C mobilized may ultimately be from stable soil C, but within each cycle, the C originates immediately from microbial biomass (Schimel 2018).

In addition to C, wet–dry cycles have also important consequences on N cycling processes. In drylands covered with BSCs, water pulses have a temperature-dependent impact: cold desert soils remain moist for longer periods of time than hot desert soils, thereby increasing the N inputs into soils (Austin et al. 2004). Since rates and duration of nitrogenase activity are mainly controlled by water availability (Hartley and Schlesinger 2000; Belnap 2002; Hartley and Schlesinger 2002; Wu et al. 2009; Zhou et al. 2016), N fixation in drylands may start within a few hours following a wetting event (Abed et al. 2010) as organic matter increases due to primary production (Wierenga et al. 1987). For example, nitrogenase activity was detected within 2 h after wetting dry cyanobacteria-dominated crusts (i.e., *Nostoc*, *Scytonema*, and *Microcoleus*) from Sahel, Niger (Malam Issa et al. 2001). In the Great Basin, nitrogenase activity was observed within 3 days after wetting cyanolichen crusts (Jeffries et al. 1992).

Wetting of dry soils also induces NO emissions, which can reach 400-fold compared to dry soils (Austin et al. 2004). Likewise, water pulses have a critical effect on denitrification where rates can increase significantly after precipitation events (Austin et al. 2004 and references thereof). Even smaller water pulses, either by rainfall, fog, or dew, following a large precipitation event can readily activate soil and BSC microorganisms to respire, photosynthesize, and perform other metabolic processes (Belnap and Lange 2001; Belnap 2003; Veste et al. 2008). N formed by metabolically active BSCs, mainly as NH_4^+, may leak to the environment increasing nutrient availability for nearby vegetation and microorganisms (reviewed in Belnap 2001b). Increased N will also lead to higher N loss through nitrification and denitrification processes as discussed above (Zaady 2005; Barger et al. 2005; Johnson et al. 2007; Strauss et al. 2012; Brankatschk et al. 2013; Liu et al. 2016).

An important aspect to consider in soil respiration is that of microbial community composition. Bacteria and fungi are ecologically and physiologically distinct groups, support different soil food-chains (De Vries et al. 2006) and differ in their C use efficiency (Sakamoto and Oba 1994). Studies have demonstrated that fungi have a higher resistance to drought and that fungal growth is largely unresponsive to drying-rewetting cycles (Bapiri et al. 2010; Evans and Wallenstein 2012; Manzoni et al. 2012; Barnard et al. 2013; Canarini et al. 2017). However, new evidence suggests that fungal communities are more strongly influenced by drying-wetting cycles than previously considered and that communities shift toward a more Ascomycota-dominated population (Hicks et al. 2019; Liu et al. 2019). In contrast, bacterial communities are more responsive to wetting pulses and may show an immediate and linear growth response following rewetting, or bacterial growth starts exponentially after a prolonged lag-period of no growth (Meisner et al. 2013, 2015).

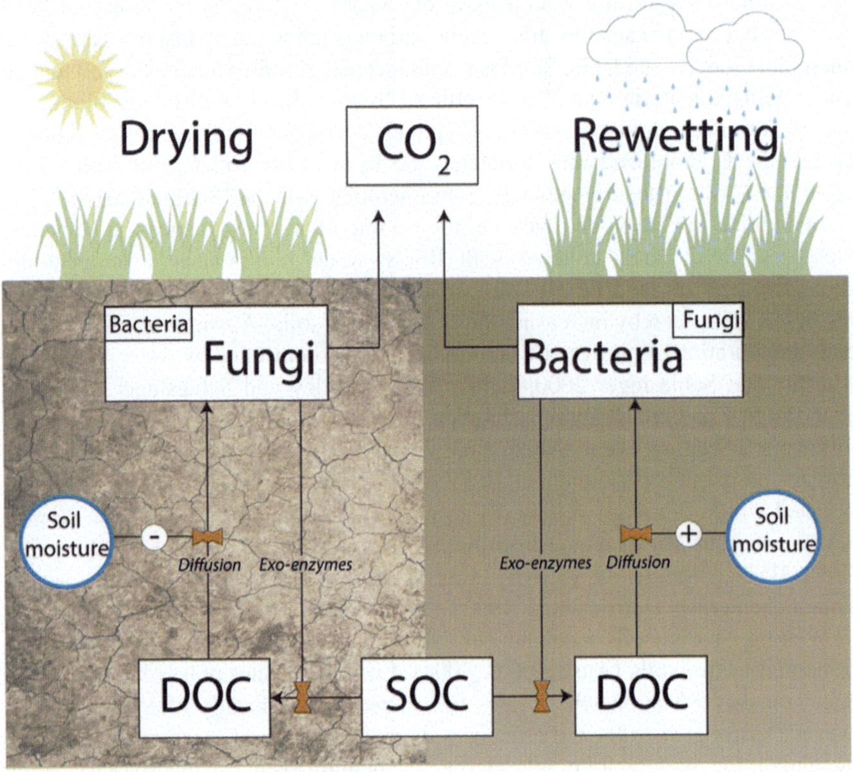

Fig. 7.1 Conceptual model of C dynamics during a drying-rewetting cycle summarizing the observed patterns obtained in a meta-analysis by Canarini et al. (2017). Fungi maintain higher respiration rates during the drying period, while bacteria show higher resilience and faster growth rates to return to pre-drought conditions when diffusion is restored (rewetting), causing a burst in respiration. Carbon-rich soils may boost this mechanism due to a higher fungal population, and by a greater release of available substrates (accumulated during the drying phase) to stimulate the bacterial community. A drought intensity threshold must be reached to initiate this response, possibly by regulating diffusion and therefore modulating the connection between available substrate and microorganisms. Adapted from Canarini et al. (2017)

This causes changes in the fungal and bacterial biomass, as well as the fungal: bacterial (F:B) ratio, which ultimately influences the C use efficiency and C loss from soils (Fig. 7.1; Canarini et al. 2017).

7.6 Impact of Climate Change on Nutrient Cycling

Climate models estimate that global warming will likely aggravate dryland expansion due to increases in evaporative demand and a global hydrological cycle with longer and more severe droughts (Schlaepfer et al. 2017; Koutroulis 2019). Of particular concern is the decrease in soil moisture that can result in: (1) changes in

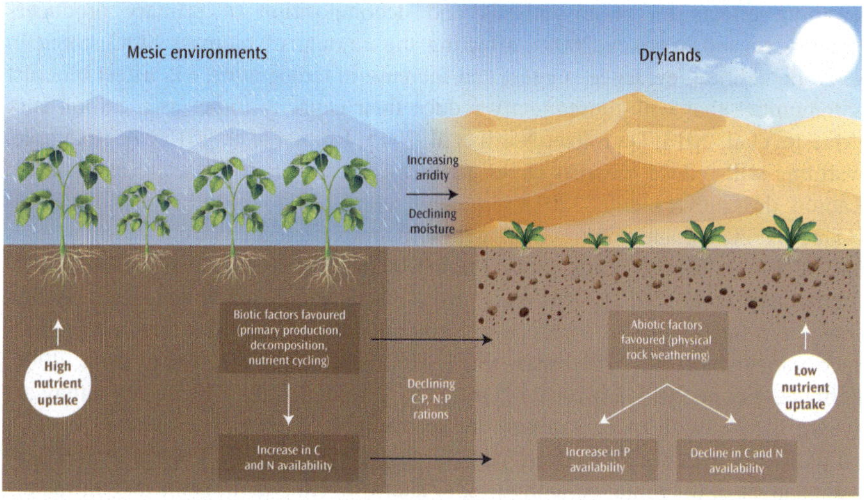

Fig. 7.2 Increase in aridity causes element cycles (C, N, and P) to decouple. Aridity is predicted to increase in many dryland ecosystems worldwide due to climate change. Delgado-Baquerizo et al. (2013) reported that, as aridity increases, available soil carbon (C) and nitrogen (N) decline, whereas available soil phosphorus (P) increases. This is a result of the impairment of biological processes that contribute to the C and N levels, and of an increase in the relative importance of abiotic processes that contribute to P availability. Adapted from Wardle (2013)

vegetation due to shifts in plant functional types (Harrison et al. 2015), woody plant mortality (Allen et al. 2010), and encroachment (D'Odorico et al. 2012), and resistance of some vegetation types (Craine et al. 2013); and (2) rise in temperatures within 3.0–4.0 °C (IPCC 2014; Huang et al. 2017; Koutroulis 2019).

Also, increase in aridity along chrono- and climosequences has been shown to cause a decline in the concentration of both the total and the most biologically available forms of C and N, but an increase in biologically available forms of P, hence distorting soil C, N, and P cycles (Finzi et al. 2011; Delgado-Baquerizo et al. 2013; Feng et al. 2016). This results in an overall decrease in C:P and N:P ratios and decoupling of C and N cycles from P (Fig. 7.2; Evans and Burke 2013; Delgado-Baquerizo et al. 2013; Wardle 2013). Such changes to the environment and uncoupling of nutrient cycles may have detrimental effects on ecosystem function (Delgado-Baquerizo et al. 2013). Reduced availability of C and N may skew C:N:P stoichiometry in soil, constraining plant and microbial activity and diversity (Finzi et al. 2011; Peñuelas et al. 2012). This may negatively impact biogeochemical reactions that control key ecosystem functions (e.g., primary production, respiration, and decomposition) and services (e.g., food production and C sequestration; (Dodds et al. 2004; Schimel and Bennett 2004; Schimel 2010; Finzi et al. 2011; Delgado-Baquerizo et al. 2013; Maestre et al. 2016; Lal 2019), that can jeopardize human well-being and nature conservation (Schlaepfer et al. 2017).

Changes in temperature, soil moisture, and vegetation structure and composition will likely modify the C sequestration capacity in high latitude drylands via direct warming effects on photosynthesis and decomposition of primary producers (Sjögersten and Wookey 2009), affecting the amount of biomass SOC stored in soils. In addition, evidence suggest that increase in temperature will affect biocrust communities that will consequently reduce their capacity to act as a carbon sink (Maestre et al. 2013; Darrouzet-Nardi et al. 2015; Escolar et al. 2015). For example, warming will reduce biocrust cover, richness, and diversity (Maestre et al. 2013; Ladrón de Guevara et al. 2018; Lafuente et al. 2018; Eldridge and Delgado-Baquerizo 2019), and communities will shift from late (i.e., dominated by mosses and lichens) toward early successional states dominated by cyanobacteria (Ferrenberg et al. 2015). This has critical implications for C sequestration and ecosystem functioning as a reduction in biocrust cover and the formation of early successional biocrusts will lessen the C storage capacity (Housman et al. 2006; Ferrenberg et al. 2015; Darrouzet-Nardi et al. 2018) of soils. This decrease may act synergistically with other warming-induced effects, such as the increase in CO_2 efflux (Maestre et al. 2013; Darrouzet-Nardi et al. 2015, 2018; Escolar et al. 2015) and changes in microbial communities, to alter C cycling in drylands and ultimately reduce soil C stocks in the mid to long term (Maestre et al. 2013).

Importantly, drylands are sensitive to land degradation and desertification which dramatically reduce the amount of SOC stored in soils (Lal 2004a; Serrano-Ortiz et al. 2012). Previous studies have estimated that the historic SOC loss from drylands due to desertification ranges between 13 and 29 Pg C (Ojima et al. 1995; Lal et al. 1999; Lal 2001). Additionally, changes in vegetation structure induced by overgrazing promote C losses that may significantly change both plant and SOC pools (Gaitán et al. 2017; Abdalla et al. 2018). As SOC has a major impact on soil physical structure and ecosystem function (e.g., nutrient retention, water storage, pollutant attenuation), its reduction can lead to reduced soil fertility and further land degradation (Abdalla et al. 2018). The magnitude of this loss underscores the importance of appropriate management systems and best strategies to restore, preserve, or increase dryland SOC to enhance ecosystem functions and services, and mitigate climate changes (Plaza et al. 2018b).

Climate change will also significantly alter N cycling processes in drylands as NO_x emissions are related to temperature (McCalley and Sparks 2009; Suddick et al. 2013). Optimal temperatures for N processes (e.g., N fixation) generally range between 20 and 30 °C (Trolldenier 1982; Montañez et al. 1995). With global temperatures rising, BNF and thus N input will consequently induce larger NO_x emissions in hot deserts where summer temperatures already exceed 40 °C (Makhalanyane et al. 2015). In addition, higher N losses will further contribute to soil infertility, therefore unable to support most plant life. Although some climate models predict higher summer precipitation events for desert areas (e.g., arid regions of central Asia), the combination of moisture and heat would greatly enhance N losses. One should also consider that most climate models only consider biological factors to predict N gases losses from soils and rarely account for abiotic impacts on the N budget (McCalley and Sparks 2009).

P cycling will be also strongly influenced by climate change, as P availability is directly and indirectly impacted by both temperature and precipitation in drylands

(Belnap 2011). Plants whose distribution is limited by available P will likely undergo spatial shits in habitat (Belnap 2011). Consequently, biological cycling of P will be reduced due to manifold factors: (1) a decline in soil moisture will not only negatively affect abiotic processes by slowing the release of bio-unavailable P, but also biotic processes (Belnap 2011). Reduced soil moisture will decrease microbial abundance and activity (Sardans et al. 2006), resulting in reduced acidification and production/excretion of enzymes, chelators, and other compounds to release bound P (Belnap 2011). Lower plant biomass, including root biomass, will further reduce soil acidification and root exudates to release bound P (Delucia et al. 1997; Li and Sarah 2003); (2) low soil moisture will also affect phosphatase activity as their effectiveness is more dependent on water availability than substrate availability (Sardans et al. 2008); (3) infrequent hydration-desiccation events will cause less P to be released from physical processes or from dead microbes (Belnap 2011); (4) severe droughts and higher temperatures will increase plant resorption of P, resulting in less P in detritus material (Killingbeck and Whitford 2001; Sardans et al. 2006; Luo et al. 2018); and (5) reduced plant cover can intensify dust emissions (Field et al. 2010; Duniway et al. 2019), resulting in substantial loss of many plant-essential nutrients including P, Na, K, and Mg from soil (Li et al. 2007; Belnap 2011; Katra et al. 2016).

7.7 Conclusion

Nutrient cycling in dryland soils and BSCs is a complex, yet active, process largely controlled by water availability (i.e., the size and frequency of precipitation events). Other abiotic factors such as temperature, C content and light, but also microbial species composition determine the rate by which nutrients are cycled. As aridity in drylands increases with climate change, the coupling between nutrient cycles will weaken and many arid environments can observe abrupt declines in organic C and total N content, while inorganic P will accumulate. Such changes are likely to have adverse effects on key ecosystem functions (e.g., primary production, respiration, and decomposition) and food services (e.g., food production and C storage). As such, comprehending the role and responses of microbial communities in drylands (soil and BSCs) can offer vital insights/predictions into future C and N fluxes, with important implications for future sustainable management and conservation policies.

Acknowledgments KJ is supported by a CONICYT-Fondecyt Postdoctorado 3190464 grant.

References

Abdalla M, Hastings A, Chadwick DR, Jones DL, Evans CD, Jones MB et al (2018) Critical review of the impacts of grazing intensity on soil organic carbon storage and other soil quality indicators in extensively managed grasslands. Agr Ecosyst Environ 253:62–81. https://doi.org/10.1016/j.agee.2017.10.023

Abed RM, Al Kharusi S, Schramm A, Robinson MD (2010) Bacterial diversity, pigments and nitrogen fixation of biological desert crusts from the Sultanate of Oman. FEMS Microbiol Ecol 72(3):418–428. https://doi.org/10.1111/j.1574-6941.2010.00854.x

Abed RM, Lam P, De Beer D, Stief P (2013) High rates of denitrification and nitrous oxide emission in arid biological soil crusts from the Sultanate of Oman. ISME J 7(9):1862–1875. https://doi.org/10.1038/ismej.2013.55

Achat DL, Bakker MR, Morel C (2009) Process-based assessment of phosphorus availability in a low phosphorus sorbing forest soil using isotopic dilution methods. Soil Sci Soc Am J 73(6): 2131–2142. https://doi.org/10.2136/sssaj2009.0009

Agarwal L, Dafale NA, Purohit HJ (2017) Microbial CO2 fixation bioprocesses and desert as future carbon sink. In: Purohit HJ (ed) Optimization and applicability of bioprocesses. Springer, Singapore

Ahlström A, Raupach MR, Schurgers G, Smith B, Arneth A, Jung M et al (2015) The dominant role of semi-arid ecosystems in the trend and variability of the land CO2 sink. Science 348(6237): 895–899. https://doi.org/10.1126/science.aaa1668

Alcamán-Arias ME, Pedrós-Alió C, Tamames J, Fernández C, Pérez-Pantoja D, Vásquez M, Díez B (2018) Diurnal changes in active carbon and nitrogen pathways along the temperature gradient in porcelana hot spring microbial mat. Front Microbiol 9(2353):1–17. https://doi.org/10.3389/fmicb.2018.02353

Allen CD, Macalady AK, Chenchouni H, Bachelet D, Mcdowell N, Vennetier M et al (2010) A global overview of drought and heat-induced tree mortality reveals emerging climate change risks for forests. Forest Ecol Manag 259(4):660–684. https://doi.org/10.1016/j.foreco.2009.09.001

Allen-Diaz B, Chapin FS, Díaz S, Howden SM, Puigdefábregas J, Stafford Smith M (1996) Rangelands in a changing climate: impacts, adaptations, and mitigation. In: Climate change 1995: impacts, adaptations and mitigation of climate change: scientific-technical analyses. Cambridge University Press, Cambridge, pp 131–158

An H, Li Q-L, Yan X, Wu X-Z, Liu R-T, Fang Y (2019) Desertification control on soil inorganic and organic carbon accumulation in the topsoil of desert grassland in Ningxia, northwest China. Ecol Eng 127:348–355. https://doi.org/10.1016/j.ecoleng.2018.12.014

Anderson G (1961) Estimation of purines and pyrimidines in soil humic acid. Soil Sci 91(3): 156–161

Anderson G, Malcolm RE (1974) The nature alkali-soluble soil organic phosphates. J Soil Sci 25(3):282–297. https://doi.org/10.1111/j.1365-2389.1974.tb01124.x

Anderson DC, Harper KT, Holmgren RC (1982) Factors influencing development of cryptogamic soil crusts in Utah deserts. J Range Manag 35(2):180–185

Aranibar JN, Anderson IC, Ringrose S, Macko SA (2003) Importance of nitrogen fixation in soil crusts of southern African arid ecosystems: acetylene reduction and stable isotope studies. J Arid Environ 54(2):345–358. https://doi.org/10.1006/jare.2002.1094

Austin AT, Vitousek PM (1998) Nutrient dynamics on a precipitation gradient in Hawai'i. Oecologia 113(4):519–529. https://doi.org/10.1007/s004420050405

Austin AT, Yahdjian L, Stark JM, Belnap J, Porporato A, Norton U et al (2004) Water pulses and biogeochemical cycles in arid and semiarid ecosystems. Oecologia 141(2):221–235. https://doi.org/10.1007/s00442-004-1519-1

Bae HS, Morrison E, Chanton JP, Ogram A (2018) Methanogens are major contributors to nitrogen fixation in soils of the Florida everglades. Appl Environ Microbiol 84(7). https://doi.org/10.1128/AEM.02222-17

Báez S, Fargione J, Moore DI, Collins SL, Gosz JR (2007) Atmospheric nitrogen deposition in the northern Chihuahuan desert: temporal trends and potential consequences. J Arid Environ 68(4): 640–651. https://doi.org/10.1016/j.jaridenv.2006.06.011

Banik S, Dey BK (1982) Available phosphate content of an alluvial soil as influenced by inoculation of some isolated phosphate-solubilizing microorganisms. Plant Soil 69(3):353–364. https://doi.org/10.1007/Bf02372456

Bapiri A, Baath E, Rousk J (2010) Drying-rewetting cycles affect fungal and bacterial growth differently in an arable soil. Microb Ecol 60(2):419–428. https://doi.org/10.1007/s00248-010-9723-5

Barger NN, Belnap J, Ojima DS (2005) NO gas loss from biologically crusted soils in Canyonlands National Park, Utah. Biogeochemistry 75:373–391. https://doi.org/10.1007/s10533-005-1378-9

Barger NN, Weber B, Garcia-Pichel F, Zaady E, Belnap J (2016) Patterns and controls on nitrogen cycling of biological soil crusts. In: Weber B, Büdel B, Belnap J (eds) Biological Soil crusts: an organizing principle in drylands. Ecological studies (analysis and synthesis). Springer, Cham, pp 257–285. https://doi.org/10.1007/978-3-319-30214-0_14

Barnard RL, Osborne CA, Firestone MK (2013) Responses of soil bacterial and fungal communities to extreme desiccation and rewetting. ISME J 7(11):2229–2241. https://doi.org/10.1038/ismej.2013.104

Barrow NJ (1983) A mechanistic model for describing the sorption and desorption of phosphate by soil. J Soil Sci 34(4):733–750. https://doi.org/10.1111/j.1365-2389.1983.tb01068.x

Bate DB, Barrett JE, Poage MA, Virginia RA (2008) Soil phosphorus cycling in an Antarctic polar desert. Geoderma 144(1–2):21–31. https://doi.org/10.1016/j.geoderma.2007.10.007

Batlle-Aguilar J, Brovelli A, Porporato A, Barry DA (2011) Modelling soil carbon and nitrogen cycles during land use change. A review. Agron Sustain Dev 31(2):251–274. https://doi.org/10.1051/agro/2010007

Baumann K, Glaser K, Mutz JE, Karsten U, Maclennan A, Hu YF et al (2017) Biological soil crusts of temperate forests: their role in P cycling. Soil Biol Biochem 109:156–166. https://doi.org/10.1016/j.soilbio.2017.02.011

Beer C, Reichstein M, Tomelleri E, Ciais P, Jung M, Carvalhais N et al (2010) Terrestrial gross carbon dioxide uptake: global distribution and covariation with climate. Science 329(5993):834–838. https://doi.org/10.1126/science.1184984

Bell C, Mcintyre N, Cox S, Tissue D, Zak J (2008) Soil microbial responses to temporal variations of moisture and temperature in a Chihuahuan Desert Grassland. Microb Ecol 56(1):153–167. https://doi.org/10.1007/s00248-007-9333-z

Bell CW, Tissue DT, Loik ME, Wallenstein MD, Acosta-Martinez V, Erickson RA, Zak JC (2014) Soil microbial and nutrient responses to 7 years of seasonally altered precipitation in a Chihuahuan Desert grassland. Glob Chang Biol 20(5):1657–1673. https://doi.org/10.1111/gcb.12418

Belnap J (2001a) Microbes and microfauna associated with biological soil crusts. In: Belnap J, Lange OL (eds) Biological soil crusts: structure, function, and management. Ecological studies (analysis and synthesis), vol 150. Springer, Berlin, pp 167–174. https://doi.org/10.1007/978-3-642-56475-8_14

Belnap J (2001b) Factors influencing nitrogen fixation and nitrogen release in biological soil crusts. In: Belnap J, Lange OL (eds) Biological soil crusts: structure, function, and management. ecological studies (analysis and synthesis). Springer, Heidelberg, pp 241–261. https://doi.org/10.1007/978-3-642-56475-8_19

Belnap J (2002) Nitrogen fixation in biological soil crusts from southeast Utah, USA. Biol Fert Soils 35(2):128–135. https://doi.org/10.1007/s00374-002-0452-x

Belnap J (2003) The world at your feet: desert biological soil crusts. Front Ecol Environ 1(4):181–189. https://doi.org/10.1890/1540-9295(2003)001[0181:Twayfd]2.0.Co;2

Belnap J (2006) The potential roles of biological soil crusts in dryland hydrologic cycles. Hydrol Process 20(15):3159–3178. https://doi.org/10.1002/hyp.6325

Belnap J (2011) Biological phosphorus cycling in dryland regions. In: Bünemann EK, Oberson A, Frossard E (eds) Phosphorus in action. biological processes in soil phosphorus cycling, vol 26. Springer, Heidelberg, pp 371–406. https://doi.org/10.1007/978-3-642-15271-9

Belnap J, Lange OL (2001) Structure and functioning of biological soil crusts: a synthesis. In: Belnap J, Lange OL (eds) Biological soil crusts: structure, function, and management. Springer, Heidelberg, pp 471–479

Belnap J, Büdel B, Lange OL (2001) Biological soil crusts: characteristics and distribution. In: Belnap J, Lange OL (eds) Biological soil crusts: structure, function, and management. Ecological studies (analysis and synthesis), vol 150. Springer, Berlin, pp 3–30. https://doi.org/10.1007/978-3-642-56475-8_1

Belnap J, Phillips SL, Duniway M, Reynolds RE (2003) Soil fertility in deserts: a review on the influence of biological soil crusts and the effect of soil surface disturbance on nutrient inputs and losses. In: Alsharhan AS, Wood WW, Goudie A, Fowler AR, Abdellatif EM (eds) Desertification in the third millennium. Swets & Zeitlinger (Balkema), Lisse, The Netherlands, pp 245–252

Belnap J, Phillips SL, Miller ME (2004) Response of desert biological soil crusts to alterations in precipitation frequency. Oecologia 141(2):306–316. https://doi.org/10.1007/s00442-003-1438-6

Billings SA, Schaeffer SM, Evans RD (2003) Nitrogen fixation by biological soil crusts and heterotrophic bacteria in an intact Mojave Desert ecosystem with elevated CO2 and added soil carbon. Soil Biol Biochem 35:643–649. https://doi.org/10.1016/S0038-0717(03)00011-7

Birch HF (1958) The effect of soil drying on humus decomposition and nitrogen availability. Plant Soil 10(1):9–31. https://doi.org/10.1007/BF01343734

Böhlke JK, Ericksen GE, Revesz K (1997) Stable isotope evidence for an atmospheric origin of desert nitrate deposits in northern Chile and southern California, U.S.A. Chem Geol 136(1–2): 135–152. https://doi.org/10.1016/S0009-2541(96)00124-6

Bowker MA, Belnap J, Davidson DW, Phillips SL (2005) Evidence for micronutrient limitation of biological soil crusts: importance to arid-lands restoration. Ecol Appl 15(6):1941–1951. https://www.jstor.org/stable/4543496

Bowker MA, Belnap J, Miller ME (2006) Spatial modeling of biological soil crusts to support rangeland assessment and monitoring. Rangeland Ecol Manag 59(5):519–529. https://doi.org/10.2111/05-179r1.1

Bowker MA, Mau RL, Maestre FT, Escolar C, Castillo-Monroy AP (2011) Functional profiles reveal unique ecological roles of various biological soil crust organisms. Funct Ecol 25(4): 787–795. https://doi.org/10.1111/j.1365-2435.2011.01835.x

Brankatschk R, Fischer T, Veste M, Zeyer J (2013) Succession of N cycling processes in biological soil crusts on a Central European inland dune. FEMS Microbiol Ecol 83(1):149–160. https://doi.org/10.1111/j.1574-6941.2012.01459.x

Buckingham SE, Neff J, Titiz-Maybach B, Reynolds RL (2010) Chemical and textural controls on phosphorus mobility in drylands of southeastern Utah. Biogeochemistry 100(1–3):105–120. https://doi.org/10.1007/s10533-010-9408-7

Butterly CR, Bünemann EK, Mcneill AM, Baldock JA, Marschner P (2009) Carbon pulses but not phosphorus pulses are related to decreases in microbial biomass during repeated drying and rewetting of soils. Soil Biol Biochem 41(7):1406–1416. https://doi.org/10.1016/j.soilbio.2009.03.018

Butusov M, Jernelöv A (2013) Phosphorus in the organic life: cells, tissues, organisms. Phosphorus: an element that could have been called Lucifer, vol 9. Springer, New York. https://doi.org/10.1007/978-1-4614-6803-5

Cable JM, Ogle K, Lucas RW, Huxman TE, Loik ME, Smith SD et al (2011) The temperature responses of soil respiration in deserts: a seven desert synthesis. Biogeochemistry 103(1–3): 71–90. https://doi.org/10.1007/s10533-010-9448-z

Canarini A, Kiaer LP, Dijkstra FA (2017) Soil carbon loss regulated by drought intensity and available substrate: a meta-analysis. Soil Biol Biochem 112:90–99. https://doi.org/10.1016/j.soilbio.2017.04.020

Caputa K, Coxson D, Sanborn P (2013) Seasonal patterns of nitrogen fixation in biological soil crusts from British Columbia's Chilcotin grasslands. Botany 91(9):631–641. https://doi.org/10.1139/cjb-2013-0014

Carreira JA, Vinegla B, Lajtha K (2006) Secondary CaCO3 and precipitation of P-Ca compounds control the retention of soil P in arid ecosystems. J Arid Environ 64(3):460–473. https://doi.org/10.1016/j.jaridenv.2005.06.003

Castillo-Monroy AP, Maestre FT, Rey A, Soliveres S, García-Palacios P (2011) Biological soil crust microsites are the main contributor to soil respiration in a semiarid ecosystem. Ecosystems 14(5). https://doi.org/10.1007/s10021-011-9449-3

Chamizo S, Canton Y, Lazaro R, Sole-Benet A, Domingo F (2012) Crust composition and disturbance drive infiltration through biological soil crusts in semiarid ecosystems. Ecosystems 15(1):148–161. https://doi.org/10.1007/s10021-011-9499-6

Chamizo S, Canton Y, Rodriguez-Caballero E, Domingo F (2016) Biocrusts positively affect the soil water balance in semiarid ecosystems. Ecohydrology 9(7):1208–1221. https://doi.org/10.1002/eco.1719

Chan Y, Lacap DC, Lau MC, Ha KY, Warren-Rhodes KA, Cockell CS et al (2012) Hypolithic microbial communities: between a rock and a hard place. Environ Microbiol 14(9):2272–2282. https://doi.org/10.1111/j.1462-2920.2012.02821.x

Chan Y, Van Nostrand JD, Zhou J, Pointing SB, Farrell RL (2013) Functional ecology of an Antarctic Dry Valley. Proc Natl Acad Sci U S A 110(22):8990–8995. https://doi.org/10.1073/pnas.1300643110

Chatli AS, Beri V, Sidhu BS (2008) Isolation and characterisation of phosphate solubilising microorganisms from the cold desert habitat of Salix alba Linn. in trans Himalayan region of Himachal Pradesh. Indian. J Microbiol 48(2):267–273. https://doi.org/10.1007/s12088-008-0037-y

Chatterjee D, Kuotsu R, Ao M, Saha S, Kumar Ray S, Ngachan SV (2019) Does rise in temperature adversely affect soil fertility, carbon fractions, microbial biomass and enzyme activities under different land uses? Curr Sci 116(12):2044–2054. https://doi.org/10.18520/cs/v116/i12/2044-2054

Chen L, Gong J, Fu B, Huang Z, Huang Y, Gui L (2007) Effect of land use conversion on soil organic carbon sequestration in the loess hilly area, loess plateau of China. Ecol Res 22(4). https://doi.org/10.1007/s11284-006-0065-1

Chen W, Zheng X, Chen Q, Wolf B, Butterbach-Bahl K, Brüggemann N, Lin S (2013) Effects of increasing precipitation and nitrogen deposition on CH_4 and N_2O fluxes and ecosystem respiration in a degraded steppe in Inner Mongolia, China. Geoderma 192:335–340. https://doi.org/10.1016/j.geoderma.2012.08.018

Cosgrove DJ (1977) Microbial transformations in the phosphorus cycle. In: Alexander M (ed) Advances in microbial ecology. Advances in microbial ecology, vol 1. Springer, Boston, MA, pp 95–134. https://doi.org/10.1007/978-1-4615-8219-9_3

Cowan DA, Sohm JA, Makhalanyane TP, Capone DG, Green TG, Cary SC, Tuffin IM (2011) Hypolithic communities: important nitrogen sources in Antarctic desert soils. Environ Microbiol Rep 3(5):581–586. https://doi.org/10.1111/j.1758-2229.2011.00266.x

Crain GM, Mclaren JR, Brunner B, Darrouzet-Nardi A (2018) Biologically available phosphorus in biocrust-dominated soils of the Chihuahuan Desert. Soil Syst 2(4):56. https://doi.org/10.3390/soilsystems2040056

Craine JM, Ocheltree TW, Nippert JB, Towne EG, Skibbe AM, Kembel SW, Fargione JE (2013) Global diversity of drought tolerance and grassland climate-change resilience. Nat Clim Change 3(1):63–67. https://doi.org/10.1038/Nclimate1634

Crits-Christoph A, Gelsinger DR, Ma B, Wierzchos J, Ravel J, Davila A et al (2016) Functional interactions of archaea, bacteria and viruses in a hypersaline endolithic community. Environ Microbiol 18(6):2064–2077. https://doi.org/10.1111/1462-2920.13259

Cross AF, Schlesinger WH (1995) A literature-review and evaluation of the Hedley fractionation—applications to the biogeochemical cycle of soil-phosphorus in natural ecosystems. Geoderma 64(3–4):197–214. https://doi.org/10.1016/0016-7061(94)00023-4

Cross AF, Schlesinger WH (2001) Biological and geochemical controls on phosphorus fractions in semiarid soils. Biogeochemistry 52(2):155–172. https://doi.org/10.1023/A:1006437504494

Cueva A, Volkmann THM, Van Haren J, Troch PA, Meredith LK (2019) Reconciling negative soil CO2 fluxes: insights from a large-scale experimental hillslope. Soil Syst 3(1):10. https://doi.org/10.3390/soilsystems3010010

Dai WH, Huang Y (2006) Relation of soil organic matter concentration to climate and altitude in zonal soils of China. Catena 65(1):87–94. https://doi.org/10.1016/j.catena.2005.10.006

Daims H, Lebedeva EV, Pjevac P, Han P, Herbold C, Albertsen M et al (2015) Complete nitrification by Nitrospira bacteria. Nature 528(7583):504–509. https://doi.org/10.1038/nature16461

Dalal RC (1977) Soil organic phosphorus. Adv Agron 29:83–117. https://doi.org/10.1016/S0065-2113(08)60216-3

Darrouzet-Nardi A, Reed SC, Grote EE, Belnap J (2015) Observations of net soil exchange of CO2 in a dryland show experimental warming increases carbon losses in biocrust soils. Biogeochemistry 126(3):363–378. https://doi.org/10.1007/s10533-015-0163-7

Darrouzet-Nardi A, Reed SC, Grote EE, Belnap J (2018) Patterns of longer-term climate change effects on CO2 efflux from biocrusted soils differ from those observed in the short term. Biogeosciences 15(14):4561–4573. https://doi.org/10.5194/bg-15-4561-2018

Delgado-Baquerizo M, Maestre FT, Gallardol A, Bowker MA, Wallenstein MD, Quero JL et al (2013) Decoupling of soil nutrient cycles as a function of aridity in global drylands. Nature 502(7473):672–676. https://doi.org/10.1038/nature12670

Delgado-Baquerizo M, Maestre FT, Eldridge DJ, Singh BK (2016) Microsite differentiation drives the abundance of soil ammonia oxidizing bacteria along aridity gradients. Front Microbiol 7: 505. https://doi.org/10.3389/fmicb.2016.00505

Delucia EH, Callaway RM, Thomas EM, Schlesinger WH (1997) Mechanisms of phosphorus acquisition for ponderosa pine seedlings under high CO2 and temperature. Ann Bot 79(2): 111–120. https://doi.org/10.1006/anbo.1996.0320

Delwiche CC, Wijler J (1956) Non-symbiotic nitrogen fixation in soil. Plant Soil 7:113–129

Denef K, Stewart CE, Brenner J, Paustian K (2008) Does long-term center-pivot irrigation increase soil carbon stocks in semi-arid agro-ecosystems? Geoderma 145(1–2):121–129. https://doi.org/10.1016/j.geoderma.2008.03.002

De Vries FT, Hoffland E, Van Eekeren N, Brussaard L, Bloem J (2006) Fungal/bacterial ratios in grasslands with contrasting nitrogen management. Soil Biol Biochem 38(8):2092–2103. https://doi.org/10.1016/j.soilbio.2006.01.008

Dodds WK, Gudder DA, Mollenhauer D (1995) The ecology of Nostoc. J Phycol 31:2–18

Dodds WK, Marti E, Tank JL, Pontius J, Hamilton SK, Grimm NB et al (2004) Carbon and nitrogen stoichiometry and nitrogen cycling rates in streams. Oecologia 140(3):458–467. https://doi.org/10.1007/s00442-004-1599-y

D'Odorico P, Okin GS, Bestelmeyer BT (2012) A synthetic review of feedbacks and drivers of shrub encroachment in arid grasslands. Ecohydrology 5(5):520–530. https://doi.org/10.1002/eco.259

Duniway MC, Pfennigwerth AA, Fick SE, Nauman TW, Belnap J, Barger NN (2019) Wind erosion and dust from US drylands: a review of causes, consequences, and solutions in a changing world. Ecosphere 10(3). https://doi.org/10.1002/ecs2.2650

Elbert W, Weber B, Burrows S, Steinkamp J, Budel B, Andreae MO, Poschl U (2012) Contribution of cryptogamic covers to the global cycles of carbon and nitrogen. Nat Geosci 5(7):459–462. https://doi.org/10.1038/NGEO1486

Eldridge DJ, Delgado-Baquerizo M (2019) The influence of climatic legacies on the distribution of dryland biocrust communities. Glob Chang Biol 25(1):327–336. https://doi.org/10.1111/gcb.14506

Eldridge DJ, Greene RSB (1994) Assessment of sediment yield by splash erosion on a semi-arid soil with varying cryptogam cover. J Arid Environ 26(3):221–232. https://doi.org/10.1006/jare.1994.1025

Escolar C, Maestre FT, Rey A (2015) Biocrusts modulate warming and rainfall exclusion effects on soil respiration in a semi-arid grassland. Soil Biol Biochem 80:9–17. https://doi.org/10.1016/j.soilbio.2014.09.019

Eswaran H, Reich PF, Kimble JM, Beinroth E, Padmanabhan E, Moncharoen P (2000) Global carbon stocks. In: Lal R (ed) Global climate change and pedogenic carbonate. CRC, Boca Raton, pp 15–26

Evans SE, Burke IC (2013) Carbon and nitrogen decoupling under an 11-year drought in the shortgrass steppe. Ecosystems 16(1):20–33. https://doi.org/10.1007/s10021-012-9593-4

Evans RD, Lange OL (2003) Biological soil crusts and ecosystem nitrogen and carbon dynamics. In: Belnap J, Lange OL (eds) Biological soil crusts: structure, function, and management. 150. Springer, Berlin, pp 263–279. https://doi.org/10.1007/978-3-642-56475-8_20

Evans SE, Wallenstein MD (2012) Soil microbial community response to drying and rewetting stress: does historical precipitation regime matter? Biogeochemistry 109(1–3):101–116. https://doi.org/10.1007/s10533-011-9638-3

Evans SE, Burke IC, Lauenroth WK (2011) Controls on soil organic carbon and nitrogen in Inner Mongolia, China: a cross-continental comparison of temperate grasslands. Glob Biogeochem Cycle 25:GB3006. https://doi.org/10.1029/2010gb003945

Ewing SA, Sutter B, Owen J, Nishiizumi K, Sharp W, Cliff SS et al (2006) A threshold in soil formation at Earth's arid–hyperarid transition. Geochimica et Cosmochimica Acta 70(21): 5293–5322. https://doi.org/10.1016/j.gca.2006.08.020

Fan J, Zhong H, Harris W, Yu G, Wang S, Hu Z, Yue Y (2008) Carbon storage in the grasslands of China based on field measurements of above- and below-ground biomass. Clim Change 86(3–4):375–396. https://doi.org/10.1007/s10584-007-9316-6

Fao (2004) Carbon sequestration in dryland soils, World soil resources reports no. 102. Rome, Italy

Feng Q, Endo KN, Cheng GD (2002) Soil carbon in desertified land in relation to site characteristics. Geoderma 106(1–2):21–43. https://doi.org/10.1016/S0016-7061(01)00099-4

Feng J, Turner BL, Lu XT, Chen ZH, Wei K, Tian JH et al (2016) Phosphorus transformations along a large-scale climosequence in arid and semiarid grasslands of northern China. Glob Biogeochem Cycle 30(9):1264–1275. https://doi.org/10.1002/2015gb005331

Fernandes VMC, Machado De Lima NM, Roush D, Rudgers J, Collins SL, Garcia-Pichel F (2018) Exposure to predicted precipitation patterns decreases population size and alters community structure of cyanobacteria in biological soil crusts from the Chihuahuan Desert. Environ Microbiol 20(1):259–269. https://doi.org/10.1111/1462-2920.13983

Ferrenberg S, Reed SC, Belnap J (2015) Climate change and physical disturbance cause similar community shifts in biological soil crusts. Proc Natl Acad Sci USA 112(39):12116–12121. https://doi.org/10.1073/pnas.1509150112

Field CB, Behrenfeld MJ, Randerson JT, Falkowski P (1998) Primary production of the biosphere: integrating terrestrial and oceanic components. Science 281(5374):237–240. https://doi.org/10.1126/science.281.5374.237

Field JP, Belnap J, Breshears DD, Neff JC, Okin GS, Whicker JJ et al (2010) The ecology of dust. Front Ecol Environ 8(8):423–430. https://doi.org/10.1890/090050

Fierer N, Allen AS, Schimel JP, and Holden PA. (2003) Controls on microbial CO2 production: a comparison of surface and subsurface soil horizons. Glob Change Biol 9(9):1322–1332. https://doi.org/10.1046/j.1365-2486.2003.00663.x

Finzi AC, Austin AT, Cleland EE, Frey SD, Houlton BZ (2011) Responses and feedbacks of coupled biogeochemical cycles to climate change: examples from terrestrial ecosystems. Front Ecol Environ 9(1):61–67. https://doi.org/10.1890/100001

Follett RF, Stewart CE, Pruessner EG, Kimble JM (2012) Effects of climate change on soil carbon and nitrogen storage in the US Great Plains. J Soil Water Conserv 67(5):331–342. https://doi.org/10.2489/jswc.67.5.331

Frossard A, Ramond JB, Seely M, Cowan DA (2015) Water regime history drives responses of soil Namib Desert microbial communities to wetting events. Sci Rep 5:12263. https://doi.org/10.1038/srep12263

Gaitán JJ, Bran DE, Oliva GE, Aguiar MR, Buono GG, Ferrante D et al (2017) Aridity and overgrazing have convergent effects on ecosystem structure and functioning in Patagonian rangelands. Land Degrad Dev 29(2):210–218. https://doi.org/10.1002/ldr.2694

Gaitán JJ, Maestre FT, Bran DE, Buono GG, Dougill AJ (2019) Biotic and abiotic drivers of topsoil organic carbon concentration in drylands have similar effects at regional and global scales. Ecosystems 22(7):1445–1456. https://doi.org/10.1007/s10021-019-00348-y

Gao Y, Tian J, Pang Y, Liu J (2017) Soil inorganic carbon sequestration following afforestation is probably induced by pedogenic carbonate formation in Northwest China. Front Plant Sci 8: 1282. https://doi.org/10.3389/fpls.2017.01282

George TS, Giles CD, Menezes-Blackburn D, Condron LM, Gama-Rodrigues AC, Jaisi D et al (2018) Organic phosphorus in the terrestrial environment: a perspective on the state of the art and future priorities. Plant Soil 427(1–2):209–211. https://doi.org/10.1007/s11104-017-3488-2

Giardina CP, Ryan MG (2000) Soil warming and organic carbon content—Reply. Nature 408(6814):790–790. https://doi.org/10.1038/35048675

Glenday J (2008) Carbon storage and emissions offset potential in an African dry forest, the Arabuko-Sokoke Forest, Kenya. Environ Monit Assess 142(1–3):85–95. https://doi.org/10.1007/s10661-007-9910-0

Gocke M, Pustovoytov K, Kuzyakov Y (2011) Carbonate recrystallization in root-free soil and rhizosphere of Triticum aestivum and Lolium perenne estimated by C-14 labeling. Biogeochemistry 103(1–3):209–222. https://doi.org/10.1007/s10533-010-9456-z

Goldberg S, Sposito G (1985) On the mechanism of specific phosphate adsorption by hydroxylated mineral surfaces—a review. Commun Soil Sci Plan 16(8):801–821. https://doi.org/10.1080/00103628509367646

Graham RC, Hirmas DR, Wood YA (2008) Large near-surface nitrate pools in soils capped by desert pavement in the Mojave Desert, California. Geology 36(3):259–262. https://doi.org/10.1130/G24343A.1

Granhall U (1981) Biological nitrogen fixation in relation to environmental factors and functioning of natural ecosystems. In: Clark FE, Rosswall T (eds) Terrestrial nitrogen cycles. Ecological bulletin. Swedish Natural Science Research Council, Stockholm, pp 131–144

Griffiths BS, Spilles A, Bonkowski M (2012) C:N:P stoichiometry and nutrient limitation of the soil microbial biomass in a grazed grassland site under experimental P limitation or excess. Ecol Process 1:6. https://doi.org/10.1186/2192-1709-1-6

Guan C, Li X, Zhang P, Chen Y (2018) Diel hysteresis between soil respiration and soil temperature in a biological soil crust covered desert ecosystem. PLoS One 13(4):e0195606. https://doi.org/10.1371/journal.pone.0195606

Guilbault MR, Matthias AD (1998) Emissions of N2O from Sonoran Desert and effluent-irrigated grass ecosystems. J Arid Environ 38(1):87–98. https://doi.org/10.1006/jare.1997.0300

Gyaneshwar P, Kumar GN, Parekh LJ, Poole PS (2002) Role of soil microorganisms in improving P nutrition of plants. Plant Soil 245(1):83–93. https://doi.org/10.1023/A:1020663916259

Haas HJ, Grunes DL, Reichman GA (1961) Phosphorus changes in Great Plains soils as influenced by cropping and manure applications. Soil Sci Soc Am Proc 25(3):214–218. https://doi.org/10.2136/sssaj1961.03615995002500030022x

Hall SJ, Huber D, Grimm NB (2008) Soil N_2O and NO emissions from an arid, urban ecosystem. J Geophys Res 113(G01016). https://doi.org/10.1029/2007JG000523

Haripal K, Sahoo S (2014) Microbial biomass carbon, nitrogen, and phosphorus dynamics along a chronosequence of abandoned tropical agroecosystems. Int J Curr Microbiol App Sci 3(9): 956–970

Harris WG (2002) Phosphate minerals. In: Dixon JB, Schulze DG (eds) Soil mineralogy with environmental applications. SSSA book series 7. Soil Science Society of America, Madison, WI, pp 637–665

Harrison SP, Gornish ES, Copeland S (2015) Climate-driven diversity loss in a grassland community. Proc Natl Acad Sci USA 112(28):8672–8677. https://doi.org/10.1073/pnas.1502074112

Hartley AE, Schlesinger WH (2000) Environmental controls on nitric oxide emission from northern Chihuahuan desert soils. Biogeochemistry 50(3):279–300. https://doi.org/10.1023/A:1006377832207

Hartley AE, Schlesinger WH (2002) Potential environmental controls on nitrogenase activity in biological crusts of the northern Chihuahuan Desert. J Arid Environ 52(3):293–304. https://doi.org/10.1006/jare.2002.1007
Hartley AE, Barger N, Belnap J, Okin GS (2007) Dryland ecosystems. In: Marschner P, Rengel Z (eds) Nutrient cycling in terrestrial ecosystems. Soil biology, vol 10. Springer, Heidelberg, pp 271–307. https://doi.org/10.1007/978-3-540-68027-7_10
He ZL, Wu J, Odonnell AG, Syers JK (1997) Seasonal responses in microbial biomass carbon, phosphorus and sulphur in soils under pasture. Biol Fert Soils 24(4):421–428. https://doi.org/10.1007/s003740050267
He N, Yu Q, Wu L, Wang Y, Han X (2008) Carbon and nitrogen store and storage potential as affected by land-use in a Leymus chinensis grassland of northern China. Soil Biol Biochem 40(12):2952–2959. https://doi.org/10.1016/j.soilbio.2008.08.018
He N-P, Wang RM, Zhang YH, Chen QS (2014) Carbon and nitrogen storage in inner Mongolian grasslands: relationships with climate and soil texture. Pedosphere 24(3):391–398. https://doi.org/10.1016/S1002-0160(14)60025-4
Hedley MJ, Stewart JWB, Chauhan BS (1982) Changes in inorganic and organic soil-phosphorus fractions induced by cultivation practices and by laboratory incubations. Soil Sci Soc Am J 46(5):970–976. https://doi.org/10.2136/sssaj1982.03615995004600050017x
Heuck C, Weig A, Spohn M (2015) Soil microbial biomass C:N:P stoichiometry and microbial use of organic phosphorus. Soil Biol Biochem 85:119–129. https://doi.org/10.1016/j.soilbio.2015.02.029
Hicks LC, Ang R, Leizeaga A, Rousk J (2019) Bacteria constrain the fungal growth response to drying-rewetting. Soil Biol Biochem 134:108–112. https://doi.org/10.1016/j.soilbio.2019.03.006
Homyak PM, Blankinship JC, Marchus K, Lucero DM, Sickman JO, Schimel JP (2016) Aridity and plant uptake interact to make dryland soils hotspots for nitric oxide (NO) emissions. Proc Natl Acad Sci U S A 113(19):E2608–E2616. https://doi.org/10.1073/pnas.1520496113
Hou EQ, Chen CR, Luo YQ, Zhou GY, Kuang YW, Zhang YG et al (2018) Effects of climate on soil phosphorus cycle and availability in natural terrestrial ecosystems. Glob Change Biol 24(8):3344–3356. https://doi.org/10.1111/gcb.14093
Housman DC, Powers HH, Collins AD, Belnap J (2006) Carbon and nitrogen fixation differ between successional stages of biological soil crusts in the Colorado Plateau and Chihuahuan Desert. J Arid Environ 66(4):620–634. https://doi.org/10.1016/j.jaridenv.2005.11.014
Hu HW, Trivedi P, He JZ, Singh BK (2017) Microbial nitrous oxide emissions in dryland ecosystems: mechanisms, microbiome and mitigation. Environ Microbiol 19(12):4808–4828. https://doi.org/10.1111/1462-2920.13795
Huang L, Zhang Z, Li X (2014) Carbon fixation and influencing factors of biological soil crusts in a revegetated area of the Tengger Desert, northern China. J Arid Environ 6(6):725–734. https://doi.org/10.1007/s40333-014-0027-3
Huang JP, Yu HP, Dai AG, Wei Y, Kang LT (2017) Drylands face potential threat under 2 degrees C global warming target. Nat Clim Change 7(6):417–425. https://doi.org/10.1038/Nclimate3275
Hügler M. and Sievert SM. (2011) Beyond the Calvin cycle: autotrophic carbon fixation in the ocean. Ann Rev Mar Sci 3:261–289. https://doi.org/10.1146/annurev-marine-120709-142712
IPCC (2014) Synthesis report. Contribution of Working Groups I, II and III to the fifth assessment report of the Intergovernmental Panel on Climate Change. Geneva, Switzerland, p 151
Jakobsen I, Leggett ME, Richardson AE (2005) Rhizosphere microorganisms and plant phosphorus uptake. In: Sims JT, Sharpley AN (eds) Phosphorus: agriculture and the environment. Agronomy monograph, vol 46. ASA, CSSA, and SSSA, Madison, WI, pp 437–494
Jeffries DL, Klopatek JM, Link SO, Bolton H (1992) Acetylene reduction by cryptogamic crusts from a blackbrush community as related to resaturation and dehydration. Soil Biol Biochem 24(11):1101–1105. https://doi.org/10.1016/0038-0717(92)90059-7

Jin Z, Zhu Y, Li X, Dong Y, An Z (2015) Soil N retention and nitrate leaching in three types of dunes in the Mu Us desert of China. Sci Rep 5:14222. https://doi.org/10.1038/srep14222

Jobbagy EG, Jackson RB (2001) The distribution of soil nutrients with depth: global patterns and the imprint of plants. Biogeochemistry 53(1):51–77. https://doi.org/10.1023/A:1010760720215

Johnson SL, Budinoff CR, Belnap J, Garcia-Pichel F (2005) Relevance of ammonium oxidation within biological soil crust communities. Environ Microbiol 7(1):1–12. https://doi.org/10.1111/j.1462-2920.2004.00649.x

Johnson SL, Neuer S, Garcia-Pichel F (2007) Export of nitrogenous compounds due to incomplete cycling within biological soil crusts of arid lands. Environ Microbiol 9(3):680–689. https://doi.org/10.1111/j.1462-2920.2006.01187.x

Johnson SL, Kuske CR, Carney TD, Housman DC, Gallegos-Graves L, Belnap J (2012) Increased temperature and altered summer precipitation have differential effects on biological soil crusts in a dryland ecosystem. Glob Chang Biol 18(8):2583–2593. https://doi.org/10.1111/j.1365-2486.2012.02709.x

Jones DL, Oburger E (2011) Solubilization of phosphorus by soil microorganisms. In: Büunemann EK, Oberson A, Frossard E (eds) Phosphorus in action. biological processes in soil phosphorus cycling. Springer, Heidelberg, pp 169–198. https://doi.org/10.1007/978-3-642-15271-9

Jungk A, Claassen N (1997) Ion diffusion in the soil-root system. Adv Agron 61:53–110. https://doi.org/10.1016/S0065-2113(08)60662-8

Kalayu G (2019) Phosphate solubilizing microorganisms: promising approach as biofertilizers. Int J Agron. https://doi.org/10.1155/2019/4917256

Katra I, Gross A, Swet N, Tanner S, Krasnov H, Angert A (2016) Substantial dust loss of bioavailable phosphorus from agricultural soils. Sci Rep 6. https://doi.org/10.1038/srep24736

Killingbeck KT, Whitford WG (2001) Nutrient resorption in shrubs growing by design, and by default in Chihuahuan Desert arroyos. Oecologia 128(3):351–359. https://doi.org/10.1007/s004420100668

Koutroulis AG (2019) Dryland changes under different levels of global warming. Sci Total Environ 655:482–511. https://doi.org/10.1016/j.scitotenv.2018.11.215

Kowalchuk GA, Stephen JR (2001) Ammonia-oxidizing bacteria: a model for molecular microbial ecology. Annu Rev Microbiol 55:485–529. https://doi.org/10.1146/annurev.micro.55.1.485

Kowalski AS, Serrano-Ortiz P, Janssens IA, Sánchez-Moral S, Cuezva S, Domingo F et al (2008) Can flux tower research neglect geochemical CO2 exchange? Agric For Meteorol 148(6–7):1045–1054. https://doi.org/10.1016/j.agrformet.2008.02.004

Kucey RMN, Janzen HH, Leggett ME (1989) Microbially mediated increases in plant-available phosphorus. Adv Agron 42:199–228

Kumar A, Kumar A, Patel H (2018) Role of microbes in phosphorus availability and acquisition by plants. Int J Curr Microbiol App Sci 7(5):1344–1347. https://doi.org/10.20546/ijcmas.2018.705.161

Kuo S (1996) Phosphorus. In: Sparks DL (ed) Methods of soil analysis part 3: chemical methods, SSSA book series 5. Soil Science Society of America, Madison, WI, pp 869–920

Lacap DC, Warren-Rhodes KA, Mckay CP, Pointing SB (2011) Cyanobacteria and chloroflexi-dominated hypolithic colonization of quartz at the hyper-arid core of the Atacama Desert, Chile. Extremophiles 15(1):31–38. https://doi.org/10.1007/s00792-010-0334-3

Lacap-Bugler DC, Lee KK, Archer S, Gillman LN, Lau MCY, Leuzinger S et al (2017) Global diversity of desert hypolithic cyanobacteria. Front Microbiol 8:867. https://doi.org/10.3389/fmicb.2017.00867

Ladrón De Guevara M, Gozalo B, Raggio J, Lafuente A, Prieto M, Maestre FT (2018) Warming reduces the cover, richness and evenness of lichen-dominated biocrusts but promotes moss growth: insights from an 8 yr experiment. New Phytol 220(3):811–823. https://doi.org/10.1111/nph.15000

Lafuente A, Berdugo M, De Guevara ML, Gozalo B, Maestre FT (2018) Simulated climate change affects how biocrusts modulate water gains and desiccation dynamics after rainfall events. Ecohydrology 11(6):e1935. https://doi.org/10.1002/eco.1935

Lajtha K, Schlesinger WH (1988) The biogeochemistry of phosphorus cycling and phosphorus availability along a desert soil chronosequence. Ecology 69(1):24–39. https://doi.org/10.2307/1943157

Lal R (2001) Potential of desertification control to sequester carbon and mitigate the greenhouse effect. Clim Change 51(1):35–72. https://doi.org/10.1023/A:1017529816140

Lal R (2002) Soil carbon dynamics in cropland and rangeland. Environ Pollut 116(3):353–362. https://doi.org/10.1016/s0269-7491(01)00211-1

Lal R (2004a) Carbon sequestration in dryland ecosystems. Environ Manag 33(4):528–544. https://doi.org/10.1007/s00267-003-9110-9

Lal R (2004b) Soil carbon sequestration impacts on global climate change and food security. Science 304(5677):1623–1627. https://doi.org/10.1126/science.1097396

Lal R (2009) Sequestering carbon in soils of arid ecosystems. Land Degrad Dev 20(4):441–454. https://doi.org/10.1002/ldr.934

Lal R (2019) Carbon cycling in global drylands. Curr Clim Change Rep 5(3):221–232. https://doi.org/10.1007/s40641-019-00132-z

Lal R, Hassan HM, Dumanski J (1999) Desertification control to sequester C and mitigate the greenhouse effect. In: Rosenberg NJ, Izaurralde RC, Malone EL (eds) Carbon sequestration in soils: science, monitoring and beyond. Battelle, Columbus, OH, pp 83–151

Lange W (1974) Chelating agents and blue-green algae. Can J Microbiol 20(10):1311–1321. https://doi.org/10.1139/m74-204

Leon-Sobrino C, Ramond JB, Maggs-Kolling G, Cowan DA (2019) Nutrient acquisition, rather than stress response over diel cycles, drives microbial transcription in a hyper-arid namib desert soil. Front Microbiol 10:1054. https://doi.org/10.3389/fmicb.2019.01054

Le Quéré C, Andrew RM, Friedlingstein P, Sitch S, Hauck J, Pongratz J et al (2018) Global carbon budget. Earth Syst Sci Data 10:2141–2194. https://doi.org/10.5194/essd-10-2141-2018

Li XZ, Sarah P (2003) Enzyme activities along a climatic transect in the Judean Desert. Catena 53(4):349–363. https://doi.org/10.1016/S0341-8162(03)00087-0

Li J, Okin GS, Hartman LJ, Epstein HE (2007) Quantitative effects of vegetation cover on wind erosion and soil nutrient loss in a desert grassland of southern New Mexico, USA. Biogeochemistry 85(3):317–332. https://doi.org/10.1007/s10533-007-9142-y

Li XD, Fu H, Guo D, Li XD, Wan CG (2010) Partitioning soil respiration and assessing the carbon balance in a Setaria italica (L.) Beauv. Cropland on the Loess Plateau, Northern China. Soil Biol Biochem 42(2):337–346. https://doi.org/10.1016/j.soilbio.2009.11.013

Li KH, Song W, Liu XJ, Shen JL (2012a) Atmospheric reactive nitrogen concentrations at ten sites with contrasting land use in an arid region of central Asia. Biogeosciences 9:4013–4021. https://doi.org/10.5194/bg-9-4013-20

Li XR, Zhang P, Su YG, Jia RL (2012b) Carbon fixation by biological soil crusts following revegetation of sand dunes in arid desert regions of China: a four-year field study. CATENA 97:119–126. https://doi.org/10.1016/j.catena.2012.05.009

Li XJ, Zhao Y, Yang HT, Zhang P, Gao YP (2018) Soil respiration of biologically-crusted soils in response to simulated precipitation pulses in the Tengger Desert, Northern China. Pedosphere 28(1):103–113. https://doi.org/10.1016/S1002-0160(17)60307-2

Lindahl BD, Finlay RD, Cairney JWG (2005) Enzymatic activities of mycelia in mycorrhizal fungal communities. In: Dighton J, White JF, Oudemans P (eds) The fungal community: its organization and role in the ecosystem, vol 23. Taylor & Francis, Boca Raton, FL, pp 331–348

Liu Q, Edwards NT, Post WM, Gu L, Ledford J, Lenhart S (2006) Temperature-independent diel variation in soil respiration observed from a temperate deciduous forest. Glob Chang Biol 12(11):2136–2145. https://doi.org/10.1111/j.1365-2486.2006.01245.x

Liu X, Qi Y, Dong Y, Peng Q, He Y, Sun L et al (2014) Response of soil N_2O emissions to precipitation pulses under different nitrogen availabilities in a semiarid temperate steppe of Inner Mongolia, China. J Arid Land 6(4):410–422. https://doi.org/10.1007/s40333-013-0211-x

Liu YR, Delgado-Baquerizo M, Trivedi P, He JZ, Singhad BK (2016) Species identity of biocrust-forming lichens drives the response of soil nitrogen cycle to altered precipitation frequency and nitrogen amendment. Soil Biol Biochem 96:128–136. https://doi.org/10.1016/j.soilbio.2016.01.021

Liu Z, Sun Y, Zhang Y, Feng W, Lai Z, Fa K, Qin S (2018) Metagenomic and ^{13}C tracing evidence for autotrophic atmospheric carbon absorption in a semiarid desert. Soil Biol Biochem 125:156–166. https://doi.org/10.1016/j.soilbio.2018.07.012

Liu D, Keiblinger KM, Leitner S, Wegner U, Zimmermann M, Fuchs S et al (2019) Response of microbial communities and their metabolic functions to drying-rewetting stress in a temperate forest soil. Microorganisms 7(5):129. https://doi.org/10.3390/microorganisms7050129

Logan JA, Prather MJ, Wofsy SC, Mcelroy MB (1981) Tropospheric chemistry: a global perspective. J Geophys Res 86(C8):7210–7254. https://doi.org/10.1029/JC086iC08p07210

Lu HB, Liu SR, Wang H, Luan JW, Schindlbacher A, Liu YC, Wang Y (2017) Experimental throughfall reduction barely affects soil carbon dynamics in a warm-temperate oak forest, central China. Sci Rep 7. https://doi.org/10.1038/s41598-017-15157-3

Luo W, Xu C, Ma W, Yue X, Liang X, Zuo X et al (2018) Effects of extreme drought on plant nutrient uptake and resorption in rhizomatous vs bunchgrass-dominated grasslands. Oecologia 188(2):633–643. https://doi.org/10.1007/s00442-018-4232-1

Luo G, Xue C, Jiang Q, Xiao Y, Zhang F, Guo S et al (2020) Soil carbon, nitrogen, and phosphorus cycling microbial populations and their resistance to global change depend on soil C:N:P stoichiometry. MSystems 5(3):e00162-00120. https://doi.org/10.1128/mSystems.00162-20

Ma WH, He JS, Yang YH, Wang XP, Liang CZ, Anwar M et al (2010) Environmental factors covary with plant diversity-productivity relationships among Chinese grassland sites. Glob Ecol Biogeogr 19(2):233–243. https://doi.org/10.1111/j.1466-8238.2009.00508.x

Ma J, Liu R, Li Y (2017) Abiotic contribution to total soil CO2 flux across a broad range of land-cover types in a desert region. J Arid Land 9(1):13–26. https://doi.org/10.1007/s40333-016-0061-4

Mackenzie FT, Vera LM, Lerman A (2002) Century-scale nitrogen and phosphorus controls of the carbon cycle. Chem Geol 190(1–4):13–32. https://doi.org/10.1016/S0009-2541(02)00108-0

Maestre FT, Bowker MA, Canton Y, Castillo-Monroy AP, Cortina J, Escolar C et al (2011) Ecology and functional roles of biological soil crusts in semi-arid ecosystems of Spain. J Arid Environ 75(12):1282–1291. https://doi.org/10.1016/j.jaridenv.2010.12.008

Maestre FT, Escolar C, De Guevara ML, Quero JL, Lazaro R, Delgado-Baquerizo M et al (2013) Changes in biocrust cover drive carbon cycle responses to climate change in drylands. Glob Chang Biol 19(12):3835–3847. https://doi.org/10.1111/gcb.12306

Maestre FT, Eldridge DJ, Soliveres S, Kefi S, Delgado-Baquerizo M, Bowker MA et al (2016) Structure and functioning of dryland ecosystems in a changing world. Annu Rev Ecol Evol Syst 47:215–237. https://doi.org/10.1146/annurev-ecolsys-121415-032311

Magid J, Nielsen NE (1992) Seasonal-variation in organic and inorganic phosphorus fractions of temperate-climate sandy soils. Plant Soil 144(2):155–165. https://doi.org/10.1007/Bf00012872

Maier S, Tamm A, Wu D, Caesar J, Grube M, Weber B (2018) Photoautotrophic organisms control microbial abundance, diversity, and physiology in different types of biological soil crusts. ISME J 12(4):1032–1046. https://doi.org/10.1038/s41396-018-0062-8

Makhalanyane TP, Valverde A, Gunnigle E, Frossard A, Ramond JB, Cowan DA (2015) Microbial ecology of hot desert edaphic systems. FEMS Microbiol Rev 39(2):203–221. https://doi.org/10.1093/femsre/fuu011

Malam Issa O, Stal LJ, Défarge C, Couté A, Trichet J (2001) Nitrogen fixation by microbial crusts from desiccated Sahelian soils (Niger). Soil Biol Biochem 33(10):1425–1428. https://doi.org/10.1016/S0038-0717(01)00046-3

Mander L, Dekker SC, Li M, Mio W, Punyasena SW, Lenton TM (2017) A morphometric analysis of vegetation patterns in dryland ecosystems. R Soc Open Sci 4(2):160443. https://doi.org/10.1098/rsos.160443

Manzoni S, Schimel JP, Porporato A (2012) Responses of soil microbial communities to water stress: results from a meta-analysis. Ecology 93(4):930–938. https://doi.org/10.1890/11-0026.1

Marion GM (1989) Correlation between long-term pedogenic CaCO3 formation rate and Modern precipitation in deserts of the America southwest. Quat Res 32(3):291–229. https://doi.org/10.1016/0033-5894(89)90095-1

Marusenko Y, Bates ST, Anderson I, Johnson SL (2013) Ammonia-oxidizing archaea and bacteria are structured by geography in biological soil crusts across North American arid lands. Ecol Process 2(9). https://doi.org/10.1186/2192-1709-2-9

Marusenko Y, Garcia-Pichel F, and Hall SJ. (2015) Ammonia-oxidizing archaea respond positively to inorganic nitrogen addition in desert soils. FEMS Microbiol Ecol 91(2):1–11. https://doi.org/10.1093/femsec/fiu023

Mayland HF, Mcintosh TH (1966) Availability of biologically fixed atmospheric nitrogen-15 to higher plants. Nature 209:421–422

Mccalley CK, Sparks JP (2008) Controls over nitric oxide and ammonia emissions from Mojave Desert soils. Oecologia 156(4):871–881. https://doi.org/10.1007/s00442-008-1031-0

Mccalley CK, Sparks JP (2009) Abiotic gas formation drives nitrogen loss from a desert ecosystem. Science 326(5954):837–840. https://doi.org/10.1126/science.1178984

Mccalley CK, Strahm BD, Sparks KL, Eller ASD, Sparks JP (2011) The effect of long-term exposure to elevated CO2 on nitrogen gas emissions from Mojave Desert soils. J Geophys Res 116(G3). https://doi.org/10.1029/2011JG001667

Mcgill WB, Cole CV (1981) Comparative aspects of cycling of organic C, N, S and P through soil organic-matter. Geoderma 26(4):267–286. https://doi.org/10.1016/0016-7061(81)90024-0

Meisner A, Baath E, Rousk J (2013) Microbial growth responses upon rewetting soil dried for four days or one year. Soil Biol Biochem 66:188–192. https://doi.org/10.1016/j.soilbio.2013.07.014

Meisner A, Rousk J, Baath E (2015) Prolonged drought changes the bacterial growth response to rewetting. Soil Biol Biochem 88:314–322. https://doi.org/10.1016/j.soilbio.2015.06.002

Meslier V, Casero MC, Dailey M, Wierzchos J, Ascaso C, Artieda O et al (2018) Fundamental drivers for endolithic microbial community assemblies in the hyperarid Atacama Desert. Environ Microbiol 20(5):1765–1781. https://doi.org/10.1111/1462-2920.14106

Michalski G, Böhlke JK, Thiemens M (2004) Long term atmospheric deposition as the source of nitrate and other salts in the Atacama Desert, Chile: new evidence from mass-independent oxygen isotopic compositions. Geochim Cosmochim Acta 68(20):4023–4038. https://doi.org/10.1016/j.gca.2004.04.009

Miller ME, Belnap J, Beatty SW, Webb BL (2006a) Effects of water additions, chemical amendments, and plants on in situ measures of nutrient bioavailability in calcareous soils of southeastern Utah, USA. Plant Soil 288(1–2):19–29. https://doi.org/10.1007/s11104-006-9014-6

Miller ME, Belnap J, Beatty SW, Reynolds RL (2006b) Performance of Bromus tectorum L. in relation to soil properties, water additions, and chemical amendments in calcareous soils of southeastern Utah, USA. Plant Soil 288(1–2):1–18. https://doi.org/10.1007/s11104-006-0058-4

Min K, Lehmeier CA, Ballantyne F, Tatarko A, Billings SA (2014) Differential effects of pH on temperature sensitivity of organic carbon and nitrogen decay. Soil Biol Biochem 76:193–200. https://doi.org/10.1016/j.soilbio.2014.05.021

Miralles I, Trasar-Cepeda C, Leiros MC, Gil-Sotres F (2013) Labile carbon in biological soil crusts in the Tabernas desert, SE Spain. Soil Biol Biochem 58:1–8. https://doi.org/10.1016/j.soilbio.2012.11.010

Miralles I, De Guevara ML, Chamizo S, Rodriguez-Caballero E, Ortega R, Van Wesemael B, Canton Y (2018) Soil CO2 exchange controlled by the interaction of biocrust successional stage and environmental variables in two semiarid ecosystems. Soil Biol Biochem 124:11–23. https://doi.org/10.1016/j.soilbio.2018.05.020

Mogul R, Vaishampayan P, Bashir M, Mckay CP, Schubert K, Bornaccorsi R et al (2017) Microbial community and biochemical dynamics of biological soil crusts across a gradient of surface coverage in the Central Mojave Desert. Front Microbiol 8:1974. https://doi.org/10.3389/fmicb.2017.01974

Montañez A, Danso SKA, Hardarson G (1995) The effect of temperature on nodulation and nitrogen fixation by five Bradyrhizobium japonicum strains. Appl Soil Ecol 2(3):165–174. https://doi.org/10.1016/0929-1393(95)00052-M

Montiel-González C, Tapia-Torres Y, Souza V, Garcia-Oliva F (2017) The response of soil microbial communities to variation in annual precipitation depends on soil nutritional status in an oligotrophic desert. PeerJ 5:e4007. https://doi.org/10.7717/peerj.4007

Mummey DL, Smith JL, Bolton H (1994) Nitrous oxide flux from a shrub-steppe ecosystem: sources and regulation. Soil Biol Biochem 26(2):279–286. https://doi.org/10.1016/0038-0717(94)90168-6

Mureva A, Ward D, Pillay T, Chivenge P, Cramer M (2018) Soil organic carbon increases in semi-arid regions while it decreases in humid regions due to woody-plant encroachment of grasslands in South Africa. Sci Rep 8. https://doi.org/10.1038/s41598-018-33701-7

Nannipieri P, Giagnoni L, Landi L, Renella G (2011) Role of phosphatase enzymes in soil. In: Bünemann E, Oberson A, Frossard E (eds) Phosphorus in action: biological processes in soil phosphorus cycling. Soil biology, vol 26. Springer, Heidelberg, pp 215–243. https://doi.org/10.1007/978-3-642-15271-9_9

Neff JC, Reynolds R, Sanford RL, Fernandez D, Lamothe P (2006) Controls of bedrock geochemistry on soil and plant nutrients in southeastern Utah. Ecosystems 9(6):879–893. https://doi.org/10.1007/s10021-005-0092-8

Nejidat A (2005) Nitrification and occurrence of salt-tolerant nitrifying bacteria in the Negev desert soils. FEMS Microbiol Ecol 52(1):21–29. https://doi.org/10.1016/j.femsec.2004.10.011

Offre P, Spang A, Schleper C (2013) Archaea in biogeochemical cycles. Annu Rev Microbiol 67: 437–457. https://doi.org/10.1146/annurev-micro-092412-155614

Ojima DS, Smith MS, Beardsley M (1995) Factors affecting carbon storage in semi-arid and arid ecosystems. In: Squires VR, Glenn EP, Ayoub AT (eds) Combating global climate change by combating land degradation. Nairobi, Kenya, UNEP, pp 93–115

Okin GS, Mahowald N, Chadwick OA, Artaxo P (2004) Impact of desert dust on the biogeochemistry of phosphorus in terrestrial ecosystems. Glob Biogeochem Cycle 18(2). https://doi.org/10.1029/2003gb002145

Orlando J, Caru M, Pommerenke B, Braker G (2012) Diversity and activity of denitrifiers of Chilean arid soil ecosystems. Front Microbiol 3:101. https://doi.org/10.3389/fmicb.2012.00101

Ou Y, Rousseau AN, Wang LX, Yan BX (2017) Spatio-temporal patterns of soil organic carbon and pH in relation to environmental factors-A case study of the Black Soil Region of Northeastern China. Agr Ecosyst Environ 245:22–31. https://doi.org/10.1016/j.agee.2017.05.003

Patra DD, Brookes PC, Coleman K, Jenkinson DS (1990) Seasonal-changes of soil microbial biomass in an arable and a grassland soil which have been under uniform management for many years. Soil Biol Biochem 22(6):739–742. https://doi.org/10.1016/0038-0717(90)90151-O

Peñuelas J, Sardans J, Rivas-Ubach A, Janssens IA (2012) The human-induced imbalance between C, N and P in Earth's life system. Glob Change Biol 18(1):3–6. https://doi.org/10.1111/j.1365-2486.2011.02568.x

Perez-Quezada JF, Delpiano CA, Snyder KA, Johnson DA, Franck N (2011) Carbon pools in an arid shrubland in Chile under natural and afforested conditions. J Arid Environ 75(1):29–37. https://doi.org/10.1016/j.jaridenv.2010.08.003

Perroni Y, Garcia-Oliva F, Tapia-Torres Y, Souza V (2014) Relationship between soil P fractions and microbial biomass in an oligotrophic grassland-desert scrub system. Ecol Res 29(3): 463–472. https://doi.org/10.1007/s11284-014-1138-1

Peterjohn WT, Schlesinger WH (1990) Nitrogen loss from deserts in the southwestern United States. Biogeochemistry 10(1):67–79. https://doi.org/10.1007/BF00000893

Plaza C, Zaccone C, Sawicka K, Mendez AM, Tarquis A, Gasco G et al (2018a) Soil resources and element stocks in drylands to face global issues. Sci Rep 8(1):13788. https://doi.org/10.1038/s41598-018-32229-0

Plaza C, Gascó G, Méndez AM, Zaccone C, Maestre FT (2018b) Soil organic matter in dryland ecosystems. In: Garcia C, Nannipieri P, Hernandez T (eds) The future of soil carbon. Its conservation and formation. Academic, London, pp 39–70. https://doi.org/10.1016/C2016-0-01797-7

Pointing SB, Belnap J (2012) Microbial colonization and controls in dryland systems. Nat Rev Microbiol 10(8):551–562. https://doi.org/10.1038/nrmicro2831

Porder S, Ramachandran S (2013) The phosphorus concentration of common rocks-a potential driver of ecosystem P status. Plant Soil 367(1–2):41–55. https://doi.org/10.1007/s11104-012-1490-2

Porporato A, D'Odorico P, Laio F, Rodriguez-Iturbe I (2003) Hydrologic controls on soil carbon and nitrogen cycles. I. Modeling scheme. Adv Water Resour 26(1):45–58. https://doi.org/10.1016/S0309-1708(02)00094-5

Posmanik R, Nejidat A, Dahan O, Gross A (2017) Seasonal and soil-type dependent emissions of nitrous oxide from irrigated desert soils amended with digested poultry manures. Sci Total Environ 1(593–594):91–98. https://doi.org/10.1016/j.scitotenv.2017.03.115

Qin Y, Li Y, Bao H, Liu F (2012) Massive atmospheric nitrate accumulation in a continental interior desert, northwestern China. Geology 40(7):623–626. https://doi.org/10.1130/G32953.1

Qin FY, Shi XZ, Xu SX, Yu DS, Wang DD (2016) Zonal differences in correlation patterns between soil organic carbon and climate factors at multi-extent. Chin Geogr Sci 26(5): 670–678. https://doi.org/10.1007/s11769-015-0736-3

Quiquampoix H, Mousain D (2005) Enzymatic hydrolysis of organic phosphorus. In: Turner BL, Frossard E, Baldwin DS (eds) Organic phosphorus in the environment. CABI, Wallingford, pp 89–112

Ramond JB, Woodborne S, Hall G, Seely M, Cowan DA (2018) Namib Desert primary productivity is driven by cryptic microbial community N-fixation. Sci Rep 8(1):6921. https://doi.org/10.1038/s41598-018-25078-4

Rasmussen C (2006) Distribution of soil organic and inorganic carbon pools by biome and soil taxa in Arizona. Soil Sci Soc Am J 70(1):256–265. https://doi.org/10.2136/sssaj2005.0118

Reynolds R, Neff J, Reheis M, Lamothe P (2006) Atmospheric dust in modern soil on aeolian sandstone, Colorado Plateau (USA): variation with landscape position and contribution to potential plant nutrients. Geoderma 130(1–2):108–123. https://doi.org/10.1016/j.geoderma.2005.01.012

Richardson AE, Simpson RJ (2011) Soil microorganisms mediating phosphorus availability. Plant Physiol 156(3):989–996. https://doi.org/10.1104/pp.111.175448

Rodriguez H, Fraga R (1999) Phosphate solubilizing bacteria and their role in plant growth promotion. Biotechnol Adv 17(4–5):319–339. https://doi.org/10.1016/S0734-9750(99)00014-2

Ronca S, Ramond JB, Jones BE, Seely M, Cowan DA (2015) Namib Desert dune/interdune transects exhibit habitat-specific edaphic bacterial communities. Front Microbiol 6:845. https://doi.org/10.3389/fmicb.2015.00845

Rosenberg NJ, Izaurralde RC, Malone EL (1999) Carbon sequestration in soils: science, monitoring and beyond. Battelle, Columbus, OH

Rotenberg E, Yakir D (2010) Contribution of semi-arid forests to the climate system. Science 327(5964):451–454. https://doi.org/10.1126/science.1179998

Runge-Metzger A (1995) Closing the cycle: obstacles to efficient P management for improved global food security. In: Scope-Scientific Committee on Problems of the Environment International Council of Scientific Unions, Indianapolis, pp 27–42

Ruttenberg KC (2003) The global phosphorus cycle. In: Schlesinger WH, Holland HD, Turekian KK (eds) Treatise on geochemistry, vol 8. Elsevier, New York, pp 585–643. https://doi.org/10.1016/B0-08-043751-6/08153-6

Ryan MG, Law BE (2005) Interpreting, measuring, and modeling soil respiration. Biogeochemistry 73(1):3–27. https://doi.org/10.1007/s10533-004-5167-7

Ryan J, Hasan HM, Baasiri M, Tabbara HS (1985) Availability and transformation of applied phosphorus in calcareous lebanese soils. Soil Sci Soc Am J 49(5):1215–1220. https://doi.org/10.2136/sssaj1985.03615995004900050029x

Safriel U, Adeel Z, Niemeijer D, Puigdefabregas J, White R, Lal R et al (2005) Dryland systems. In: Hassan R, Scholes R, Ash N (eds) Ecosystems and human well-being: current state and trends: findings of the Condition and Trends Working Group, vol 1. Island Press, Washington, DC, pp 623–662

Sahrawat KL (2003) Importance of inorganic carbon in sequestering carbon in soils of the dry regions. Curr Sci 84(7):864–865

Sakamoto K, Oba Y (1994) Effect of fungal to bacterial biomass ratio on the relationship between CO2 evolution and total soil microbial biomass. Biol Fert Soils 17(1):39–44. https://doi.org/10.1007/Bf00418670

Sancho LG, Belnap J, Colesie C, Raggio J, Weber B (2016) Carbon budgets of biological soil crusts at micro-, meso-, and global scales. In: Weber B, Büdel B, Belnap J (eds) Biological soil crusts: an organizing principle in drylands. Ecological studies (analysis and synthesis), vol 226. Springer, Cham, pp 287–304. https://doi.org/10.1007/978-3-319-30214-0_15

Sanderman J (2012) Can management induced changes in the carbonate system drive soil carbon sequestration? A review with particular focus on Australia. Agr Ecosyst Environ 155:70–77. https://doi.org/10.1016/j.agee.2012.04.015

Sardans J, Penuelas J, Estiarte M (2006) Warming and drought alter soil phosphatase activity and soil P availability in a Mediterranean shrubland. Plant Soil 289(1–2):227–238. https://doi.org/10.1007/s11104-006-9131-2

Sardans J, Penuelas J, Ogaya R (2008) Experimental drought reduced acid and alkaline phosphatase activity and increased organic extractable P in soil in a Quercus ilex Mediterranean forest. Eur J Soil Biol 44(5–6):509–520. https://doi.org/10.1016/j.ejsobi.2008.09.011

Sardans J, Rivas-Ubach A, Penuelas J (2012) The elemental stoichiometry of aquatic and terrestrial ecosystems and its relationships with organismic lifestyle and ecosystem structure and function: a review and perspectives. Biogeochemistry 111(1–3):1–39. https://doi.org/10.1007/s10533-011-9640-9

Schaeffer SM, Evans RD (2005) Pulse additions of soil carbon and nitrogen affect soil nitrogen dynamics in an arid Colorado Plateau shrubland. Oecologia 145(3):425–433. https://doi.org/10.1007/s00442-005-0140-2

Scharpenseel HW, Mtimet A, Freytag J (2000) Soil inorganic carbon and global change. In: Lal R, Kimble JM, Eswaran H, Stewart BA (eds) Global climate change and pedogenic carbonates. CRC, Boca Raton, FL, pp 27–42

Schimel DS (2010) Drylands in the earth system. Science 327(5964):418–419. https://doi.org/10.1126/science.1184946

Schimel JP (2018) Life in dry soils: effects of drought on soil microbial communities and processes. Annu Rev Ecol Evol Syst 49:409–432. https://doi.org/10.1146/annurev-ecolsys-110617-062614

Schimel JP, Bennett J (2004) Nitrogen mineralization: challenges of a changing paradigm. Ecology 85(3):591–602. https://doi.org/10.1890/03-8002

Schlaepfer DR, Bradford JB, Lauenroth WK, Munson SM, Tietjen B, Hall SA et al (2017) Climate change reduces extent of temperate drylands and intensifies drought in deep soils. Nat Commun 8. https://doi.org/10.1038/ncomms14196

Schlesinger WH (1985) The formation of caliche in soils of the Mojave Desert, California. Geochimica et Cosmochimica Acta 49(1):57–66. https://doi.org/10.1016/0016-7037(85)90191-7

Schlesinger WH (2002) Inorganic carbon and the global C cycle. In: Lal R (ed) Encyclopedia of soil science. Marcel Dekker, New York, pp 706–708

Schlesinger WH (2006) Inorganic carbon and the global C cycle. In: Lal R (ed) Encyclopedia of soil science, 3rd edn. CRC, Boca Raton, FL. https://doi.org/10.4324/9781315161860

Schlesinger WH, Pippen JS, Wallenstein MD, Hofmockel KS, Klepeis DM, Mahall BE (2003) Community composition and photosynthesis by photoautotrophs under quartz pebbles, Southern Mojave Desert. Ecology 84(12):3222–3231. https://doi.org/10.1890/02-0549

Schwabedissen SG, Lohse KA, Reed SC, Aho KA, Magnuson TS (2017) Nitrogenase activity by biological soil crusts in cold sagebrush steppe ecosystems. Biogeochemistry 134:57–76. https://doi.org/10.1007/s10533-017-0342-9

Scurlock JMO, Hall DO (1998) The global carbon sink: a grassland perspective. Glob Chang Biol 4:229–233. https://doi.org/10.1046/j.1365-2486.1998.00151.x

Selmants PC, Hart SC (2010) Phosphorus and soil development: does the Walker and Syers model apply to semiarid ecosystems? Ecology 91(2):474–484. https://doi.org/10.1890/09-0243.1

Serrano-Ortiz P, Sánchez-Cañete EP, Oyonarte C (2012) The carbon cycle in drylands. In: Lal R, Lorenz K, Hüttl R, Schneider B, Von Braun J (eds) Recarbonization of the biosphere. Springer, Dordrecht, pp 347–368. https://doi.org/10.1007/978-94-007-4159-1_15

Shoun H, Fushinobu S, Jiang L, Kim SW, Wakagi T (2012) Fungal denitrification and nitric oxide reductase cytochrome P450nor. Philos Trans R Soc Lond B Biol Sci 367(1593):1186–1194. https://doi.org/10.1098/rstb.2011.0335

Siebers N, Sumann M, Kaiser K, Amelung W (2017) Climatic effects on phosphorus fractions of native and cultivated North American grassland soils. Soil Sci Soc Am J 81(2):299–309. https://doi.org/10.2136/sssaj2016.06.0181

Sims JT, Pierzynski GM (2005) Chemistry of phosphorus in soils. Chemical processes in soils. SSSA book series, vol 8. SSSA, Madison, WI, pp 151–192

Sinsabaugh RS (1994) Enzymic analysis of microbial pattern and process. Biol Fert Soils 17(1): 69–74. https://doi.org/10.1007/BF00418675

Sinsabaugh RL, Lauber CL, Weintraub MN, Ahmed B, Allison SD, Crenshaw C et al (2008) Stoichiometry of soil enzyme activity at global scale. Ecol Lett 11(11):1252–1264. https://doi.org/10.1111/j.1461-0248.2008.01245.x

Sjögersten S, Wookey PA (2009) The impact of climate change on ecosystem carbon dynamics at the Scandinavian mountain birch forest-tundra heath ecotone. Ambio 38(1):2–10. https://www.jstor.org/stable/25515792

Smith KR, Waring BG (2019) Broad-scale patterns of soil carbon (C) pools and fluxes across semiarid ecosystems are linked to climate and soil texture. Ecosystems 22(4):742–753. https://doi.org/10.1007/s10021-018-0299-0

Sollins P, Robertson GP, Uehara G (1988) Nutrient mobility in variable-charge and permanent-charge soils. Biogeochemistry 6(3):181–199. https://doi.org/10.1007/Bf02182995

Spohn M, Kuzyakov Y (2013) Phosphorus mineralization can be driven by microbial need for carbon. Soil Biol Biochem 61:69–75. https://doi.org/10.1016/j.soilbio.2013.02.013

Stark JM, Firestone MK (1995) Mechanisms for soil-moisture effects on activity of nitrifying bacteria. Appl Environ Microbiol 61(1):218–221

Stein LY (2019) Insights into the physiology of ammonia-oxidizing microorganisms. Curr Opin Chem Biol 49:9–15. https://doi.org/10.1016/j.cbpa.2018.09.003

Stevenson FJS, Cole MA (1999) Cycles of soils: carbon, nitrogen, phosphorus, sulfur, micronutrients, 2nd edn. Wiley, New York

Strauss SL, Day TA, Garcia-Pichel F (2012) Nitrogen cycling in desert biological soil crusts across biogeographic regions in the Southwestern United States. Biogeochemistry 108(1/3):171–182. https://doi.org/10.1007/s10533-011-9587-x

Stursova M, Crenshaw CL, Sinsabaugh RL (2006) Microbial responses to long-term N deposition in a semiarid grassland. Microb Ecol 51(1):90–98. https://doi.org/10.1007/s00248-005-5156-y

Su Y-G, Zhao X, Li A-X, Li X-R, Huang G (2011) Nitrogen fixation in biological soil crusts from the Tengger desert, northern China. Eur J Soil Biol 47(3):182–187. https://doi.org/10.1016/j.ejsobi.2011.04.001

Suddick EC, Whitney P, Townsend AR, Davidson EA (2013) The role of nitrogen in climate change and the impacts of nitrogen–climate interactions in the United States: foreword to thematic issue. Biogeochemistry 114:1–10. https://doi.org/10.1007/s10533-012-9795-z

Tan WF, Zhang R, Cao H, Huang CQ, Yang QK, Wang M-K, Koopal LK (2014) Soil inorganic carbon stock under different soil types and land uses on the Loess Plateau region of China. Catena 121:22–30. https://doi.org/10.1016/j.catena.2014.04.014

Thomey ML, Ford PL, Reeves MC, Finch DM, Litvak ME, Collins SL (2014) Climate change impacts on future carbon stores and management of warm deserts of the United States. Rangelands 36(1):16–24. https://doi.org/10.2111/RANGELANDS-D-13-00045.1

Tian JH, Boitt G, Black A, Wakelin S, Condron LM, Chen LJ (2017) Accumulation and distribution of phosphorus in the soil profile under fertilized grazed pasture. Agr Ecosyst Environ 239:228–235. https://doi.org/10.1016/j.agee.2017.01.022

Tiessen H, Moir JO (1993) Characterization of available p by sequential extraction. In: Carter MR (ed) Soil sampling and methods of analysis. Lewis, Boca Raton, FL, pp 75–86

Tiessen H, Stewart JWB, Cole CV (1984) Pathways of phosphorus transformations in soils of differing pedogenesis. Soil Sci Soc Am J 48(4):853–858. https://doi.org/10.2136/sssaj1984.03615995004800040031x

Trolldenier G (1982) Effect of soil temperature on nitrogen fixation on roots of rice and reed. Plant Soil 68(2):217–222. https://doi.org/10.1007/BF02373707

Tucker CL, Reed SC (2016) Low soil moisture during hot periods drives apparent negative temperature sensitivity of soil respiration in a dryland ecosystem: a multi-model comparison. Biogeochemistry 128(1–2):155–169. https://doi.org/10.1007/s10533-016-0200-1

Turner BL, Mckelvie ID, Haygarth PM (2002) Characterisation of water-extractable soil organic phosphorus by phosphatase hydrolysis. Soil Biol Biochem 34(1):27–35. https://doi.org/10.1016/S0038-0717(01)00144-4

Turner BL, Cade-Menun BJ, Westermann DT (2003a) Organic phosphorus composition and potential bioavailability in semi-arid arable soils of the western United States. Soil Sci Soc Am J 67(4):1168–1179. https://doi.org/10.2136/sssaj2003.1168

Turner BL, Baxter R, Ellwood NTW, Whitton BA (2003b) Seasonal phosphatase activities of mosses from Upper Teesdale, northern England. J Bryol 25:189–200. https://doi.org/10.1179/037366803235001670

Turner BL, Condron LM, Richardson SJ, Peltzer DA, Allison VJ (2007) Soil organic phosphorus transformations during pedogenesis. Ecosystems 10(7):1166–1181. https://doi.org/10.1007/s10021-007-9086-z

Van Amstel AR, Swart RJ (1994) Methane and nitrous oxide emissions: an introduction. Fertil Res 37:213–225. https://doi.org/10.1007/BF00748940

Van Kessel MA, Speth DR, Albertsen M, Nielsen PH, Op Den Camp HJ, Kartal B et al (2015) Complete nitrification by a single microorganism. Nature 528(7583):555–559. https://doi.org/10.1038/nature16459

Veste M, Littmann T, Breckle SW, Yair A (2001) The role of biological soil crusts on desert sand dunes in the Northwestern Negev, Israel. In: Breckle SW, Veste M, Wucherer W (eds) Sustainable land use in deserts. Springer, Berlin, pp 357–367. https://doi.org/10.1007/978-3-642-59560-8_38

Veste M, Heusinkveld BG, Berkowicz SM, Breckle SW, Littmann T, Jacobs AFG (2008) Dew formation and activity of biological soil crusts. In: Breckle SW, Yair A, Veste M (eds) Arid dune ecosystems. Ecological studies (analysis and synthesis). Springer, Heidelberg

Vikram S, Guerrero LD, Makhalanyane TP, Le PT, Seely M, Cowan DA (2016) Metagenomic analysis provides insights into functional capacity in a hyperarid desert soil niche community. Environ Microbiol 18(6):1875–1888. https://doi.org/10.1111/1462-2920.13088

Vitousek PM, Porder S, Houlton BZ, Chadwick OA (2010) Terrestrial phosphorus limitation: mechanisms, implications, and nitrogen-phosphorus interactions. Ecol Appl 20(1):5–15. https://doi.org/10.1890/08-0127.1

Walker TW, Syers JK (1976) The fate of phosphorus during pedogenesis. Geoderma 15(1):1–19. https://doi.org/10.1016/0016-7061(76)90066-5

Walvoord MA, Phillips FM, Stonestrom DA, Evans RD, Hartsough PC, Newman BD, Striegl RG (2003) A reservoir of nitrate beneath desert soils. Science 302(5647):1021–1024. https://doi.org/10.1126/science.1086435

Wang ZY, Sun G, Luo P, Mou CX, Wang J (2013) A study of soil-dynamics based on a simulated drought in an alpine meadow on the Tibetan Plateau. J Mt Sci 10(5):833–844. https://doi.org/10.1007/s11629-013-2697-2
Wang B, Zha TS, Jia X, Wu B, Zhang YQ, Qin SG (2014a) Soil moisture modifies the response of soil respiration to temperature in a desert shrub ecosystem. Biogeosciences 11(2):259–268. https://doi.org/10.5194/bg-11-259-2014
Wang C, Wang X, Liu D, Wu H, Lu X, Fang Y et al (2014b) Aridity threshold in controlling ecosystem nitrogen cycling in arid and semi-arid grasslands. Nat Commun 5:4799. https://doi.org/10.1038/ncomms5799
Wang F, Michalski G, Seo J-H, Ge W (2014c) Geochemical, isotopic, and mineralogical constraints on atmospheric deposition in the hyper-arid Atacama Desert, Chile. Geochim Cosmochim Ac 135:29–48. https://doi.org/10.1016/j.gca.2014.03.017
Wang M, Su YZ, Yang X (2014d) Spatial distribution of soil organic carbon and its influencing factors in desert grasslands of the Hexi Corridor, Northwest China. PLoS One 9(4). https://doi.org/10.1371/journal.pone.0094652
Wang X, Wang J, Xu M, Zhang W, Fan T, Zhang J (2015) Carbon accumulation in arid croplands of northwest China: pedogenic carbonate exceeding organic carbon. Sci Rep 5:11439. https://doi.org/10.1038/srep11439
Wardle DA (2013) Drivers of decoupling in drylands. Nature 502(7473):628–629. https://doi.org/10.1038/502628a
Warren SD (2003) Synopsis: influence of biological soil crusts on arid land hydrology and soil stability. In: Belnap J, Lange OL (eds) Biological soil crusts: structure, function and management. Springer, Heidelberg, pp 349–360
Weber B, Wu D, Tamm A, Ruckteschler N (2015) Biological soil crusts accelerate the nitrogen cycle through large NO and HONO emissions in drylands. PNAS 112(50):15384–15389. https://doi.org/10.1073/pnas.1515818112
Wei ST, Lacap-Bugler DC, Lau MC, Caruso T, Rao S, De Los Rios A et al (2016) Taxonomic and functional diversity of soil and hypolithic microbial communities in Miers Valley, McMurdo Dry Valleys, Antarctica. Front Microbiol 7:1642. https://doi.org/10.3389/fmicb.2016.01642
Weil RR, Brady NC (2017) The nature and properties of soils, 15th edn. Pearson, Boston
White CS, Moore DI, Craig JA (2004) Regional-scale drought increases potential soil fertility in semiarid grasslands. Biol Fert Soils 40(1):73–78. https://doi.org/10.1007/s00374-004-0744-4
White DA, Welty-Bernard A, Rasmussen C, Schwartz E (2009) Vegetation controls on soil organic carbon dynamics in an arid, hyperthermic ecosystem. Geoderma 150(1–2):214–223. https://doi.org/10.1016/j.geoderma.2009.02.011
Whitelaw MA (2000) Growth promotion of plants inoculated with phosphate-solubilizing fungi. Adv Agron 69:99–151. https://doi.org/10.1016/S0065-2113(08)60948-7
Whitford WG (1999) Comparison of ecosystem processes in the Nama-karoo and other deserts. In: Deand WRJ, Milton SJ (eds) The Karoo: ecological patterns and processes. Cambridge University Press, Cambridge, pp 291–313
Whitford WG (2002) Ecology of desert systems. Academic, San Diego, CA
Wierenga PJ, Hendrickx JMH, Nash MH, Ludwig JA, Daugherty LA (1987) Variation of soil and vegetation with distance along a transect in the Chihuahuan Desert. J Arid Environ 13(1):53–63
Wiesmeier M, Barthold F, Blank B, Kögel-Knabner I (2011) Digital mapping of soil organic matter stocks using Random Forest modeling in a semi-arid steppe ecosystem. Plant Soil 340(1–2):7–24. https://doi.org/10.1007/s11104-010-0425-z
Wild A, Oke OL (1966) Organic phosphate compounds in calcium chloride extracts of soils: identification and availability to plants. J Soil Sci 17(2):356–371. https://doi.org/10.1111/j.1365-2389.1966.tb01479.x
Williams EJ, Guenther A, Fehsenfeldi FC (1992) An inventory of nitric oxide emissions from soils in the United States. J Geophys Res 97(D7):7511–7519. https://doi.org/10.1029/92JD00412
Wilske B, Burgheimer J, Karnieli A, Zaady E, Andreae MO, Yakir D, Kesselmeier J (2008) The CO2 exchange of biological soil crusts in a semiarid grass-shrubland at the northern transition

zone of the Negev desert, Israel. Biogeosciences 5(5):1411–1423. https://doi.org/10.5194/bg-5-1411-2008
Wilske B, Burgheimer J, Karnieli A, Zaady E (2009) Modeling the variability in annual carbon fluxes related to biological soil crusts in a Mediterranean shrubland. Biogeosci Discus 6:7295–7324. https://doi.org/10.5194/bgd-6-7295-2009
Wohlfahrt G, Fenstermaker LF, Arnone JA (2008) Large annual net ecosystem CO2 uptake of a Mojave Desert ecosystem. Glob Chang Biol 14:1475–1487. https://doi.org/10.1111/j.1365-2486.2008.01593.x
Woomer PL, Touré M, Sall M (2004) Carbon stocks in Senegal's Sahel transition zone. J Arid Environ 59(9):499–510. https://doi.org/10.1016/j.jaridenv.2004.03.027
Wu JS, Huang M, Xiao HA, Su YR, Tong CL, Huang DY, Syers JK (2007) Dynamics in microbial immobilization and transformations of phosphorus in highly weathered subtropical soil following organic amendments. Plant Soil 290(1–2):333–342. https://doi.org/10.1007/s11104-006-9165-5
Wu N, Zhang YM, Downing A (2009) Comparative study of nitrogenase activity in different types of biological soil crusts in the Gurbantunggut Desert, Northwestern China. J Arid Environ 73(9):828–833. https://doi.org/10.1016/j.jaridenv.2009.04.002
Wynn JG, Bird MI, Vellen L, Grand-Clement E, Carter J, Berry SL (2006) Continental-scale measurement of the soil organic carbon pool with climatic, edaphic, and biotic controls. Glob Biogeochem Cycle 20(1):GB1007. https://doi.org/10.1029/2005gb002576
Xu XF, Thornton PE, Post WM (2013) A global analysis of soil microbial biomass carbon, nitrogen and phosphorus in terrestrial ecosystems. Glob Ecol Biogeogr 22(6):737–749. https://doi.org/10.1111/geb.12029
Yahdjian L, Sala O (2010) Size of precipitation pulses controls nitrogen transformation and losses in an arid Patagonian ecosystem. Ecosystems 13(4):575–585. https://doi.org/10.1007/s10021-010-9341-6
Yan D, Li J, Pei J, Cui J, Nie M, Fang C (2017) The temperature sensitivity of soil organic carbon decomposition is greater in subsoil than in topsoil during laboratory incubation. Sci Rep 7(1):5181. https://doi.org/10.1038/s41598-017-05293-1
Yang X, Post WM (2011) Phosphorus transformations as a function of pedogenesis: a synthesis of soil phosphorus data using Hedley fractionation method. Biogeosciences 8(10):2907–2916. https://doi.org/10.5194/bg-8-2907-2011
Yang YH, Mohammat A, Feng JM, Zhou R, Fang JY (2007) Storage, patterns and environmental controls of soil organic carbon in China. Biogeochemistry 84(2):131–141. https://doi.org/10.1007/s10533-007-9109-z
Yang K, Zhu JJ, Zhang M, Yan QL, Sun OJ (2010) Soil microbial biomass carbon and nitrogen in forest ecosystems of Northeast China: a comparison between natural secondary forest and larch plantation. J Plant Ecol 3(3):175–182. https://doi.org/10.1093/jpe/rtq022
Yang J, Kang Y, Sakurai K, Ohnishi K (2017) Fixation of carbon dioxide by chemoautotrophic bacteria in grassland soil under dark conditions. Acta Agric Scand B Soil Plant Sci 67(4):362–371. https://doi.org/10.1080/09064710.2017.1281433
Yan-Gui S, Xin-Rong L, Ying-Wu C, Zhi-Shan Z, Yan L (2013) Carbon fixation of cyanobacterial–algal crusts after desert fixation and its implication to soil organic carbon accumulation in desert. Land Degrad Dev 24(4):342–349. https://doi.org/10.1002/ldr.1131
Yao J, Liu H, Huang J, Gao Z, Wang G, Li D et al (2020) Accelerated dryland expansion regulates future variability in dryland gross primary production. Nat Commun 11(1):1665. https://doi.org/10.1038/s41467-020-15515-2
Yu J, Glazer N, Steinberger Y (2014) Carbon utilization, microbial biomass, and respiration in biological soil crusts in the Negev Desert. Biol Fertil Soils 50(2):285–293. https://doi.org/10.1007/s00374-013-0856-9
Yuan BC, Yue DX (2012) Soil microbial and enzymatic activities across a chronosequence of chinese pine plantation development on the Loess Plateau of China. Pedosphere 22(1):1–12. https://doi.org/10.1016/S1002-0160(11)60186-0

Yuan H, Ge T, Chen C, O'Donnell AG, Wu J (2012) Significant role for microbial autotrophy in the sequestration of soil carbon. Appl Environ Microbiol 78(7):2328–2336. https://doi.org/10.1128/AEM.06881-11

Yue P, Cui X, Wu W, Gong Y, Li K, Misselbrook T, Liu X (2020) Are annual nitrous oxide fluxes sensitive to warming and increasing precipitation in the Gurbantunggut Desert? Land Degrad Dev. https://doi.org/10.1002/ldr.3636

Zaady E (2005) Seasonal change and nitrogen cycling in a patchy Negev Desert: a review. Arid Land Res Manag 19(2):111–124. https://doi.org/10.1080/15324980590916512

Zaady E, Arbel S, Barkai D, Sarig S (2013) Long-term impact of agricultural practices on biological soil crusts and their hydrological processes in a semiarid landscape. J Arid Environ 90:5–11. https://doi.org/10.1016/j.jaridenv.2012.10.021

Zehr JP, Jenkins BD, Short SM, Steward GF (2003) Nitrogenase gene diversity and microbial community structure: a cross-system comparison. Environ Microbiol 5(7):539–554

Zhang P, Shao M (2014) Spatial variability and stocks of soil organic carbon in the Gobi desert of Northwestern China. PLoS One 9(4):e93584. https://doi.org/10.1371/journal.pone.0093584

Zhang XD, Amelung W, Ying YA, Zech W (1998) Amino sugar signature of particle-size fractions in soils of the native prairie as affected by climate. Soil Sci 163(3):220–229. https://doi.org/10.1097/00010694-199803000-00007

Zhang ZS, Li XR, Nowak RS, Wu P, Gao YH, Zhao Y et al (2013) Effect of sand-stabilizing shrubs on soil respiration in a temperate desert. Plant Soil 367(1–2):449–463. https://doi.org/10.1007/s11104-012-1465-3

Zhang Y, Xiao X, Guanter L, Zhou S, Ciais P, Joiner J et al (2016a) Precipitation and carbon-water coupling jointly control the interannual variability of global land gross primary production. Sci Rep 6:39748. https://doi.org/10.1038/srep39748

Zhang ZS, Zhao Y, Dong XJ, Shi YF, Chen YL, Hu YG (2016b) Evolution of soil respiration depends on biological soil crusts across a 50-year chronosequence of desert revegetation. Soil Sci Plant Nutr 62(2):140–149. https://doi.org/10.1080/00380768.2016.1165597

Zhao MS, Heinsch FA, Nemani RR, Running SW (2005) Improvements of the MODIS terrestrial gross and net primary production global data set. Remote Sens Environ 95(2):164–176. https://doi.org/10.1016/j.rse.2004.12.011

Zhou X, Smith H, Giraldo Silva A, Belnap J, Garcia-Pichel F (2016) Differential responses of dinitrogen fixation, diazotrophic cyanobacteria and ammonia oxidation reveal a potential warming-induced imbalance of the N-cycle in biological soil crusts. PLoS One 11(10):e0164932. https://doi.org/10.1371/journal.pone.0164932

Zhu SH, Li CF, Shao H, Ju WM, Lv NN (2019) The response of carbon stocks of drylands in Central Asia to changes of CO2 and climate during past 35 years. Sci Total Environ 687:330–340. https://doi.org/10.1016/j.scitotenv.2019.06.089

Zscheischler J, Reichstein M, Von Buttlar J, Mu M, Randerson JT, Mahecha MD (2014) Carbon cycle extremes during the 21st century in CMIP5 models: future evolution and attribution to climatic drivers. Geophys Res Lett 41:8853–8861. https://doi.org/10.1002/2014GL062409

Zuo R, Jin S, Chen M, Guan X, Wang J, Zhai Y et al (2018) In-situ study of migration and transformation of nitrogen in groundwater based on continuous observations at a contaminated desert site. J Contam Hydrol 211:39–48. https://doi.org/10.1016/j.jconhyd.2018.03.003

Diversity and Plant Growth-Promoting Properties of Microbiomes Associated with Plants in Desert Soils

8

Ramona Marasco, Maria J. Mosqueira, Ameur Cherif, and Daniele Daffonchio

Abstract

Plants inhabiting desert ecosystems, also known as xerophytes, exhibits morphological and physiological adaptations to resist the abiotic stresses, such as drought and salinity. They also exploit the ecological services provided by the microbial communities naturally associated with their organs and tissues. In this chapter, we provide a critical review of the bacterial and fungal communities associated with xerophytic plants living in hot desert ecosystems, as well as the plant-growth-promoting (PGP) microorganisms they include. We debate the composition, structure, and functionality of the microbial communities inferred by molecular analyses, along with the roles and potential exploitation of PGP microorganisms as demonstrated by studies that apply cultivation approaches. PGP microorganisms influence several aspects of the plant response to drought and salinity, inducing morphological, physiological, and biochemical changes associated to water uptake and reduction of water losses. A deeper understanding of the population dynamics and organismal interactions in the complex associations of desert plants and microorganisms may contribute to optimize the exploitation of the plant-microbial partnership. The implementation of sustainable biotechnological approaches for agriculture production in arid lands, based on the microbial resource, is also considered as a measure to counteract the constant reduction of crop land due to the ongoing climate change and land desertification.

R. Marasco (✉) · M. J. Mosqueira · D. Daffonchio (✉)
Biological and Environmental Sciences and Engineering Division, King Abdullah University of Science and Technology, Thuwal, Kingdom of Saudi Arabia
e-mail: ramona.marasco@kaust.edu.sa; daniele.daffonchio@kaust.edu.sa

A. Cherif
University of Manouba, ISBST, BVBGR-LR11ES31, Biotechpole Sidi Thabet, Ariana, Tunisia

J.-B. Ramond, D. A. Cowan (eds.), *Microbiology of Hot Deserts*, Ecological Studies 244, https://doi.org/10.1007/978-3-030-98415-1_8

Keywords

Plant-microbiota · Desert plant · PGP microorganisms · Microbial diversity · Arid ecosystem

8.1 Introduction

Around one-fifth of the world's surface is classified as desert (Laity 2009). Deserts are biomes characterized by low mean annual precipitation and high annual evaporation (Sandquist 2014). Besides their distinctive aridity, deserts usually present low nutrient availability, elevated temperatures and sun irradiation, high salinity, and strong wind erosion (Sandquist 2014; Lester et al. 2007; Stomeo et al. 2013). Such environmental pressures drive the diversity and ecological adaptation strategies of the organisms living in the desert (Noy-Meir 1974; Gutterman 2002; Makhalanyane et al. 2015a). Vegetation is an important component determining the desert biome. The first and most important challenges for desert plants are aridity and water scarcity, along with extreme pH and high salinity, which also negatively affect plants' water uptake (Thorup 1969; Aroca et al. 2012). In deserts, mesophytic plants, those adapted to temperate climates (Maximov 1931), would consume more water than that they could uptake from the soil, leading to wilting and death. On the contrary, specialized xerophytic plants exhibit specific adaptations to survive in such water-limiting conditions (Fahn and Cutler 1992), including morphological and physiological changes in both above- and below-ground compartments (Sandquist 2014; Danin 1996; Danin 2000; Wickens 1998). Morphological adaptations are mainly directed toward the optimization of water management by reducing the amount of water lost by evapotranspiration, favouring conditions to economize water uptake, improving the water-holding capacity of the soil, and by storing water in the plant tissues (Wickens 1998). For instance, succulent plants store water in their stems and leaves (*e.g.*, the *Cactaceae* family; (Bárcenas et al. 2011)), while shrubs and bushes growing in the sandy soils of the dunes develop specialized and deep root systems (*e.g.*, the rhizosheath of speargrasses of the *Poaceae* family; (Danin 1996)) that favour retention and long storage of water. Additional physiological adaptations in halophytes, such as salt secretion and osmoregulation (Flowers and Colmer 2008; Flowers and Colmer 2015), allow the plants to resist high salinity typical of many desert/arid soils.

Along with the multiple morphological and physiological adaptations, xerophytic plants exploit positive interactions with their associated microorganisms (Rolli et al. 2015; Marasco et al. 2012; Fonseca-García et al. 2016; Marasco et al. 2016; Camarena-Pozos et al. 2019). Indeed, plants have been recently re-defined as "metaorganisms" composed of the host plant and its associated microbiota, both acting synergically as a single entity undergoing evolutionary selection (Zilber-Rosenberg and Rosenberg 2008; Daffonchio et al. 2015). Besides plant pathogens, a higher number of microorganisms are beneficial to the plant host and successfully cooperate under the nutrient-poor conditions of deserts to optimize resource utilization. They provide valuable ecological services to the metaorganism, including growth promotion of the plant and protection against abiotic and biotic stresses (Marasco et al. 2012; Daffonchio et al. 2015; Bashan and Holguin 2002; Vigani et al. 2018). Beneficial bacteria and fungi, known as plant growth-promoting (PGP) microorganisms, benefit the plant metaorganism through both direct and indirect mechanisms (Spaepen et al. 2009). Although far from being fully elucidated, these mechanisms include the ability to (1) fix atmospheric nitrogen, (2) solubilize mineral phosphates and increase the bioavailability of other macronutrients, such as magnesium, calcium and potassium, and micronutrients (*e.g.*, iron, manganese, copper, and zinc) (Lemanceau et al. 2009; Santi et al. 2013; Balloi et al. 2010), (3) synthesize antibiotics, enzymes, and/or antimicrobial compounds (Olanrewaju et al. 2017), and (4) produce hormone-like substances interfering with the physiological hormone balance of the plant hosts (Spaepen et al. 2009; Glick et al. 2007a).

In this context, understanding the association of desert plants and microorganisms may improve efforts to comprehend the overall ecological processes taking place at the interface between sand, rocky-gravel and salty substrates/sediments and the plant root system (the rhizosphere and root endosphere). Thus, here we assess the available knowledge on the composition, structure, functionality, and biotechnological potential of the microbial communities associated with plants living in hot deserts with the goal of facilitating the wise management of this plant-complementary biome. We focus our chapter on bacterial and fungal communities associated with plants in hot desert ecosystems. We first categorize plants according to three main adaptations that are common in deserts: (1) changes in leaf/stem morphology, using, as an example, succulent plants; (2) changes in the root system structure, focusing on the rhizosheath of desert shrubs and grasses; and (3) salinity tolerance of halophytic desert plants. In addition, we give venue to (4) an "iconic desert-crop", the date palm (*Phoenix dactylifera*), cultivated under desert farming conditions in oasis agroecosystems.

8.2 Plant-Associated Microorganisms in Hot Desert Biomes

Microbial communities associated with plant compartments are generally a subset of the microbial pool of the surrounding soil/sand (Van Der Heijden and Schlaeppi 2015; Marasco et al. 2018a; Marasco et al. 2018b), which is thus the primary source of the plant microbiome (Lundberg et al. 2012; Philippot et al. 2013). Microorganisms colonizing the plant tissues can be as well vertically inherited by seeds (Berg and Raaijmakers 2018; Jooste et al. 2019) and further transmitted from one plant generation to the next (Nelson 2018; Mitter et al. 2017), as the culmination of a complex process of microbial interactions mediated by plant throughout its life cycle (Johnston-Monje et al. 2016; Rodríguez et al. 2018; Cavazos et al. 2018).

Microbial communities of desert soils have been proven to be different from those present in other soils and to have lower levels of both functional and phylogenetic diversity (Fierer et al. 2012a). One of the most common bacterial phyla found in desert ecosystems globally is Actinobacteria (Chap. 4), especially *Rubrobacter*,

Arthrobacter, *Thermopolyspora*, and *Streptomyces* (Makhalanyane et al. 2015a; Drees et al. 2006; Makhalanyane et al. 2013; Santhanam et al. 2013). The prevalence and ubiquitous occurrence and survival in arid soils of these genera are explained by their metabolic versatility, tolerance to desiccation and solute stress, and multiple UV repair mechanisms (Chater and Chandra 2006; Gao and Garcia-Pichel 2011; Neilson et al. 2012; Stevenson and Hallsworth 2014; Mohammadipanah and Wink 2016). Similarly, Proteobacteria, Gemmatimonadetes, Firmicutes, and Cyanobacteria (Makhalanyane et al. 2013; Bahl et al. 2011; Fierer et al. 2012b; Lacap et al. 2011), which are ubiquitously distributed in other soils, have also been found in desert soils in high abundances (Fierer et al. 2012b; Köberl et al. 2011; Delgado-Baquerizo et al. 2018). In this contest, Proteobacteria have a versatile physiology ranging from symbiotic to free-living lifestyles and including N fixation, photosynthesis (Boldareva-Nuianzina et al. 2013; Raymond 2008), spore formation, and desiccation and UV resistance capacities (Harel et al. 2004; Singh et al. 2010; Singh et al. 2013; Guesmi et al. 2019). Similarly, several taxa of Bacteroidetes (family *Flavobacteriales* and genera *Adheribacter* and *Pontibacter*; (Prestel et al. 2013; Subhash et al. 2014)) are highly abundant, probably due to their ability to grow under the alkaline and oligotrophy conditions common in desert soils (Lauber et al. 2009).

Fungal diversity in desert soils is also relatively high [(Taylor-George et al. 1983; Murgia et al. 2018; Ranzoni 1968); Chap. 4]. Despite the limitations of cultivation approaches (Degnan and Ochman 2012), members of Ascomycota (*Cladosporium*, *Cladophialophora*, *Aspergillus*, *Penicillium*, *Alternaria*, *Leptosphaerulina*, *Phoma*, *Aureobasidium*, among others) and Basidiomycota, which included thermophilic and thermotolerant members (Zeikus 1979; Mouchacca 2007; Moustafa 1978), have been isolated from both vegetated and barren soils (Grishkan and Nevo 2010).

Plants are able to attract and sustain their associated microbial communities from the surrounding soil (Van Der Heijden and Schlaeppi 2015; Marasco et al. 2018b; Hirsch and Mauchline 2012) by means of rhizodeposition of carbon sources released by the roots into the rhizospheric-soil (Jones et al. 2009). The root is a key driver in the microbial recruitment process: a subset of the microbial communities of the soil microbial pool is enriched in the rhizosphere and further selected by interaction with the rhizoplane, which acts as a physical barrier allowing only certain microbial types, named endophytes, to colonize the internal root tissues, the endophytes (Van Der Heijden and Schlaeppi 2015; Marasco et al. 2018b). Each plant compartment is an "ecological niche" in which specific microbial communities are hosted (Coleman-Derr et al. 2016; Beckers et al. 2017). Here, we summarize the current knowledge on plant-associated microorganisms in desert environments. We also consider the beneficial functional potential of microorganisms for the host plant.

8.2.1 Microorganisms Associated with Succulent Plants in Arid Ecosystems

Succulents are xerophytic plants with fleshy and relatively thickened tissues. Succulence is present in 30 plant orders and encompasses great morphological and physiological diversity ((Nyffeler and Eggli 2010); Fig. 8.1). Agaves (*Agavaceae* family, 10 genera; (Verhoek 1998)) and cacti (*Cactaceae* family, 174 genera; (Bárcenas et al. 2011)) are among the most recognizable succulent plants, mainly distributed in arid and semi-arid systems of the American continent (Hernández-Hernández et al. 2014). They generally present shallow roots and fleshy stems (*cactus*; (Barthlott and Hunt 1993)) or succulent leaves (*agave*; (Verhoek 1998)) that act as water-storage systems during periods of water scarcity (Willert et al. 1990; Eggli and Nyffeler 2009; Cui and Nobel 2006). These plants have crassulacean acid metabolism (CAM photosynthesis)–as opposed to classical photosynthesis–that allows them to fix carbon (CO_2) during the night and keep the leaf stomata closed during the hot day to reduce water loss due to evapotranspiration (Sandquist 2014).

Several culture-independent and culture-based studies revealed complex microbial communities associated with cacti (*Mammillaria carnea*, *Opuntia pilifera,* and *Stenocereus stellatus*) and agaves (*Agave tequilana*, *A. salmiana,* and *A. deserti*), mainly composed by bacteria and fungi (Fonseca-García et al. 2016; Camarena-Pozos et al. 2019; Coleman-Derr et al. 2016; Aguirre-Garrido et al. 2012; Fonseca-García et al. 2018; Puente et al. 2004). The compositions of bacterial communities were substantially different in each compartment (*i.e.*, rhizosphere, root, leaves) of these succulent plants; a portion of microorganisms in the soil surrounding the root colonize the different below- and above-ground plant compartments in a niche-partitioning process of selection (Coleman-Derr et al. 2016; Fonseca-García et al. 2018; Desgarennes et al. 2014; Flores-Núñez et al. 2020; López-Lozano et al. 2020). Despite the recognized effect of the host species, the site, and the season on the microbial community assembly, these factors played a minor role in these plants (Coleman-Derr et al. 2016; López-Lozano et al. 2020), while water proved to be an important driver (Kavamura et al. 2018; Karray et al. 2020). The relative abundance of water in the system (for instance, due to precipitation) affected the composition of the microbial community and the functionality patterns of the succulent plant *Cereus*

Fig. 8.1 Examples of succulent plants from desert environments. Species of succulent plant, (**a**) including agave and (**a–c**) cacti, growth in the botanic garden *Jardin Majorelle* in Marrakesh, Marocco

jamacaru and *Opuntia ficus-indica* (Kavamura et al. 2018; Karray et al. 2020); water scarcity reduced soil pore connectivity and microbial movement, thus favouring spatial isolation of communities and the increment of bacterial diversity (Carson et al. 2010).

Proteobacteria (orders *Pseudomonadales* and *Enterobacteriales*), Actinobacteria (*Actinomycetales*), Firmicutes (*Bacillales*), and Bacteroidetes (*Flavobacteriales*) represent the dominant prokaryotic phyla in the different plant compartments of agaves and cacti (Coleman-Derr et al. 2016; Fonseca-García et al. 2018), which share core beneficial taxa (*Azotobacter*, *Pseudomonas*, *Enterobacter*, *Bacillus,* and *Citrobacter*) involved in nitrogen fixation, auxin and siderophore production, phosphate solubilization and other bioweathering activities (Aguirre-Garrido et al. 2012; Puente et al. 2004; Desgarennes et al. 2014; Lopez et al. 2011; Puente et al. 2009; Lopez and Bacilio 2020; de la Torre-Hernández et al. 2020).

Nitrogen is the most limiting factor of biological activity after water scarcity (McCalley and Sparks 2009), and bacterial rock-weathering is considered essential to increase soil fertility (Mapelli et al. 2012). Bacteria associated with the root system of cacti growing in rocky substrates improve the availability of significant amounts of nutrients (P, K, Mg, and Mn) from the substrate (Puente et al. 2004; Puente et al. 2009; Lopez and Bacilio 2020). Inoculation of these bacteria in a rock mineral soil promoted the growth of giant cardon cactus (*Pachycereus pringlei*) seedlings; non-inoculated plants grew less vigorously and some died (Puente et al. 2009). These microbial symbionts are vertically transmitted to seeds and their elimination from the seed by antibiotic treatment blocked seedling development, confirming the importance of the symbiotic relationship between bacteria and their succulent hosts under the harsh environmental conditions of the desert (Fonseca-García et al. 2016; Puente et al. 2009).

A recent study by Camarena-Pozos and co-authors (Camarena-Pozos et al. 2019) revealed that approximately 90% of the bacterial strains isolated from the endosphere (roots, leaves, and seeds), rhizosphere, and phyllosphere of agaves and cacti positively affected the host fitness of both the original and of non-native hosts (*i.e.*, *Arabidopsis thaliana* and *Nicotiana benthamiana*) via the emission of volatile organic compounds (VOCs). Similarly, endophytic bacteria (*Bacillus* spp.) of *Euphorbia trigona*–a cactus species in the arid regions of Cameroon–were able to colonize exogenous tomato plants and support their resistance to water stress (Eke et al. 2019).

Xerophytic plants also exploit endophytic and epiphytic fungi and their secondary metabolites to increase their fitness (Loro et al. 2012). In the fungal communities of succulent plants (agaves and cacti), a limited number of phyla (*i.e.*, *Pleosporales*, *Sordariales*, *Chaetothyriales*, *Capnodiales,* and *Dothideales*) of Ascomycota represent up to 90% of the community, followed by Basidiomycota such as *Agaricales* (Fonseca-García et al. 2018; Suryanarayanan et al. 2005; Bezerra et al. 2013). In general, the composition of the fungal community is less variable than that of the bacterial community, and it is mainly influenced by the plant species and its biogeography (Fonseca-García et al. 2016; Coleman-Derr et al. 2016). Notably, under extreme environmental conditions, fungal species typically known to be pathogenic, such as Ascomycota *Fusarium oxysporum*, can convert their lifestyle from free-living or pathogenic to mutualistic (Redman et al. 2001). It is also possible that such fungi (*F. oxysporum*) may prefer an endophytic lifestyle in an arid environment as they are often isolated as endophytic symbionts from succulent agaves and cacti (Bezerra et al. 2013; Pieterse et al. 2018). Fungal endophytes of the succulent *Agave victoria-reginae* provide beneficial services to the host (Obledo et al. 2003). In vivo tests showed that *F. oxysporum* determined increases in root length, leaf stomata number, photosynthetic efficiency, and plant development (Obledo et al. 2003).

Overall, studies on succulent plants revealed a large variety of associated microorganisms, some of which define shared core microbiome between agaves and cacti. Several of these taxa provide beneficial services to the host plant and might be important for plant survival and success in the desert ecosystem. However, considering the vast diversity of succulent plants, further efforts are warranted to enlarge our view on the symbiotic associations between these iconic xerophytes and microorganisms.

8.2.2 Microorganisms Associated with the Rhizosheath-Root Systems of Xerophytic Desert Plants

Changes in the root system morphology and the development of specialized root functional traits is one of the first strategies that plants use to adapt to the conditions of the desert (Danin 1996). Usually, these root adaptations occur with the aim of increasing water uptake and preventing water loss (Sandquist 2014). One example of a root adaptation is the formation of sandy-rhizosheath in xerophytic desert plants, such as speargrasses (Fig. 8.2a and b). The rhizosheath is defined as the encasing cylinder, made of root hairs, fungal hyphae, and sand particles, that surrounds the root system of certain plants (Fig. 8.2b–e; (Brown et al. 2017; Pang et al. 2017)). Global screenings of the roots of several desert plants concluded that rhizosheath occur across different genera throughout the angiosperms (Danin 1996; Brown et al. 2017; Smith et al. 2011), particularly in several members of the *Poaceae* and *Cyperaceae* families grown in desert sand (Marasco et al. 2018b; Hegazi et al. 2009; Roth-Nebelsick et al. 2012; York et al. 2016; Liu et al. 2011); for example, 107 of 130 South African grass species can form rhizosheath (Bailey and Scholes 1997).

It is proposed that bioadhesive mucilage components of exudates by roots and bacteria, along with root hairs and fungal hyphae, are important factors in the formation of the compact cylinders of sand grains around the roots (Fig. 8.2c–e; (Wullstein and Pratt 1981; Moreno-Espíndola et al. 2007)). This unique structure enables plants to sustain and increase nutrient and water uptake from the soil (Brown et al. 2017; Pang et al. 2017; Young 1995; North and Nobel 1997; Benard et al. 2016; Galloway et al. 2019; Marasco et al. 2022), provides mechanical protection to the root in dry and hardened soils (Albalasmeh and Ghezzehei 2014), increases adaptation to salt stress (Ashraf and Harris 2004; Ashraf 2006), and facilitates water

Fig. 8.2 Rhizosheath structure of desert speargrasses. (**a** and **b**) *Stipagrostis sabulicola* growing in the Namib desert dune and its rhizosheath-root system. (**c**) Rhizosheath-root system fragments collected from *S. sabulicola*. (**d**) Visualization at the stereomicroscope of rhizosheath structure. (**e**) Scanning Electron Microscopy magnification of the rhizosheath' external part, showing the root hairs entrapping sand grains. (**f**) Microorganisms associated to the rhizosheath-root system of speargrasses colonizing both sand grains and root hairs

movement across the root-soil interface (Young 1995; North and Nobel 1997; Othman et al. 2004; Bhatnagar and Bhatnagar 2005; Bergmann et al. 2009). During extended drought periods, some grass species increase the thickness of their rhizosheath (Harnett et al. 2013), mainly stimulating the production of mucilage (*i.e.*, xyloglucan) so as to enhance water infiltration and aeration (Galloway et al. 2019).

The environmental conditions generated inside the rhizosheath provide a more favourable ecological niche with respect to the desert sand, where soil microorganisms can find resources and improved conditions to survive and flourish (Fig. 8.2f; (Danin 1996; Marasco et al. 2018b; Othman et al. 2003a; Othman et al. 2003b; Wullstein et al. 1979; Marasco et al. 2022)). Microorganisms associated with the rhizosheath-root system are recruited from the surrounding bulk sand and are significantly enriched in the rhizosheath compartments, doubling their number per unit of solid substrate (Marasco et al. 2018b). By using cultivation-independent techniques (*e.g.*, next-generation sequencing, denaturing gradient gel electrophoresis), further selection mediated by the root rhizoplane was detected, revealing specific microbial community in each rhizosheath-root system compartment (Marasco et al. 2018b; Marasco et al. 2022), as already observed in other desert plants (Marasco et al. 2012; Coleman-Derr et al. 2016; Mapelli et al. 2020; Marasco et al. 2021). Independently of the host species (*e.g.*, *Stipagrostis sabulicola*, *Stipagrostis seelyae*, and *Cladoraphis spinosa*, (Marasco et al. 2018b)), the formation of the rhizosheath-root system microbial community occurs by neutral assembly processes in which the limited microbial diversity available in the bulk sand is randomly reassembled by the plant hosts. These communities are mainly dominated by desert-adapted Actinobacteria and Alphaproteobacteria (*e.g.*, *Lechevalieria*, *Streptomyces,* and *Microvirga*) as well as by saprophytic Ascomycota fungi (*e.g.*, *Curvularia*, *Aspergillus,* and *Thielavia*).

Cultivation-dependent studies have revealed that N-fixing bacteria, such as members of the *Bacillus*, *Paenibacillus*, *Enterobacter*, *Agrobacterium,* and *Chryseomonas* genera, are important components of the recruited isolates (Hegazi et al. 2009; Othman et al. 2004; Tian et al. 2019). The high relative abundance of N-fixing bacteria has been confirmed in the rhizosheath of both annual (*Bromus madritensis*, *Hordeum murinum*, *Lolium perenne*, *Trachynia distachya*, among others; (Hegazi et al. 2009)) and perennial speargrasses (*e.g.*, *Aristida purpurea*, *Elymus lanceolatus*, *Oryzopsis hymenoides*, *Stipa comata*, *Panicum turgidum*, *Stipagrostis scoparia*, *S. pennata*; (Othman et al. 2004; Bergmann et al. 2009; Othman et al. 2003b; Tian et al. 2019; Akbari et al. 2015; Hanna et al. 2013; Moreno-Espíndola et al. 2013). Both symbiotic and free-living N-fixing microorganisms represent a main source of fixed nitrogen, and their enrichment is crucial for providing–directly and indirectly–sufficient organic nitrogen to the xerophytic plants that thrive in the nitrogen-poor desert soil (Tian et al. 2019; Köberl et al. 2016; Klironomos and Hart 2001; Shelef et al. 2013; Bar-Shmuel et al. 2018). Along with diazotrophic bacteria, the rhizosheath favours beneficial interactions with microorganisms that might contribute to plant nutrition (P solubilizers and siderophore producers) and growth promotion (auxin and 1-aminocyclopropane-1-carboxylate [ACC] deaminase producers; (Akbari et al. 2015)), as well as to the water-holding capacity of the rhizosheath-root system throughout the production of exopolysaccharide (EPS; (Moreno-Espíndola et al. 2007; Watt et al. 1993; Marasco et al. 2022)). PGP bacteria such as Bacilli, Pseudomonads, and Rhizobia synthesize a wide spectrum of multifunctional polysaccharides in extreme environments (Ali et al. 2014; Raddadi et al. 2018; Upadhyay et al. 2011; Qurashi and Sabri 2012). Under water-limiting conditions, inoculation with EPS-producing bacteria significantly increases the root-adhering soil-per-root tissue ratio (Alami et al. 2000), alleviating the negative effect of drought on plants (Vanhaverbeke et al. 2003). Notably, a recent study conducted by Zhang and colleagues (Zhang et al. 2020) demonstrated how a pronounced rhizosheath can be formed in rice (*Oryza sativa*) only in response to a moderate soil drying conditions. The analysis of the bacterial community associated to the newly formed rhizosheath revealed that the associated bacteria contribute to the rhizosheath formation by mechanisms involving the ethylene response of plant. Indeed, an enrichment of enterobacters that carry ACC deaminase activity was observed; in vivo experiment conducted using the wild-type *Enterobacter* isolate resulted in rhizosheath formation, while inoculation with an ACC deaminase-deficient mutant strain did not, confirming that the ACC deaminase activity is crucial for enhanced rhizosheath formation during drought (Zhang et al. 2020). A holistic understanding of the rhizosheath structure as a plant adaptation to desert conditions requires a comprehensive study of the microbial communities associated with this structure and a comparative evaluation with respect to other plant compartments. As most of the available information focusing on this structure, and specifically on rhizosheath found in desert grasses and shrubs, comes from cultivation-based techniques and in vitro functionality tests, more efforts should be dedicated to characterize the structure and functionality of the overall associated microbial communities by using cultivation-independent methods, as well as to disentangle the ecological drivers affecting such community assembly.

8.2.3 Microorganisms Associated with Halophytes Inhabiting Arid Ecosystems

Soil salinity is a common characteristic of arid and semi-arid soils due to high evaporation rates surpassing water input from precipitation, which results in the build-up and concentration of solutes in the soil as water evaporates (Chap. 1). Furthermore, groundwater basins, which in theory would increase water availability in desert soils, tend to have high concentrations of solutes (Sandquist 2014). Salinity produces different effects on plant growth and development through different mechanisms, including (1) osmotic imbalances that eventually induce a drought-like water deficit, (2) changes in the physical structure and mechanical properties of the soil, which reduce water permeability, and (3) enhancement of ion and ROS toxicity, which leads to oxidative damage and metabolic and physiological dysfunction (Daffonchio et al. 2015; Sharma et al. 2012). Plants have physiological mechanisms to counteract the negative effects of salinity (Zhu 2001). Halophytes, are able to thrive in saline soils due to their capacity to control the cell balance of ionic concentrations through (1) the accumulation of compatible solutes in vacuoles, (2) the compartmentalization of ions, or (3) the excretion of cellular salt through the

Fig. 8.3 Halophytic plants (**a**) *Salicornia* plant growing in Tunisian hypersaline soils. (**b**) Acclimation of *S. strobilacea* plantlets in pots containing sterile marine water before the treatment with *gfp*-labelled bacteria. (**c**) Rhizocompetence and recolonization ability of *S. strobilacea* rhizospheric strains after 96 h by *gfp*-labelled *Pseudomonas* sp. strains SR7-77 (Marasco et al. 2016). The red colour represents the auto-fluorescence of root tissues when excited by the UV laser while the green colour the *gfp* fluorescence of labelled bacterial cells

stem or leaves (Sandquist 2014). Examples of desert "salt accumulators" are the perennial shrubs *Allenrolfea* and *Salicornia* (Fig. 8.3a) and some members of *Atriplex* (saltbushes), while desert "salt excretors" include the salt cedar tree *Tamarix* (Sandquist 2014) and salt grass *Distichlis* (Pessarakli et al. 2012).

A body of research on desert halophyte microbial communities has used cultivation-based methods with the aim of exploring the PGP traits carried by microorganisms able to cope with both drought and salinity stresses (Marasco et al. 2016; Mapelli et al. 2013; Jenkins 2003; Kaplan et al. 2013; Fterich et al. 2011). *Halomonas*, *Oceanobacillus*, *Marinobacter,* and *Bacillus*, dominating the rhizosphere of several halophytes (among others, *Salicornia* spp.; (Marasco et al. 2016; Mapelli et al. 2013; Gontia et al. 2011)), are genera typical of saline environments (such as mangrove sediments; (Booth et al. 2019a; Booth et al. 2019b; Sefrji et al. 2021, 2022)) and have several biotechnological potentials (Marasco et al. 2016; Dias et al. 2009; El-Tarabily 2008; Soldan et al. 2019). Bacterial isolates from the salt-accumulating plant *Atriplex halimus* L. confirmed the prevalence of halotolerant and halophilic bacterial taxa in the rhizosphere (Kaplan et al. 2013), including Actinobacteria (genera *Arthrobacter*, *Streptomyces*, *Microbacterium,* and *Kocuria*), Proteobacteria (*Pseudomonas*, *Halomonas* and *Stenotrophomonas*) and Firmicutes (*Bacillus* and *Oceanobacillus*). On the other hand, *Fusarium* and *Penicillium* appear to be the main genera within the fungal community (Kaplan et al. 2013). Notably, passing from the below-ground (*i.e.*, root and rhizosphere) to the above-ground (*i.e.*, leaves) compartments of the salt-excreting desert tree *Tamarix*, the most abundant taxa among bacteria and fungi were *Oceanospirillales* (genus *Halomonas*) and Ascomycota (*Lecanoromycetes* and *Onygenales*). The presence of *Halomonas* in the phyllosphere of salt-excreting trees is not surprising, as most known members of this genus are able to grow in conditions of up to 20% salinity (El Hidri et al. 2013; Belkin and Qvit-Raz 2010). Sequencing of the hypervariable regions of 16S and 18S rRNA genes revealed that the recruitment, association, and assembly of these bacterial and fungal taxa, respectively, are mainly driven by the geography of the plants (Finkel et al. 2011) and the salt concentration (Zhang et al. 2019; Szymańska et al. 2018).

Desert-halophytic microbial partners of plants can reduce salt stress in their hosts (Marasco et al. 2016; Abu-Ghosh et al. 2014) and help plants to resist oligotrophy stress and enhance root activity in poor soils (Etesami and Beattie 2018; Hassen et al. 2018; Halo et al. 2015). They are mainly involved in biofertilization processes, including nitrogen fixation (*e.g.*, *Rhizobium*, *Bradyrrhizobium*, *Mesorrhizobium,* and *Ensifer* associated with salt-tolerant trees *Prosopis*; (Jenkins 2003; Fterich et al. 2011)), ammonia production (Goswami et al. 2014), and the release of nutrients (*e.g.*, iron and phosphate; (Kaplan et al. 2013; Goswami et al. 2014)). Some genera, such as *Sphingomonas* and *Pseudomonas*, secrete the plant growth hormone auxin (Asaf et al. 2018), stably colonize the root system and can move to the aerial parts as well (Fig. 8.3b–c; (Marasco et al. 2016)). These microorganisms provide to the host plant fundamental ecological services supporting adaptation, growth, and development in the saline desert soils (Marasco et al. 2016; Daffonchio et al. 2015).

In this contest, the elucidation of the mechanisms of action of halophilic PGP microorganisms can expand their applications in different crops (Daffonchio et al. 2015; Soussi et al. 2016).

8.2.4 Microorganisms Associated with Oasis Date Palm, a Model Crop in Desert Agro-Ecosystems

The conversion of arid and semi-arid landscapes into arable land is a method to combat the effects of climate change and land desertification (Clery 2011). Desert agro-ecosystems usually rely on irrigation and are characterized by the effect of human intervention and land management (*i.e.*, agricultural practices), which can significantly influence both the soil and the plant-associated microbial communities (Philippot et al. 2013). Several studies investigated the ecology and role of microbial communities under desert farming (Marasco et al. 2012; Köberl et al. 2011; Soussi et al. 2016; Cherif et al. 2015; Ferjani et al. 2015; Köberl et al. 2013; Mosqueira et al. 2019; Kazerooni et al. 2017; Marasco et al. 2021). The most renowned agro-ecosystem exploiting desert farming is the oasis (Fig. 8.4a and b). Oases are unique and highly productive rural desert agro-ecosystems that provide agricultural products and social services to humans living in the desert (de Grenade et al. 2016; Chao and Krueger 2007). Oases also contribute to the preservation of crop diversity in arid regions (Achtak et al. 2010). In desert oases, date palms (*Phoenix dactylifera* L.) create the highest and widest canopies, which recondition the desert climate and produce favourable conditions for cropping. Date palms provide shade, decrease air temperature, and enhance air moisture below the canopy (de Grenade 2013). The resulting microclimate allows the cultivation of different crops below the date palm's canopy, including olive, almond and pomegranate trees, cereals such as barley and rice, leguminous and forage plants (de Grenade et al. 2016; de Grenade 2013). In this context, date palms in the desert oasis agro-ecosystem represent an excellent model plant for understanding the diversity and functionality of the microbial communities in a multispecies agro-ecosystem where the conditions are opposite to the monoculture because of the simultaneous and continuous coexistence of multiple crop species.

Fig. 8.4 Oasis agro-ecosystem. (**a** and **b**) Oasis agro-ecosystems in Saudi Arabia and Tunisia, respectively, dominated by *Phoenix dactylifera*. (**c**) *P. dactylifera* root sampling

Oases are embedded within the harsh and resource-scarce desert soil, which, *per se*, is characterized by low phylogenetic and functional microbial diversity (Köberl et al. 2011; Köberl et al. 2013; Fierer et al. 2009; Makhalanyane et al. 2015b). In the oasis date palms select a PGP core microbiota in their root system, which is taxonomically and functionally specific, conserved and co-evolved with the host plant. In the desert oasis ecosystem, the date palm, rather than soil type or location, is the main driver responsible for the selection of the bacterial communities associated with the root system (Mosqueira et al. 2019; Yaish et al. 2016). This is the opposite pattern from that observed in conventional agro-ecosystems (Philippot et al. 2013). Studies conducted in different oases in Tunisia and Oman evidenced that, regardless of the edaphic conditions and geographic location, the bacterial communities in the date palm root system (*i.e.*, root tissues and rhizosphere; Fig. 8.4c) enriched *Gammaproteobacteria* (*Pseudomonas*, *Enterobacter,* and *Rahnella* genera), Actinobacteria (*Streptomyces*, *Arthrobacter* and *Microbacterium*), and Firmicutes (*Bacillus* and *Staphylococcus*), and there was a lower abundance of *Betaproteobacteria* (*Achromobacter*) and Bacteroidetes (*Flavobacterium*) (Cherif et al. 2015; Ferjani et al. 2015; Mosqueira et al. 2019; Yaish et al. 2016; Yaish et al. 2015).

Several isolates from the date palm root system showed PGP potential properties capable of favouring abiotic stress tolerance to drought and salinity in the host plant (Cherif et al. 2015; Yaish and Kumar 2015). Numerous endophytic and rhizospheric strains of bacteria and fungi have been isolated from the date palm root system (Cherif et al. 2015; Ferjani et al. 2015; Yaish and Kumar 2015; Al-Yahya'ei et al. 2011; Thennarasu et al. 2019) that were shown to be capable of PGP activities in vitro and *in planta* involved in plant growth promotion and protection from drought and salinity (Cherif et al. 2015; Yaish and Kumar 2015). These bacteria may facilitate plant growth in a variety of ways, by improving the availability of important nutrients (N, P, K, Fe, and Ca) or modulating hormone homeostasis through the synthesis of phytohormones or enzymes, or by producing EPS involved in water retention and root protection (Cherif et al. 2015; Ferjani et al. 2015; Yaish and Kumar 2015; Glick and Glick 2012). For instance, biofertilizer containing PGP bacteria to provide nitrogen (*Azotobacter chroococcum* and *Azospirillum brasilense*), phosphorus (*Bacillus megaterium*), and potassium (*Bacillus circulans*) significantly improved the date palm nutrient uptake under stress, favouring the vegetative growth of the plant (El-sharabasy et al. 2018). Similarly, endophytic biopromoter strains (*Pseudomonas* spp.) increased date palm biomass under repeated drought periods, possibly by affecting the hormonal pathway of the plant (Cherif et al. 2015).

Other important components of the beneficial date palm microbiome are the arbuscular mycorrhizal fungi (AMF), which establish a symbiosis with the plant by penetrating the cortical cells and colonizing the root tissues (Jeffries et al. 2003; Zougari-Elwedi and Issami 2019; Zougari-Elwedi et al. 2016; Al-Karaki 2013). In traditional oases in Oman, the date palm AMF were dominated by the genus *Glomus*, followed by *Scutellospora* and *Racocetra* (Al-Yahya'ei et al. 2011). AMF increase plant tolerance to various biotic and abiotic stresses by enhancing the root uptake of nutrients (P, N, Cu and Fe) and water (Al-Karaki 2013; El Kinany et al. 2019). Greenhouse experiments showed that inoculation with spores of *Glomus mosseae* isolated from the rhizosphere of date palm cultivated in arid/saline regions of Tunisia improved soil fertility and plant growth, with a significant increase of soluble P and N (Zougari-Elwedi and Issami 2019). Similarly, the commercial strain *Glomus iranicum* significantly promoted the growth and nutrition of micropropagated date palm (cv. *Feggous*) in greenhouse conditions (El Kinany et al. 2019).

AMF also contributed to the formation of stable soil aggregates, building up a macro porous structure of soil that allows penetration of water and air (Al-Karaki 2013). All these beneficial effects (i.e., plant growth and fitness) revealed how AMF can increase agricultural productivity both reducing the impact of date palm cultivation on ecosystems and maximizing dates yield. Thus, the use of AMF represents a sustainable solution for the functioning of agricultural ecosystems in arid (Al-Karaki 2013).

8.3 PGP Microorganisms Associated with Desert-Adapted Plants as a Biotechnological Resource

Under the ongoing climate change, drought and soil salinity have been raised as major concerns for agriculture in arid and semi-arid regions, where temperatures are elevated, rainfalls scarce, and evaporation rates high (Daffonchio et al. 2015; Shabani et al. 2015; Shabani et al. 2012; Berdugo et al. 2020). Abiotic stresses not only induce strong reductions in plant productivity but also cause specific, as well as unspecific reactions and damages at the cellular level, resulting in tissue and organ necrosis (Parida and Das 2005).

Several studies consolidated the evidence that beneficial PGP microorganisms represent a potentially powerful tool to increase abiotic stress tolerance in agriculture crops in a sustainable manner (Aroca et al. 2012; Hassen et al. 2018; Grover et al. 2011; Trivedi et al. 2020; Alsharif et al. 2020). However, to exploit PGP microorganisms as biofertilizers and biopromoters, it is crucial to follow a rationale that secures the proper selection of the bacterial inoculant for the final host by considering its PGP recruitment strategies, the environmental setting, and its compatibility with the autochthonous microbial communities.

8.3.1 Where to Select Valuable PGP Microorganisms Against Drought and Their Essential Features

Environmental and geo-climatic factors are crucial in shaping the diversity and functionality of plant-associated microbial communities. In arid and semi-arid climates, these factors are of particular importance in determining the microbial community assembly and, ultimately, the stability and functionality of the ecosystem (Marasco et al. 2012). Plants that thrive in arid and semi-arid environments act as reservoirs of stress-tolerant microorganisms able to confer beneficial services to the

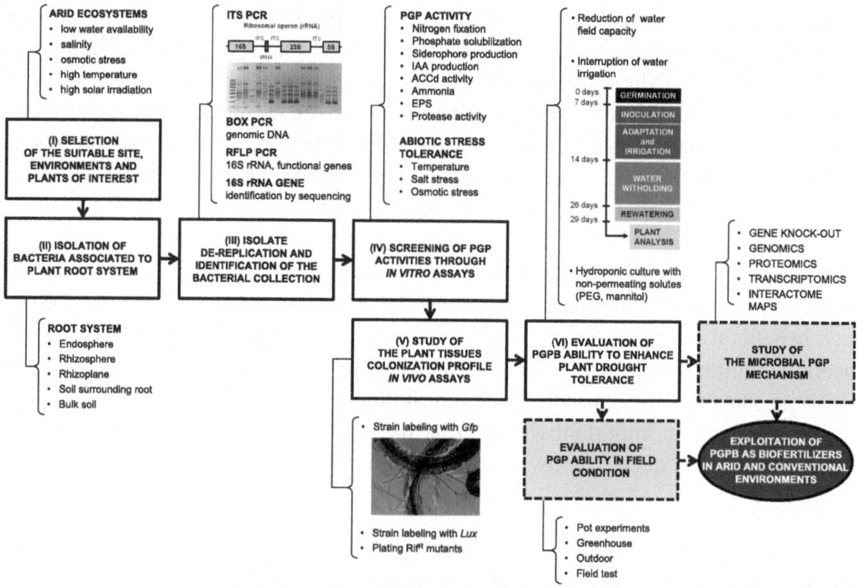

Fig. 8.5 Flow chart of the proposed methodological pipeline for the selection of PGP microorganisms involved in the enhancement of plant resistance to drought stress. Proposed rational includes: (i) selection of valuable source, (ii) sampling and isolation of microorganisms from plant tissues and rhizosphere, (iii) reduction of genomic redundancy and identification of microbial isolates, (iv) screening for in vitro PGP traits and stress tolerance, and eventual selection of potential PGP microorganisms, (v) assessment of root colonization capacity of the isolates using labelled mutant derivatives and selection of efficient colonizers, (vi) screening for in vivo activity under stress to evaluate the effectiveness of the beneficial traits (including both, plant growth chamber and greenhouse-scale experiments). Although specific to drought stress, same steps are transferable to other types of abiotic stress

host plant, contributing to its tolerance to drought and salt stresses (Marasco et al. 2012; Marasco et al. 2016; Mapelli et al. 2020; Ferjani et al. 2015). Thus, plants that are adapted to the arid climates can serve as a novel model to understand and elucidate the factors determining plant growth promotion and plant survival under abiotic stresses (Marasco et al. 2018b; Köberl et al. 2011).

To increase the chances that PGP microorganisms efficiently cope with drought stress and deliver that protection to plants, the isolates must be able to (1) efficiently colonize the rhizosphere or the endosphere of the targeted plant and (2) respond rapidly to environmental fluctuations and thrive under the stress conditions of the targeted environment (Fig. 8.5).

1. *Efficient Colonization.* The ideal source to search for efficient microorganisms able to survive the competition in the rhizosphere and further colonize the root system is the original environment in which they exert their beneficial services in a mutualistic association with the plant; that is, the plant root system (rhizosphere and root tissue). It has been proved that the interactions among plant and PGP

microorganisms can have a limited level of specificity: PGP microorganisms were able to cross-colonize others plant hosts than the one of the origin, at least on a short term (Marasco et al. 2013).

2. *Stress Response*. Microorganisms inhabiting stressful environments such as those in the desert have evolved stress-adaptive traits that allow them to survive under fluctuating environmental conditions, water limitation, salinity, osmotic and heat stresses, and desiccation (Harel et al. 2004; Makhalanyane et al. 2015b).

Even though all plant species are colonized by PGP microbial communities in their rhizosphere and internal root tissues, the composition and functional role of such beneficial microorganisms change according to the plant host and the environmental conditions (Marasco et al. 2018a; Hartmann et al. 2009; Peiffer et al. 2013). Thus, to sustain the growth and development of plants cultivated in arid environments, it is important to select as the source the root and rhizosphere of plants naturally adapted to grow in dry and saline soils.

8.3.2 How to Isolate Powerful PGP Microorganisms

Once the plant model and the sampling environment have been selected, a methodological pipeline needs to be defined to isolate and screen microorganisms carrying out beneficial PGP activities (Fig. 8.5). An efficient strategy should allow the study of a relatively small fraction of strains in the system with large chances of presenting the beneficial activities. This strategy should include procedures for: (1) sampling and collection of plant tissues and rhizosphere; (2) isolation of cultivable microorganisms using suitable media (including selective media and enrichment procedures); (3) reduction of genomic redundancy and identification of microbial isolates; (4) screening for in vitro PGP traits and stress tolerance and selection of potential PGP microorganisms (those positive for several traits); (5) assessment of root colonization capacity of the isolates using labelled mutant derivatives and selection of efficient colonizers; (6) screening for in vivo activity under stress to evaluate the effectiveness of the beneficial traits (including plant growth chamber, greenhouse-scale experiments and field trial); and finally, (7) selection of the most promising strains and establishment of the final PGP collection of microorganisms (examples in Rolli et al., 2017).

Efforts should be increased to learn "*which microorganisms are there*" and garner information on "*what they are doing*", understand the relevance of diversity and whether there is a minimal core microbiome providing all the essential services for specific crops, and optimize breeding approaches to enhance the interactions with the associated microorganisms. Such a comprehensive analysis would provide useful knowledge about the mechanisms involved in plant protection and growth promotion, as well as the conditions needed by bacteria to exert their beneficial services.

8.3.3 The Functional Roles Exerted by Microorganisms Associated to Desert Plants

Microorganisms activate different mechanisms that enhance abiotic stress tolerance in plants, including protection from mechanical, osmotic, and oxidative stresses, as well as manipulation of hormone homeostasis (Rolli et al. 2015; Daffonchio et al. 2015; Vigani et al. 2018). For instance, the ability of microorganisms to produce EPS has a direct effect on the soil particles and overall physical conditions of the rhizospheric soil, thus protecting the root against physical damage (Amellal et al. 1998) and favouring the water-holding capacity of the rhizosphere/rhizosheath (Raddadi et al. 2018). Microorganisms can also protect the plant against stress by producing osmoprotectants (compatible solutes such as glycine betaine, proline, trehalose, and spermidine, among others) that are involved in osmotic adjustment (Boch et al. 1994; Csonka and Hanson 1991), as well as stress-related enzymes involved in ROS detoxification (Farr and Kogoma 1991). Another capacity of PGP microorganisms is their ability to produce phytohormone-like molecules (*e.g.*, auxin). Patten and Glick (2002) estimated that 80% of bacteria isolated from the rhizosphere display this activity. PGP microorganisms can also affect plant growth through regulation of phytohormone levels (Glick et al. 1998), for instance, by lowering the levels of ethylene in plants through the production of the enzyme 1-aminocyclopropane-1-carboxylate (ACC) deaminase able to hydrolyze ACC–the immediate biosynthetic precursor of ethylene (Glick et al. 2007b; Glick 2014). Ethylene is the stress hormone synthesized by plants as a direct consequence of stressful condition, causing reduction of plant growth. The expression of ACC deaminase by plant-associated bacteria could decrease the concentration of ethylene, releasing the plant from stress and preserving normal growth. This activity is one of the main studied PGP's mechanism to improve abiotic stress tolerance in plants (Glick 2014). Endophytic and rhizospheric bacteria capable of producing ACC deaminase were isolated from several plants and encompass a wide phylogenetic diversity; among the most studied are members of the genera *Pseudomonas, Bacillus, Paenibacillus, Burkholderia, Citrobacter, Enterobacter, Azospirillum, Rhizobium*, and *Agrobacterium* (Glick and Santoyo 2020). Their ability to sustain plant growth under abiotic stress, by determining increases both in the fresh and dry biomass of the plants, has been demonstrated under microcosm conditions with different model plants such as pepper, tomato, wheat, grapevine, and date palm, among others (Rolli et al. 2015; Marasco et al. 2012; Cherif et al. 2015; Lim and Kim 2013). Several examples of successful use of microorganisms isolated from desert plants for promoting plant growth under stressful conditions have been described. Bacterial strains isolated from the root system of the date palm cultivated in Tunisian oases were able to promote date palm growth and protect the plant from drought stress (Cherif et al. 2015). Two endophytic isolates, *Pseudomonas frederiksbergensis* E102 and *P. brassicacearum* E141, with a range of PGP traits *in vitro* including EPS production, inorganic phosphate solubilization, auxin and ammonia production, nitrogen fixation, and ACC deaminase activity, promoted date palm growth under drought, significantly increasing root and shoot biomass (Cherif et al. 2015). Similarly, other studies performed on canola suggested that endophytes isolated from date palms cultivated in Oman increased plant abiotic stress tolerance, altering ethylene and auxin levels in the host plant (Yaish et al. 2015). *Bacillus licheniformis* strain A2, isolated from the rhizosphere of the salt shrub *Suaeda fructicosa*, presents in vitro PGP activities (*i.e.*, auxin, siderophore and ammonia production, phosphate solubilization) and promotes the growth of the groundnut *Arachis hypogaea* under saline conditions (Goswami et al. 2014).

These examples demonstrate that PGP microorganisms of desert-associated plants exhibit an important protective/promoting potential in various ecosystems and under diverse climatic conditions (Hassen et al. 2018).

8.4 Conclusions

The ecology of plant-microorganism associations in desert ecosystems can be considered an incipient field of study as we are still beginning to understand the microbial diversity associated to desert plants. From the "metaorganism" point of view, as observed for plants growing in temperate ecosystems, xerophytic vegetation inhabiting desert biomes also harbours different and heterogeneous microbial communities. The harsh conditions of the desert strongly affect the way in which microbial communities are assembled and associated with the plant, determining a consistent presence of both taxonomical and functional core microbiomes. Indeed, since the low-resource properties of the desert soil (*i.e.*, water, nutrient, biodiversity), plants enrich and select the microorganisms available in the surrounding soil to survive in such extreme ecosystems, determining a consistent and co-evolved core microbiome across plant species. Among these microorganisms, a high number of beneficial microorganisms (*i.e.*, PGP) successfully interact with plants, accomplishing essential functions and ecological services complementary to the physiological features encoded by the host plant.

These beneficial PGP microorganisms not only act as biopromoter and bioprotector for their original host, but–as many recent studies have demonstrated–they are a powerful tool to increase abiotic stress tolerance in agricultural crops in a sustainable manner, motivating the interest in study the microbial communities associated with desert vegetation. Overall, knowledge of the microbial ecology of plant-associated communities will guide our efforts to understand the complex association of plants and microorganisms in desert biomes in order to improve agricultural management and conservation practices under the ongoing climate change scenario.

Acknowledgements DD acknowledges the financial support of King Abdullah University and Technology (KAUST) through the baseline research fund.

References

Abu-Ghosh S, Droby S, Korine C (2014) Seasonal and plant-dependent variations in diversity, abundance and stress tolerance of epiphytic yeasts in desert habitats. Environ Microbiol Rep 6: 373–382

Achtak H, Ater M, Oukabli A, Santoni S, Kjellberg F, Khadari B (2010) Traditional agroecosystems as conservatories and incubators of cultivated plant varietal diversity: The case of fig (Ficus carica L.) in Morocco. BMC Plant Biol 10:28

Aguirre-Garrido JF, Montiel-Lugo D, Hernández-Rodríguez C, Torres-Cortes G, Millán V, Toro N et al (2012) Bacterial community structure in the rhizosphere of three cactus species from semi-arid highlands in central Mexico. Antonie Van Leeuwenhoek 101:891–904

Akbari DL, Akbari LF, Bhadania Roshani A, Parkhiya MV, Golakiya BA (2015) Plant growth promoting traits and in vitro antagonism of drought tolerant endophytic bacteria isolated from grasses of kutch. J Cell Tissue Res 15:5241–5246

Alami Y, Achouak W, Marol C, Heulin T (2000) Rhizosphere soil aggregation and plant growth promotion of sunflowers by an exopolysaccharide-producing Rhizobium sp. strain isolated from sunflower roots. Appl Environ Microbiol 66:3393–3398

Albalasmeh AA, Ghezzehei TA (2014) Interplay between soil drying and root exudation in rhizosheath development. Plant Soil 374:739–751

Ali SZ, Sandhya V, Venkateswar RL (2014) Isolation and characterization of drought-tolerant ACC deaminase and exopolysaccharide-producing fluorescent Pseudomonas sp. Ann Microbiol 64: 493–502

Al-Karaki GN (2013) Application of mycorrhizae in sustainable date palm cultivation. Emirates J Food Agric 25:854–862

Alsharif W, Saad MM, Hirt H (2020) Desert microbes for boosting sustainable agriculture in extreme environments. Front Microbiol 11:1666

Al-Yahya'ei MN, Oehl F, Vallino M, Lumini E, Redecker D, Wiemken A et al (2011) Unique arbuscular mycorrhizal fungal communities uncovered in date palm plantations and surrounding desert habitats of Southern Arabia. Mycorrhiza 21:195–209

Amellal N, Burtin G, Bartoli F, Heulin T (1998) Colonization of wheat roots by an exopolysaccharide-producing Pantoea agglomerans strain and its effect on rhizosphere soil aggregation. Appl Environ Microbiol 64:3740–3747

Aroca R, Porcel R, Ruiz-Lozano JM (2012) Regulation of root water uptake under abiotic stress conditions. J Exp Bot 63:43–57

Asaf S, Khan AL, Khan MA, Al-Harrasi A, Lee I-J (2018) Complete genome sequencing and analysis of endophytic Sphingomonas sp. LK11 and its potential in plant growth. 3. Biotech 8: 389

Ashraf M (2006) Bacterial exo-polysaccharides and productivity of the salt affected soils. II. Effect of exo-polysaccharides (EPS) producing bacterial inoculation on growth of roots of wheat (Triticum aestivum L.) plants grown in a salt-affected soil. Int J Environ Sci Technol 3:45–53

Ashraf M, Harris PJC (2004) Potential biochemical indicators of salinity tolerance in plants. Plant Sci 166:3–16

Bahl J, Lau MCY, Smith GJD, Vijaykrishna D, Cary SC, Lacap DC et al (2011) Ancient origins determine global biogeography of hot and cold desert cyanobacteria. Nat Commun 2:163

Bailey C, Scholes M (1997) Rhizosheath occurrence in South African grasses. South African J Bot 63:484–490

Balloi A, Rolli E, Marasco R, Mapelli F, Tamagnini I, Cappitelli F et al (2010) The role of microorganisms in bioremediation and phytoremediation of polluted and stressed soils. Agrochimica. 54:353–369

Bárcenas RT, Yesson C, Hawkins JA (2011) Molecular systematics of the Cactaceae. Cladistics 27: 470–489

Bar-Shmuel N, Rogovin E, Rachmilevitch S, Friedman A-L-L, Shelef O, Hoffmann I et al (2018) Tripartite symbiosis of plant-weevil-bacteria is a widespread phenomenon in the Negev Desert. Sci Rep 8:2420

Barthlott W, Hunt DR (1993) Cactaceae. In: Kubitzki K, Rohwer JG, Bittrich V (eds) Flowering plants dicotyledons. Springer, Berlin, Heidelberg, pp 161–197

Bashan Y, Holguin G (2002) Plant growth-promoting bacteria: a potential tool for arid mangrove reforestation. Trees - Struct Funct 16:159–166

Beckers B, Op De Beeck M, Weyens N, Boerjan W, Vangronsveld J, De Beeck MO et al (2017) Structural variability and niche differentiation in the rhizosphere and endosphere bacterial microbiome of field-grown poplar trees. Microbiome 5:25

Belkin S, Qvit-Raz N (2010) Life on a leaf: Bacterial epiphytes of a salt-excreting desert tree. Springer, Dordrecht, pp 393–406

Benard P, Kroener E, Vontobel P, Kaestner A, Carminati A (2016) Water percolation through the root-soil interface. Adv Water Resour 95:190–198

Berdugo M, Delgado-Baquerizo M, Soliveres S, Hernández-Clemente R, Zhao Y, Gaitán JJ et al (2020) Global ecosystem thresholds driven by aridity. Science 367(6479):787–790. https://doi.org/10.1126/science.aay5958

Berg G, Raaijmakers JM (2018) Saving seed microbiomes. ISME J 12:1167–1170

Bergmann D, Zehfus M, Zierer L, Smith B, Gabel M (2009) Grass rhizosheaths: associated bacterial communities and potential for nitrogen fixation. West North Am Nat 69:105–114

Bezerra JDP, Santos MGS, Barbosa RN, Svedese VM, Lima DMM, Fernandes MJS et al (2013) Fungal endophytes from cactus Cereus jamacaru in Brazilian tropical dry forest: a first study. Symbiosis 60:53–63

Bhatnagar A, Bhatnagar M (2005) Microbial diversity in desert ecosystems. Curr Sci 89:91–100

Boch J, Kempf B, Bremer E (1994) Osmoregulation in Bacillus subtilis: Synthesis of the osmoprotectant glycine betaine from exogenously provided choline. J Bacteriol 176:5364–5371

Boldareva-Nuianzina EN, Bláhová Z, Sobotka R, Koblížek M (2013) Distribution and origin of oxygen-dependent and oxygen-independent forms of mg-protoporphyrin monomethylester cyclase among phototrophic proteobacteria. Appl Environ Microbiol 79:2596–2604

Booth JM, Fusi M, Marasco R, Michoud G, Fodelianakis S, Merlino G et al (2019a) The role of fungi in heterogeneous sediment microbial networks. Sci Rep 9:7537

Booth JM, Fusi M, Marasco R, Mbobo T, Daffonchio D (2019b) Fiddler crab bioturbation determines consistent changes in bacterial communities across contrasting environmental conditions. Sci Rep 9:3749

Brown LK, George TS, Neugebauer K, White PJ (2017) The rhizosheath – a potential trait for future agricultural sustainability occurs in orders throughout the angiosperms. Plant Soil 418:115–128

Camarena-Pozos DA, Flores-Núñez VM, López MG, López-Bucio J, Partida-Martínez LP (2019) Smells from the desert: microbial volatiles that affect plant growth and development of native and non-native plant species. Plant Cell Environ 42:1368–1380

Carson JK, Gonzalez-Quiñones V, Murphy DV, Hinz C, Shaw JA, Gleeson DB (2010) Low pore connectivity increases bacterial diversity in soil. Appl Environ Microbiol 76:3936–3942

Cavazos BR, Bohner TF, Donald ML, Sneck ME, Shadow A, Omacini M et al (2018) Testing the roles of vertical transmission and drought stress in the prevalence of heritable fungal endophytes in annual grass populations. New Phytol 219:1075–1084

Chao CCT, Krueger RR (2007) The date palm (Phoenix dactylifera L.): overview of biology, uses, and cultivation. HortScience 42:1077–1082

Chater KF, Chandra G (2006) The evolution of development in Streptomyces analysed by genome comparisons. FEMS Microbiol Rev 30:651–672

Cherif H, Marasco R, Rolli E, Ferjani R, Fusi M, Soussi A et al (2015) Oasis desert farming selects environment-specific date palm root endophytic communities and cultivable bacteria that promote resistance to drought. Environ Microbiol Rep 7:668–678

Clery D (2011) Greenhouse-power plant hybrid set to make Jordan's desert bloom. Science (80-) 331:136–136

Coleman-Derr D, Desgarennes D, Fonseca-Garcia C, Gross S, Clingenpeel S, Woyke T et al (2016) Plant compartment and biogeography affect microbiome composition in cultivated and native Agave species. New Phytol 209:798–811

Csonka LN, Hanson AD (1991) Prokaryotic Osmoregulation. Annu Rev Microbiol 45:569–606

Cui M, Nobel PS (2006) Nutrient status, water uptake and gas exchange for three desert succulents infected with mycorrhizal fungi. New Phytol 122:643–649

Daffonchio D, Hirt H, Berg G (2015) Principles of plant-microbe interactions. Springer International Publishing, Cham

Danin A (1996) Plant adaptations to environmental stresses in desert dunes. In: Cloudsley-Thompson J, Punzo F (eds) Adaptations of desert organisms. Springer, Berlin

Danin A (2000) Plants of desert dunes. In: Cloudsley-Thompson J (ed) Adaptations of desert organisms. Springer, Berlin

de Carmen Orozco-Mosqueda M, Glick BR, Santoyo G (2020) ACC deaminase in plant growth-promoting bacteria (PGPB): An efficient mechanism to counter salt stress in crops. Microbiol Res 235:126439

de Grenade R (2013) Date palm as a keystone species in Baja California peninsula, Mexico oases. J Arid Environ 94:59–67

de Grenade R, Nabhan GP, Cariño OM (2016) Oases of the Baja California peninsula as sacred spaces of agrobiodiversity persistence. Agric Human Values 33:455–474

de la Torre-Hernández ME, Salinas-Virgen LI, Aguirre-Garrido JF, Fernández-González AJ, Martínez-Abarca F, Montiel-Lugo D et al (2020) Composition, structure, and PGPR traits of the rhizospheric bacterial communities associated with wild and cultivated Echinocactus platyacanthus and Neobuxbaumia polylopha. Front Microbiol 11:1424

Degnan PH, Ochman H (2012) Illumina-based analysis of microbial community diversity. ISME J 6:183–194

Delgado-Baquerizo M, Oliverio AM, Brewer TE, Benavent-González A, Eldridge DJ, Bardgett RD et al (2018) A global atlas of the dominant bacteria found in soil. Science (80-) 359:320–325

Desgarennes D, Garrido E, Torres-Gomez MJ, Peña-Cabriales JJ, Partida-Martinez LP (2014) Diazotrophic potential among bacterial communities associated with wild and cultivated Agave species. FEMS Microbiol Ecol 90:844–857

Dias ACF, Costa FEC, Andreote FD, Lacava PT, Teixeira MA, Assumpção LC et al (2009) Isolation of micropropagated strawberry endophytic bacteria and assessment of their potential for plant growth promotion. World J Microbiol Biotechnol 25:189–195

Drees KP, Neilson JW, Betancourt JL, Quade J, Henderson DA, Pryor BM et al (2006) Bacterial community structure in the hyperarid core of the Atacama Desert, Chile. Appl Environ Microbiol 72:7902–7908

Eggli U, Nyffeler R (2009) Living under temporarily arid conditions - succulence as an adaptive strategy. Bradleya 27:13–36

Eke P, Kumar A, Sahu KP, Wakam LN, Sheoran N, Ashajyothi M et al (2019) Endophytic bacteria of desert cactus (Euphorbia trigonas Mill) confer drought tolerance and induce growth promotion in tomato (Solanum lycopersicum L.). Microbiol Res 228:126302

El Hidri D, Guesmi A, Najjari A, Cherif H, Ettoumi B, Hamdi C et al (2013) Cultivation-Dependant Assessment, Diversity, and Ecology of Haloalkaliphilic Bacteria in Arid Saline Systems of Southern Tunisia. Biomed Res Int 2013:1–15

El Kinany S, Achbani E, Faggroud M, Ouahmane L, El Hilali R, Haggoud A et al (2019) Effect of organic fertilizer and commercial arbuscular mycorrhizal fungi on the growth of micropropagated date palm cv, Feggouss. J Saudi Soc Agric Sci 18:411–417

El-sharabasy SF, Orf HOM, Abotaleb HH, Abdel-galeil LM, Saber TY (2018) Effect of Plant Growth Promoting Rhizobacteria (PGPR) on growth and leaf chemical composition of date palm plants cv. Bartamuda under salinity stress. Middle East J Agric Res 7:618–624

El-Tarabily KA (2008) Promotion of tomato (Lycopersicon esculentum Mill.) plant growth by rhizosphere competent 1-aminocyclopropane-1-carboxylic acid deaminase-producing streptomycete actinomycetes. Plant Soil 308:161–174

Etesami H, Beattie GA (2018) Mining halophytes for plant growth-promoting halotolerant bacteria to enhance the salinity tolerance of non-halophytic crops. Front Microbiol 9:148
Fahn A, Cutler DF (1992) Xerophytes. Gebrüder Borntraeger, Berlin
Farr SB, Kogoma T (1991) Oxidative stress responses in Escherichia coli and Salmonella typhimurium. Microbiol Rev 55:561–585
Ferjani R, Marasco R, Rolli E, Cherif H, Cherif A, Gtari M et al (2015) The date palm tree rhizosphere is a niche for plant growth promoting bacteria in the oasis ecosystem. Biomed Res Int 2015:1–10
Fierer N, Strickland MS, Liptzin D, Bradford MA, Cleveland CC (2009) Global patterns in belowground communities. Ecol Lett 12:1238–1249
Fierer N, Lauber CL, Ramirez KS, Zaneveld J, Bradford M, a, Knight R. (2012a) Comparative metagenomic, phylogenetic and physiological analyses of soil microbial communities across nitrogen gradients. ISME J 6:1007–1017
Fierer N, Leff JW, Adams BJ, Nielsen UN, Bates ST, Lauber CL et al (2012b) Cross-biome metagenomic analyses of soil microbial communities and their functional attributes. Proc Natl Acad Sci 109:21390–21395
Finkel OM, Burch AY, Lindow SE, Post AF, Belkin S (2011) Geographical location determines the population structure in phyllosphere microbial communities of a salt-excreting desert tree. Appl Environ Microbiol 77:7647–7655
Flores-Núñez VM, Fonseca-García C, Desgarennes D, Eloe-Fadrosh E, Woyke T, Partida-Martínez LP (2020) Functional signatures of the epiphytic prokaryotic microbiome of agaves and cacti. Front Microbiol 10:1–13
Flowers TJ, Colmer TD (2008) Salinity tolerance in halophytes*. New Phytol 179:945–963
Flowers TJ, Colmer TD (2015) Plant salt tolerance: adaptations in halophytes. Ann Bot 115:327–331
Fonseca-García C, Coleman-Derr D, Garrido E, Visel A, Tringe SG, Partida-Martínez LP (2016) The cacti microbiome: interplay between habitat-filtering and host-specificity. Front Microbiol 7:150
Fonseca-García C, Desgarennes D, Flores-Núñez VM, Partida-Martínez LP (2018) The Microbiome of desert CAM Plants: lessons from amplicon sequencing and metagenomics. In: Metagenomics. Elsevier, Amsterdam, pp 231–254
Fterich A, Mahdhi M, Caviedes MA, Pajuelo E, Rivas R, Rodriguez-Llorente ID et al (2011) Characterization of root-nodulating bacteria associated to Prosopis farcta growing in the arid regions of Tunisia. Arch Microbiol 193:385–397
Galloway AF, Knox P, Krause K (2019) Sticky mucilages and exudates of plants: putative microenvironmental design elements with biotechnological value. New Phytol 225:1461–1469
Gao Q, Garcia-Pichel F (2011) Microbial ultraviolet sunscreens. Nat Rev Microbiol 9:791–802
Glick BR (2014) Bacteria with ACC deaminase can promote plant growth and help to feed the world. Microbiol Res 169:30–39
Glick BR, Glick BR (2012) Plant growth-promoting bacteria: mechanisms and applications. Scientifica (Cairo) 2012:1–15
Glick BR, Penrose DM, Li J (1998) A model for the lowering of plant ethylene concentrations by plant growth-promoting bacteria. J Theor Biol 190:63–68
Glick BR, Zhenyu C, Czarny J (2007a) Promotion of plant growth by ACCdeaminase-producing soil bacteria. New Perspect Approaches Plant Growth-Promoting Rhizobacteria Res 119:329–339
Glick BR, Cheng Z, Czarny J, Duan J (2007b) Promotion of plant growth by ACC deaminase-producing soil bacteria. Eur J Plant Pathol 119:329–339
Gontia I, Kavita K, Schmid M, Hartmann A, Jha B (2011) Brachybacterium saurashtrense sp. nov., a halotolerant root-associated bacterium with plant growth-promoting potential. Int J Syst Evol Microbiol 61:2799–2804

Goswami D, Dhandhukia P, Patel P, Thakker JN (2014) Screening of PGPR from saline desert of Kutch: Growth promotion in Arachis hypogea by Bacillus licheniformis A2. Microbiol Res 169: 66–75

Grishkan I, Nevo E (2010) Spatiotemporal distribution of soil microfungi in the Makhtesh Ramon area, central Negev desert, Israel. Fungal Ecol 3:326–337

Grover M, Ali SZ, Sandhya V, Rasul A, Venkateswarlu B (2011) Role of microorganisms in adaptation of agriculture crops to abiotic stresses. World J Microbiol Biotechnol 27:1231–1240

Guesmi S, Chouchane H, Neifar M, Hosni F, Cherif A, Sghaier H (2019) Radiation-inducible radioprotective exopolysaccharides of Bacillus siamensis CV5 from irradiated roots of Cistanche violacea to decrease free radical damage produced by ionizing radiation. Int J Radiat Biol 95:1552–1563

Gutterman Y (2002) The desert biome, survival adaptations and strategies of annual plant species. In: Survival strategies of annual desert plants. Springer, Berlin, pp 1–36

Halo BA, Khan AL, Waqas M, Al-Harrasi A, Hussain J, Ali L et al (2015) Endophytic bacteria (Sphingomonas sp. LK11) and gibberellin can improve Solanum lycopersicum growth and oxidative stress under salinity. J Plant Interact 10:117–125

Hanna AL, Youssef HH, Amer WM, Monib M, Fayez M, Hegazi NA (2013) Diversity of bacteria nesting the plant cover of north Sinai deserts, Egypt. J Adv Res 4:13–26

Harel Y, Ohad I, Kaplan A (2004) Activation of photosynthesis and resistance to photoinhibition in cyanobacteria within biological desert crust. Plant Physiol 136:3070–3079

Harnett AC, Wilson GWT, Ott JP, Setshogo M (2013) Variation in root system traits among African semi-arid savanna grasses: Implications for drought tolerance. Austral Ecol 38:383–392

Hartmann M, Lee S, Hallam SJ, Mohn WW (2009) Bacterial, archaeal and eukaryal community structures throughout soil horizons of harvested and naturally disturbed forest stands. Environ Microbiol 11:3045–3062

Hassen W, Neifar M, Cherif H, Najjari A, Chouchane H, Driouich RC et al (2018) Pseudomonas rhizophila S211, a new plant growth-promoting rhizobacterium with potential in pesticide-bioremediation. Front Microbiol 9:1–17

Hegazi N, Fayez M, Hanna A, Youssef H, Amer W, Monib M. Spore-forming bacteria are potential residents in the root spheres of rhizo-sheathed xerophytic plants of north Sinai deserts. 4 th Conference on Recent Technologies in Agriculture 2009. pp 870–884.

Hernández-Hernández T, Brown JW, Schlumpberger BO, Eguiarte LE, Magallón S (2014) Beyond aridification: multiple explanations for the elevated diversification of cacti in the new world succulent biome. New Phytol 202:1382–1397

Hirsch PR, Mauchline TH (2012) Who's who in the plant root microbiome? Nat Biotechnol 30: 961–962

Jeffries P, Gianinazzi S, Perotto S, Turnau K, Barea J-M (2003) The contribution of arbuscular mycorrhizal fungi in sustainable maintenance of plant health and soil fertility. Biol Fertil Soils 37:1–16

Jenkins MB (2003) Rhizobial and bradyrhizobial symbionts of mesquite from the Sonoran Desert: salt tolerance, facultative halophily and nitrate respiration. Soil Biol Biochem 35:1675–1682

Johnston-Monje D, Lundberg DS, Lazarovits G, Reis VM, Raizada MN (2016) Bacterial populations in juvenile maize rhizospheres originate from both seed and soil. Plant Soil 405: 337–355

Jones DL, Nguyen C, Finlay RD (2009) Carbon flow in the rhizosphere: carbon trading at the soil–root interface. Plant Soil 321:5–33

Jooste M, Roets F, Midgley GF, Oberlander KC, Dreyer LL (2019) Nitrogen-fixing bacteria and Oxalis – evidence for a vertically inherited bacterial symbiosis. BMC Plant Biol 19:441

Kaplan D, Maymon M, Agapakis CM, Lee A, Wang A, Prigge BA et al (2013) A survey of the microbial community in the rhizosphere of two dominant shrubs of the Negev Desert highlands, Zygophyllum dumosum (Zygophyllaceae) and Atriplex halimus (Amaranthaceae), using cultivation-dependent and cultivation-independent methods. Am J Bot 100:1713–1725

Karray F, Gargouri M, Chebaane A, Mhiri N, Mliki A, Sayadi S (2020) Climatic aridity gradient modulates the diversity of the rhizosphere and endosphere bacterial microbiomes of opuntia ficus-indica. Front Microbiol 11:1622

Kavamura VN, Taketani RG, Ferreira C, de Melo IS, Mendes R (2018) The role of species turnover in structuring bacterial communities in a local scale in the cactus rhizosphere. Plant Soil 425: 101–112

Kazerooni EA, Maharachchikumbura SSN, Rethinasamy V, Al-Mahrouqi H, Al-Sadi AM (2017) Fungal diversity in tomato rhizosphere soil under conventional and desert farming systems. Front Microbiol 8:1–8

Klironomos JN, Hart MM (2001) Animal nitrogen swap for plant carbon. Nature 410:651–652

Köberl M, Müller H, Ramadan EM, Berg G (2011) Desert farming benefits from microbial potential in arid soils and promotes diversity and plant health. PLoS One 6:e24452

Köberl M, Schmidt R, Ramadan EM, Bauer R, Berg G (2013) The microbiome of medicinal plants: Diversity and importance for plant growth, quality, and health. Front Microbiol 4:1–9

Köberl M, Erlacher A, Ramadan EM, El-Arabi TF, Müller H, Bragina A et al (2016) Comparisons of diazotrophic communities in native and agricultural desert ecosystems reveal plants as important drivers in diversity. FEMS Microbiol Ecol 92:fiv166

Lacap DC, Warren-Rhodes KA, McKay CP, Pointing SB (2011) Cyanobacteria and chloroflexi-dominated hypolithic colonization of quartz at the hyper-arid core of the Atacama Desert, Chile. Extremophiles 15:31–38

Laity JJ (2009) Deserts and desert environments. John Wiley & Sons, Hoboken

Lauber CL, Hamady M, Knight R, Fierer N (2009) Pyrosequencing-based assessment of soil pH as a predictor of soil bacterial community structure at the continental scale. Appl Environ Microbiol 75:5111–5120

Lemanceau P, Bauer P, Kraemer S, Briat J-F (2009) Iron dynamics in the rhizosphere as a case study for analyzing interactions between soils, plants and microbes. Plant Soil 321:513–535

Lester ED, Satomi M, Ponce A (2007) Microflora of extreme arid Atacama Desert soils. Soil Biol Biochem 39:704–708

Lim JH, Kim SD (2013) Induction of drought stress resistance by multi-functional PGPR *Bacillus licheniformis* K11 in pepper. Plant Pathol J 29:201–208

Liu W, Zhou X, Li G, Li L, Kong L, Wang C et al (2011) Multiple plant surface signals are sensed by different mechanisms in the rice blast fungus for appressorium formation. PLoS Pathog 7: e1001261

Lopez BR, Bacilio M (2020) Weathering and soil formation in hot, dry environments mediated by plant–microbe interactions. Biol Fertil Soils 56:447–459

Lopez BR, Bashan Y, Bacilio M (2011) Endophytic bacteria of Mammillaria fraileana, an endemic rock-colonizing cactus of the southern Sonoran Desert. Arch Microbiol 193:527–541

López-Lozano NE, Echeverría Molinar A, Ortiz Durán EA, Hernández Rosales M, Souza V (2020) Bacterial diversity and interaction networks of Agave lechuguilla rhizosphere differ significantly from bulk soil in the oligotrophic sasin of Cuatro Cienegas. Front Plant Sci 11:1–14

Loro M, Valero-Jiménez CA, Nozawa S, Márquez LM (2012) Diversity and composition of fungal endophytes in semiarid Northwest Venezuela. J Arid Environ 85:46–55

Lundberg DS, Lebeis SL, Paredes SH, Yourstone S, Gehring J, Malfatti S et al (2012) Defining the core Arabidopsis thaliana root microbiome. Nature 488:86–90

Makhalanyane TP, Valverde A, Lacap DC, Pointing SB, Tuffin MI, Cowan DA (2013) Evidence of species recruitment and development of hot desert hypolithic communities. Environ Microbiol Rep 5:219–224

Makhalanyane TP, Valverde A, Velázquez D, Gunnigle E, Van Goethem MW, Quesada A et al (2015a) Ecology and biogeochemistry of cyanobacteria in soils, permafrost, aquatic and cryptic polar habitats. Biodivers Conserv 24:819–840

Makhalanyane TP, Valverde A, Gunnigle E, Frossard A, Ramond JBJ-BJ-BB, Cowan DA (2015b) Microbial ecology of hot desert edaphic systems. FEMS Microbiol Rev 39:203–221

Mapelli F, Marasco R, Balloi A, Rolli E, Cappitelli F, Daffonchio D et al (2012) Mineral–microbe interactions: biotechnological potential of bioweathering. J Biotechnol 157:473–481

Mapelli F, Marasco R, Rolli E, Barbato M, Cherif H, Guesmi A et al (2013) Potential for plant growth promotion of rhizobacteria associated with Salicornia growing in Tunisian hypersaline soils. Biomed Res Int 2013:1–13

Mapelli F, Riva V, Vergani L, Choukrallah R, Borin S (2020) Unveiling the microbiota diversity of the xerophyte Argania spinosa L. skeels root system and residuesphere. Microb Ecol 80: 822–836

Marasco R, Rolli E, Ettoumi B, Vigani G, Mapelli F, Borin S et al (2012) A drought resistance-promoting microbiome is selected by root system under desert farming. PLoS One 7:e48479

Marasco R, Rolli E, Vigani G, Borin S, Sorlini C, Ouzari H et al (2013) Are drought-resistance promoting bacteria cross-compatible with different plant models? Plant Signal Behav 8:e26741

Marasco R, Mapelli F, Rolli E, Mosqueira MJ, Fusi M, Bariselli P et al (2016) Salicornia strobilacea (Synonym of Halocnemum strobilaceum) grown under different tidal regimes selects rhizosphere bacteria capable of promoting plant growth. Front Microbiol 7:1286

Marasco R, Rolli E, Fusi M, Michoud G, Daffonchio D (2018a) Grapevine rootstocks shape underground bacterial microbiome and networking but not potential functionality. Microbiome 6:3

Marasco R, Mosqueira MJ, Fusi M, Ramond J, Merlino G, Booth JM et al (2018b) Rhizosheath microbial community assembly of sympatric desert speargrasses is independent of the plant host. Microbiome 6:215

Marasco R, Fusi M, Rolli E, Ettoumi B, Tambone F, Borin S et al (2021) Aridity modulates belowground bacterial community dynamics in olive tree. Environ Microbiol 23 (10):6275–6291. https://doi.org/10.1111/1462-2920.15764

Marasco R, Fusi M, Mosqueira M, Booth JM, Rossi F, Cardinale M et al (2022) Rhizosheath–root system changes exopolysaccharide content but stabilizes bacterial community across contrasting seasons in a desert environment. Environ Microbiome 17(1):1–19. https://doi.org/10.1186/s40793-022-00407-3

Maximov N (1931) The physiological significance of the xeromorphic structure of plants. J Ecol 19: 273–282

McCalley CK, Sparks JP (2009) Abiotic gas formation drives nitrogen loss from a desert ecosystem. Science (80-) 326:837–840

Mitter B, Pfaffenbichler N, Flavell R, Compant S, Antonielli L, Petric A et al (2017) A new approach to modify plant microbiomes and traits by introducing beneficial bacteria at flowering into progeny seeds. Front Microbiol 8:1–10

Mohammadipanah F, Wink J (2016) Actinobacteria from arid and desert habitats: diversity and biological activity. Front Microbiol 6:1–10

Moreno-Espíndola IP, Rivera-Becerril F, de Jesús F-GM, De León-González F (2007) Role of root-hairs and hyphae in adhesion of sand particles. Soil Biol Biochem 39:2520–2526

Moreno-Espíndola PI, Jesús FGM, Fernando DLG, Facundo RB, Diego GH (2013) Comunidad bacteriana cultivable asociada a la rizocoraza de Amaranthus hypochondriacus. Terra Latinoam 31:57–69

Mosqueira MJ, Marasco R, Fusi M, Michoud G, Merlino G, Cherif A et al (2019) Consistent bacterial selection by date palm root system across heterogeneous desert oasis agroecosystems. Sci Rep 9:4033

Mouchacca J (2007) Heat tolerant fungi and applied research: addition to the previously treated group of strictly thermotolerant species. World J Microbiol Biotechnol 23:1755–1770

Moustafa AF (1978) A supplementary annolated list of the fungi of Kuwait. J Univ Kuwait 5:61–82

Murgia M, Fiamma M, Barac A, Deligios M, Mazzarello V, Paglietti B et al (2018) Biodiversity of fungi in hot desert sands. Microbiologyopen 8(1):e00595

Neilson JW, Quade J, Ortiz M, Nelson WM, Legatzki A, Tian F et al (2012) Life at the hyperarid margin: novel bacterial diversity in arid soils of the Atacama Desert, Chile. Extremophiles 16: 553–566

Nelson EB (2018) The seed microbiome: origins, interactions, and impacts. Plant Soil 422:7–34

North GGBG, Nobel PPS (1997) Drought-induced changes in soil contact and hydraulic conductivity for roots of Opuntia ficus-indica with and without rhizosheaths. Plant Soil 191:249–258

Noy-Meir I (1974) Desert Ecosystems: Higher trophic levels. Annu Rev Ecol Syst 5:195–214

Nyffeler R, Eggli U (2010) An up-to-date familial and suprafamilial classification of succulent plants. Bradleya 28:125–144

Obledo EN, Barragán-Barragán LB, Gutiérrez-González P, Ramírez-Hernández BC, Ramírez JJ, Rodríguez-Garay B (2003) Increased photosyntethic efficiency generated by fungal symbiosis in Agave victoria-reginae. Plant Cell Tissue Organ Cult 74:237–241

Olanrewaju OS, Glick BR, Babalola OO (2017) Mechanisms of action of plant growth promoting bacteria. World J Microbiol Biotechnol 33:197

Othman A, Shawky M, Amer W, Fayez M, Monib M, Hegazi N (2003a) Biodiversity of microorganisms in semi-arid soils of north sinai deserts. Arch Agron Soil Sci 49:241–260

Othman AA, Amer WM, Fayez M, Monib M, Hegazi NA (2003b) Biodiversity of diazotrophs associated to the plant cover of north sinai deserts: Biodiversität diazotropher assoziiert mit der pflanendecke der wüsten nordsinais. Arch Agron Soil Sci 49:683–705

Othman AA, Amer WM, Fayez M, Hegazi N (2004) Rhizosheath of sinai desert plants is a potential repository for associative diazotrophs. Microbiol Res 159:285–293

Pang J, Ryan MH, Siddique KHMM, Simpson RJ (2017) Unwrapping the rhizosheath. Plant Soil 418:129–139

Parida AK, Das AB (2005) Salt tolerance and salinity effects on plants: a review. Ecotoxicol Environ Saf 60:324–349

Patten CL, Glick BR (2002) Role of Pseudomonas putida indoleacetic acid in development of the host plant root system. Appl Environ Microbiol 68:3795–3801

Peiffer JA, Spor A, Koren O, Jin Z, Tringe SG, Dangl JL et al (2013) Diversity and heritability of the maize rhizosphere microbiome under field conditions. Proc Natl Acad Sci 110:6548–6553

Pessarakli M, Harivandi MA, Kopec DM, Ray DT (2012) Growth responses and nitrogen uptake by saltgrass (Distichlis spicata L.), a halophytic plant species, under salt stress, using the 15 N technique. Int J Agron 2012:1–9

Philippot L, Raaijmakers JM, Lemanceau P, van der Putten WH (2013) Going back to the roots: the microbial ecology of the rhizosphere. Nat Rev Microbiol 11:789–799

Pieterse Z, Aveling TAS, Jacobs A, Cowan DA (2018) Seasonal variability in fungal endophytes from Aizoaceae plants in the Succulent Karoo biodiversity hotspot, South Africa. J Arid Environ 156:19–26

Prestel E, Regeard C, Salamitou S, Neveu J, DuBow MS (2013) The bacteria and bacteriophages from a mesquite flats site of the death valley desert. Antonie Van Leeuwenhoek 103:1329–1341

Puente ME, Li CY, Bashan Y (2009) Rock-degrading endophytic bacteria in cacti. Environ Exp Bot 66:389–401

Puente ME, Bashan Y, Li CY, Lebsky VK (2004) Microbial populations and activities in the rhizoplane of rock-weathering desert plants. I. Root colonization and weathering of igneous rocks. Plant Biol 6:629–642

Qurashi AW, Sabri AN (2012) Bacterial exopolysaccharide and biofilm formation stimulate chickpea growth and soil aggregation under salt stress. Brazilian J Microbiol 43:1183–1191

Raddadi N, Giacomucci L, Marasco R, Daffonchio D, Cherif A, Fava F (2018) Bacterial polyextremotolerant bioemulsifiers from arid soils improve water retention capacity and humidity uptake in sandy soil. Microb Cell Fact 17:83

Ranzoni F (1968) Fungi Isolated in culture from soils of the Sonoran Desert. Mycologia 63: 157–168

Raymond J (2008) Coloring in the tree of life. Trends Microbiol 16:41–43

Redman RS, Dunigan DD, Rodriguez RJ (2001) Fungal symbiosis from mutualism to parasitism: who controls the outcome, host or invader? New Phytol 151:705–716

Rodríguez CE, Mitter B, Barret M, Sessitsch A, Compant S (2018) Commentary: seed bacterial inhabitants and their routes of colonization. Plant Soil 422:129–134

Rolli E, Marasco R, Vigani G, Ettoumi B, Mapelli F, Deangelis ML et al (2015) Improved plant resistance to drought is promoted by the root-associated microbiome as a water stress-dependent trait. Environ Microbiol 17:316–331

Rolli E, Marasco R, Saderi S, Corretto E, Mapelli F, Cherif A et al (2017) Root-associated bacteria promote grapevine growth: from the laboratory to the field. Plant Soil 410:369–382

Roth-Nebelsick A, Ebner M, Miranda T, Gottschalk V, Voigt D, Gorb S et al (2012) Leaf surface structures enable the endemic Namib desert grass Stipagrostis sabulicola to irrigate itself with fog water. J R Soc Interface 9:1965–1974

Sandquist DR (2014) Plants in deserts. In: Monson RK (ed) Ecology and the environment. Springer, New York, pp 297–326

Santhanam R, Rong X, Huang Y, Andrews BA, Asenjo JA, Goodfellow M (2013) Streptomyces bullii sp. nov., isolated from a hyper-arid Atacama Desert soil. Antonie Van Leeuwenhoek 103: 367–373

Santi C, Bogusz D, Franche C (2013) Biological nitrogen fixation in non-legume plants. Ann Bot 111:743–767

Sefrji FO, Marasco R, Michoud G, Seferji KA, Merlino G, Daffonchio D (2021) Kaustia mangrovi gen. nov., sp. nov. isolated from Red Sea mangrove sediments belongs to the recently proposed Parvibaculaceae family within the order Rhizobiales. Int J Syst Evol Microbiol 71(5). https://doi.org/10.1099/ijsem.0.004806

Sefrji FO, Marasco R, Michoud G, Merlino G, Daffonchio D (2022) Insights into the cultivable bacterial fraction of sediments from the Red Sea mangroves and physiological, chemotaxonomic, and genomic characterization of Mangrovibacillus cuniculi gen. nov., sp. nov., a novel member of the Bacillaceae family. Front Microbiol 391. https://doi.org/10.3389/fmicb.2022.777986

Shabani F, Kumar L, Taylor S (2012) Climate change impacts on the future distribution of date palms: a modeling exercise using CLIMEX. PLoS One 7:e48021

Shabani F, Kumar L, Taylor S (2015) Distribution of date palm in the Middle East based on future climate scenarios. Exp Agric 51:244–263

Sharma P, Jha AB, Dubey RS, Pessarakli M (2012) Reactive oxygen species, oxidative damage, and antioxidative defense mechanism in plants under stressful conditions. J Bot 2012:1–26

Shelef O, Helman Y, Friedman ALL, Behar A, Rachmilevitch S (2013) Tri-party underground symbiosis between a weevil, bacteria and a desert plant. PLoS One 8:e76588

Singh H, Fernandes T, Apte SK (2010) Unusual radioresistance of nitrogen-fixing cultures of Anabaena strains. J Biosci 35:427–434

Singh H, Anurag K, Apte SK (2013) High radiation and desiccation tolerance of nitrogen-fixing cultures of the cyanobacterium Anabaena sp. strain PCC 7120 emanates from genome/proteome repair capabilities. Photosynth Res 118:71–81

Smith RJ, Hopper SD, Shane MW (2011) Sand-binding roots in Haemodoraceae: global survey and morphology in a phylogenetic context. Plant Soil 348:453–470

Soldan R, Mapelli F, Crotti E, Schnell S, Daffonchio D, Marasco R et al (2019) Bacterial endophytes of mangrove propagules elicit early establishment of the natural host and promote growth of cereal crops under salt stress. Microbiol Res 223–225:33–43

Soussi A, Ferjani R, Marasco R, Guesmi A, Cherif H, Rolli E et al (2016) Plant-associated microbiomes in arid lands: diversity, ecology and biotechnological potential. Plant Soil 405: 357–370

Spaepen S, Vanderleyden J, Okon Y (2009) Chapter 7 plant growth-promoting actions of rhizobacteria. In: Advances in botanical research, 1st edn. Elsevier Ltd, Amsterdam

Stevenson A, Hallsworth JE (2014) Water and temperature relations of soil Actinobacteria. Environ Microbiol Rep 6:744–755

Stomeo F, Valverde A, Pointing SB, McKay CP, Warren-Rhodes KA, Tuffin MI et al (2013) Hypolithic and soil microbial community assembly along an aridity gradient in the Namib Desert. Extremophiles 17:329–337

Subhash Y, Sasikala C, Ramana CV (2014) Pontibacter ruber sp. nov. and Pontibacter deserti sp. nov., isolated from the desert. Int J Syst Evol Microbiol 64:1006–1011

Suryanarayanan TS, Wittlinger SK, Faeth SH (2005) Endophytic fungi associated with cacti in Arizona. Mycol Res 109:635–639

Szymańska S, Borruso L, Brusetti L, Hulisz P, Furtado B, Hrynkiewicz K (2018) Bacterial microbiome of root-associated endophytes of Salicornia europaea in correspondence to different levels of salinity. Environ Sci Pollut Res 25:25420–25431

Taylor-George S, Palmer F, Staley JT, Borns DJ, Curtiss B, Adams JB (1983) Fungi and bacteria involved in desert varnish formation. Microb Ecol 9:227–245

Thennarasu S, Natarajan E, Muthukumar B (2019) Novel plant growth promoting rhizobacteria from rhizospheres of date palm (phoenix dactylifera): 16S rRNA typing and phylogenetic analyses reveal their distinct identity. Res Rev A J Microbiol Virol 9:11–25

Thorup JT (1969) pH effect on root growth and water uptake by plants. Agron J 61:225–227

Tian Y, Ma X, Li Y, Cheng C, Ge F, An D (2019) Relationship between microbial diversity and nitrogenase activity of Stipagrostis pennata rhizosheath. J Cell Biochem 120:13501–13508

Trivedi P, Leach JE, Tringe SG, Sa T, Singh BK (2020) Plant–microbiome interactions: from community assembly to plant health. Nat Rev Microbiol 18(11):607–621

Upadhyay SK, Singh JS, Singh DP (2011) Exopolysaccharide-producing plant growth-promoting rhizobacteria under salinity condition. Pedosphere 21:214–222

Van Der Heijden MGAA, Schlaeppi K (2015) Root surface as a frontier for plant microbiome research. Proc Natl Acad Sci U S A 112:2299–2300

Vanhaverbeke C, Heyraud A, Mazeau K (2003) Conformational analysis of the exopolysaccharide from Burkholderia caribensis strain MWAP71: Impact on the interaction with soils. Biopolymers 69:480–497

Verhoek S (1998) Agavaceae. In: Kubitzki K (ed) Flowering plants monocotyledons. Springer, Berlin, Heidelberg, pp 60–70

Vigani G, Rolli E, Marasco R, Dell'Orto M, Michoud G, Soussi A et al (2018) Root bacterial endophytes confer drought resistance and enhance expression and activity of a vacuolar H + -pumping pyrophosphatase in pepper plants. Environ Microbiol. https://doi.org/10.1111/1462-2920.14272

Watt M, McCully ME, Jeffree CE (1993) Plant and bacterial mucilages of the maize rhizosphere: Comparison of their soil binding properties and histochemistry in a model system. Plant Soil 151:151–165

Wickens G (1998) Anatomical and morphological adaptations. In: Ecophysiology of economic plants in arid and semi-arid lands. Springer, Berlin, pp 145–160

Willert DJ, Eller BM, Werger MJA, Brinckmann E (1990) Desert succulents and their life strategies. Vegetatio 90:133–143

Wullstein LH, Pratt SA (1981) Scanning electron microscopy of rhizosheaths of Oryzopsis hymenoides. Am J Bot 68:408–419

Wullstein LH, Bruening ML, Bollen WB (1979) Nitrogen fixation associated with sand grain root sheaths (Rhizosheaths) of certain xeric grasses. Physiol Plant 46:1–4

Yaish MW, Kumar PP (2015) Salt tolerance research in date palm tree (Phoenix dactylifera L.), past, present, and future perspectives. Front Plant Sci 6:1–5

Yaish MW, Antony I, Glick BR (2015) Isolation and characterization of endophytic plant growth-promoting bacteria from date palm tree (Phoenix dactylifera L.) and their potential role in salinity tolerance. Antonie Van Leeuwenhoek 107:1519–1532

Yaish MW, Al-Harrasi I, Alansari AS, Al-Yahyai R, Glick BR (2016) The use of high throughput DNA sequence analysis to assess the endophytic microbiome of date palm roots grown under different levels of salt stress. Int Microbiol 19:143–155

York LM, Carminati A, Mooney SJ, Ritz K, Bennett MJ (2016) The holistic rhizosphere: integrating zones, processes, and semantics in the soil influenced by roots. J Exp Bot 67: 3629–3643

Young IIM (1995) Variation in moisture contents between bulk soil and the rhizosheath of wheat (Triticum aestivum L. cv. Wembley). New Phytol 130:135–139

Zeikus JG (1979) Thermophilic bacteria: ecology, physiology and technology. Enzyme Microb Technol 4:243–252

Zhang K, Shi Y, Cui X, Yue P, Li K, Liu X et al (2019) Salinity is a key determinant for soil microbial communities in a desert ecosystem. mSystems 4:1–11

Zhang Y, Du H, Xu F, Ding Y, Gui Y, Zhang J et al (2020) Root-Bacteria associations boost rhizosheath formation in moderately dry soil through ethylene responses. Plant Physiol 183: 780–792

Zhu J-K (2001) Plant salt stress. In: Encyclopedia of life sciences. John Wiley & Sons, Ltd, Chichester, pp 2–4

Zilber-Rosenberg I, Rosenberg E (2008) Role of microorganisms in the evolution of animals and plants: the hologenome theory of evolution. FEMS Microbiol Rev 32:723–735

Zougari-Elwedi B, Issami W (2019) Arbuscular mycorrhizal fungi (Glomus mosseae) selection by date palm root system: The clue to a sustainable fertile soil in Jerid region of Tunisia. Int J Rural Dev Environ Heal Res 3:113–122

Zougari-Elwedi B, Issami W, Msetra A, Sanaa M, Yolande D, Lounes-Haj sahraoui A. (2016) Monitoring the evolution of the arbuscular mycorrhizal fungi associated with date palm. J New Sci Agric Biotechnol 31:1822–1831

Insights of Extreme Desert Ecology to the Habitats and Habitability of Mars

9

K. Warren-Rhodes, M. Phillips, A. Davila, and C. P. McKay

Abstract

Desert ecosystems are a key repository for important Mars analog habitats and the extant or extinct life within them. We provide an overview of four main desert habitat types—soils, sediments, salts, and rocks—and the extreme microbiology living within them, with a particular focus on the hyperarid Atacama Desert and Dry Valleys of Antarctica, the driest and coldest limits for life on Earth. We construct habitat maps of Mars from an ecological perspective and the first estimates of study sample sizes of key habitats from historical and recent Mars orbiter and lander imagery and data. We review the lessons that can be drawn for the search for life on Mars from decades of microbial ecology work in end-member terrestrial deserts.

Keywords

Mars · Habitat · Ecology · Atacama · Desert · Microbial · Analog · Hypolith · Endolith · Hyperarid

Supplementary Information The online version contains supplementary material available at [https://doi.org/10.1007/978-3-030-98415-1_9].

K. Warren-Rhodes (✉)
NASA Ames Research Center, Moffett Field, CA, USA

SETI Institute, Science Council, Mountain View, CA, USA
e-mail: krhodes@seti.org

M. Phillips
Johns Hopkins, Applied Physics Laboratory, Laurel, MD, USA

A. Davila · C. P. McKay
NASA Ames Research Center, Moffett Field, CA, USA

J.-B. Ramond, D. A. Cowan (eds.), *Microbiology of Hot Deserts*, Ecological Studies 244, https://doi.org/10.1007/978-3-030-98415-1_9

9.1 Introduction

Mars is a desert planet and during its evolutionary history, hydrological and geochemical forces have forged environments reminiscent of deserts on Earth, including evaporite deposits, hypersaline lakes, alluvial plains, and volcanic desert pavements. Like Mars, many terrestrial "analog" desert environments are inhospitable due to extreme temperatures, high solar irradiation, oligotrophic conditions, and/or low to near zero water availability. And yet, terrestrial deserts support a wealth of unique and hardy microorganisms. Our knowledge of the ecology of many of the world's most extreme hot and cold deserts has increased exponentially over the past several decades. In particular, desert regions once thought to be too extreme even for microbial life are now regarded as restrictive yet habitable environments, where specialized microbial communities eke out a living with minimal nutrients and liquid water. Here, we review and distill data from two of the Earth's end–member deserts—the coldest, namely the Antarctic Dry Valleys, and the driest, the Atacama Desert in Chile, with a view to generating working habitat maps of Mars and furnishing insights from an ecosystem ecology perspective of the potential and strategies for the search for life on Mars.

9.2 Desert Habitats and Ecosystems on Earth

Habitability and Life

In this paper we define habitability as the "potential for an environment to support life, be it on planet-wide or microscopic scales" Hays et al. (2015). Habitability nomenclature primarily originates from the fields of biology and ecology, where a "habitat" is the physical place where a species/organism/community *may* live. As Cockell et al. (2011) explained with their "habitability triad," virtually any environment in the Universe can be classified as either "uninhabitable" or "habitable." There are underlying issues with both concepts (Supplementary Material Box 1), as *uninhabitable space* is defined based on current theoretical principles of habitability, which are subject to change with new discoveries, and therefore its boundaries are conceptual. On the other hand, *habitable space* is entirely empirically derived, based on the habitability envelope defined by the known limits of life on Earth, i.e., a planetary data point of one. Herein, to be "inhabited," we assume life is *extant* (metabolically viable) and either actively metabolizing, growing and/or reproducing, or *dormant*, i.e., in a "reversible state of low or no metabolic activity" (Wilhelm et al. 2018; Goordial et al. 2016). We acknowledge that inhabited environments may have (1) existed in the past but within which life is now *extinct*, i.e., dead, the remains of which are biochemically or physically preserved (Hays et al. 2017); or (2) be *vacant*, i.e., no life or trace of life remains.

Desert Habitats and Ecosystems

An ecosystem is a biological community of living organisms and their physical environment. Ecosystems are comprised of both abiotic and biotic components and can exist at multiple scales and in hierarchical arrangements of units, from a single speck of dirt to a planetary body (Belnap et al. 2003a; Velland 2017). As an example, a hypersaline lake defined by its many physical and chemical parameters (e.g., salinity, pH, soluble ion composition) is a potential habitat, as is a salt dome or quartz rock. When a hypersaline lake, salt dome, or quartz rock supports extant microorganisms, it is an ecosystem. A habitat or ecosystem may change over time. A hypersaline lake subject to drought can evaporate and leave behind salt deposits with a record of the water history, entombed organic matter, and/or dead microorganisms, becoming a record of the extinct ecosystem.

On the Earth, roughly six distinct biomes/large-scale ecosystems exist. Amongst them, deserts cover roughly one-quarter of the Earth's land surface (33 million km^2) and are defined "climatologically as the sum of the arid and hyper-arid areas of the globe; biologically, as the ecoregions that contain plants and animals adapted for survival in arid environments; and, physically, as large contiguous areas with ample extensions of bare soil and low vegetation cover" (UNEP 2006; Makhalanyane et al. 2015). Further distinctions between deserts are commonly based on an Aridity Index (AI, Thornwaite 1948; UNEP 1997), calculated as the ratio of mean annual precipitation (MAP) to potential evapotranspiration (PET). Deserts are defined as having an AI of <1, and further subdivided into subhumid (AI $=$ 0.5–0.65), semi-arid (AI $=$ 0.2–0.5), arid (AI $=$ 0.05–0.2), and hyperarid (AI <0.05) (Barrow 1992; Makhalanyane et al. 2015). With an AI of ~0.2, rainfall in arid deserts, which constitute roughly 12% of land on Earth, supplies a maximum of 20% of the water needed for optimum plant growth; in contrast, in hyperarid deserts, which cover about 8% of the earth's surface, rainfall supplies at most 5% (Barrow 1992; UNEP 1997).

Desert biomes encompass a significant diversity of habitats, which may be classified in several ways (Chap. 1), for example by discipline (ecology, geology), temporal constraints (e.g., ephemeral, perennial), and/or by spatial scale (e.g., landscape, micro-scale). For this paper, we considered three classification systems (Table 9.1), each relevant to Martian habitability and Earth analogs: (1) geomorphic units or ecological habitat types; (2) key desert substrates; and (3) Mars classifications, as circumscribed by contemporary or past mission and technology data (e.g., orbiter, spectrometer). We chose the second option as the most parsimonious, and thus organized the paper, although not rigidly, around four desert substrates/habitats: (1) soils and regolith; (2) lithic substrates; (3) sediments; (4) and salts. These categories reflect primarily a microbial ecology and habitats perspective, rather than strictly geological or mineralogical considerations. Furthermore, because of the relevance and utility of other habitat classifications, throughout the paper, we refer to and combine these substrate types with units from the first and third categories (Table 9.1) to better illustrate and complement our descriptions. It is important to remember that, although in this paper we partition substrates and habitats, this is a simplified approach, as all of the above can be or are intimately

Table 9.1 Classification systems of habitats

Terrestrial desert ecological habitat/ Geomorphic unit	Substrate types	Mars database classification (spectral, S; geomorphic, G)
Alluvial fans/desert pavement, gravel plains, plateaus, permafrost landscapes **(BSC, biofilms or mats)**	**Soils and regolith**	Rock abundance (G); permafrost regions (G)
Alluvial fans, outcrops, desert pavement, hillslopes, gullies, cliffs, drainage channels, gravel, polygonal patterned ground, permafrost landscapes **(lithic, lichens, moss, soil communities)**	**Lithic substrates** (rocks, boulders) - quartz, granite, sandstone, calcite ignimbrite	RSLs (G); polygonal networks, polygonal patterned ground (G)
Playas (dried lake beds, ephemeral), polygonal pattern ground; paleolakes, gullies, outcrops, drainage channels **(BSC, biofilms or mats, planktonic communities)**	**Sediments (clastic)** (clays, mudstones)	Hydrous minerals (S) Clay weathering sequences (S)
Salars, salt deposits (nodules, polygons, domes, polygonal networks), salt lakes (lagunas), outcrops, drainage channels **(lithic, lichens, soil and salt communities, BSC, biofilms or mats, planktonic communities)**	**Salts (chemical sediments, evaporites)** – Halites, sulfates	Chlorides (S) Sulfates (S) Hydrous minerals (S) Clay weathering sequences (S)

BSC biological soil crust

interconnected across space and time to, and nested within, the larger ecosystem context. Likewise, despite a large variety of habitats in deserts we are unable to cover all habitats in equal breadth and focus on the most ubiquitous ones or those of highest relevance to Mars (For a full inventory, see Dohm et al. 2011).

9.2.1 Soils

In terms of surface coverage, one of the most common substrates in terrestrial deserts is soil. The term soil refers to natural material on the immediate surface of the Earth composed of unconsolidated organic and inorganic solids, liquid, and gases altered by biological, chemical, and/or physical agents (Soil Survey Staff 1999; Certini and Ugolini 2013). Terrestrial soils are a biologically active medium, distinguished from regolith—a layer of loose, heterogeneous superficial deposits covering solid rock that can include dust, soil, broken rock, and other materials—by the presence of organic matter and the potential to support plants and/or microbial life. The microbial ecology of hot desert soils is the subject of Chap. 3. Soils also differ from sediments, which are formed when rock is weathered, eroded, and then transported elsewhere. Once rock particles are deposited somewhere else, they become

Fig. 9.1 (**a**) Soils of an alluvial fan/desert pavement near Domeyko range, Atacama Desert. Note rover for scale; (**b**) Lithic substrates of a quartz desert pavement near Aguas Calientes in the Atacama Desert (rocks are ~1–5 cm in size). (**c**) Playa sediments in background versus dark, mafic desert pavement of paleo-playa in foreground, near Domeyko range, Atacama Desert. (**d**) Colonized halite salt nodules in the Salar Grande, Atacama Desert (cm ruler in bottom of photo for scale)

sediments (Prothero and Schwab 2004, socratic.org), which often also lack the structured layers (or distinct horizons) commonly developed in soils.

Soils in deserts are often mantled by an armor of gravel and bedrock debris, forming a so-called desert pavement (Fig. 9.1a, b). Because bare soils and lithic substrates of desert pavement tend to be inhabited by distinct microbial consortia (Pointing et al. 2009; Makhalanyane et al. 2015; Goordial et al. 2017), we treat lithic substrates below as a separate ecological unit. Surface and subsurface soils in deserts form from a variety of processes, namely chemical weathering, biological mechanisms, and other geological forces. Soil formation in deserts is more limited than in other biomes—due to low rainfall (which limits downward movement) and comparatively lower plant and microbial biomass—and much of the soil development and biogeochemical activity is restricted to the near-surface, where water is most prevalent. Indeed, the atmosphere-soil interface and first few millimeters to centimeters of most desert soils ("duricrust"), especially in hyperarid environments, constitute a significant ecological boundary (Belnap et al. 2003a; Davis et al. 2010). Near-surface soils typically function as the engine for biogeochemical processes and activity in hyperarid deserts and represent a fundamental transition and structuring agent for microbial community abundance, distribution, and diversity (Belnap et al. 2003a; Lester et al. 2007; Warren-Rhodes et al. 2019).

When compared with more arid regimes, hyperarid soils possess unique soil taxonomic features: little to no leaching of the most soluble salts, such that halites, sulfates, and nitrates can build up to very high levels in heavily-enriched indurated subsurface horizons; minimal to no plant cover; few pedogenic carbonates; and extremely low levels of organic carbon and/or biotic mixing (Rech et al. 2003; Ewing et al. 2006, 2008; Quade et al. 2006; Amundson et al. 2012; Crits-Christoph et al. 2013; Bockheim and Haus 2014; Finstad et al. 2017; Goordial et al. 2017). In hyperarid soils in the driest portions of the Atacama Desert, major biogeochemical cycles, including carbon and nitrogen, reach critical thresholds and can be largely absent (Ewing et al. 2006, 2008), such that new and more relevant classifications (petrosalic, nitric horizons, and Petrosalids group in the Aridisols order) have been proposed (Finstad et al. 2017).

In both hot and cold deserts, polygonal patterned ground in soils is also a prominent geomorphological feature (Fig. 9.2). Polygons form through thermal

a b

Fig. 9.2 (**a**) Polygonal (~20 m wide) patterned ground covered by mafic desert pavement near Lake Fryxell, Dry Valleys, Antarctica. Photo credit: J. Goordial. (**b**) Phoenix Sol 1 image of polygonal patterned ground (~5–10 m wide) covered by dark desert pavement shaped by subsurface ice in the northern high latitudes of Mars (Photo credit: NASA; Smith et al. 2009; Mellon et al. 2009a)

(thermal contraction polygons, frost-sorted, sorted patterned ground, ice-wedge networks) and/or desiccation processes across space and time (Bockheim 1997, 2002; Barrett et al. 2004; Bockheim and Haus 2014). The presence and morphology of polygonal patterned ground often reflect the occurrence and influence of aqueous processes and salts in soils (Levy et al. 2008; Lousada et al. 2018). In warmer deserts, patterned ground and polygonal networks arise from ephemeral and/or seasonal cycles of wetting after precipitation events, subsequent drying, and evaporation, and these geomorphological features occur in soils, playas, and hypersaline environments such as dried salt lakes (salars) (*see below*).

In contrast, in cold, dry deserts on Earth, such as the Antarctic Dry Valleys, thermal contraction polygons are widespread and created from peri-glacial processes such as seasonal thermal (freeze-thaw) and moisture or micro-climate changes in ice-cemented soils (McKay et al. 1998; Levy et al. 2008). In the McMurdo Dry Valleys, dry permafrost terrain—ground that never warms above 0 °C and has negligible ice content (McKay et al. 2019)—also has polygonal structures, with mineral soils mantled by desert pavement (sandstone, quartz, granite; Pointing et al. 2009). Indeed, one of the few places on Earth where both ice-cemented and dry permafrost concurrently exist are the McMurdo Dry Valleys (e.g., University Valley, Goordial et al. 2017)—making them analogs for the dry permafrost in the polar regions of Mars (Goordial et al. 2017; McKay et al. 2019).

Microbial Communities in Soils

Most bare, pavement and permafrost desert soils support vegetation as well as edaphic microbial communities (Gómez-Silva 2018; Pointing et al. 2009; Lee et al. 2012; Stomeo et al. 2013; Wei et al. 2016). Edaphic communities in less arid hot and cold deserts can form complex biological soil crusts (BSC; for more details, the readers are referred Chap. 2 of this book), where photosynthetic and non-photosynthetic microorganisms coexist within and bind the topsoil matrix (Belnap et al. 2003a; Belnap et al. 2003b; Mogul et al. 2017; Baumann et al. 2018; Lehnert et al. 2018; Jung et al. 2019). BSC can be taxonomically diverse and include cyanobacteria, lichens, mosses, algae, microfungi, and bacteria (Belnap et al. 2003b). Up to 70% of living soil cover in arid landscapes can be BSC, and they supply critical ecosystems services (photosynthesis/primary production, soil stabilization, erosion/desertification prevention, nutrient, and soil organic matter supply, moisture retention; Belnap et al. 1994; Elbert et al. 2012; Williams et al. 2012). BSC are intimately linked to their geochemical and physical environment, with their presence or absence and biodiversity most strongly controlled by water, soil texture (Kidron et al. 2012; Williams et al. 2012), orientation, light, temperature, pH, and salt content (Nash et al. 1979; Eldridge 1996; Eldridge and Koen 1998; Eldridge et al. 2000; Lalley and Viles 2005; Cereceda et al. 2008; Kidron et al. 2010, 2011, 2014; Baumann et al. 2018; Chap. 2).

With increasing dryness, vegetation and BSC in desert soils become fragmented and eventually disappear, and hyperarid soils gradually become more regolith-like (Navarro-Gonzalez et al. 2010; Wilhelm et al. 2017, 2018). In hot and cold deserts with higher water availability (e.g., Mojave Desert), and within the driest deserts at sites with more moisture (e.g., southern, coastal, and higher altitude Atacama, coastal Antarctica), soils can be remarkably diverse and localized (Lee et al. 2012; Wei et al. 2016). In these locations, edaphic microorganisms still play central roles in ecosystem function, altering biogeochemistry (mainly through action via pH changes), catalyzing weathering, and cycling essential nutrients such as carbon, nitrogen, and sulfur (Belnap et al. 2003b; Makhalanyane et al. 2015; Mogul et al. 2017).

However, as conditions become drier and less conducive to life, microbial communities likewise become increasingly depauperate, with lower abundance and diversity (or compositional changes, e.g., less cyanobacteria or archaea versus bacteria: Kidron et al. 2010; Niederberger et al. 2015; Lee et al. 2018) and spatial distributions that are rare, patchy and strongly linked to water availability, salt content/conductivity, pH, nutrients and/or temperature (Bagaley 2006; Bowker et al. 2005, 2006; Büdel et al. 2009; Wong et al. 2010; Lee et al. 2012; Stomeo et al. 2012; Crits-Christoph et al. 2013; Garcia-Pichel and Belnap 1996, 2003, Garcia-Pichel et al. 2014; Kidron et al. 2011, 2014; Wilhelm et al. 2018; Goordial et al. 2017; Schulze-Makuch et al. 2018; Warren-Rhodes et al. 2019). At the cold-arid limits, Goordial et al. (2017) found oligotrophic permafrost soils with primarily dormant or absent populations characterized by significantly lower biomass, diversity, and activity compared to those from lower elevation sites with warmer climates and greater water availability.

Nonetheless, it is critical to reiterate that, even in the driest and coldest soils on Earth, desiccation-resistant and cold-adapted microorganisms live within these edaphic habitats, with dominant phyla being the *Actinobacteria* and *Proteobacteria* but also *Chloroflexi, Acidobacteria*, and *Cyanobacteria* (Navarro-Gonzalez et al. 2003; Gilchinsky et al. 2007; Cary et al. 2010; Lee et al. 2012; Neilson et al. 2017;

Crits-Christoph et al. 2013; Robinson et al. 2013; Wei et al. 2016; Schulze-Makuch et al. 2018; Goordial et al. 2017; Fernández-Martínez et al. 2019; Warren-Rhodes et al. 2019). Viruses also appear to be present, particularly the family Caudovirales in the Atacama Desert (Schulze-Makuch et al. 2018) and other hyperarid soils (e.g., Namib Desert, Vikram et al. 2016). Whether these soil microbial communities at the dry and cold/dry limits are primarily indigenous or allochthonous, active or dormant, and persistent or transient continues to be debated, with intriguing implications for potential life on Mars.

In addition to surface soils, in both hot and cold deserts on Earth substantial differences are also apparent for subsurface microbial communities (Steven et al. 2013), with shallow surface (~0–20 cm), mid-depth (~30–50 cm), and deep (~80 cm) soil horizons often distinguishable from one another, as has been demonstrated across multiple locations and environmental conditions (Steven et al. 2013; Schulze-Makuch et al. 2018; Fernández-Martínez et al. 2019; Warren-Rhodes et al. 2019). Soils forming on alluvial fans have been observed to be structured compositionally and decrease in biomass and diversity with depth, with variations explained by moisture content and geochemistry (Lester et al. 2007; Crits-Christoph et al. 2013; Robinson et al. 2013; Schulze-Makuch et al. 2018; Fernández-Martínez et al. 2019; Warren-Rhodes et al. 2019; Chaps. 3 and 10).

In both hot and cold deserts, polygonal networks and patterned ground soils are likewise key units for ecological scaling and structure (Barrett et al. 2004). Antarctic polygon interiors were shown to harbor higher biomass and diversity than polygon edges, due to more favorable fine-scale soil properties (i.e., higher soil moisture and organic matter, Barrett et al. 2009). In contrast, at the rock-soil interface, cyanobacterial community colonization, for example, was greater at polygon edges versus interiors in the Arctic (Cockell and Stokes 2006), where freeze–thaw movement of rocks opens up light in soils at rock edges, enhancing habitability by providing new niches for photosynthetic communities, even on rocks with low translucence (Cockell and Stokes 2006). Likewise, halite polygon trough or cracked edges in deep subsurface ancient deposits had greater habitability, as evidenced by significantly higher mineral and lipid content (likely a result of greater wind deposit and infill; Cockell et al. 2020).

Also relevant to Mars, surface and subsurface ice and soils of polygonal networks in the cold dry valleys of Antarctica support abundant and diverse—much more so than originally anticipated—soil (and lithic) microbial communities (Pointing et al. 2009; Cary et al. 2010; Lee et al. 2012; Wei et al. 2016). However, in the coldest and driest permafrost, soil microbial communities may be functionally dormant, with "permanently sub-freezing temperatures, low water activity, oligotrophy and age" (Goordial et al. 2017) essentially precluding metabolism and growth, i.e., a viable ecosystem. These dry permafrost (overlaying ice-cemented permafrost) soils may be remarkably good terrestrial microbial analogs for the dry permafrost characterizing the polar regions of Mars, although it is important to note that in these same Antarctic locations, colonization of lithic habitats such as sandstone cliffs remains widespread (Archer et al. 2017).

9.2.2 Lithic Substrates

Lithic substrates are another ubiquitous habitat in deserts, particularly within desert pavement and rocky outcrops (Fig. 9.1b). Desert pavements are indicative of relatively old, stable soils where wind erosion is the primary mechanism of landscape change. They can cover alluvial fans, plains, discharge channels or gullies, lava fields, plateaus, and hillslopes. Desert pavements primarily form when wind strips fines from surface soils, leaving behind larger rock fragments (mm to cm). Significant water transport events, lava flows and gravity can also play a role in the origin, size, and composition of desert pavements. Desert pavements may thus also contain larger (meter-scale) rocks and boulders originally deposited (and occasionally transported) by powerful yet infrequent (decadal, century) flash flood or mudflow events (Pointing et al. 2009; Archer et al. 2017). Depending on geographical location and geological history, desert pavements can be composed of dark, mafic materials ("black gobi," i.e., basalts, ignimbrites, rhyolites), or more translucent rocks and minerals ("white gobi," i.e., quartz, granites, sandstones, limestones, gypsum, halite), or a combination of both.

Microbial Communities of Lithic Substrates
In most deserts, microorganisms inhabit the bulk of available rock and soil space. Lithic microbial communities colonize the surface (epilithic), undersides (hypolithic), and/or interiors (endolithic) of rocks, boulders, outcrops, and cliffs (Golubic et al. 1981), and they have been described for deserts of all types and on every continent (Bahl et al. 2011; Lacap-Bugler et al. 2017; Gómez-Silva 2018; Nienow 2019). Similar to plants, lithic habitats function as resource islands, with elevated nutrients, water holding capacity, and microbial activity compared to bare soils (Ewing et al. 2007). Given their ubiquity across the world's deserts and their comparatively simple community structure, lithic substrates are important ecological and astrobiological model systems (Warren-Rhodes et al. 2007; Caruso et al. 2011; Wong et al. 2010; Chan et al. 2012; Wierzchos et al. 2015; Meslier and DiRuggiero 2019; Qu et al. 2020).

Lithic microorganisms may be partly recruited from adjacent soils and/or from nearby plant "islands"—an example of "nested microbiomes" (Amend et al. 2019). However, most recent studies suggest soils and lithic desert microbial communities are distinct from one another (Yung et al. 2014; Pointing et al. 2009; Wong et al. 2010; Stomeo et al. 2012, Makhalanyane et al. 2013; Van Goethem et al. 2016; Wei et al. 2014, 2016; Lacap-Bugler et al. 2017; Garido-Benavent et al. 2020). In contrast with soils, which primarily support *Actinobacteria* and other cosmopolitan taxa (Lee et al. 2012; Niederberger et al. 2012; Stomeo et al. 2012; Archer et al. 2017), lithic communities, like BSCs, are dominated by photosynthetic algae and/or cyanobacteria that form mutualistic associations with a diversity of heterotrophic taxa. In hyperarid deserts, these consortia often function as the principal, if not sole, primary producers and play important roles in weathering and biogeochemical cycling (Friedmann and Galun 1974; Golubic et al. 1981; Cockell and Stokes

2006; Warren-Rhodes et al. 2006; Omelon 2008; Yung et al. 2014; Makhalanyane et al. 2015; Wei et al. 2016; Ramond et al. 2018).

Key phyla/genera in lithic photosynthetic communities include *Cyanobacteria* (e.g., *Chroococcidiopsis, Phormidium, Halothece*-salt rocks), as well as *Chloroflexi, Thermi/Deinococci, Actinobacteria, Acidobacteria, Proteobacteria, and Bacteroidetes*—i.e., microorganisms exceptionally resistant to ionizing radiation, temperature extremes and desiccation (Billi et al. 2000; Pointing et al. 2007; Wong et al. 2010; Stomeo et al. 2013; Meslier et al. 2018; Yung et al. 2014; Lacap et al. 2011; Lacap-Bugler et al. 2017; Cockell et al. 2020). In wetter desert environments, archaeal, fungal, algal, and moss components also occur (Wong et al. 2010). Viruses may also be important predator community components of lithic substrates, although specific predator-prey interactions remain poorly understood (Valverde et al. 2015; Bezuidt et al. 2020; Qu et al. 2020).

While each of the three main types of lithic microbial communities shares many common members, adaptive strategies, and ecological trends (e.g., lower diversity with water scarcity), each of these niches are distinct. Thus, we treat them separately below, highlighting some of their commonalities and differences.

Epilithic Communities Epilithic microbial communities cover vast areas of the world's arid, semiarid and polar deserts, colonizing almost any rock surface (sandstones, quartz pebbles, gravels, granites, basalts) within key ecological units such as gravel plains, desert pavement, and outcrops/cliffs (Eldridge 1996; Lalley and Viles 2005; Lalley et al. 2006; de los Ríos et al. 2014). Epiliths often display comparatively rich biological diversity, including cyanobacteria, algae, moss, fungi, and particularly lichens. Even in the driest hot and cold deserts, while their numbers may be quite restricted, complex epilithic communities survive, for example, on rock surfaces in foggy areas or within desert varnish (i.e., nanostratigraphic iron and manganese oxide rock surface coatings; Kuhlman et al. 2008; Espisito et al. 2015).

Epilithic communities, in particular, are distinguished from other types of lithic consortia by their morphological and physiological adaptations that help them scavenge and imbibe water from sources other than rainfall, namely fog, dew, water vapor, and snowmelt (Rundel 1978; Cereceda et al. 2008; Kaseke 2009; Kidron et al. 2011). These adaptations provide a competitive advantage over both vascular plants and other lithic microbiota, allowing them to colonize areas with exceedingly little precipitation. The presence and spatial distribution of epilithic life has been linked most strongly to these unique water sources (Rundel 1978; Nash et al. 1979; Eldridge and Greene 1994; Eldridge and Koen 1998; Lalley and Viles 2005; Cereceda et al. 2008; Kidron et al. 2002, 2005, 2010, 2014). Additionally, the abundance, diversity, and spatial distribution of epilithic communities in arid and hyperarid deserts is also intimately tied to their substrate properties and physico-chemical micro-environmental parameters, for example with micro-topography dictating not only fog dispersal and/or dew capture but also light and temperature, and thus microbial spatial pattern (Lalley and Viles 2005; Bowker et al. 2006; Lalley et al. 2006; Sun 2013).

Hypolithic Communities Hypoliths, coined by Vogel (1955) as "window-pane" algae for their greenhouse-like habitats, colonize the undersides of translucent rocks (e.g., quartz, limestone, granite) which protect them from intense solar radiation and desiccation (Friedmann et al. 1967; Billi et al. 2000; Cockell and Stokes 2006; Billi et al. 2011). Estimates suggest only ~1 mm of rock substrate in even the driest deserts is sufficient to block excessive solar irradiance, dampen high evaporation, and shield lithic communities from lethal harm (Cockell et al. 2008; Wierzchos et al. 2015). Hypolithic communities occupy cold and hot desert rock habitats ranging from drainage channels and alluvial fans in the Atacama Desert and quartz pavement in the Antarctic Dry Valleys to thousands of kilometers of gobi gravel plains in the Middle East, Mongolia, western China, Namibia, and the US (Cameron and Blank 1965; Friedmann and Galun 1974; Broady and Kibblewhite 1991; Büdel and Wessels 1991; Schlesinger et al. 2003; Cowan et al. 2010; Traci et al. 2010; Khan et al. 2011; Lacap-Bugler et al. 2017; Warren-Rhodes et al. 2006, 2013; Wei et al. 2016; Ekwealor and Fisher 2020).

Hypolithic abundance, measured as colonization frequency of suitable rock substrates, is one ecological parameter that has been strongly linked to macro-scale water availability (total mean annual precipitation, MAP; total/metabolic hrs year^{-1}). At global scales, colonization drops exponentially, from 100% at MAP of roughly ≥80 mm year^{-1} in "wet" hot deserts such as the Mojave (Schlesinger et al. 2003) to ≤5% in the driest deserts on Earth, such as the Atacama or in western China (Warren-Rhodes et al. 2006, 2007). Abundance has also been linked to "growing season" (# days when photosynthesis is possible, Lacap-Bugler et al. 2017) At regional/within desert (km) scales, colonization generally falls linearly with MAP for the driest deserts, and also exhibits moderate to significant variation at landscape scales (within a site, m-to-10s m) (Warren-Rhodes et al. 2006; 2014).

At fine-scales (cm to mm), hypolithic colonization is likewise patchy but non-random, and tied to micro-scale physical controls (larger rocks, orientations that preserve early morning moisture, microtopography) and water availability (Warren-Rhodes et al. 2007). As with epilithic systems, fog, dew, and snowmelt can be important sources of moisture for hypoliths, with access to these sources boosting colonization to higher levels and richer community diversity in coastal sites in the Atacama (100%, Azúa-Bustos et al. 2011), Negev (McKay 2016) and Namib Deserts (100%, Kaseke et al. 2012; Warren-Rhodes et al. 2013), in the Antarctic coastal and dry valleys (Pointing et al. 2009; Wei et al. 2016) and in high altitude sites across the Qinghai-Tibetan plateau (Warren-Rhodes et al. 2007; Wong et al. 2010). Similarly, in polar deserts, hypolithic colonization, quantified within polygon patterned ground covered by desert pavement, ranges from ~5% (McKelvey Valley, quartz, Dry Valleys) to 100% (opaque rocks, Maritime sites), but is lower (36%, quartz) in high-altitude tundra desert locations in Tibet (Wong et al. 2010).

In contrast to abundance, hypolithic community diversity cannot be explained solely by water availability; temperature effects (contemporary and ancient environmental legacies) must also be invoked (Pointing et al. 2007; Bahl et al. 2011). Overall, species richness and community diversity tend to be lower in polar/cold than non-polar deserts (Caruso et al. 2011), and "distinct community shifts occur between hot and cold and dry and wet hyperarid deserts" (Pointing et al. 2007; Lacap-Bugler et al. 2017). Dominant cyanobacteria in warm desert hypoliths are typically coccoid members of *Pleuroscapsales* order (esp. genus *Chroococcidiopsis*), versus filamentous *Oscillatoriales* (esp. genus *Phormidium*) in cold/polar desert environments (Chan et al. 2012; Makhalanyane et al. 2013; Lacap-Bugler et al. 2017). Globally, communities in the Antarctic Dry Valley are the most distinct from all other deserts (Lacap-Bugler et al. 2017).

Endolithic Communities In contrast to epilithic or hypolithic microorganisms, which are exposed to air and soil/rock interfaces, endolithic communities seek refuge millimeters to centimeters beneath the surface of rocks, boulders, outcrops, and cliffs within sandstone, granite, quartz, gypsum, ignimbrite, and carbonate substrates (de los Ríos et al. 2005; Walker and Pace 2007; Wierzchos et al. 2011, 2012b, 2015, 2018; DiRuggiero et al. 2013; Meslier et al. 2018; Chap. 4). Endoliths also commonly populate salt habitats (Fig. 9.1d) and siliceous sinters (e.g., extinct hot springs, Walker et al. 2005). Endolithic communities occupying the fissures and cracks of rocks are called chasmo(endo)lithic, whereas those that live in the pore spaces are called cryptoendolithic or hypoendolithic. Here we will refer to all types generally as endolithic.

Like hypoliths, endolithic habitats are widespread in desert biomes across the globe, and well-described, including for the Atacama Desert and Antarctic Dry Valleys (Friedmann 1980, 1982; Omelon 2016; Archer et al. 2017; Meslier et al. 2018). A similar cast of characters to hypoliths comprises the hyperarid endolithic microbial community, i.e., predominantly *Cyanobacteria, Actionobacteria, Chloroflexi*, and *Proteobacteria*, but their component members and structure are highly specific to the particular rock type they inhabit (Archer et al. 2017; Meslier et al. 2018). Gypsum and calcite substrates tend to have higher community diversity, for example, than ignimbrite or granite substrates, which support higher relative abundances of *Cyanobacteria* ($\geq$70% versus 40–60%, respectively) as well as *Deinococcus*, *Gemmatiomonadetes*, *Bacteroidetes*, and *Armatimonadetes* (Meslier et al. 2018). Endoliths can also be quite different both from one another across temporal and spatial scales (Archer et al. 2017; Garrido-Benavent et al. 2020), between substrates (Archer et al. 2017; Meslier et al. 2018), aspects/orientations (Archer et al. 2017) and from hypolithic systems (e.g., Pointing et al. 2009). Globally, endolithic community diversity differs most strikingly between polar and hot deserts, driven by macroclimate, especially temperature, rather than solely by geographic distance at these large scales (Qu et al. 2020).

An ecological lifestyle feature distinguishing endoliths from their epi- or hypolithic counterparts is the organization of endolithic communities into distinct colored 'biofilm' bands or layers (encapsulated in extrapolymeric substances, EPS), which comprise unique microbial consortia and/or the same members but in many (but not all) cases protected by different compositions of photosynthetic pigments (chlorophyll, beta-carotene, scytonemin; Gao and Garcia-Pichel 2011; Archer et al. 2017). These eco-engineered layers have been linked to adaptive survival strategies

(UV protection, water conservation) and subtleties of the rock substrate micro-architecture (Wierzchos et al. 2015; Archer et al. 2017).

Endolithic colonization has been quantified in a few Antarctic locations, especially water-rich coastal and hyporheic zones, to be at least as frequent (1–3% hypoliths versus 5% endoliths in weathered sandstones, Pointing et al. 2009; 30–100% of weathered granite substrates, Yung et al. 2014) as hypolithic colonization, although comparative data remains limited. Also similar to hypolithic systems, the overarching driver for endolithic community structure and composition is water availability, primarily rainfall but also snowmelt and other novel water sources (Friedmann and McKay 1985; Friedmann et al. 1987; Davila et al. 2008a, b; Büdel et al. 2008, 2009; Omelon 2008; Wierzchos et al. 2012a; Sun 2013). In contrast to hypolithic communities, however, at the micro-scale, the architecture of the rock, i.e., "the space available for colonization and its physical structure, linked to water retention capabilities" (Meslier et al. 2018) appears to be more important for endolithic communities, which are confined within the structure itself, than for hypoliths. Thus, endoliths are likely to be more constrained by Goldilocks conditions where light, salinity, temperature, and rock geochemistry dictate more strictly their existence, spatial distribution, abundance, and/or diversity (Friedmann and Mckay 1985; Nienow et al. 1988; McKay et al. 1993; Pointing et al. 2009; Sun 2013). For example, at the cold/dry limit in the Antarctic Dry Valleys, endolithic community composition of available substrate is extremely different depending on microclimate (e.g., aspect/orientation, Yung et al. 2014) or available rock type, e.g., with cyanobacterial-dominated communities in sandstone and granite boulders versus more biodiverse and heterogeneous lichen endoliths in weathered sandstone cliffs. These variations are explained by micro-sale rock properties, including porosity and salt content (Pointing et al. 2009; Wong et al. 2010; Cámara et al. 2014; Archer et al. 2017).

9.2.3 Sediments

Despite low precipitation and high evapotranspiration rates, sedimentary systems still develop in desert environments, which are capable of sustaining relatively diverse and complex communities of organisms. There are two general types of sedimentary deposits: (1) Clastic sediments composed predominantly of broken pieces or clasts of older weathered and eroded rocks; and (2) Chemical sediments formed by the precipitation of minerals from water (commonly called salts or evaporites, see Sect. 9.2.4).

Across the world's deserts, sediments constitute the bulk of over 50,000 large and small playas (i.e., dry ephemeral lakes, Figs. 9.1c, 9.3a–b). Playas' distinct fluvial-transported compositions consist primarily of fine-grained, laminated clastic sediments such as clays, together with chemical sediments that precipitate in situ (e.g., sulfates, halites; Droste 1961; Gutiérrez 2005). Situated in endorheic topographic lows, playas are defined by transitory water dynamics (Rosen 1994; Briere 2000; Navarro et al. 2009; Shaw and Bryant 2011; Kidron et al. 2014). Following

Fig. 9.3 (**a**) Sun reflects off clay sediments in a dry playa in the Atacama Desert. Note trucks in left on playa for scale. Credit: K. Warren-Rhodes. (**b**) Halite salars with contemporary laguna (~300 m length) in the Altiplano, Chile. Credit: K. Warren-Rhodes

precipitation and flash-flooding events, water pools in a playa and is retained for significant periods of time, slowly infiltrating to subsurface layers (Bowler et al. 1986; Davis et al. 2010; Hang et al. 2015; Goldstein et al. 2017; McKenna and Sala 2018). Even with repeated wetting, desiccation, and evaporation cycles, playas do not form thick salt deposits (see "salars" below). Instead, fragile thin crusts of biotic or abiotic origin and/or hard-packed surfaces (often polygonally patterned) are generated after rainfall (Rosen 1994; Navarro et al. 2009). An important facet of playa crusts is their vulnerability upon desiccation to aeolian deflation and dispersal. Indeed, playa sediments can be the key source of dust circulating within and from deserts at local to global scales, with concomitant transport of biological (e.g., disease vectors) material and pollutants (Cahill et al. 1996; Reheis 1997; Mahowald et al. 2003; Griffin 2004; Reynolds et al. 2009).

Both contemporary and ancient lakes of a wide variety exist within hot and cold deserts on Earth. While a detailed treatment of these ecological habitats and their biology is beyond the scope of this review, lake sediments constitute an important reservoir of biological diversity in desert systems and in the paleo-record. Chemically, the majority of lakes in arid environments are hypersaline (see lagunas and salars in next section) or alkaline (Demergasso et al. 2004; Bowler et al. 1986) but acidic and ice-covered lakes are also common (Escudero et al. 2015), particularly in cold hyperarid deserts in Antarctica (Imhoff et al. 1979; Bowman et al. 2016), Chile/Bolivia (altiplano) and China (Qinghai/Tibet). There is widespread terminology, but limited consensus on, the boundaries distinguishing desert playas from lakes (Gutiérrez 2005). In some cases, temporal aspects differentiate water bodies in closed basins (playa = wet $\leq$25% of year; playa lake, wet 25–75% of the year) and lakes (wet = 100% of year) or ecological features, such as water depth (lacustrine, $\geq$ 2 m) (Gutierrez 2005). Regardless of their ephemeral, perennial, or permanent nature, or varying sizes and origins, topographical depressions in deserts are incredibly numerous, collecting scarce—and in some cases the only—available water outside of limited precipitation events, making them critical ecological habitats.

Microbiology of Desert Playa and Lake Sediments

Playas and lake/paleo-lake sediments constitute rich microbiological repositories in deserts, including the Atacama and Antarctic Dry Valleys (Albarracín et al. 2015; Lee et al. 2018), hosting abundant microbial communities in both the water column, in sediment-associated mats and dry sediments. Playa and microbial desert lacustrine ecosystems are geochemically and biologically quite diverse, supporting *Chloroflexi, Actinobacteria, Proteobacteria* (esp *Alphaproteobacteria*) but also *Bacteroidetes, Firmicutes, Gemmatimonadetes, Cyanobacteria* (usually low abundances unless after a bloom), archaea, fungi, and diatoms (Navarro et al. 2009; Sirisena et al. 2018; Fernández-Martínez et al. 2019).

Playa and lacustrine sediment communities are often highly specialized and structured by spatial location (e.g., distance from water source) or by depth, reflecting adaptation to particular microenvironments, especially gradients in climate and geochemistry (e.g., aerobic communities near-surface vs. anaerobic communities at depth; Finstad et al. 2017; Fernández-Martínez et al. 2019; Warren-Rhodes et al. 2019). Surface and near-surface (<1 m) dry sediment communities may be dominated by *Actinobacteria* and *Alphaproteobacteria* (halotolerant/alkalotolerant/spore-forming/desiccation-resistant), as well as anoxygenic phototrophic *Chloroflexi*. Clear shifts in diversity, related to concomitant shifts in soil moisture, nutrients (e.g., phosphorus), and/or salinity, often occur in deeper playa sediments, which support less diverse but more halotolerant communities that also include primarily *Actinobacteria* and *Proteobacteria* but different orders/families and/or specialized deep taxa (e.g., Sirisena et al. 2018; *Methylobacterium radiotolerans*, Fernández-Martínez et al. 2019; Warren-Rhodes et al. 2019).

As well as spatial structuring, temporal shifts in playa and lacustrine desert sediment communities can also occur at multiple scales. In most deserts, even within the Atacama's hyperarid core, microbial populations can experience rapid ecological succession after stochastic precipitation events, flourishing in surface waters as biofilms grow and die on sediments, followed by decay, disintegration, and dispersal into the wider landscape via wind and/or water action (unpublished data, K. Warren-Rhodes; Belnap and Gillette 1998; Fernández-Martínez et al. 2019; Qu et al. 2020).

9.2.4 Salts

Scant rainfall and high evapotranspiration favor the development of evaporitic desert ecosystems and the long-term preservation of chemical sediments (primarily soluble salts such as sulfates, halides, and nitrates, hereafter referred simply to as salts or evaporites). Salt habitats include large-scale hypersaline lakes (*lagunas*, i.e., actively forming, and/or paleolakes), briny ponds, ice-covered brines, and salt-encrusted landscapes, such as salt-covered playas or gypsum outcrops. Much of the discussion that follows pertains to temperate deserts such as the Atacama, where salt habitats are more commonly found. Salt habitats in cold and in polar deserts are comparatively rare, although they represent a unique opportunity to study the combined effects of low temperature and high salt concentration on microbial communities. A notable example is the perennially cold brine of Lake Vida, in the Antarctic Dry Valleys, which forms beneath a thick layer of ice, and appears to sustain microbial communities despite perennial subfreezing temperatures and the absence of light (Murray et al. 2012). Salt habitats can range from regional (tens to hundreds of kilometers) to fine-scale (hundreds of meters to centimeters) and include evaporite deposits, salt polygons, nodules and domes, salt crusts, amongst many others (Stoertz and Ericksen 1974; Chong 1984, 1988; Ericksen and Salas 1990; Wierzchos et al. 2006; Szynkiewicz et al. 2010; Artieda et al. 2015; Finstad et al. 2016, 2017; Pfeiffer et al. 2018).

Microbiology of Salt Habitats

The habitability of salt-rich environments depends on a complex interplay of physical and chemical parameters, including salt composition and content, pH, temperature, and radiation (Oren 2002a, b). Despite saline environments being often considered extreme for life, many contemporary salt habitats host abundant and often unique microbial communities, well-adapted to survival amidst the higher salinity, ultraviolet radiation, and extreme temperature fluctuations characterizing these habitats. Salt habitats can host a variety of communities, including microbial crusts and mats, lithic communities, sediment microbiomes (see sediments above), and microbialites, whose remnants can persist over geological eras as microfossils within these hypersaline systems (Noffke et al. 2001; Rothschild and Mancinelli 2001; Demergasso et al. 2004; Vogel et al. 2009; Murray et al. 2012; Robinson et al. 2013; Farias et al. 2013, 2014; Fernández-Martínez et al. 2019; Rasuk 2014; McKay 2016; Sirisena et al. 2018). However, there are also a few natural environments with high salt content that appear to preclude biological processes (e.g., Belilla et al. 2019).

While salt habitats share many microbial taxa with the soils, rocks, and sediments elaborated above, they also foster communities markedly different from those of other substrates, particularly distinguished by higher abundances and diversity of archaea and also *Halobacteriales*, *Bacteroidetes*, algae (*Chlorophyta*), *Proteobacteria* and *Cyanobacteria* (especially *Halothece*, *Euhalothece*, and/or *Chroococcidiopsis*), as these habitats require special adaptations (radiation and desiccation-resistance) to overcome high salt content (Rothschild et al. 1994; Oren 2002a, b; Oren 2013; Gramain et al. 2011; Wierzchos et al. 2015). Evaporite communities in the Atacama Desert have been well-studied, particularly the wide-spread lagunas and extensive salars (salt-encrusted playas) that dot the Altiplano and Central Depression, typically forming in low-lying topographical depressions (e.g., the Salar de Atacama, Demergasso et al. 2010; Dorador et al. 2010). Overall, these studies have revealed microbial communities in halite and gypsum salts often share common inhabitants but can be quite distinct from one another (Robinson et al. 2013). Even within the same salt environment, for example, abundance and biodiversity can differ significantly, both by habitat type (e.g., microbial mats versus dry halite crusts of a laguna or salar) and across moisture and salinity gradients (Rasuk 2014; Finstad et al. 2017). As with lithic microbial communities, in general evaporite systems exhibit a reduction in biodiversity and biomass with decreasing water availability, and conversely with increasing salinity. Finstad et al. (2017) and de los Ríos et al. (2010), for example, both quantified decreasing complexity in these microbial parameters for halite crusts and nodules (see below) as moisture and salinity stress intensified.

Halite nodules that are part of salt-encrusted playas represent a unique type of salt habitat that so far has only been described in the Atacama Desert (Fig. 9.1d). The salt nodules are approximately 50–70 cm in diameter and 10–20 cm thick and are found scattered over dry Miocene lake basins along the western flank of the Atacama's hyperarid core. The existence of subaerial salt nodules in a landscape that is millions of years old is a testimony to the extreme and long-lasting dryness in the region. The interior of the nodules is porous and relatively friable, and it is home to specialized, salt-adapted communities that rely on atmospheric inputs of water such as fog and deliquescence (Wierzchos et al. 2006, 2012a; Davila et al. 2008a, b). In fog-rich regions, such as Salar Grande and Salar Llamara, halite nodules support high biomass and biodiversity. These active microbial communities form distinct colonized layers visible by their green (e.g., chlorophyll) and sometimes pink/red/orange-colored photosynthetic pigments (e.g., carotenoids, scytonmin; Wierzchos et al. 2006, 2015; Robinson et al. 2013; Davila et al. 2013), and they include rich consortiums of cyanobacteria, algae, a large variety of heterotrophic bacteria (*Bacteroidetes*), halophilic archaea (*Halorhabdus, Haloarcula*) and, to a lesser extent, fungi (Robinson et al. 2013; Finstad et al. 2017). These substrates are still strongly salt-selected, dominated by a widely distributed but closely related single cyanobacterial cluster (*Aphanothece-Halothece-Euhalothece*) that appears uniquely adapted to especially high salinity-high desiccation micro-habitats (Robinson et al. 2013; Finstad et al. 2017). Moreover, these contemporary halite communities are distinct from ancient evaporite deposits in the region (Gramain et al. 2011; Parro et al. 2011). Viruses appear to be active in some halite nodules in Salar Grande and seem to primarily target *Halobacteria* and *Salinibacter* hosts for infection (Uritskiy et al. 2019).

In much drier settings in the Atacama, halite nodules are some of the last refugia to support primary productivity (Davila et al. 2008a, b, 2010; de los Ríos et al. 2010; Wierzchos et al. 2006, 2012a; Robinson et al. 2013). In these regions, the hygroscopic nature of the salt substrate facilitates microbial activity by generating deliquescent brines when the ambient air relative humidity reaches a threshold value of 75% (Davila et al. 2008a, b, 2013; Wierzchos et al. 2012a). Indeed, the least diverse halite microbial systems on Earth have been documented for halite nodules in the hyperarid core at Yungay (see below, Wierzchos et al. 2006; Davila et al. 2008a, b; de los Ríos et al. 2005; Robinson et al. 2013; Finstad et al. 2017). There, colonization is largely constrained to a single, near-surface endolithic layer, supporting only three main groups of microorganisms: a single phylotype of cyanobacterium (*Halothece*), mutualistic heterotrophic bacteria (10% of bacterial abundance, mainly *Bacteroidetes* such as *Salinibacter*), and archaea (mainly an unidentified *Halobacteriacae sp. and Halococcus sp*). In comparison to Salar Grande and other wetter sites, cyanobacterial primary producers in the Yungay nodules comprise a much greater total relative abundance (89% vs 28%), photosynthetic algae are absent and less diverse archaeal taxa survive. The Yungay communities are also distinct from salty soils only a few cm away, which lack cyanobacteria and contain more *Proteobacteria* (e.g., *Halomonas, Shewanella, Acinetobacter*) and yet other archaeal taxa (e.g., *Haloterrigena*, Robinson et al. 2013). The ecological explanation for why microbial communities inside salt nodules can survive and thrive resides with the water-retention properties of the salt substrate after a wetting event, which allows the nodule interiors to remain wet for days or even weeks (compared to hours for nearby quartz rock habitats), even under hyperarid conditions.

Similarly, at coastal and inland Atacama sites with higher precipitation levels, gypsum crusts and outcrops host relatively more complex populations than drier locations, including algal, cyanobacterial, lichen, and fungal components, some of which survive on fog-rich inputs of atmospheric water vapor for photosynthetic activity (Wierzchos et al. 2011, 2013). Lower abundance and diversity also accompanies lower water content in gypsum salt habitats, with the apparent limit reached at Yungay in the hyperarid core, where gypsic crusts are widely available but virtually uncolonized (~1028 hrs at $\geq$60% air RH vs ~2961 hrs at Salar Grande; Wierzchos et al. 2011). Similar to halites and lithic substrates, gypsum hypersaline habitats (crusts, outcrops, domes, lagunas, microbial mats) in the Atacama support epilithic, hypolithic, and endolithic colonization modes dominated by similar key Phlya, namely *Cyanobacteria* (~20–90% of total abundance) *Actinobacteria*, *Proteobacteria* and *Chloroflexi* (Hughes and Lawley 2003; Wierzchos et al. 2015; Dong et al. 2007; Meslier et al. 2018). However, they too can also harbor distinct photosynthetic taxa distinguished by the particular salt or lithic substrate they inhabit (Robinson et al. 2013; Meslier et al. 2018).

Finally, as described above for clay sediment playas, evaporite microbial ecosystems can also be subject to intensely rapid and prolific ecological blooms and successional cycles following precipitation events (Uritskiy et al. 2019; Azúa-Bustos et al. 2018), catalyzing abundant microbial growth and surface photosynthetic processes and carbon, nitrogen, and sulfur cycling (e.g., sulfate fermentation and reduction, sulfide phototrophy and chemolithotrophy, Navarro et al. 2009; Fernández-Martínez et al. 2019).

9.3 Insights from Extreme Desert Ecology

Several notable insights recur in our ecological review of microbial life in terrestrial end-member deserts. First is a general trend toward ecological extinction with greater xeric stress. In this extinction sequence, as water declines, macro- and micro-fauna are the first to disappear from deserts; next comes flora, then microbial surface life, and then eukaryotes. Finally, all that remains in the most extreme end-member deserts are simple cryptic, mostly prokaryotic microbial systems. Basic ecological principles such as succession and convergence in ecosystem structure and survival strategies are also hallmarks of this extinction trend. Second is a shift from the macro- to the micro- scale for key determinants of the abundance, diversity, distribution, and function of microbial communities. Based on our review, these overarching trends and processes are a common thread for microbial life across extreme desert soils, sediments, rocks, and salt substrates on the Earth. Given their predictability and possible universality, an understanding and recognition of the ecological principles underpinning and organizing terrestrial desert analogs may thus hold important lessons for the search for ancient and/or extant microbial systems on other planets, including Mars.

9.3.1 Trends of Ecological Change in Deserts

Declines to Ecological Extinction

As evident from the review above, desert ecosystems respond to increasing dryness in multiple, convergent ways. We refer to this (often serial) process as "ecological extinction," a continuum that charts first the disappearance of flora and fauna and then the progressive loss of microbial ecosystems. Formerly diverse and abundant microbial consortia are replaced by (1) rare microbial communities that (2) inhabit fewer niches and are comprised of (3) less diverse (e.g., loss of algae, filamentous taxa, Jung et al. 2019) (4) nimbler (e.g., rapid "switching" of metabolic activities on/off as a response to water or temperature changes, Sun 2013; Goordial et al. 2017) and (5) hardier members (dormancy/cryptobiosis, radiation- or desiccation resistance; Belnap et al. 2003a; Lee et al. 2018; Shen et al. 2021). The end of the continuum is widely available but vacant unoccupied habitats (Fig. 9.4).

From our review, it is clear that abundant water availability even in the driest deserts on Earth supports higher plants and animals, and buffers microbial communities from the hardships of abiotic constraints, such that most habitats (soils, rocks, clastic/chemical sediments) are suitable for survival and growth. Thus, in "wet" desert environments (soils ~25% GWC; ~100 mm MAP; rocks $\geq$1500 hrs liquid water year^{-1}), virtually all available substrate is colonized with high microbial diversity, and "typical" biotic and ecological processes (predation, competition, biogeochemical cycling) control microbial destinies (Warren-Rhodes et al. 2006, 2013; Lee et al. 2018; Davila et al. 2019). Lithic photosynthetic communities, for example, in wetter locations of Antarctica, the Arctic, Iceland, the Qinghai-Tibetan Plateau, Negev, and Atacama, occupy an expanded territory of potential substrates from translucent stones to dark, mafic rocks—i.e. those typically unsuitable in drier sites—such as basalts, rhyolitic glass, ignimbrites and opaque dolomitic rocks (Cockell and Stokes 2006; Herrera et al. 2009; Olsson-Francis et al. 2012, 2019; Brady et al. 2017).

As water availability declines, soil and surface habitats become increasingly untenable (soils threshold at ~8% GWC, Pointing et al. 2009; Makhalanyane et al. 2015; Neilson et al. 2017; Lee et al. 2018; Chap. 10). As xeric stress intensifies, even lithobiontic photosynthetic communities begin to struggle (~ $\leq$ 25 mm MAP) and retreat to fewer and fewer niches and substrate types, i.e., only those habitats with optimal features (high translucency, higher porosity, larger-sized rocks) that enhance

Fig. 9.4 Vacant empty habitat. Thousands of suitable translucent rock habitats exist at this site in the core of the Atacama Desert for endolithic or hypolithic photosynthetic communities. Yet <1% of this habitat is colonized. Large quartz rocks in the bottom foreground are ~30 cm wide. Photo Credit: K. Warren-Rhodes

water capture and retention (Wierzchos et al. 2006; Warren-Rhodes et al. 2007; Davila et al. 2008a, b, 2019). Mafic and dark rocks are mostly uninhabited, although desert varnish communities can sometimes persist but their extent remains unknown (Warren-Rhodes 2006, 2007; Cockell and Stokes 2006; Kuhlman et al. 2008; Pointing et al. 2009).

Microbial desert ecosystems also become less diverse with decreasing amounts of water, losing BSC and epiliths such as lichens (unless there is fog) at the arid-hyperarid transition (Pointing et al. 2009; Palzelt et al. 2014). Lithic communities can lose filamentous and most eukaryotic taxa at this stage (e.g, Jung et al. 2019), while surface and subsurface soil communities likewise experience similar transitions to lower diversity, with patchy distributions dominated by desiccation and radiation-resistant bacteria (e.g., hardy spore-forming soil taxa *Actinobacteria*, but also *Proteobacteria, Chloroflexi and Firmicutes*; (Bagaley 2006; Pointing et al. 2009; Crits-Christoph et al. 2013; Schulze-Makuch et al. 2018; Warren-Rhodes et al. 2019). In the Atacama Desert, for example, soil heterotrophic biomass can be up to four times lower in hyperarid versus semi-arid locations, and diversity declines in both surface and subsurface soils (Bagaley 2006; Crits-Christoph et al. 2013). These declining trends are not unique to soil bacteria or lithobiontic ecosystems, but appear to be common across substrates, niches, and a wide variety of deserts, having been similarly demonstrated for lipid biodiversity and abundance, soil microbial activity (Wilhelm et al. 2018), and for salt-dwelling endolithic microorganisms, amongst others (e.g., Robinson et al. 2013; Qu et al. 2020). With declines in microbial abundance, diversity and activity, biogeochemical cycling and function also slow, paralleling the marked reductions in abiotic processes as water scarcity deepens, such as soil formation (Ewing et al. 2006, 2008; Finstad et al. 2014; Goordial et al. 2017).

In the driest parts of the Atacama Desert soil microbial abundance and diversity is the lowest found anywhere on Earth and lithobiontic photosynthetic community numbers and biodiversity decline precipitously to $<<1\%$ of available substrate (Warren-Rhodes et al. 2006; Pointing et al. 2009; Robinson et al. 2013; Schulze-Makuch et al. 2018). At this stage (~1 mm year^{-1} MAP, $\leq$200 hrs liquid water year^{-1}), biogeochemical cycling and soil processes typically cease or slow to geological timescales (e.g., carbon cycling, Ewing et al. 2006; Azúa-Bustos et al. 2017; Pfeiffer et al. 2018) and another microbial ecological shift toward extinction occurs, where even the hardiest endolithic communities become restricted to a single type of refugia substrate, namely deliquescing halite nodules, which possess efficient water capture and retention properties (Wierzchos et al. 2006; Davila et al. 2008a, b).

In the coldest habitats of the Antarctic Dry Valleys, similar transitions as those described for hot deserts in terms of trends to ecological extinction (in biomass and diversity of surface soils and lithic niches) also occur, with significantly higher values at locations with more moisture close to the coast but declining inland and upward at more desiccated sites (e.g., Cowan et al. 2002; Pointing et al. 2009; Lee et al. 2012; Yung et al. 2014; Wei et al. 2016). Here too, surface expression of microbial life eventually disappears, as microorganisms retreat to protected endolithic niches and community structure simplifies to include only the exceptionally hardy microbial survivalists, e.g., more coccoid (esp., *Chroococcidiopsis, Halothece*) rather than filamentous (Wierzchos et al. 2015) and more desiccation/radiation/cold-adapted taxa versus those less well-adapted to severe environmental stress (Yung et al. 2014; Wierzchos et al. 2015; Goordial et al. 2017; Lacap-Bugler et al. 2017). Like in the driest portions of the Atacama Desert, in the coldest Antarctic Dry Valley locations, microbial life becomes exceedingly rare, mostly transient and/or entirely dormant (i.e., metabolically and reproductively inactive), as it attempts to survive at the coldest and driest limits cataloged on Earth (Goordial et al. 2016).

Notwithstanding the above convergent trends, however, it is important to consider that many questions remain unanswered about even the most well-studied microbial ecosystems in the driest hot and cold end-member deserts on Earth. Are these systems static, i.e., once conditions change do we see permanent or ephemeral ("succession") phase shifts back toward greater abundance and diversity? Do these systems harbor significant repositories of dormant and diverse 'dark matter' microbial worlds, ready to spring to life as conditions ease? Are surviving microbial communities 'simpler' in terms of their interactions? Are they mutualists, prey, predators, and/or competitors, and how do these fundamental relationships play out under changing conditions? Addressing these and other knowledge gaps will continue to hold remarkable insights into the search for and detection of life on other desert worlds.

The Criticality of Scale

The abundance, distribution, function, and diversity of microbial communities in desert substrates are strongly linked to the physical, geochemical, and climate regimes within which they survive. These factors operate at, and may be dominant drivers of, microbial ecology at multiple and varying spatial scales. Macro-scale climate and other abiotic determinants of microbial abundance, spatial distribution and diversity tend to be drivers at global to continental scales (e.g., between deserts), whereas micro-scale factors appear to take over at regional and smaller scales (e.g, within deserts, within a single site).

Water, Temperature and Substrate Properties Control Microbial Ecology but Effects Vary by Scale and System

For soil, BSC, hypolithic, and some hypersaline systems, water availability is the overarching control on desert microbial abundance and biodiversity at macro to micro-scales (Warren-Rhodes et al. 2006, 2007; Büdel et al. 2008, 2009; Finstad et al. 2017; Jung et al. 2019). In environments where xeric stress is the ultimate constraint on life, the timing, intensity, frequency, duration, and source of water explains most, if not all, microbial ecology outcomes—from global to single rock scales. This can be easily understood if one considers the interconnectivity of habitats and ecosystems with the larger regional context (Belnap et al. 2005, Belnap 2006; Warren-Rhodes et al. 2006; Turnbull et al. 2012; Garcia-Pichel et al. 2014). As air, dust, and especially water move through the larger desert landscape—from hillslopes, across desert pavement terrain, via drainage channels/gullies, down to topographic lows in playas and salt lakes—each of the key desert habitats are linked to one another in an integrated ecosystem whole (Fig. 9.5). At larger spatial scales, macro-climate and hydrological systems supply the particular sources (fog, snow, rain) of water, and physical conditions such as light and temperature constrain, influence and channel water flow and availability across the broad landscape to individual rock and soil micro-scales. This nested spatial connectivity also likely explains the spatially nested nature of some ecosystem microbiomes (Amend et al. 2019). Once water limitations are eased, other secondary factors may come into play as dominant controls, such as temperature, light, nutrients (carbon, phosphorus, nitrogen, sulfur), pH, and/or biotic interactions like predation, competition, resource

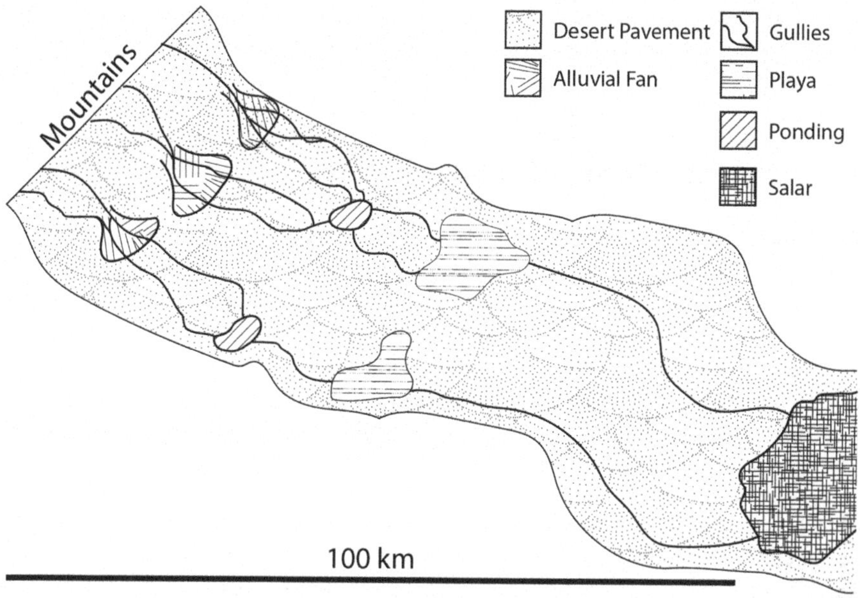

Fig. 9.5 The interconnectivity of habitats in the Yungay Valley ecosystem, Atacama Desert, Chile

partitioning, and habitat differentiation (Ettema and Wardle 2002; Lee et al. 2018; Ekwealor and Fisher 2020).

In contrast, for other microbial systems such as endoliths, temperature and light conditions can be of more importance than water to abundance and/or community diversity. Sun (2013) showed light levels, not snowmelt availability, explained Antarctic endolithic colonization, while research on endolithic and soil communities demonstrated macro-scale temperature is more critical than water in controlling biodiversity at global and continental scales, with specificity in climate adaptabilities of particular endolithic community members, namely phototrophic cyanobacterial taxa, explaining some observed differences (Qu et al. 2020).

At fine-scales, microenvironmental factors like temperature, light availability, and rock properties also show deterministic effects, although they still may be linked to the availability and retention of water (Berner and Evenari 1978; Bagaley 2006; Omelon et al. 2006; Warren-Rhodes et al. 2006; Kidron 2007; Warren-Rhodes et al. 2007; Cowan et al. 2011; Crits-Christoph et al. 2013; Archer et al. 2017; Yung et al. 2014; Meslier et al. 2018). For instance, micro-scale habitat properties (rock type, aspect/orientation, light, size, porosity) have been hypothesized to affect hypolithic colonization and orientation at the individual rock or smaller scales (Warren-Rhodes et al. 2007) and, more recently, shown to be the key drivers of microbial communities' existence and composition, spatial pattern, survival and activity for endolithic systems at fine-scales (Meslier et al. 2018; Qu et al. 2020). Scale thus matters profoundly in ecology, and the interconnection between habitats and their larger environmental context are critical for understanding the potential types and distribution of past or contemporary life and biosignatures on both the Earth and other planets.

9.4 Potential Desert Habitats on Mars

One look at Mars and a terrestrial desert ecologist feels right at home. As far as the (rover) eye can see are familiar sights: wide expanses of (1) desert pavement mantling hillslopes carved by (2) drainage channels and gullies bisecting (3) broad alluvial fans and plateaus draining into (4) ancient lakebeds and playas. In the polar highlands, dark "rock-strewn" desert pavement stretches across thousands of km of frigid hyperarid landscapes composed of salty permafrost and polygonal ice-cemented soils (Figs. 9.2b and 9.6, Arvidson et al. 2009; Hecht et al. 2009; Smith et al. 2009).

To date, a spatial assessment of the varied types and areas of desert habitats on Mars is unavailable. Furthermore, even for the larger landscapes mapped by previous or current orbiters (Ody, ME, MRO, MAVEN, ExoMars TGO), landers (Viking 1 and 2, Phoenix, InSight) and rovers (Pathfinder, Spirit, Opportunity, Curiosity), a complete ecological analysis of habitats and mission sampling data has yet to be performed. Given the ambitious goals and success of planetary science and exploration missions to Mars, such data and analytical gaps are understandable. Nevertheless, with past (and future) science in mind, below we examine Mars from an ecologist's perspective to (1) map, where possible, the extent of key terrestrial-like desert analog habitats, (2) generate summary ecological sampling statistics, and (3) highlight gaps and potential questions that could be addressed with future ecological analyses and data.

9.4.1 Soils and Regolith

Viking

The 1975 Viking landers were the first to sample Martian soils/regolith (hereafter, we use the words soils and regolith interchangeably for Mars; Moore et al. 1987; Arvidson et al. 1989). Both landers operated surface samplers on "rock-strewn patterned ground" (Mutch 1976; Mutch et al. 1977) for roughly 600 sols (639 lander 1, 596 lander 2) and sampled a total of roughly 3 m^2 trenched surface soils from 'nominal sampling fields' of ~10 m^2 each. Roughly 59 trenches, including 3 "deep holes," were dug and sampled from two types of ecological/geomorphic units: 1) dark, mafic (basalts) desert pavement (Lander 1 and 2 sites), including permafrost polygons (Lander 2 site); and 2) "dune-like" drifts (Lander 1 site). Three main categories of soils were accessed (i.e., 86% of the surface sampler arm workspace, Moore and Jakosky 1989): "fine-grained drift," "blocky," and "crusty to cloddy" soils, with the first two occurring at Lander 1 site, and the third comprising most of Lander site 2 on the opposite side of the planet. Desert pavement surfaces had

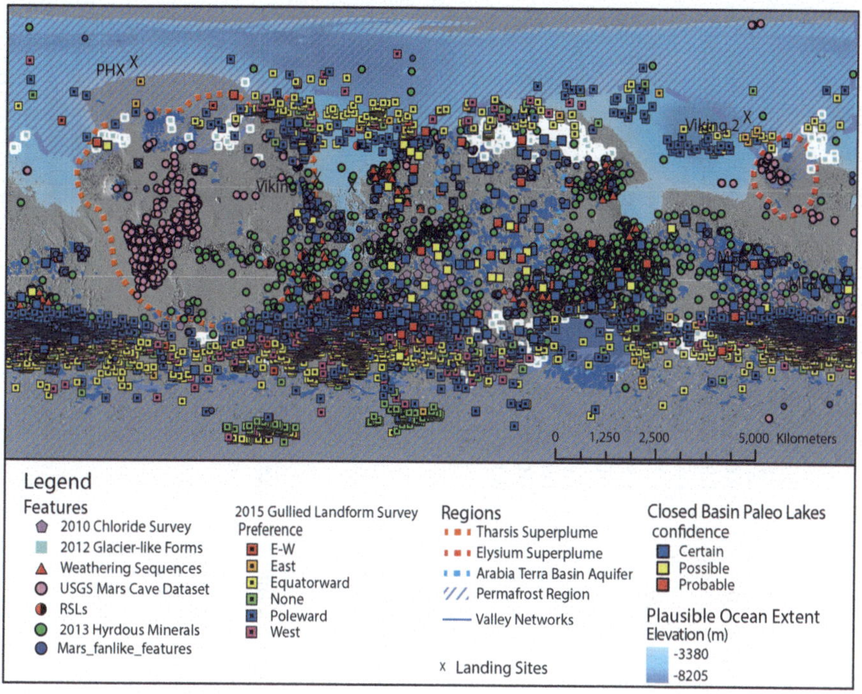

Fig. 9.6 Map of all features that could plausibly be associated with habitats (past or present) on Mars. See associated supplementary table for references and full descriptions of the units (descriptions from JMARS). *2010 Chloride Survey*: 642 deposits of potential chloride salt-bearing material, mostly within southern highland Noachian and Early Hesperian terrains. *2012 Glacier-like Forms*: 1309 features identified using CTX images indicate geologically recent water ice deformation. *Weathering Sequences*: 140 clay weathering sequences, including Fe/Mg- and Al-clays, identified with CRISM. *USGS Mars Cave Dataset*: GIS-based catalog of over 1200 cave-entrance candidates. *RSLs*: regions where RSL sightings have been confirmed with HiRISE. *2013 Hydrous Minerals*: hydrous mineral detections with OMEGA and CRISM. *Mars Fanlike Features*: Global catalog of mars fanlike features, actively being updated (Mondro et al. LPSC 2019 [#2739]). *2015 Gullied Landform Survey*: mapped using CTX, global distribution of gullied landforms with associated orientation preference. *Closed Basin Paleo Lakes*: candidate impact-crater hosted closed-basin paleo lakes

duricrusts several centimeters thick covering softer materials (Arvidson et al. 2009). One duricrust sample was acquired, in addition to several samples from underneath rocks of the desert pavement that had been nudged away. Samples were taken from both the surface and subsurface (~2–12 cm). "Very little" information on rocks at the sites was obtained (Arvidson et al. 2009), and no rocks were found to have fragile or weathered rinds—a key observation as most terrestrial colonized lithic habitats coincide with weathered substrates.

A total of 60 samples were analyzed by the Viking landers (Table 9.2). The biological instruments, BIO (Labeled Release, Pyrolytic Release, and Gas Exchange) were each provided four samples at each Viking site. For each of the eight biological experiments that successfully executed analysis (BIO, four at each

Table 9.2 Summary of Potential (Paleo) Habitats' Sample and Analysis Statistics for Viking, Phoenix and Curiosity Lander and Rover Missions through May 2020

	Potential (paleo/ modern) habitats sampled	Area sampled	No. samples run	Quantity and type of samples	Key results
Viking (1975): SOILS ~600+ sols	Eolian deposits (dust); dark desert pavement (polygonal ground), mafic (basalt) rocks	**~10 m^2 per site, ~20 m^2 total work space (dark DP, permafrost soils)**; **~3 m^2 total trench area**; mean trench area ~ 600 cm^2, mean trench volume ~ 4200 cm^3; ~60 trenches; 3 deep holes; range depth below surface ~2–12 cm; mean depth: 4.7 cm;	**~60 samples soils** 24 biological = BIO samples (0.1 g each), 4 GCMS analyzed (~0.1 g) 31 XRFS analyzed (~0.1 g)	**~ 6 g total soils** BIO: **2.4 g BIO** • 9 samples dune/drift • 3 samples blocky soil • 12 samples cloddy to crusty soils (permafrost, water ice? See Phoenix below) **GCMS: 0.4 g; XRFS: 3.1 g**	• soils: Basalts, iron-rich; 1200–1800 g/cm^3 bulk density • little learned about rocks (no weathered rind) • water content of soils 0.1 to <1% **-detection (reinterpreted; Navarro-Gonzalez et al.** 2010) **of low parts per billion (ppb) levels of organics** -soil chemical reactivity when wetted -no conclusive evidence of life
Phoenix (2007): SOILS 152 sols 30,000 images;	~0.6 Ga permafrost soils (water-ice-bearing soils; polygonal ground); dark desert pavement, mafic (basalt) rocks	**~3 m^2 workspace,** 12 trenches, max depth 18.3 cm; 30 samples acquired by scoop *and delivered* to instruments	**18 permafrost soil samples analyzed (0.052–1 g per sample)** 7 surface (~0–2.5 cm) 4 subsurface (2.5-5 cm) 6 scrape piles above ice; 1 soils under rocks and 2 blanks	**~7 g in total permafrost soils analyzed**	• definitive water ice signature (~1% water by weight); pure ice and pore ice in soils • detection of perchlorates and calcium carbonates (Heet et al. 2009, Smith et al. 2009) • **first in-situ ice detected on Mars**, ice table ~3–15 cm below soil surface • soils most similar to

					crusty to cloddy sols at Viking Lander 2 site
Curiosity (2012-May 8 2020, MSL): SEDIMENTS, ROCKS ~2758 sols+, ~ 22.1 km; 674,758 images	Paleolake (~3–3.5 Ga lacustrine sediments); lithic habitats (sandstone, clay-rich sediments, mudstones, sulfates, eolian sands)	**25 drills, 6 scoops; drill accessible workspace ~ 2 m^2 maximum; drill depth ~ 7 cm**	**Total: ~ 93 solid samples analyzed; 51 mudstone samples, 24 sandstone samples, 17 eolian sand samples; 1 heterolithic sample (~0.065 g per sample,** range 0.05 to 0.1 g per sample); **376.2 g** total excavated mass SAM (to 2018): **2.135 g,** Dune material: 2 (0.27 g) Sandstone: 4 (0.54 g) Mudstone: 14 (1.125 g) Eolian sand/dust: 4 (0.2 g)	**~9.64** g total solid samples analyzed (**7.53 g drilled** and **2.12 g scooped**; 4.36 g SAM +0.76 g CheMIN = **5.12 g mudstones**, 1.745 g SAM + 0.0595 g CheMIN = **2.34 g sandstones**; 1.79 g SAM + 0.33 CheMIN = **2.12 g eolian sands**; **0.07 g heterolithic** sample. SAM: 7.895 g sample analyzed; avg. ~0.24 g per drill; CheMIN: 1.75 g sample analyzed; ~0.05 g per drill	• discovered fine-grained sedimentary rocks, inferred to represent ancient lake suited to support chemolithoautotrophy • **detection of organic matter** (>50 nm organic carbon); preserved organics possible to support heterotrophy (Freissinet et al. 2015; Eigenbrode et al. 2018)
Total				**22.64 g total solid samples analyzed;** ~13 g permafrost soils, ~5.12 g mudstones, ~2.34 g sandstones, ~2.12 g eolian sands	

The table includes only science missions that acquired and delivered physical samples via drill or scoop to analytical instruments, including those that had the capability to detect the presence of organic compounds. It also tallies only samples that were successfully analyzed. Potential habitats refer to both modern and/or ancient habitats site), 0.1 g of soil was used, for a total of 24 samples and 2.4 g regolith analyzed (Levin and Straat 1976; Moore et al. 1987). All Lander 1 site BIO samples (n = 12) were of dune-like drift materials, and all Lander 2 samples (n = 12) were of blocky materials. For GCMS, the objective was to search for organic molecules and inorganic volatiles, and four in different environments were successfully analyzed: two samples from drift material (Sandy Flats, GCMS 1) and blocky material (Rocky Flats, GCMS 3) at Lander Site 1 and two samples from crusty to cloddy material (Bonneville Salt Flats, GCMS 1) and a surface originally under rock 3 (Badger) at Lander Site 2. For each GCMS, a maximum of 60-mm^3 was analyzed, or ~ 0.4 g soil total. For XRFS, ~0.1 g was analyzed per sample, for a total of ~3.1 g. In all, thirty-one XFRS and four GCMS soil samples revealed the presence of basalts, iron-rich clays and elements Fe, Ca, Si, Al, and Ti. Eighteen physical properties measurements were also completed, and atmospheric and ground surface meteorological data were collected (Moore et al. 1987). Water vapor saturated the atmosphere and snowfall was detected at Lander site 2 in wintertime. Thus, in total, Viking analyzed roughly **~6 g of martian soil**, from which no signs of extant or extinct life or organic matter were conclusively detected at the time (Klein 1979) (Table 9.2). Recent reinterpretations, however, of Viking GCMS data point to the successful detection of Martian organic material (Guzman et al. 2018).

Phoenix

More recent lander and orbiter data also reveal Mars to be a "permafrost dominated planet" (Levy et al. 2008, Fig. 9.7). As on Earth, Mars' polygonal terrain is quite varied. Small-scale polygons (few meters diameter) have been observed, for example, at the 2008 Phoenix landing site and linked to thermal contractions and water/ice dynamics (Smith et al. 2009). This polygonal patterned ground, similar to that studied in Antarctica, signals the "presence of widespread, shallow cohesive icy soil" that cooled and warmed seasonally, creating high-shouldered, round-centered polygon interiors bounded by 20–50 cm deep trough bounding edges (Smith et al. 2009; Levy et al. 2008). Large-scale thermal contraction (HiRISE (PSP_002070_2250) and clastic polygons tens to hundreds of meters in diameter have also been mapped and modeled for Mars, for example near Utopia Planitia (Yoshikawa 2003; Lefort et al. 2009) and Lyot Crater (NASA/JPL/Univ. Arizona, HiRISE ESP_016985_2315), respectively, and possible patterned ground has been imaged for the Elysium Planitia region (HiRISE (PSP_004072_1845; Washburn 1956; Black 1976; Maloof et al. 2002).

The Phoenix mission, like Viking, sampled surface and sub-surface materials in the desert pavement of the northern plains of Mars, including geomorphic units associated with water-ice-rich permafrost soils and polygonal ground (Mellon et al. 2009a; Smith et al. 2009). The desert pavement unit was created from aeolian stripping and ice soil sublimation and the polygonal ground may reflect widespread cohesive icy soils that have undergone seasonal or longer-term cooling and cracking by elastic failure. In all, Phoenix acquired 31 samples from a total of 12 trenches: nine surface (~0–2.5 cm), four subsurface (~2.5–5 cm), and seven soils scraped from piles above ice; four icy soils; and two soils beneath rocks (Arvidson et al. 2009;

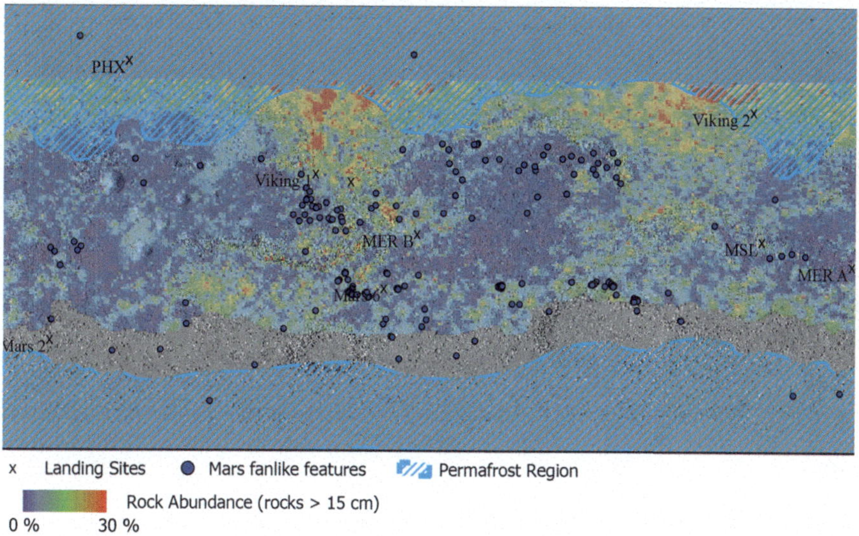

Fig. 9.7 Map of plausible permafrost soils and lithic substrate habitats on Mars. Base map is MOLA shaded relief, overlain by TES Rock Abundance data (Nowicki and Christensen 2007) for rocks >15 cm in diameter. Alluvial fan-like features (Di Achille and Hynek 2010; Mondro et al. 2019 and references therein) are indicated by the blue dots. Permafrost regions were derived from Neutron Spectrometer epithermal neutron counts (< 6 counts/second). Mission landing sites are denoted with x's and labeled

Mellon et al. 2009b). Icy-soil interfaces occurred at a mean depth of 4.6 cm (maximum depth excavated was 18.3 cm, Mellon et al. 2009b) and the Phoenix soils were "similar in appearance and properties to the weakly cohesive crusty and cloddy soils" characterized by the Viking 2 lander. Phoenix soils contained ~1% water content (% weight, Smith et al. 2009). In total, **~7 g of regolith** was analyzed by Phoenix, and it was the first mission to definitively detect water ice in soils on Mars (Table 9.2).

9.4.2 Lithic Substrates

Mars is a primarily basalt-covered world that has weathered in a near-exclusively polar-desert climate for billions of years (Fig. 9.7, Christensen et al. 2001; McSween et al. 2003; McSween et al. 2009; McSween 2015). Multiple landed missions have revealed the widespread presence of desert pavement consisting of lithic material (boulders, rocks, and pebbles), including abundant sandstone and mudstone outcrops in Gale Crater and rocky terrain throughout Gusev Crater and Meridiani Planum. Despite this, there is currently no available estimate of the total area on Mars comprised of desert pavement or its larger ecological units, including hillslopes, permafrost plains and plateaus, although general estimates of rock abundance have been quantified (Fig. 9.7).

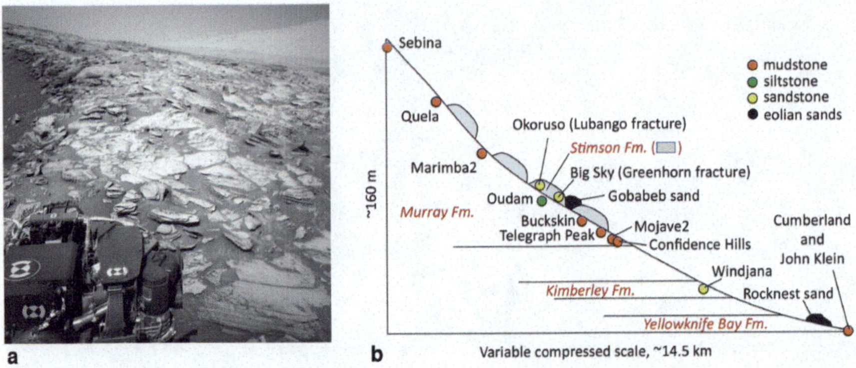

Fig. 9.8 (**a**) Mastcam image MSL Curiosity Rover of sandstone lithic habitats at Greenheugh Pediment (sol 2698, March 6, 2020). (**b**) ChemMIN sample types and locations (Vaniman et al. 2018)

As described above, Viking and Phoenix landers extensively investigated desert pavement-soil habitats, but lithic substrates either were not a main focus and/or were unsuccessfully interrogated (e.g., no RAT/DRT). However, in several other landed and rover missions, mineralogical features and geological composition of hundreds of soils, rocks, and outcrops across vast swaths of pavement terrain have been imaged and analyzed spectrally by remote instruments (e.g., 700 distinct APXS measurements by MER and MSL through 2014, Yen et al. 2014). While detailed geological reviews are beyond the scope of this chapter, we note that neither extensive quartz fields nor significant numbers of lithic substrates that typically support endolithic colonization on Earth, such as granites, ignimbrites, and calcite outcrops, have been detected and/or subsequently analyzed for the presence of organics or biosignatures on Mars, although opaline sinters (Gusev Crater) and possible lacustrine carbonates (Jezero Crater) have been located from orbit and/or by rovers (Squyres et al. 2008; Ruff et al. 2011; Brown et al. 2019; Horgan et al. 2019).

One recent exception has been the sandstone outcrops at Gale Crater (Fig. 9.8). During Curiosity's primary mission, the first drill (Treiman et al. 2016) into sandstone bedrock occurred in the Kimberley formation, with a surface exposure age of ~54 Myr, relatively young compared to the Sheepbed formation and Murray Buttes mudstones (Sect. 9.4.3, Fig. 9.8). By May 2020 (sol 2768), roughly 24 sandstone samples (~2.34 g total) had been drilled and analyzed by the SAM and CheMIN instruments on Curiosity. APXS and ChemCam data show that the bulk of the sandstone samples' composition is basalt-derived sediments. The SAM instrument did not detect organics in sandstones (through 2018; Sutter et al. 2017). While sandstones appear to be abundant, from 4 rock samples (through 2018), neither fossils nor biosignatures have been detected, although it is important to note that these imaging and organic analysis capabilities were not as detailed as would be done on laboratories on Earth.

9.4.3 Sediments of Potential Paleoplayas/Paleolakes on Mars

As elaborated in Sect. 9.2.3, ephemeral playas and lakes in terrestrial deserts are analogs for the ubiquitous (candidate) paleolakes and widespread paleo-fluvial activity observed on the surface of Mars (Fig. 9.9, Wharton et al. 1995; Cabrol and Grin 1999, 2001, 2003; Baldridge et al. 2004; Goudge et al. 2015). Terrestrial playas are rich in phyllosilicates (e.g., clays, mudstones) and salts, such as chlorides and hydrated sulfates.

Visual and geochemical characterizations of sedimentary rocks on Mars have in the past few decades supplied direct evidence of ancient aqueous environments reminiscent of Darwin's "warm little pond" (Squyres et al. 2004, 2008) and of seafloor hydrothermal settings like those home to some of the earliest life forms on Earth (Weiss et al. 2016; Michalski et al. 2017). Orbital data indicate the formation of standing bodies of water during the early Amazonian (2–3 billion years ago) and perhaps during the Late Amazonian (<2 billion years ago) (Fassett et al. 2010). The Eridania basin in the southern highlands of Mars was likely once an inland sea host to a variety of habitable environments from sub-aqueous

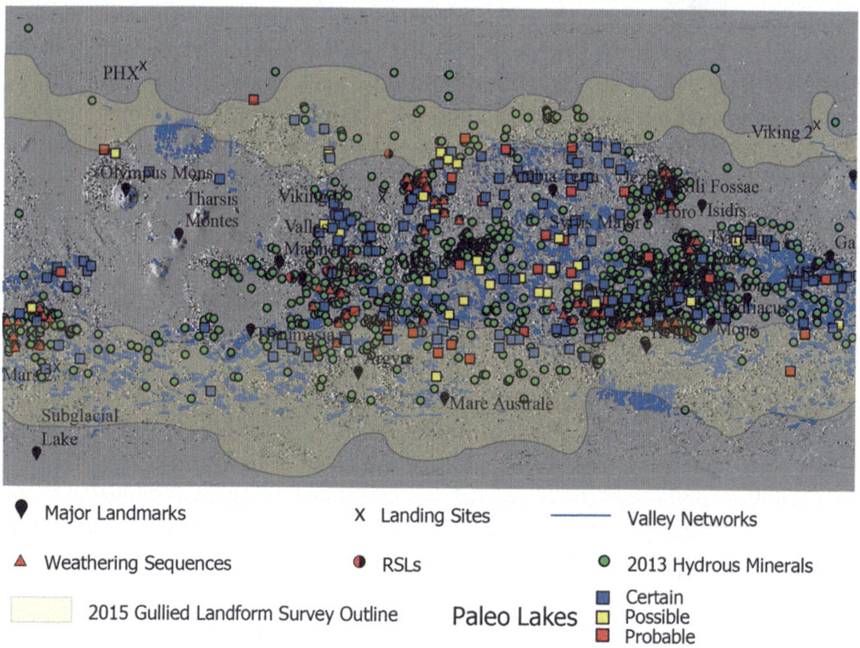

Fig. 9.9 Plausible martian habitats within sediments. Valley networks (Hynek et al. 2010), clay weathering sequences (Carter et al. 2013), RSLs (McEwen et al. 2014; Ojha et al. 2014), hydrous minerals (Carter et al. 2013), closed basin paleo lakes (Forsythe and Blackwelder 1998; Cabrol and Grin 1999; Ori et al. 2000; Grant and Parker 2002; Malin and Edgett 2003; Fassett and Head 2007; Mangold et al. 2007; Di Achille and Hynek 2010; Hauber et al. 2013; Goudge et al. 2015), and regions with gullied landforms (Harrison et al. 2015) are shown

hydrothermal vents to playa-lake-type deposits along its margins as the sea dried (Michalski et al. 2017). Opportunity and Curiosity rovers also collected data on martian aqueous paleo-playa environments. Finely layered paleo-playa deposits, for example, were observed by Opportunity in Meridiani Planum (sulfate salts, weathered basalts, phyllosilicates, hematites), and these environments likely have an origin controlled by a regionally extensive and fluctuating groundwater table marked by ephemeral recharge/evaporation/upwelling (Andrews-Hanna et al. 2010).

The accumulated material in Gale Crater, currently being mapped and investigated by the Curiosity rover, too had (at least in part) an aqueous origin, containing remnant paleolake deposits that accumulated over ten thousand to millions of years during Mars' more clement conditions >3 Gyr ago (Grotzinger et al. 2015). Such lakes may have been capable of sustaining life if it existed on Mars during the Hesperian period, as late as 3.5 Gyr ago (Grotzinger et al. 2014, 2015; Hurowitz et al. 2017).

Curiosity

Curiosity (2012 to present) is the first rover to have successfully drilled into sediments (mudstones) in an aqueous environment on Mars interpreted to be a paleolake habitat. During MSL's pioneering primary mission in Gale Crater, two full-depth drill holes (John Klein and Cumberland, ~7 cm below surface) and one mini-drill (~ 2.5 cm below surface, John Klein) were completed in mudstones in the Sheepbed formation (Fig. 9.8, Yellowknife Bay, Gale Crater, Abbey et al. 2019). APXS and ChemCam data revealed abundant clay minerals likely formed in a low energy, aqueous freshwater (slightly neutral or alkaline) environment (lakes, playas) suitable for earth-like microbes (Grotzinger et al. 2014). Notably, the second Cumberland drill in the Sheepbed mudstones confirmed the first definitive detection of organic molecules on the surface of Mars (Freissinet et al. 2015).

By May 2020 (sol 2768), roughly 51 drills in mudstones had been acquired along the 22 km traverse, for a total of ~5.12 g of mudstone samples (A. Vasavada, pers. comm.). In total MSL SAM and CheMIN have analyzed **~9.64 g** of solid samples (53.1% mudstones, 24.3% sandstones; A. Vasavada, pers. comm.). From 13 samples (two at JK, seven from CB, and one each at CH, MJ, TP, and BK; through Buckskin in the ~3.5 Ga Murray formation, Fig. 9.8) analyzed successfully, 5 detected organic matter (>50 nmoles organic carbon in ~50 mg sample) from the near-surface ($\leq$ 7 cm) of Mars and raised the possibility of ancient heterotrophic microbial systems in the Gale Crater paleolake system (Freissinet et al. 2015; Sutter et al. 2017; Eigenbrode et al. 2018).

9.4.4 Salts

Evaporitic minerals have been associated with many geologic features on Mars, including sediment outcrops (Meridiani Planum, Valles Marineris), depressions, and impact craters (Nili Fossae, Mawrth Vallis, Terra Sirenum) (Poulet et al. 2005;

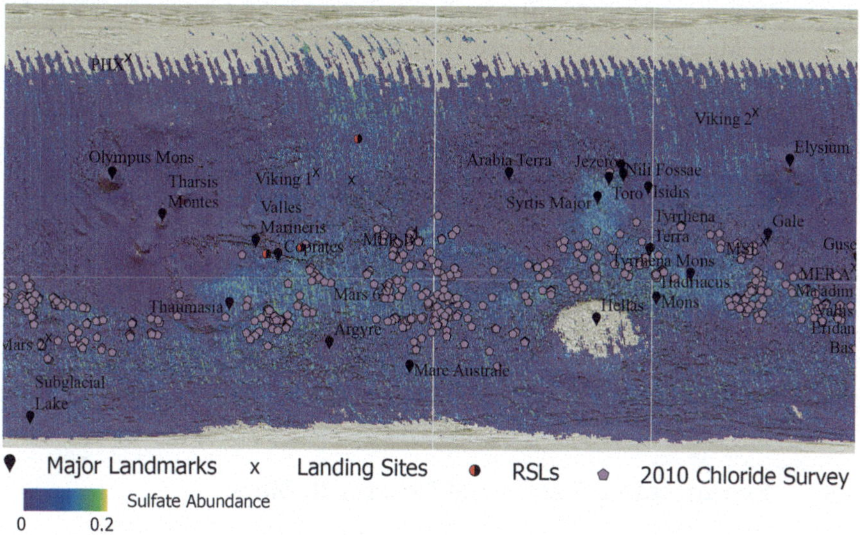

Fig. 9.10 Map of plausible salt habitats on Mars. Base map is global sulfate abundance (Bandfield 2002; Bandfield et al. 2003). Chloride deposits from Osterloo et al. (2010) mapped with lilac pentagons. Recurring Slope Lineae (RSL, McEwen et al. 2014; Ojha et al. 2014) are plotted with red and black dots. Major geologic features are indicated for reference, and mission landing sites are denoted with x's

Bibring et al. 2006; Mustard et al. 2008; Osterloo et al. 2008, 2010; Dohm et al. 2011; Ehlmann et al. 2011; Fairén et al. 2018; Sirisena et al. 2019).

Sulfates and magnesium, calcium and sodium perchlorates have been detected in the regolith by orbiters, landers (Phoenix), and/or rovers (Opportunity, Curiosity) (Squyres et al. 2008; Heet et al. 2009; Fisher et al. 2010; Yen et al. 2014). Calcium sulfate minerals are widespread and gypsum, basanite, and anhydrite were recently discovered to be abundant in Gale Crater along Curiosity's traverse (Vaniman et al. 2018). However, no signs of colonization or microfossils have been identified.

Chloride salts have also been detected on the surface of Mars, mainly in dark-toned Noachian units of the southern highlands and have been reported based on their distinctive featurelessness in thermal infrared (TIR) wavelengths (Fig. 9.10, Osterloo et al. 2008). Chloride deposits occur in local topographic lows and are most easily explained as evaporite deposits based on their geologic context, geomorphologic characteristics, spectral nature (at both TIR and VSWIR wavelengths), and chemical composition (Osterloo et al. 2008, 2010; Glotch et al. 2016). The deposits are typically small and discontinuous but are dispersed across most of the southern highlands of Mars, indicating that they were once part of a more globally expansive unit of which only scattered erosional remnants remain.

Chloride deposits are confined to Noachian and early Hesperian units, indicating they formed in the first ~1 Ga of Mars geologic history. Further geomorphological analysis suggests that chloride deposits may have undergone desiccation, evidenced by polygonal cracking, and may encase or mantle other sedimentary features like dunes (Osterloo et al. 2008, 2010). Spectral observations of chloride salts in the VSWIR by Murchie et al. (2009) and Wray et al. (2009) using the CRISM instrument onboard the Mars Reconnaissance Orbiter (MRO) support the interpretation of Osterloo et al. (2008) that the deposits are comprised of a chloride salt. El-Maarry et al. (2013) also suggest that many of these chloride-bearing features "have a lacustrine origin and a geologic setting similar to playas on Earth." The implications of ephemeral lakes, chloride salt pans and playas comprised of, for example, phyllosilicates and halite (similar to terrestrial analogs described above in the Atacama Desert) on the surface of Mars has clear astrobiological implications because chloride salts are postulated to be the final refuge for life in the Earth's most extreme arid-desert environments (Davila et al. 2010; Davila and Schulze-Makuch 2016). No direct physical samples of chloride salt crusts or nodules have been accessed by any landed or rover mission to Mars.

9.5 Extrapolating Terrestrial Deserts to Mars

One objective of this chapter has been to apply an ecological perspective to the mapping of potential habitats on Mars and to the analysis of two central questions in Astrobiology: (1) What is the potential for life on Mars? and (2) What are the lessons terrestrial desert ecology can impart to the robotic search for potential life and biosignatures on Mars?

McKay (2020) recently reviewed the most promising locations for life-detection missions based on five prerequisites: liquid water, carbon, biologically available nitrogen, energy and organic material. From potential candidates, he identified equatorial lake sediments and polar (salty) ice-cemented ground as two of the most promising targets on Mars. Our review of terrestrial end-member deserts concurs with the importance of these two habitats and raises additional implications for future robotic missions that search for biosignatures and life.

Evidence Suggests Extant Surface Life Is Untenable

A preponderance of data from hot and cold desert ecology in the last several decades points to a fundamental winnowing of the habitability of suitable substrates as temperature and water availability limits are approached on Earth. As evident from our review above, in wetter desert environments, complex microbial consortia live within a wide variety of habitats, but as dryness and/or cold intensify the overall trend is ultimately toward ecological extinction. This natural telescoping of habitability has several implications for Mars.

First, the data suggest that the likelihood of extant surface life on Mars is small and would require exhaustive sampling. In the driest and coldest environments in the Atacama Desert and Antarctic Dry Valleys, microbial communities exhibit the lowest biomass and activity and least diversity of any on Earth. Moreover, they are not uniformly distributed in space but occur in patchy, non-random spatial patterns with rare hotspots amidst large swaths of suitable yet uncolonized habitat (Fig. 9.4). Furthermore, such hotspots typically support a "community of survivors," scratching out an existence or in a state of dormancy, i.e., not actively metabolizing, growing, or reproducing (Wilhelm et al. 2018; Goordial et al. 2017). Based on the sparsest distributions yet mapped for terrestrial desert communities, maximum colonization likelihoods of roughly <<0.1%–1% of suitable habitat might be expected—the proverbial needle in a haystack. Mars, however, is ~100 times drier than the driest hyperarid core of the Atacama Desert, with a mean annual air temperature of −60 °C versus −30 °C for the coldest Antarctic Dry Valleys (Davila et al. 2019). These probabilities must be kept in mind for even the most promising targeted surface habitats on Mars.

Inventory of Samples on Mars

Over the past four decades of exploration on Mars by orbiters, landers and rovers, thousands of landscapes, outcrops, rocks, soils and sediments have been interrogated by remote or contact instruments (spectral, XRD/XRF, LIBS). However, during only three Mars missions (Viking, Phoenix, Curiosity) were discrete physical samples (drill, scoop) analyzed for life or organics detection. In total from these three missions, ~69 permafrost soil samples, ~51 mudstone, ~26 eolian sand/dune/drift, ~24 sandstone, and 1 heterolithic samples were completed (Table 9.2). These samples sum up to total of **~22.64 g** of soil and rock material. No conclusive signs of biosignatures for extant or extinct life have been found for any soil, sand/dune, heterolithic, or sandstone samples. To put this in methodological perspective, however, in the most extreme deserts on Earth, this total quantity of soil samples is generally what might be collected *for a single study at a single site*. Given the totals and individual aliquots to instruments of a *maximum* < 0.1 g, it is unlikely that detection (DNA, lipids) in terrestrial extreme deserts such as the Atacama or certain Antarctic Dry Valleys would have occurred.

Table 9.3 highlights some of the main techniques, targeted molecules, advantages and disadvantages (bias) and sample quantities required for detection in both the Atacama Desert and Dry Valleys, Antarctica. As Table 9.3 shows, low sample quantities can be problematic for unambiguous detection even in terrestrial settings, depending upon the substrate, technique, and signature analyzed. Well-known discrepancies between culture-independent and cultivation methods, for example, have led to one approach in the same desert region finding rare to no microbial presence versus a different approach revealing high-diversity soils populated by microbial "dark matter" biospheres (Hugenholtz et al. 1998; Idris et al. 2017). Regardless of technique, oligotrophic and extremely low biomass soils pose significant technical obstacles for the detection of life even in terrestrial settings. Low biomass soils in the Atacama Desert, for example, have presented challenges in extracting sufficient DNA, lipids, and ATP for analysis in samples that otherwise are known to contain cells. Extraction efficiencies are generally low with extracted compounds often at undetectable levels (or "signals at the level of the blanks," e.g., Skelley et al. 2007; Lynch et al. 2014).

Thus, it is enlightening to consider the ~51 mudstone samples (~5 g) analyzed by Curiosity SAM and ChemMin on Mars (Leshin et al. 2013; Freissinet et al. 2015; Eigenbrode et al. 2018; Stern et al. 2015; Sutter et al. 2017; Franz et al. 2020; Szopa

Table 9.3 Some standard terrestrial biological assessment techniques, targeted molecules, sample quantities, and methodology advantages and disadvantages

Target/Method	Atacama Hyperarid Core - surface	Dry Valleys Antarctica – surface	Advantages/Disadvantages for biosignature detection
16S rRNA gene-based qPCR (DNA sequencing)	0.5–1 g	0.5–15 g	**Advantages**: Very sensitive; ubiquitous **Disadvantages:** Complex sample preparation; earth-centric (universal 16S rRNA gene); unstable over geologic times even under extreme cold/dryness
Metagenomics-sequencing	15 g	2 g	**Advantages**: Community-wide genetic map; function/metabolic profiles; **Disadvantages**: Complex sample preparation; earth-centric; requires relatively large amounts of extracted DNA; unstable over geologic times even under extreme cold/dryness
ATP-fluorescence	6 g	100 mg	**Advantages**: Quantitative (biomass); fast, simple, robust **Disadvantages**: Earth-centric; questionable accuracy; unstable over geologic times even under extreme cold/dryness
Lipids-mass spectroscopy	50–200 g	100 g	**Advantages**: Generic, source-diagnostic (abiotic vs biotic); resistant to degradation over long geologic times (>10^8 years) **Disadvantages**: Complex sample preparation
Peptides-immunoassays	1–3 g	1–3 g	**Advantages**: Sensitive; ubiquitous; resistant to degradation over geologic times (>10^6 years) under extreme cold/dryness **Disadvantages**: Earth-centric; complex sample preparation
Amino acids-fluorescence/mass spectroscopy	1.5 g	0.4 g	**Advantages**: Generic, source-diagnostic (abiotic vs biotic); resistant to degradation over geologic times (>10^6–10^7 years) under extreme cold/dryness; note chirality is lost before amino acids are degraded **Disadvantages:** Complex sample preparation
Pigments-spectroscopy	n/a	n/a	**Advantages**: Non-contact **Disadvantages:** Earth-centric; unstable over geologic times even under extreme cold/dryness

Volcanic oligotrophic mineral soils required 60*0.25 g = 15 g for standard DNA extractions to achieve the quantity necessary (Lynch et al. 2014). Yungay alluvial duricrust soils shielded below rocks; exposed duricrust yielded signals at or below the signal of the blanks (Skelley et al. 2007). Goordial et al. 2017: for low biomass soils, 10 g of soil per extraction per sample (for community DNA 2 g of soil ×5 extractions pooled; for cryptoendoliths in sandstone, 100 mg crushed rock per sample; No RNA detectable after extraction for 2 g permafrost soils ×10 extractions per sample = 20 g per sample, pooled). PLFA, 50 g Yungay alluvial surface soils (Lester et al. 2007). Recovery of DNA ranged from 1.06–8.44 ng, with bio-rich valley floor soils in Antarctica (mean 229 ng DNA recovered/g soil) highly variable across the coastal valley floor (Archer et al. 2017). Antarctic Dry Valley soils had undetectable but quantifiable DNA in 0.75 g soil and ultra-low biomass high-altitude wind-blown sediments samples contained 0.02 ng/m^3 air averaged over all samples, with a total volume of air of 72 m^3 (Archer et al. 2017). Lipids 50–300 g Yungay alluvial surface soils (Wilhelm et al. 2017, 2018). Bound (polymeric) amino acids 3 g Yungay alluvial surface soils (Wilhelm et al. 2018). “Free” amino acids. For amino acids, Antarctica, 0.4 g soils were used for each of three replicates (Noell et al. 2018). For ATP, 100 mg soil per assay with replicates required for accurate values (Cowan et al. 2002; Cowan and Casanueva 2007) and 6 g in the Yungay hyperarid core of the Atacama Desert (Schulze-Makuch et al. 2018) et al. 2020). Notably, five of thirteen mudstone samples (28% of ~1.125 g total sample analyzed, CB-3, CB-5, CB-6, CH and MJ-2, Fig. 9.9), or roughly 38.5%, showed organic matter detections (pers. comm., B. Sutter; Freissinet et al. 2015; Eigenbrode et al. 2018; Szopa et al. 2020)—a higher success rate than might be predicted based on our review. The origin of the organic matter (geological, biological, or meteoritic) remains unknown, but these data may signal potentially significant gaps in the applicability of terrestrial colonization and preservation rates and/or the need for additional studies to quantify comparative numbers for terrestrial desert permafrost, sediments, salts, and lithic substrates, as well as those in other extreme environments, particularly aqueous ones (lakes, deltas, oceans, hot springs).

Second, terrestrial desert analogs demonstrate that only a restricted set of suitable substrates and geochemical environments support even the hardiest life under the harshest conditions. At the limits for life on Earth, only those desert habitats with *exactly* the right "Goldilocks" suite of physical features support microorganisms, with deliquescent salts as one last probable refugia. As Mars transformed from a wetter world to an increasingly cold and dry planet over its geological history, terrestrial extreme desert microbial ecology points to salt habitats such as halite or gypsum deposits as a likely repository for surface life. To date, no samples of such salt niches have been directly analyzed by life detection or organics sampling instruments. Salts thus should remain a top possible priority for exploration on Mars, in addition to other surface substrates (e.g., lacustrine sediments) for locating biosignatures (Davila and Schulze-Makuch 2016).

Subsurface Oases May Exist

Should surface habitats be eschewed as an unlikely proposition for biosignature or life-detection missions, our review next points to potential subsurface oases. Shielded below from harmful radiation, dry permafrost soils and pockets of ice, and possibly even subsurface brines and thin films, may exist or have existed in geologically recent times on Mars (Davila and Schulze-Makuch 2016; Goordial et al. 2017; Davila et al. 2019). In the coldest and driest deserts on Earth, life is often in a state of dormancy (cryptobiosis; "metabolocally inactive reversible rest period"). Yet, ample evidence demonstrates that even in these habitats, when stochastic events ease conditions even momentarily (e.g., ephemeral liquid water, warmer temperatures) metabolism and reproduction can reignite. It is conceivable that pockets of "sleeping beauty" microbial life survive in such habitable subsurface oases on Mars, frozen in a state of perpetual hibernation/stasis but ready to awake upon resuscitation after thousands to millions of years of slumber (Cano and Borucki 1995; Davila et al. 2019; Shen et al. 2021).

If liquid water (of suitable salinity) is available to subsurface life on Mars, other essential requirements, namely energy, key elements (CHNOPS), and trace nutrients (Na, Ca, Fe, etc.) could circumscribe their survival (McKay 2020). While non-desert terrestrial chemolithotrophic ecosystems, characterized by a continuous and sufficient source of water availability ("deep, dark energy biospheres") demonstrate microbial life can persist in the deep subsurface (caves, salt mines, hydrothermal vents, deep aquifers; Fisk and Giovannoni 1999; Chapelle et al. 2002; Colwell and

D'Hondt 2013; Farmer 2018), cold and exceedingly dry desert subsurface soils in both the Atacama Desert and the Dry Valleys point to a less favorable picture. Ramping up and down from habitable periods is costly in terms of energy, and to our knowledge, most extant "oasis" subsurface communities scavenge carbon to carry out these functions from either surface photosynthetic production (washed down by precipitation or warming/melting processes) and/or obtain carbon from relic sources. Chemolithotrophy alone thus may not be sufficient to cover the high energy costs needed to eke out a living in a hibernation model for life in exceedingly dry and/or icy permafrost subsurface settings where unfavorable conditions have persisted for thousands to millions, or even billions, of years. Certainly, this viewpoint is quite speculative, but it points to the need for new "hibernation/metabolic arrest" microbial ecosystem models and long-term study of both deep, dark, and dry permafrost and hyperarid rocks and soils. There, the focus is not on typical active microbial ecosystems and processes but rather on those with long dormancy periods and brief, punctuated periods of sporadic activity, microbial succession, and innovative metabolism and reproductive capabilities.

Terrestrial end-member deserts also shed valuable insights into the processes of xero- and/or cryo-preservation of microfossils and biosignatures (e.g., lipids and organic carbon degradation products such as kerogen). Quantitatively comparing the preservation of organic matter in various substrates in terrestrial deserts should be a key objective for astrobiology. Experiences from the Atacama core, in particular, suggest extreme desiccation may represent one of the best possible means to delay degradation of organic material and preserve biosignatures over prolonged time frames (Wilhelm et al. 2017). Data from the Atacama Desert (and altiplano for salts and high altitude hot springs) and Antarctic Dry Valleys also confirm that ice, clay, hydrated minerals, and evaporite minerals (e.g., opaline silica, phyllosilicates, carbonates, halides, sulfates, Wierzchos and Ascaso 2001, 2002; Ruff and Farmer 2016; Wilhelm et al. 2017) are all effective substrates for entombment/preservation, and a search for these materials in the "last outposts" of extremely dry and cold desert habitats on Mars (Davila and Schulze-Makuch 2016) is a continuing strategy for future Mars biosignature detection missions.

Microbial Spatial Pattern in Terrestrial Analog Environments as a Search Template

Finally, studies in the Earth's end-member deserts also highlight the important considerations of microbial succession and spatial pattern in the preservation and search for biosignatures. Whereas current assumptions—and recent positive results from Curiosity's mudstone samples—suggest that the highest priority analog habitats may be paleo-aquatic ecosystems (abundant biomass, fast burial, and high preservation potential), it is important to acknowledge the possibilities of overprinting by secondary colonizers in certain habitats (e.g., endoliths that may obliterate evidence of the original microbial inhabitants once hot springs or hypersaline lakes dry up) and non-random colonization and preservation patterns that may point to more effective targeted life detection approaches within these larger promising search spaces.

To that end, quantitatively mapping the microbial world on Earth with consistent and comparative techniques across all proposed high priority potential target habitats (e.g, Yingst et al. 2016, 2020) and at the scales most relevant to microbial spatial pattern is a key logistical step forward. Creating such microbial maps, and teasing out the factors that control these patterns, could furnish the signposts and templates for more effective search for habitable zones and life detection at the micro-scale (Farmer 2018). A quantitative recognition of how microorganisms—dead or alive—are distributed in space, interact with one another and their environment (i.e., "co-evolve"; e.g., Cabrol 2018) and are preserved temporally across multiple analog habitats could more rapidly narrow the proposed search space within priority target landscapes. In that vein, an understanding of the nested hierarchical patterns of terrestrial life may significantly improve the likelihood of detecting extant or ancient biosignatures across vast areas of uninhabited territory on other planets such as Mars.

Acknowledgments The authors gratefully appreciate access to the data and insights for the Curiosity samples at Gale Crater from Ashwin Vasavada, Brad Sutter, and David Blake. We also kindly thank Stephen B. Pointing, Stephen Archer, Mary Beth Wilhelm, Chris Greening, and Charles Cockell for sharing their technical expertise on various aspects of the chapter.

References

Abbey W, Anderson R, Beegle L, Hurowitz J, Williford K, Peters G, Morookian J, Collins C, Feldman J, Kinnett R, Jandura L, Limonadi D, Logan C, McCloskey S, Melko J, Okon A, Robinson M, Roumeliotis C, Seybold C, Singer J, Warner N (2019) A look back: the drilling campaign of the curiosity rover during the Mars science Laboratory's prime Mission. Icarus 319:1–13

Albarracín VH et al (2015) High-up: a remote reservoir of microbial extremophiles in central Andean wetlands. Front Microbiol 6:1404

Amend A, Cobian G, Laruson A, Remple K, Tucker S, Poff K, Antaky C, Boraks A, Jones C, Kuehu D, Lensing B, Pejhanmehr M, Richardson D, Riley P (2019) Phytobiomes are compositionally nested from the ground up. PeerJ 7:e6609. https://doi.org/10.7717/peerj.6609

Amundson R, Dietrich W, Bellugi D, Ewing S, Nishiizumi K, Chong G, Owen J, Finkel R, Heimsath A, Stewart B (2012) Geomorphic evidence for the late Pliocene onset of hyperaridity in the Atacama Desert. Geol Soc Am Bull 124:1048–1070

Andrews-Hanna J, Zuber M, Arvidson R, Wiseman S (2010) Early Mars hydrology: Meridiani playa deposits and the sedimentary record of Arabia Terra. JGR 115:E06002. https://doi.org/10.1029/2009JE003485

Archer S, de los Ríos A, Lee K, Niederberger T, Cary C, Coyne K, Douglas S, Lacap-Bugler D, Pointing S (2017) Endolithic microbial diversity in sandstone and granite from the McMurdo dry valleys, Antarctica. Polar Biol 40:997–1006. https://doi.org/10.1007/s00300-016-2024-9

Artieda O et al (2015) Surface evolution of salt-encrusted playas under extreme and continued dryness. Earth Surf Process Landf 40(14):1939–1950

Arvidson RE, Gooding JL, Moore H (1989) The Martian surface as imaged, sampled, and analyzed by the Viking landers. Rev Geophys 27:39–60. https://doi.org/10.1029/RG027i001p00039

Arvidson R, Bonitz R, Robinson M, Carsten J, Volpe R, Trebi-Ollennu A, Mellon M, Chu P, Davis K, Wilson J, Shaw A, Greenbergre R, Siebach K, Stein T, Cull S, Goetz W, Morris R, Ming D, Keller H, Lemmon M, Sizemore H, Mehta M (2009) Results form the Mars Phoenix Lander robotic arm experiment. JGR 114:E00#02. https://doi.org/10.1029/2009JE003408

Azúa-Bustos A, Gonzálaz-Silva C, Mancilla RA, Salas L, Gómez-Silva B, McKay CP, Vicuña R (2011) Hypolithic cyanobacteria supported mainly by fog in the coastal range of the Atacama Desert. Micro Ecol 61:568–581. https://doi.org/10.1007/s00248-010-9784-5

Azúa-Bustos A, González-Silva C, Corsini G (2017) The Hyperarid Core of the Atacama Desert, an extremely dry and carbon deprived habitat of potential interest for the field of carbon science. Front Microbiol 8:993. https://doi.org/10.3389/fmicb.2017.00993

Azúa-Bustos A, Fairén AG, González-Silva C, Ascaso C, Carrizo D, Fernández-Martínez MA, García-Descalzo L, García-Villadangos M, Martin-Redondo M, Sánchez-García L, Wierzchose J, Parro V (2018) Unprecedented rains decimate surface microbial communities in the hyperarid core of the Atacama Desert. Sci Rep 8:16706. https://doi.org/10.1038/s41598-018-35051-w

Bagaley D (2006) Uncovering bacterial diversity on and below the surface of a hyper-arid environment, the Atacama Desert, Chile. Thesis, Louisana State University, Baton Rouge, 156 pp

Bahl J, Lau M, Smith G, Vijaykrishna D, Cary SC, Lacap D (2011) Ancient origins determine global biogeography of hot and cold desert cyanobacteria. Nat Commun 2:163. https://doi.org/10.1038/ncomms1167

Baldridge A, Farmer J, Moersch J (2004) Mars remote sensing analog studies in the Badwater Basin, Death Valley, California. JGR 109:E12006. https://doi.org/10.1029/2004JE002315

Bandfield JL (2002) Global mineral distributions on Mars. J Geophys Res Planets 107(E6):9–1

Bandfield JL, Glotch TD, Christensen PR (2003) Spectroscopic identification of carbonate minerals in the Martian dust. Science 301:1084–1087

Barrett J, Virginia R, Wall D, Parsons A, Powers L, Burkins M (2004) Variation in biogeochemistry and soil biodiversity across spatial scales in a polar desert ecosystem. Ecology 85(11): 3105–3118

Barrett J, Ball B, Simmons B (2009) Standard procedures for soil research in the McMurdo dry valleys LTER, pp. 1–32

Barrow C (1992) In: Middleton N, Thomas D (eds) World atlas of desertification, vol 3. Edward Arnold, London, p 249

Baumman K, Jung P, Samolov E, Lehnert L, Büdel B, Karsten U, Bendix J, Achilles S, Schermer M, Matus F, Oses R, Osses P, Morshedizad M, Oehlschläger C, Hu Y, Klysubun W, Leinweber P (2018) Biological soil crusts along a climatic gradient in Chile: richness and imprints of phototrophic microorganism in phosphorus biogeochemical cycling. Soil Biol Biochem 127:286–300. https://doi.org/10.1016/j.soilbio.2018.09.035

Belilla J, Moreira D, Jardillier L, Reboul G, Benzerara K, López-García JM, Bertolino P, López-Archilla AI, López-García P (2019) Hyperdiverse archaea near life limits at the polyextreme geothermal Dallol area. Nat Ecol Evol. https://doi.org/10.1038/s41559-019-1005-0

Belnap J (2006) The potential roles of biological soil crusts in dryland hydrologic cycles. Hydrol Process 20:3159–3178. https://doi.org/10.1002/hyp.6325

Belnap J, Gillette D (1998) Vulnerability of desert biological soil crusts to wind erosion: the influences of crust development, soil texture and disturbance. J Arid Environ 39:133–142

Belnap J, Harper K, Warren S (1994) Surface disturbance of cryptobiotic soil crusts: Nitrogenase activity, chlorophyll content and chlorophyll degradation. Arid Soil Res Rehab 8(1):1–8. https://doi.org/10.1080/15324989309381373

Belnap J, Hawkes C, Firestone M (2003a) Boundaries in miniature: two examples from soil. Bioscience 53(8):739–749

Belnap J, Büdel B, Lange O (2003b) In: Belnap J, Lange OL (eds) Biological soils crusts: characteristics and distribution. Biological soil crusts: structure, function and management. Spring-Verlap, Berlin, pp 3–31

Belnap J, Welter J, Grimm N, Barger N, Ludwig J (2005) Linkages between microbial and hydrological processes in arid and semiarid watersheds. Ecology 86(2):298–307

Berner T, Evenari M (1978) The influence of temperature and light penetration on the abundance of the hypolithic algae in the Negev Desert of Israel. Oecologia 33:255–260

Bezuidt O, Lebre P, Pierneef R, León-Sobrino C, Adriaenssens E, Cowan D, Van de Peer Y, Makhalanyane T (2020) Phages actively challenge nice communities in Antarctic soils. mSystems 5:e00234-20. https://doi.org/10.1128/mSystems.00234-20

Bibring J, Langevin Y, Mustard J, Poulet F, Arvidson R, Gendrin A, Gondet B, Mangold N, Pinet P, Forget F, Omega team (2006) Global mineralogical and aqueous Mars history derived from OMEGA/Mars express data. Science 312:400–404

Billi D, Friedmann EI, Hofer KG, Caiola MG, Ocampo-Friedmann R (2000) Ionizing-radiation resistance in the desiccation-tolerant cyanobacterium Chroococcidiopsis. Appl Environ Microbiol 66:1489–1492. https://doi.org/10.1128/aem.66.4.1489-1492.2000

Billi D, Viaggiu E, Cockell CS, Rabbow E, Horneck G, Onofri S (2011) Damage escape and repair in dried Chroococcidiopsis spp. from hot and cold deserts exposed to simulated space and Martian conditions. Astrobiology 11:65–73. https://doi.org/10.1089/ast.2009.0430

Black RF (1976) Periglacial features indicative of permafrost: ice and soil wedges. Quat Res 6:3–26

Bockheim J (1997) Properties and classification of cold desert soils from Antarctica. J Soil Sci Soc America 61:224–231

Bockheim J (2002) Landform and soil development in the McMurdo dry valleys, Antarctica: a regional synthesis. Arct Antarct Alp Res 34:308–317

Bockheim J, Haus N (2014) Distribution of organic carbon in the soils of Antarctica. In: Hartemink A, McSweeney K (eds) Soil carbon, Progress in soil science. Springer, Cham, pp 373–380

Bowker M, Belnap J, Daivdson D, Phillips S (2005) Evidence for micronutrient limitation of biological soil crusts: important to arid-lands restoration. Ecol Appl 15(6):1941–1951

Bowker M, Belnap J, Davidson D, Goldstein H (2006) Correlates of biological soil crust abundance across a continuum of spatial scales: support for a hierarchical conceptual model. J Appl Ecol 43:152–163

Bowler J, Qi H, Kezao C, Head M, Baoyin Y (1986) Radiocarbon dating of playa-lake hydrologic changes: examples from northwestern China and Central Australia. Palaeogeogr Palaeoclimatol Palaeoecol 54:241–260. https://doi.org/10.1016/0031-0182(86)90127-6

Bowman JS et al (2016) Microbial community dynamics in two polar extremes: the lakes of the McMurdo Dry Valleys and the West Antarctic Peninsula marine ecosystem. Bioscience 66 (10):829–847

Brady A, Slater G, Gibbons E, Kobs Nawotniak S, Hughes S, Payler S, Stevens A, Cockell C, Haberle C, Sehlke A, Elphic R, Lim D (2017). Detection of microbial organic biomarkers in terrestrial basalts in a Mars analogue environment. Astrobiological Science Conference, Arizona; #3492. LPI Contrib. No 1965

Briere PR (2000) Playa, playa lake, sabkha: proposed definitions for old terms. J Arid Environ 45: 1–7. https://doi.org/10.1006/jare.2000.0633

Broady P, Kibblewhite A (1991) Morphological characterization of Oscillatoriales (cyanobacteria) from Ross Island and southern Victoria land. Antarct Antarct Sci 3(1):35–45. https://doi.org/10.1017/S095410209100007X

Brown A, Viviano C, Goudge T (2019) Olivine-carbonate mineralogy of the Jezero crater region. JGR Planets 125:e2019JE006011. https://doi.org/10.1029/2019JE006011

Büdel B, Wessels DCJ (1991) Rock inhabiting blue-green algae/cyanobacteria from hot arid regions. Algological Studies/Archiv für Hydrobiologie, Supplement Volumes:385–398

Büdel B, Bendix J, Bicker FR, Allan Green TG (2008) Dewfall as a water source frequently activates the endolithic cyanobacterial communities in the granites of Taylor Valley, Antarctica. J Phycol 44:1415–1424

Büdel B, Darienko T, Deutschewitz K, Dojani S, Friedl T, Mohr K (2009) Southern African biological soil crusts are ubiquitous and highly diverse in drylands, being restricted by rainfall frequency. Microbiol Ecol 57:229–247. https://doi.org/10.1007/s00248-008-9449-9

Cabrol N (2018) The coevoluation of life and environment on Mars: an ecosystem perspective on the robotic exploration of biosignatures. Astrobiology 18(1):1–27. https://doi.org/10.1089/ast.2017.1756

Cabrol N, Grin E (1999) Distribution, classification and ages of Martian impact crater lakes. Icarus 142:160–172

Cabrol N, Grin E (2001) The evolution of lacustrine environments on Mars: is Mars only hydrologically dormant? Icarus 149:291–328

Cabrol N, Grin E (2003) Overview on the formation of paleolakes and ponds on Mars. Glob Planet Chang 35(3–4):199–219. https://doi.org/10.1016/50921-8181(02)00127-3

Cahill T, Gill T, Reid J, Gearhart E, Gillette D (1996) Saltating particles, playa crusts and dust aerosols at Owens (dry) Lake, California. Earth Surf Process Landf 21(7):621–639

Cámara B, Suzuki S, Nealson K, Wierzchos J, Ascaso C, Artieda O, de los Ríos A (2014) Ignimbrite textural properties as determinants of endolithic colonization patterns from hyper-arid Atacama Desert. Int Microbiol 17:235–247. https://doi.org/10.2436/20.1501.01.226

Cameron RE, Blank GB (1965) Soil studies-microflora of desert regions. VIII. Distribution and abundance of desert microflora. Space Programs Summary 37–34 4:193–202

Cano RJ, Borucki MK (1995) Revival and identification of bacterial spores in 25-to 40-million-year-old Dominican amber. Science 268(5213):1060–1064

Caruso T, Chan Y, Lacap D, Lau M, McKay CP, Pointing SB (2011) Stochastic and deterministic processes interact in the assembly of desert microbial communities on a global scale. ISME J 5: 1406–1413. https://doi.org/10.1038/ismej.2011.21

Carter J, Poulet F, Bibring JP, Mangold N, Murchie S (2013) Hydrous minerals on Mars as seen by the CRISM and OMEGA imaging spectrometers: updated global view. J Geophys Res Planets 118:831–858

Cary SC et al (2010) On the rocks: the microbiology of Antarctic Dry Valley soils. Nat Rev Microbiol 8(2):129–138

Cereceda P, Larrain H, Osses P, Farías M, Egaña I (2008) The climate of the coast and fog zone in the Tarapacá region, Atacama Desert, Chile. Atmos Res 87:301–311

Certini G, Ugolini FC (2013) An updated, expanded, universal definition of soil. Geoderma 192: 378–379

Chan Y, Lacap DC, Lau MCY, Ha KY, Warren-Rhodes KA, Cockell CS et al (2012) Hypolithic microbial communities: between a rock and a hard place. Environ Microbiol 14:2272–2282. https://doi.org/10.1111/j.1462-2920.2012.02821

Chapelle F, O'Neill K, Bradley P, Methé B, Clufo S, Knobel L, Lovely D (2002) A hydrogen-based subsurface microbial community dominated by methanogens. Nature 415:312–315

Chong G (1984) Die Salare in Nordchile: Geologie, Struktur und Geochimie. Goetekt Forsch 67:1–146

Chong G (1988) The Cenozoic saline deposits of the Chilean Andes between18° and 27° south. In: Bahlburg H, Breitkreuz C, Giese P (eds) The southern Central Andes. Springer-Verlag, Heidelberg, pp 135–151

Christensen P, Bandfield J, Hamilton V, Ruff S, Kieffer H, Titus T, Malin M, Morris R, Lane M, Clrk R, Jakosky B, Mellon M, Pearl J, Conrath B, Smith M, Clancy R, Kuzmin R, Roush T, Mehall G, Gorelick N, Bender K, Murray K, Dason S, Greene E, Silverman S, Grenfield M (2001) Mars Global Surveyor Thermal Emission Spectrometer experiment: Investigation description and surface science results. J Geophys Res 106(E10):28823–23871. https://doi.org/10.1029/2000JE001370.4010.1002

Cockell C, Stokes D (2006) Hypolithic colonization of opaque rocks in the Arctic and Antarctic polar desert. Arct Antarct Alp Res 38(3):335–342

Cockell C, McKay CP, Warren-Rhodes KA, Horneck G (2008) Ultraviolet radiation-induced limitation to epilithic microbial growth in arid deserts- Dosimetric experiments in the hyperarid core of the Atacama Desert. J Photochem Photobiol 90:79–87

Cockell C, Balme M, Bridges J, Davila A, Schwenzer S (2011) Uninhabited Habitats on Mars. Icarus. https://doi.org/10.1016/j.icarus.2011.10.025

Cockell C, Wilhelm MB, Perl S, Wadsworth J, Payler S, Paling S, Edwards T, McMahon S (2020) 0.25 Ga salt deposits preserve geological signatures of habitable conditions and ancient lipids. Astrobiology 20(7):864–877. https://doi.org/10.1089/ast.2019.2053

Colwell F, D'Hondt S (2013) Nature and extent of the deep biosphere. Rev Mineral Geochem 75: 547–574. https://doi.org/10.2138/rmg.2013.75.17
Cowan D, Casanueva A (2007) Stability of ATP in Antarctic desert soils. Polar Biol 30:1599–1603
Cowan D, Russell N, Mamais A, Sheppard D (2002) Antarctic Dry Valley mineral soils contain unexpectedly high levels of microbial biomass. Extremophiles 6:431–436
Cowan DA, Khan N, Pointing S, Cary SC (2010) Diverse hypolithic refuge communities in Antarctic dry valleys. Antarct Sci 22:714–720
Cowan DA, Pointing SB, Stevens MI, Cary SC, Stomeo F, Tuffin IM (2011b) Distribution and abiotic influences on hypolithic communities in an Antarctic Dry Valley. Polar Biol 34:307–311
Crits-Christoph A, Robinson C, Barnum T, Fricke F, Davila A, Jedynak B, McKay C, DiRuggiero J (2013) Colonization patterns of soil microbial communities in the Atacama Desert. Microbiome 1:28. https://doi.org/10.1186/2049-2618-1-28
Davila A, Schulze-Makuch D (2016) The last possible outposts for life on Mars. Astrobiology 16: 159–168
Davila AF, Gómez-Silva B, de los Rios A (2008a) Facilitation of endolithic microbial survival in the Hyperarid Core of the Atacama Desert by mineral deliquescence. J Geophys Res 113: G01028
Davila A, Gómez-Silva B, de los Ríos A, Ascaso C, Olivares H, McKay C, Wierzchos J (2008b) Facilitation of endolithic microbial survival in the hyperarid core of the Atacama Desert by mineral deliquescence. J Geophys Res 113:G01028. https://doi.org/10.1029/2007JG000561
Davila AF, Duport L, Melchiorri R, Jänchen J, Valea S, de los Rios A, Fairén A, Möhlmann D, McKay C, Ascaso C, Wierzchos J (2010) Hygroscopic salts and the potential for life on Mars. Astrobiology 10:617–628
Davila A, Hawes I, Ascaso C, Wierzchos J (2013) Salt deliquescence drives photosynthesis in the hyperarid Atacama Desert. Environ Microbiol Rep 5:583–587. https://doi.org/10.1111/1758-2229.12050
Davila A, Kahre M, Quinn R, Des Marais D (2019) The biological potential of present day Mars. In: Meadows V et al (eds) Planetary astrobiology. Univ. of Arizona, Tucson
Davis W, de Pater I, McKay CP (2010) Rain infiltration and crust formation in the extreme environment arid zone of the Atacama Desert. Chile Planet Space Sci 58(4):616–622
de los Ríos A, Wierzchos J, Sancho L, Green T, Ascaso C (2005) Ecology of endolithic lichens colonizing granite in continental Antarctica. Lichenologist 37:383
de los Ríos A, Valea S, Ascaso C, Davila A, Kastovsky J, CP MK, Gómez-Silva B, Wierzchos J (2010) Comparative analysis of the microbial communities inhabiting halite evaporites of the Atacama Desert. Int Microbiol 13:79–89. https://doi.org/10.2436/20.1501.01.113
de los Ríos A, Wierzchos J, Ascaso C (2014) The lithic microbial ecosystems of Antarctica's McMurdo dry valleys. Antarct Sci 26:459–477
Demergasso C, Casamayor E, Chong G, Galleguillos P, Escudero L, Pedros-Alio C (2004) Distribution of prokaryotic genetic diversity in athalssohaline lakes of the Atacama Desert, northern Chile. FEMS Microbiol Ecol 48:57–69
Demergasso C, Dorador C, Menese D, Blamey J, Cabrol N, Escudero L, Chong G (2010) Prokaryotic diversity pattern in high-altitude ecosystems of the Chilean Altiplano. J Geophys Res Biogeosci 115:G2
Di Achille G, Hynek BM (2010) Ancient ocean on Mars supported by global distribution of deltas and valleys. Nat Geosci 3(7):459–463
DiRuggiero J, Wierzchos J, Robinson C, Souterre T, Ravel J, Artieda O (2013) Microbial colonization of chasmoendolithic habitats in the hyper-arid zone of the Atacama Desert. Biogeosciences 10:1171–1179. https://doi.org/10.1038/s41396-018-0049-5
Dohm J, Miyamoto H, Ori G, Fairén A, Davilia A, Komatsu G, Mahaney W, Illiams J, Joye S, Achille G, Oehler D, Marzo G, Schluze-Makuch D, Acocella V, Glmoclija M, Pondrelli M, Boston P, Hart K, Anderson R, Baker V, Fink W, Kelleher B, Furfaro R, Gross C, Hare T, Frazer A, Ip E, Allen C, Kim K, Muruyama S, McGuire P, Netoff D, Parnell J, Wendt I, Wheelock S, Steele A, Hancock R, Havics R, Costa P, Krinsely D (2011) An inventory of

potentially hapitable environments on Mars: geological and biological perspectives. Geol Soc Am Spec Pap 483:317–347

Dong H, Rech J, Jiang H, Sun H, Buck B (2007) Endolithic cyanobacteria in soil gypsum: occurrences in Atacama (Chile), Mojave (United States) and Al-Jafr Basin (Jordan) deserts. JGR Biogeosci 112:G2. https://doi.org/10.1029/2006JG000385

Dorador C, Vila I, Remonsellez F, Imhoff JF, Witzel KP (2010) Unique clusters of Archaea in Salar de Huasco, an athalassohaline evaporitic basin of the Chilean Altiplano. FEMS Microbiol Ecol 73(2):291–302

Droste J (1961) Clay mineral composition of sediments in some desert lakes in Nevada, California, and Oregon. Science 133:1928

Ehlmann B, Mustard J, Murchie S, Bibring JP, Meunier A, Fraeman A, Langevin Y (2011) Subsurface water and clay mineral formation during the early history of Mars. Nature 479: 53–60

Eigenbrode JL, Summons RE, Steele A, Freissinet C, Millan M, Navarro-González R, Sutter B, McAdam AC, Franz HB, Glavin DP, Archer PD, Mahaffy PR, Conrad PG, Hurowitz JA, Grotzinger JP, Gupta S, Ming DW, Sumner DY, Szopa C et al (2018) Organic matter preserved in 3-billion-year-old mudstones at Gale crater. Science, Mars. https://doi.org/10.1126/science.aas9185

Ekwealor J, Fisher K (2020) Life under quartz: Hypolithic mosses in the Mojave Desert. PLoS One 15(7):e0235928. https://doi.org/10.1371/journal.pone.0235928

Elbert W, Weber B, Burrows S, Steinkamp J, Büdel B, Andreae M, Pöschl U (2012) Contribution of cryptogamic covers to the global cycles of carbon and nitrogen. Nat Geosci 5:459–462

Eldridge DJ (1996) Distribution and floristics of terricolous lichens in soil crusts in arid and semi-arid New South Wales, Australia. Austr J Bot 44:581–599

Eldridge DJ, Greene RSB (1994) Microbiotic soil crusts: a review of their roles in soil and ecological processes in the rangelands of Australia. Aust J Soil Res 32:389–415. https://doi.org/10.1071/SR9940389

Eldridge DJ, Koen TB (1998) Cover and floristics of microphytic soil crusts in relation to indices of landscape health. Plant Ecol 137:1010–1114

Eldridge DJ, Zaady E, Shachak M (2000) Infiltration through three contrasting biological soil crusts in patterned landscapes in the Negev, Israel. Catena 40:323–336. https://doi.org/10.1016/S0341-8162(00)00082-5

El-Maarry M, Pommerol A, Thomas N (2013) Analysis of polygonal cracking patterns in chloride-bearing terrains on Mars: indicators of ancient playa settings. JGR Planets 118:2263–2278. https://doi.org/10.1002/2013JE004463

Ericksen G, Salas R (1990) Geology and resources of salars in the Central Andes. In: Ericksen GE, Cañas MT, Reinemund JA (eds) Geology of the Andes and its relation to hydrocarbon and mineral resources, Earth science series, vol 11. Circum-Pacific Council for Energy and Mineral Resources, Houston, Texas, pp 165–172

Escudero L, Bijman J, Mariela G, Pueyo Mur J, Chong G, Demergasso C (2015) Organotrophic and mixotrophic sulfur oxidation in an acidic salt flat in northern chile. Adv Mater 1130:63–66. https://doi.org/10.4028/www.scientific.net/amr.1130.63

Espisito A, Ahmed E, Ciccazzo S, Sikorski J, Overmann J, Holmström S, Brusetti L (2015) Comparison of rock varnish bacterial communities with surrounding non-varnished rock surfaces: taxon-specific analysis and morphological description. Microbial Ecology 70:741–750. https://doi.org/10.1007/s00248-015-0617-4

Ettema C, Wardle D (2002) Spatial soil ecology. Trends Ecol Evol 17:177–183. https://doi.org/10.1016/S0169-5347(02)02496-5

Ewing S, Sutter B, Owen J, Nishizumi K, Sharp W, Cliff S, Perry K, Dietrich B, McKay C, Amundson R (2006) A threshold in soil formation at Earth's arid–hyperarid transition. Geochim Cosmochim Acta 70(21):5293–5322. https://doi.org/10.1016/j.gca.2006.08.020

Ewing S, Southard R, Macalady J, Hartshorn A, Johnson M (2007) Soil microbial fingerprints, carbon and nitrogen in a Mojave Desert creosote-bush ecosystem. Soil Sci Soc Am J 71(2): 469–475

Ewing S, Macalady J, Warren-Rhodes K, McKay C, Amundson R (2008) Changes in the soil C cycle at the arid-hyperarid transition in the Atacama Desert. J Geophys Res 113(G2):G02S90. https://doi.org/10.1029/2007JG000495

Fairén A, Parro V, Schulze-Makuch D, Whyte L (2018) Is searching for Martian life a priority for the Mars community? Astrobiol 18(2):101–107. https://doi.org/10.1089/ast.2017.1772

Farias ME, Rascovan N, Toneatti DM, Albarracin VH, Flores MR, Poire DG, Collavino MM, Aguilar OM, Vazquez MP, Polerecky L (2013) The discovery of stromatolites developing at 3570 m above sea level in a high-altitude volcanic lake Socompa, Argentinean Andes. PLoS One 8(1):e53497. https://doi.org/10.1371/journal.pone.0053497

Farias ME, Contreras M, Rasuk M, Kurth D, Flores M, Poiré D, Novoa F, Visscher P (2014) Characterization of bacterial diversity associated with microbial mats, gypsum evaporites and carbonate microbialites in thalassic wetlands: Tebenquiche and La Brava, Salar de Atacama, Chile. Extremophiles 18:311–329. https://doi.org/10.1007/s00792-013-0617-6

Farmer J (2018) Habitability as a tool in astrobiological explortion. In: From habitability to life on Mars. Elsevier Press, Cambridge, pp 1–12. https://doi.org/10.1016/B978-0-12-809935-3.00002-5

Fassett CI, Head JW III (2007) Layered mantling deposits in northeast Arabia Terra, Mars: Noachian-Hesperian sedimentation, erosion, and terrain inversion. J Geophys Res Planets 112 (E8)

Fassett C, Dickson J, Head J, Levy J, Marchant D (2010) Supraglacial and proglacial valleys on Amazonian Mars. Icarus 208(1):86–100. https://doi.org/10.1016/j.icarus.2010.0.021

Fernández-Martínez MA, Severino R, Moreno-Paz M, Gallardo-Carreño I, Blanco Y, Warren-Rhodes K, García-Villadangos M, Ruiz-Bermejo M, Barberán A, Wettergreen D, Cabrol N, Parro V (2019) Prokaryotic community structure and metabolisms in shallow subsurface of Atacama Desert playas and alluvial fans after heavy rains: repairing and preparing for next dry period. Front Microbiol 10:1641. https://doi.org/10.3389/fmicb.2019.01641

Finstad K, Pfeiffer M, Amundson R (2014) Hyperarid soils and the soil taxonomy. Soil Sci Soc America J 78(6):1845. https://doi.org/10.2136/sssaj2014.06.0247

Finstad K, Pfeiffer M, McNicol G, Barnes J, Demergasso C, Chong G, Amundson R (2016) Rates and geochemical processes of soil and salt crust formation in Salars of the Atacama Desert, Chile. Geoderma 284:57–72

Finstad K, Probst A, Thomas B, Anderson G, Demergasso C, Echeverria A, Amundson R, Banfield J (2017) Microbial community structure and the persistence of cyanobacterial populations in salt crusts of the hyperarid atacama desert from genome-resolved metagenomics. Front Microbiol 8. https://doi.org/10.3389/fmicb.2017.01435

Fisher DA, Hecht MH, Kounaves SP, Catling DC (2010) A perchlorate brine lubricated deformable bed facilitating flow of the north polar cap of Mars: possible mechanism for water table recharging. J Geophys Res Planets 115(E2)

Fisk M, Giovannoni S (1999) Sources of nutrients and energy for a deep biosphere on Mars. JGR 104(E5):11805–11815

Forsythe RD, Blackwelder CR (1998) Noachian Martian salars (closed drainage salt pans) and sabkhas as lander/sample return mission target. In: Mars Surveyor 2001 Landing Site Workshop, pp 76–77

Franz H, Mahaffy P, Webster C, Flesch G, Raaen E, Freissinet C, Atreya S, House C, McAdam A, Knudson C, Archer P, Stern J, Steele A, Sutter B, Eigenbrode J, Glavin D, Lewis J, Malespin C, Millan M, Ming D, Navarro-González R, Summons R (2020) Indigenous and exogenous organics and surface-atmosphere cycling inferred from carbon and oxygen isotopes at Gale crater. Nature Astronomy 4:526–532. https://doi.org/10.1038/s41550-019-0990-x

Freissinet C, Glavin D, Mahaffy P, Miller K, Eigenbrode J, Summons R, Brunner A, Buch A, Szopa C, Archer P, Franz H, Atreya S, Brinkerhoff W, Cabane M, Coll P, Conrad P, Des

Marais D, Dworkin J, Fairén A, Francois P, Grotzinger J, Kashyap S, ten Kate I, Leshin L, Malespin C, Martin M, Martin-Torres F, McAam A, Ming D, Navarro-González R, Pavlov A, Prats B, Squyres S, Steele A, Stern J, Sumner D, Sutter B, Zorzano M, MSL Science Team (2015) Organic molecules in the Sheepbed Mudstone, Gale Crater. Mars JGR 120(3):495–514. https://doi.org/10.1002/2014JE004737
Friedmann I (1980) Endolithic microbial life in hot and cold deserts. Orig Life 10:223–235. https://doi.org/10.1007/bf00928400
Friedmann EI (1982) Endolithic microorganisms in the antarctic cold desert. Science 215(4536): 1045–1053
Friedmann EI, Galun M (1974) Desert algae, lichens and fungi. In: Brown G (ed) Desert Biology. Academic Press, New York, pp 166–204
Friedmann EI, McKay CP (1985) A method for continuous monitoring of snow: application to the cryptoendolithic microbial community of Antarctica. Antarct J US 20:179–181
Friedmann I, Lipkin Y, Ocampo-Paus R (1967) Desert algae of the Negev (Israel). Phycologia 6: 185–200. https://doi.org/10.2216/i0031-8884-6-4-185.1
Friedmann EI, McKay CP, Nienow JA (1987) The cryptoendolithic microbial environment in the Ross Desert of Antarctica: satellite-transmitted continuous nanoclimate data, 1984 to 1986. Polar Biol 7(5):273–287
Gao Q, Garcia-Pichel F (2011) Microbial ultraviolet sunscreens. Nat Rev Microbiol 9:791–802
Garcia-Pichel F, Belnap J (1996) Microenvironments and microscale productivity of cyanobacterial desert crusts. J Phycol 32:774–782
Garcia-Pichel F, Belnap J (2003) Small-scale environments and distribution of biological soil crusts. In: Belnap J, Lange OL (eds) Biological soil crusts: structure, function, and management, vol 2. Springer, Berlin, pp 193–202
Garcia-Pichel F, Loza V, Marusenko Y, Mateo P, Potrafka R (2014) Temperature drives the continental-scale distribution of microbes in topsoil communities. Science 340(6140): 1574–1577. https://doi.org/10.1126/science.1236404
Garrido-Benavent I, Pérez-Ortega S, Durán, Ascaso C, Pointing S, Rodriguez-Cielos R, Navarro F, de los Ríos A (2020) Differential colonization and succession of microbial communities in rock and soil substrates on a maritime Antarctic glacier forefield. Front Microbiol. https://doi.org/10.3389/fmicb.2020.00126
Glotch TD, Bandfield JL, Wolff MJ, Arnold JA, Che C (2016) Constraints on the composition and particle size of chloride salt-bearing deposits on Mars. J Geophys Res: Planets 121(3):454–471. https://doi.org/10.1002/2015JE004921
Goldstein H, Breit G, Reynolds R (2017) Controls on the chemical composition of saline surface crusts and emitted dust from a wet playa in the Mojave Desert (USA). J Arid Environ 140:50–66. https://doi.org/10.1016/j.jaridenv.2017.01.010
Golubic S, Friedmann I, Schneider J (1981) The lithobiontic ecological niche, with special reference to microorganisms. J Sediment Petrol 51:475–478
Gómez-Silva B (2018) Lithobiontic life: "Atacama rocks are well and alive". Antonie van Leeuwenhoek. J Microbiol 102(1). https://doi.org/10.1007/s10482-018-1033-9
Goordial J, Davila A, Lacelle D, Lacelle D, Pollard W, Marinova M, Greer C, DiRuggiero J, McKay CP, Whyte LG (2016) Nearing the cold-arid limits of microbial life in permafrost of an upper dry valley, Antarctica. ISME J 10:1613–1624. https://doi.org/10.1038/ismej.2015.239
Goordial J, Davila A, Greer C, Cannam R, DiRuggerio J, McKay CP, Whyte LG (2017) Comparative activity and functional ecology of permafrost soils and lithic niches in a hyper-arid polar desert. Environ Microbiol 19:443–458. https://doi.org/10.1111/1462-2920.13353
Goudge T, Mustard J, Head J, Fassett C, Wiseman S (2015) Assessing the mineralogy of the watershed and fan deposits of the Jezero crater paleolake system, Mars. JGR Planets 120:775–808. https://doi.org/10.1002/2014JE004782
Gramain A, Díaz GC, Demergasso C, Lowestein T, McGenity T (2011) Archaeal diversity along a subterranean salt core from the Salar Grande (Chile). Environ Microbiol 13:2105–2121

Grant JA, Parker TJ (2002) Drainage evolution in the Margaritifer Sinus region, Mars. J Geophys Res Planets 107:4–1. https://doi.org/10.1029/2001JE001678

Griffin D (2004) Dust storms and their impact on ocean and human health: dust in Earth's Atmosphere. EcoHealth 1:284–295. https://doi.org/10.1007/s10393-004-0120-8

Grotzinger J, Sumner D, Kah L, Gupta S, Edgar L, Rubin D, Lewis K, Shiebr J, Mangold N, Milliken R, Conrad P, Des Marais D, Farmer J, Siebach K, Calef F, Hurowtiz J, McLennan S, Ming D, Vaniman D, Crisp J, Vasavada A, Edgett K, Malin M, Blake D, Gellert R, Mahaffy P, Wiens R, Maurie S, Grant J, Wilson S, Anderson R, Beegle L, Arvidson R, Hallet B, Sletten R, Rice M, Bell J III, Griffes J, Ehlmann B, Anderson R, Bristow T, Dietrich W, Dromart G, Eigenbrode J, Fraeman A, Hardgrove C, Herkenhoff K, Jandura L, Kocurek G, Lee S, Leshin L, Leveille R, Limonadi D, Maki J, McCloskey S, Meyer M, Minitti M, Newson H, Oehler D, Okon A, Palucis M, Parker T, Rowland S, Schmidt M, Squyres S, Steele A, Stolper E, Summons R, Treiman A, Williams R, Yingst A, MSL Science Team (2014) A habitable fluvio-lacustrine environment at yellowknife bay, gale crater, Mars. Science 343:6169. https://doi.org/10.1126/science.1242777

Grotzinger J, Gupta S, Malin M, Rubin D, Schieber J, Siebach K, Sumner D, Stack K, Vasavada A, Arvidson R, Calef F, Edgar L, Fischer W, Grant J, Griffes J, Kah L, Lamb M, Lewis K, Mangold N, Minitti M, Palusis M, Rice M, Williams R, Yingst R, Blake D, Blaney D, Conrad P, Crisp J, Dietrich W, Dromart G, Edgett K, Ewing R, Gellert R, Hurowitz J, Kocurek G, Mahaffy P, McBride M, McLennan S, Mischna M, Ming D, Milliken R, Newsom H, Oehler D, Parker T, Vaniman D, Wiesn R, Wilson S (2015) Deposition, exhumation and paleoclimate of an ancient lake deposit, Gale crater, Mars. Science 350:6257. https://doi.org/10.1126/science.aac7575

Gutiérrez M (2005) Desert lakes: playa lakes and sebkhas. Clim Geomorphol:383–405

Guzman M, McKay CP, Quinn RC, Szopa C, Davila AF, Navarro-González R, Freissinet C (2018) Identification of chlorobenzene in the viking gas chromatograph-mass spectrometer data sets: reanalysis of viking mission data consistent with aromatic organic compounds on Mars. J Geophys Res: Planets. https://doi.org/10.1029/2018JE005544

Hang C, Nadeau DF, Jensen DD, Hoch SW, Pardyjak E (2015) Playa soil moisture and evaporation dynamics during the MATERHORN field program. Bound.-Layer Meteorol 159:521–538. https://doi.org/10.1007/s10546-015-0058-0

Harrison TN, Osinski GR, Tornabene LL, Jones E (2015) Global documentation of gullies with the Mars Reconnaissance Orbiter Context Camera and implications for their formation. Icarus 252:236–254

Hauber E, Platz T, Reiss D, Le Deit L, Kleinhans MG, Marra WA, de Haas T, Carbonneau P (2013) Asynchronous formation of Hesperian and Amazonian-aged deltas on Mars and implications for climate. J Geophys Res Planets 118(7):1529–1544

Hays L, Archenbach L, Bailey J, Barnes R, Barros J, Bertka C, Boston P, Boyd E, Cable M, Chen I (2015) NASA astrobiology strategy. NASA, Washington, DC

Hays L, Graham H, Des Marais D, Hausrath E, Horgon B, McCollom T, Parenteau M, Potter-McIntyre S, Williams A, Lynch K (2017) Biosignature preservation and detection in Mars analog environments. Astrobiology 17:363–400

Hecht M, Kounaves S, Quinn R, West S, Youg M, Ming D, Catling D, Clark B, Boynton W, Hoffman J, DeFlores L, Gospodinova K, Kapit J, Smith P (2009) Detection of perchlorate and the soluble chemistry of martian soil at the phoenix lander site. Science 325:64–67. https://doi.org/10.1126/science.1172466

Heet T, Arvidson R, Cull S, Mellon M, Seelos K (2009) Geomorphic and geologic settings of the Phoenix Lander mission landing site. JGR 114:E00E04. https://doi.org/10.1029/2009JE003416

Herrera A, Cockell C, Self S, Blaxter M, Reitner J, Thorsteinsson T, Arp G, Dröse W, Tindle A (2009) A cryptoendolithic community in volcanic glass. Astrobiology 9(4):369–381. https://doi.org/10.1089/ast.2008.0278

Horgan B, Anderson R, Dromart G, Amador E, Rice M (2019) The mineral diversity of Jezero crater: Evidence for possible lacustrine carbonates on Mars. Icarus. https://doi.org/10.1016/j.icarus.2019.113526

Hugenholtz P, Boebel B, Pace N (1998) MiniReview: impact of culture-independent studies on the emergingn phylogenetic view of bacterial diversity. J Bacteriol 180(18):4764–4775

Hughes KA, Lawley B (2003) A novel Antarctic microbial endolithic community within gypsum crusts. Environ Microbiol 5(7):555–565

Hurowitz JA et al (2017) Redox stratification of an ancient lake in Gale crater, Mars. Science 356(6341):6849. https://doi.org/10.1126/science.aah6849

Hynek BM, Beach M, Hoke MR (2010) Updated global map of Martian valley networks and implications for climate and hydrologic processes. J Geophys Res Planets 115(E9)

Idris H, Goodfellow M, Sanderson R, Asenjo J, Bull A (2017) Actionbacterial rare biospheres and dark matter revealed in habitats of the Chilean Atacama. Sci Rep 7:8373

Imhoff J, Sahl H, Soliman G, Trüper H (1979) The Wadi Natrun: chemical composition and microbial mass developments in alkaline brines of eutrophic desert lakes. Geomicrobiology 1(3):219–234

Jung P, Schermer M, Briegel-Williams L, Baumann K, Leinweber P (2019) Water availability shapes edaphic and lithic cyanobacterial communities in the Atacama Desert. J Phycol 55:1306–1318. https://doi.org/10.1111/jpy.12908

Kaseke K (2009) Non-rainfall Atmospheric Water in Arid Soil Microhydrology and Ecology, MSc thesis, Univ. of Stellenbosch, Stellenbosch, South Africa

Kaseke K, Mills A, Henschel J, Seely M, Brown R (2012) The effects of desert pavments on soil microhydrology. Pure Appl Geophys 169:873–880

Khan N, Tuffin M, Stafford W, Cary C, Lacap D, Pointing S, Cowan D (2011) Hypolithic microbial communities of quartz rocks from Miers Valley, McMurdo Dry Valleys, Antarctica. Polar Biol 34:1657–1668

Kidron GJ (2005) Angle and aspect dependent dew precipitation in the Negev Desert, Israel. J Hydrol 301:66–74

Kidron GJ (2007) Millimeter-scale microrelief affecting runoff yield over microbiotic crusts within a dune ecosystem in the Negev Desert. Catena 70:266–273. https://doi.org/10.1016/j.catena.2006.08.010

Kidron GJ, Herrnstadtw I, Barzilay E, Campus R (2002) The role of dew as a moisture source for sand microbiotic crusts in the Negev Desert, Israel. J Arid Environ 52(4):517–533

Kidron GJ, Vonshak A, Dor I, Barinova S, Abeliovich A (2010) Properties and spatial. distribution of microbiotic crusts in the Negev Desert. Catena 82:92–101

Kidron GJ, Temina M, Starinsky A (2011) An investigation of the role of water (rain and dew) in controlling the growth form of lichens on cobbles in the Negev Desert. Geomicrobiol J 28:335–346

Kidron GJ, Barinova S, Vonshak A (2012) The effects of heavy winter rains and rare summer rains on biological soil crusts in the Negev Desert. Catena 95:6–11. https://doi.org/10.1016/j.catena.2012.02.021

Kidron GJ, Starinksky A, Yaalon D (2014) Cyanobacteria are confined to dewless habitats within a dew desert: implications for past and future climate change for lithic microorganisms. J Hydrol 519:3606–3614

Klein HP (1979) The Viking mission and the search for life on Mars. Rev Geophys 17(7):1655. https://doi.org/10.1029/RG017i007p01655

Kuhlman K, Venkat P, La Duc M, Kuhlman G, Mckay C (2008) Evidence of a microbial community associated with rock varnish at Yungay, Atacama Desert, Chile. J Geophys Res 113:G04022. https://doi.org/10.1029/2007JG000677

Lacap D, Warren-Rhodes K, McKay CP, Pointing SB (2011) Cyanobacteria and chloroflexi-dominated hypolithic colonization of quartz at the hyper-arid core of the Atacama Desert, Chile. Extremophiles 15:31–38. https://doi.org/10.1007/s00792-010-0334-3

Lacap-Bugler DC, Lee KK, Archer S, Gillman LN, Lau MCY, Leuzinger S et al (2017) Global diversity of desert hypolithic cyanobacteria. Front Microbiol 8:867. https://doi.org/10.3389/fmicb.2017.00867

Lalley J, Viles H (2005) Terricolous lichens in the northern Namib Desert of Namibia: distribution and community composition. The Lichenologist 37(1):77–91. https://doi.org/10.1017/S0024282904014203

Lalley J, Viles H, Copeman N, Cowley C (2006) The influence of multi-scale environmental variables on the distribution of terricolous lichens in a fog desert. J Veg Sci 17:831–838

Lee K, Barbier B, Bottos E, McDonald I, Cary SC (2012) The inter-valley soil comparative survey: the ecology of Dry Valley edaphic communities. ISME J 6:1046–1057. https://doi.org/10.1038/ismej.2011.170

Lee K, Caruso T, Archer S, Gillman L, Lau M, Cary SC, Lee C, Pointing SB (2018) Stochastic and deterministic effects of a moisture gradient on soil microbial communities in the McMurdo Dry Valleys of Antarctica. Frontiers Microbiol 9:2619. https://doi.org/10.3389/fmicb.2018.02619

Lefort A, Russell P, Thomas N, McEwen A, Dundas C, Kirk R (2009) Observations of periglacial landforms in Utopia Planitia with the High Resolution Imaging Science Experiment (HiRISE). J Phys Res 114:E040005. https://doi.org/10.1029/2008JE003264

Lehnert L, Thies B, Trachte K, Achilles S, Osses P, Baumann K, Schmidt J, Samolov E, Jung P, Leinweber P, Karsten U, Büdel B, Bendix J (2018) A case study on fog/low stratus occurrence at Las Lomitas, Atacama Desert (Chile) as a water source for biological soil crusts. Aerosol Air Quality Res 18:254–269. https://doi.org/10.4209/aaqr.2017.01.0021

Leshin LA, Mahaffy PR, Webster CR, Cabane M, Coll P, Conrad PG, Archer PD Jr, Atreya SK, Brunner AE, Buch A, Eigenbrode JL (2013) Volatile, isotope, and organic analysis of martian fines with the Mars Curiosity rover. Science 341(6153):1238937

Lester ED, Satomi M, Ponce A (2007) Microflora of extreme arid atacama desert soils. Soil Biol Biochem 39(2):704–708

Levin G, Straat P (1976) Viking labeled release biology experiment: interim results. Science 194: 1322–1329

Levy J, Head J, Marchant D (2008) The role of thermal contraction crack polygons in cold-desert fluvial systems. Antarctic Science 20(6):565–579. https://doi.org/10.1017/S0954102008001375

Lousada M, Pina P, Vierira G, Bandeira L, Mora C (2018) Evaluation of the use of very high resolution aerial imagery for accurate ice-wedg polygon mapping (Adventdalen, Svalbard). Sci Total Environ 615:1574–1583. https://doi.org/10.1016/j.scitotenv.2017.09.153

Lynch RC, Darcy JL, Kane NC, Nemergut DR, Schmidt SK (2014) Metagenomic evidence for metabolism of trace atmospheric gases by high-elevation desert Actinobacteria. Front Microbiol 5:698

Mahowald N, Bryant R, Corral J, Steinberger L (2003) Ephemeral lakes and desert dust sources. Geophys Res Lett 30(2):1074. https://doi.org/10.1029/2002GL016041

Malin MC, Edgett KS (2003) Evidence for persistent flow and aqueous sedimentation on early Mars. Science 302(5652):1931–1934

Makhalanyane TP, Valverde A, Lacap DC, Pointing SB, Tuffin MI, Cowan DA (2013) Evidence of species recruitment and development of hot desert hypolithic communities. Environ Microbiol Rep 5:219–224. https://doi.org/10.1111/1758-2229.12003

Makhalanyane T, Valverde A, Gunnigle E, Frossard A, Raymond JB, Cowan D (2015) Microbial ecology of hot desert edaphic systems. FEMS Microbiol Rev 39:203–221. https://doi.org/10.1093/femsre/fuu011

Maloof A, Kellogg J, Anders A (2002) Neoproterozoic sand wedges: crack formation in frozen soils under diurnal forcing during a snowball Earth. Earth Planetary Sci Letters 204:1–15

Mangold N, Poulet F, Mustard JF, Bibring JP, Gondet B, Langevin Y, Ansan V, Masson P, Fassett C, Head JW III, Hoffmann H (2007) Mineralogy of the Nili Fossae region with OMEGA/Mars Express data: 2. Aqueous alteration of the crust. J Geophys Res Planets 112(E8)

McEwen AS, Dundas CM, Mattson SS, Toigo AD, Ojha L, Wray JJ, Chojnacki M, Byrne S, Murchie SL, Thomas N (2014) Recurring slope lineae in equatorial regions of Mars. Nat Geosci 7(1):53–58

McKay CP (2016) Water sources for cyanobacteria below desert rocks in the Negev Desert determined by conductivity. Glob Ecol Conserv 6:145–151. https://doi.org/10.1016/S0012-821X(02)00960-3

McKay CP (2020) What is life—and when do we search for it on other worlds. Astrobiology 20(2): 163–166. https://doi.org/10.1089/ast.2019.2136

McKay CP, Nienow J, Meyer M, Friedmann EI (1993) Continuous nanoclimate data (1985–1988) from the Ross Desert (McMurdo Dry Valley) cryptoendolithic microbial ecosystem. In: Antarctic Meteorology and Climatology: Studies Based on Automatic Weather Stations. American Geophysical Union 61:201–207

McKay C, Mellon M, Friedmann E (1998) Soil temperatures and stability of ice-cemented ground in the McMurdo Dry Valleys, Antarctica. Antarct Sci 10:31–38

McKay C, Balaban E, Abrahams S, Lewis N (2019) Dry permafrost soil over ice-cemented ground at Elephant Head, Ellsworth Land, Antarctica. Antarct Sci 31(5):263–270

McKenna OP, Sala OE (2018) Groundwater recharge in desert playas: current rates and future effects of climate change. Environ Res Lett 13:014025. https://doi.org/10.1088/1748-9326/aa9eb6

McSween H (2015) Petrology on Mars. American Mineralogist 100(11–12):2380–2395. https://doi.org/10.2138/am-2015-5257

McSween H, Grove T, Wyatt M (2003) Constraints on the composition and petrogenesis of the Martian crust. JGR Planets 108:E12. https://doi.org/10.1029/2003JE002175

McSween H, Taylor G, Wyatt M (2009) Elemental composition of the Martian crust. Science 324: 736. https://doi.org/10.1126/science.1165870

Mellon M, Malin M, Arvidson R, Searls M, Sizemore H, Heet T, Lemmon M, Keller HU, Marshall J (2009a) The periglacial landscape at the Phoenix landing site. JGR 114:E00E06. https://doi.org/10.1029/2009JE003418

Mellon M, Arvidson R, Sizemore H, Searls M, Blaney D, Cull S, Hecht M, Heet T, Keller HU, Lemmon M, Markiewicz W, Ming D, Morris R, Pike T, Zent A (2009b) Ground ice at the Phoenix landing site: Stability state and origin. JGR 114:E00E07. https://doi.org/10.1029/2009JE003417

Meslier V, DiRuggiero J (2019) Endolithic microbial communities as model systems for ecology and astrobiology. In: Seckbach J, Rampelotto PH (eds) Model ecosystems in extreme environments. Elsevier, pp 145–168

Meslier V et al (2018) Fundamental drivers for endolithic microbial community assemblies in the hyperarid Atacama Desert. Environ Microbiol 20(5):1765–1781

Michalski J, Dobrea E, Niles P, Cuadros J (2017) Ancient hydrothermal seafloor deposits in Eridania basin on Mars. Nature Commun 8:15978. https://doi.org/10.1038/ncomms15978

Mogul R, Vaishampayan P, Bashir M, McKay CP, Schubert K, Bornaccorsi R, Gomez E, Tharayil S, Payton G, Capra J, Andaya J, Bacon L, Bargoma E, Black D, Boos K, Brant M, Chabot M, Chau D, Cisneros J, Chu G, Curnutt J, DiMizio J, Engelbrecht C, Gott C, Harnoto R, Hovanesian R, Johnson S, Lavergne B, Martinez G, Mans P, Morales E, Oei A, Peplow G, Piaget R, Ponce N, Renteria E, Rodriguez V, Rodriguez J, Santander M, Sarmiento K, Scheppelmann A, Schroter G, Sexton D, Stephenson J, Symer K, Russo-Tait T, Weigel B, Wilhelm MB (2017) Microbial community and biochemical dynamics of biological soil crusts across a gradient of surface coverage in the Central Mojave Desert. Front Microbiol 8:1974. https://doi.org/10.3389/fmicb.2017.01974

Mondro CA, Fedo CM, Moersch JE (2019) Identification and morphologic characterization of small and medium alluvial fans on mars using the ctx global mosaic [2739]: lunar and planetary science conference. https://www.hou.usra.edu/meetings/lpsc2019/pdf/2739.pdf

Moore H, Jakosky B (1989) Viking landing sites, remote-sensing observations and physical properties of Martian surface materials. Icarus 81(1):164–184. https://doi.org/10.1016/0019-1035(89)90132-2

Moore H, Hutton R, Clow G, Spitzer C (1987) Physical properties of the surface materials at the Viking landing sites. US Geological Survey Professional Paper 1389. US Government Printing Office, Washington DC

Murchie S, Mustard J, Ehlmann B, Milliken R, Bishop J, McKeown N, Dobrea E, Seeelos F, Buczkowski D, Wiseman S, Arvidson R, Wray J, Swayze G, Clark R, Des Marais D, McEwen A, Bibring JP (2009) A synthesis of Martian aqueous mineralogy after 1 Mars year of observations from the Mars Reconnaisssance Orbiter. JGR Planets 114:E2. https://doi.org/10.1029/2009JE003342

Murray AE, Kenig F, Fritsen CH, McKay CP, Cawley KM, Edwards R, Kuhn E, McKnight DM, Ostrom NE, Peng V, Ponce A, Priscu JC, Samarkin V, Townsend AT, Wagh P, Young SA, Yung PT, Doran PT (2012) Microbial life at −13 C in the brine of an ice-sealed Antarctic lake. Proc Nat Acad Sci 109(50):20,626–20,631. https://doi.org/10.1073/pnas.1208607109

Mustard JF, Murchie SL, Pelkey SM, Ehlmann B, Milliken R, Grant J, Bibring J, Poulet F, Bishop J, Dobrea E, Roach L, Seelos F, Arvidson R, Wiseman S, Green R, Hash C, Humm D, Malaret E, McGovern J, Seelos K, Clancy T, Clark R, Des Marais D, Izenberg N, Knudson A, Langevin Y, Martin T, McGuire P, Morris R, Robinson M, Roush T, Smith M, Swayze G, Taylor H, Titus T, Wolff M (2008) Hydrated silicate minerals on mars observed by the mars reconnaissance orbiter CRISM instrument. Nature 454:305–309

Mutch T (1976) The surface of mars: the view from the viking 2 lander'. Science 194(4271): 1277–1283

Mutch TA, Arvidson RE, Guinness EA, Binder AB, Morris EC (1977) The geology of the Viking Lander 2 site. J Geophys Res 82:4452–4467. https://doi.org/10.1029/JS082i028p04452

Nash T, Nebecker G, Moser T, Reeves T (1979) Lichen vegetation gradients in relation to the Pacific coast of Baja California: the maritime influence. Madroño 26:149–163

Navarro JB, Moser DP, Flores A, Ross C, Rosen MR, Dong H et al (2009) Bacterial succession within an ephemeral hypereutrophic Mojave Desert playa Lake. Microb Ecol 57:307–320. https://doi.org/10.1007/s00248-008-9426-3

Navarro-Gonzalez R, Rainey F, Molina P, Bagaley D, Hollen B, de la Rosa J, Small A, Quinn R, Grunthaner F, Cáceres L, Gomez-Sílva B, McKay CP (2003) Mars-like soils in the Atacama Desert, Chile and the dry limit of microbial life. Science 302(5647):1018–1021

Navarro-Gonzalez R, Vargas E, de la Rosa H, Raga A, McKay CP (2010) Reanalysis of the Viking results suggests perchlorate and organics at midlatitudes on Mars. JGR Planets 115:E12010. https://doi.org/10.1029/2010JE003599

Neilson JW, Califf K, Cardona C, Copeland A, van Treuren W, Josephson KL et al (2017) Significant impacts of increasing aridity on the arid soil microbiome. mSystems 2:e00195-e00116. https://doi.org/10.1128/mSystems.00195-16

Niederberger T, Sohm J, Gunderson T, Parker A, Tirindelli J, Capone D, Carpenter E, Cary SC (2015) Microbial community composition of transiently wetted Antarctic Dry Valley soils. Front Microbiol 6:9. https://doi.org/10.3389/fmicb.2015.00009

Nienow J (2019) Extremophiles: dry environments (including Cryptoendoliths). In: Schmidt T (ed) Encyclopedia of microbiology, 4th edn. Elsevier, Oxford, pp 239–262

Nienow J, McKay CP, Friedmann EI (1988) The cryptoendolithic microbial environment in the Ross Desert of Antarctica: light in the photosynthetically active region. Microbiol Ecol 16:271–289. https://doi.org/10.1007/BF02011700

Noell A, Fisher A, Fors-Francis K, Sherritt S (2018) Subcritical water extraction of amino acids from Mars analog soils. Electrophoresis 39:2854–3863

Noffke N, Gerdes G, Klenke T, Krumbein W (2001) Microbially induced sedimentary structures—a new category within the classification of primary sedimentary structures. J Sediment Res 71: 649–656

Nowicki SA, Christensen PR (2007) Rock abundance on Mars from the thermal emission spectrometer. J Geophys Res Planets 112:E5

Ojha L, McEwen A, Dundas C, Byrne S, Mattson S, Wray J, Masse M, Schaefer E (2014) HiRISE observations of recurring slope lineae (RSL) during southern summer on Mars. Icarus 231:365–376

Olsson-Francis K, Simpson A, Wolff-Boenisch D, Cockell C (2012) The effect of rock composition on cyanobacterial weathering of crystalline basalt and rhyolite. Geobiology 10:434–444. https://doi.org/10.1111/j.1472-4669.2012.00333

Olsson-Francis R, Onofri K, Ott S, Pacelli S, Podolich C, Rabbow O, Reitz G, Rettberg P, Reva O, Rothschild L, Sancho L, Shulze-Makuch D, Selbmann L, Serrano P, Szewzyk U, Verseux C, Wadsworth J, Wagner D, Westall F, Wolteer D, Zucconi L (2019) Limits of life and the habitability of Mars: The ESA space experiment BIOMEX on the ISS. Astrobiology 19(2): 145–157. https://doi.org/10.1089/ast.2018.1897

Omelon C (2008) Endolithic microbial communities in polar desert habitats. Geomicrobiol J 25(7–8):404–414. https://doi.org/10.1080/01490450802403057

Omelon C (2016) Endolithic microorganisms and their habitats. In: Hurst C (ed) Their world: a diversity of microbial environments. Advances in environmental microbiology, vol 1. Springer, Cham, pp 171–201

Omelon CR, Pollard WH, Ferris FG (2006) Environmental controls on microbial colonization of high Arctic cryptoendolithic habitats. Polar Biol 30:19–29. https://doi.org/10.1007/s00300-006-0155-0

Oren A (2002a) Molecular ecology of extremely halophilic Archaea and Bacteria. FEMS Microbiol Ecol 39(1):1–7. https://doi.org/10.1111/j.1574-6941.2002.tb00900.x

Oren A (2002b) Diversity of halophilic microorganisms: Environments, phylogeny, physiology, and applications. J Ind Microbiol Biotechnol 28(1):56–63. https://doi.org/10.1038/sj/jim/7000176

Oren A (2013) Life at high salt concentrations, intracellular KCl concentrations and acidic proteomes. Front Microbiol 4:315

Ori GG, Marinangeli L, Baliva A (2000) Terraces and Gilbert-type deltas in crater lakes in Ismenius Lacus and Memnonia (Mars). J Geophys Res Planets 105(E7):17629–17641

Osterloo MM et al (2008) Chloride-bearing materials in the southern highlands of Mars. Science 319(5870):1651–1654

Osterloo MM et al (2010) Geologic context of proposed chloride-bearing materials on Mars. J Geophys Res 115(E10):1–29

Palzelt J, Hodac L, Friedl T (2014) Biodversity of soil cyanobacteria in the hyper-arid Atacama Desert. J Phycol 50(4):698–710

Parro V, Diego-Castilla G, Moreno-Paz M, Blanco Y, Cruz-Gill P, Rodríguez-Manfredi J, Fernández-Remolar D, Gómez F, Gómez M, Rivas L, Demergasso C, Echeverría A, Urtuvia V, Ruiz-Bermejo M, García-Villadangos M, Postigo M, Sánchez-Román M, Chong-Díaz GC, Gómez-Elvira J (2011) A microbial oasis in the hypersaline Atacama subsurface discovered by the Life Detector Chip: Implications for the search for life on Mars. Astrobiology 11(10):1–20

Pfeiffer M, Latorre C, Santoro C, Gayo E, Rojas R, Carrevedo M, McRostic V, Finstad KM, Heimsath A, Jungers M, de Pol-Holz R, Amundson R (2018) Chronology, stratigraphy and hydrological modelling of extensive wetlands and paleolakes in the hyperarid core of the Atacama Desert during the late Quaternary. Quat Sci Rev 197:224–245. https://doi.org/10.1016/j.quascirev.2018.08.001

Pointing S, Warren-Rhodes KA, Lacap D, Rhodes K, McKay CP (2007) Hypolithic community shifts occur as a result of liquid water availability along environmental gradients in China's hot and cold hyperarid deserts. Environ Microbiol 9:414–424. https://doi.org/10.1111/j.1462-2920

Pointing SB et al (2009) Highly specialized microbial diversity in hyper-arid polar desert. Proc Natl Acad Sci 106(47):19964–19969

Poulet F, Bibring J, Mustard J, Gendrin A, Mangold N, Langevin Y, Arvidson R, Gondet B, Gomez C, the Omega team (2005) Phyllosilicates on Mars and implications for early Martian climate. Nature 438:623–627

Prothero DR, Schwab F (2004) Sedimentary geology. An introduction to sedimentary rocks and stratigraphy. W. H. Freeman and Co., New York

Qu E, Omelon C, Oren A, Meslier V, Cowan D, Maggs-Kölling G, DiRuggiero J (2020) Trophic selective pressures organize the composition of endolithic microbial communities from global deserts. Frontiers Microbiol 10:2952. https://doi.org/10.3389/fmicb.2019.02952

Quade J, Rech J, Latorre C, Betancourt J, Glesson E, Kalin M (2006) Soils at the hyperarid margin: the isotopic composition of soil carbonate from the Atacama Desert, Northern Chile. Geochemica et Cosmochemica Acta 71(15):3772–3795. https://doi.org/10.1016/j.gca.2007.02.016

Ramond JB, Woodborne S, Hall G, Seely M, Cowan D (2018) Namib Desert primary productivity is driven by cryptic microbial community N-fixation. Scientific Reports 8:6921

Rasuk MC (2014) Microbial characterization of microbial ecosystems associated to evaporites domes of gypsum in Salar de Llamara in Atacama Desert. Microb Ecol 68:483–494

Rech J, Quade A, Hart W (2003) Isotopic evidence for the source of Ca and S in soil gypsum, anhydrite and calcite in the Atacama Desert, Chile. Geochimica et Cosmochemica Acta 67:575–586

Reheis M (1997) Dust deposition downwind of Owens (dry) lake, 1991–1994: Preliminary findings. JGR Atmospheres 102:25999–26008. https://doi.org/10.1029/97JD01967

Reynolds R, Bogle R, Vogel J, Goldstein H, Yount J (2009) Dust emission at Franklin Lake Playa, Mojave Desert (USA): response to meteorological and hydrological changes 2005–2008. Natural Resources and Environmental Issues V15- Saline Lakes Around the World: Unique Systems with Unique Values: Article 18. http://digitalcommons.usu.edu/nrei/vol15/iss1/18.

Robinson CK, Wierzchos J, Black C, Crits-Christoph A, Ma B, Ravel J, Ascaso C, Artieda O, Valea S, Roldán M, Gómez-Silva B, Diruggiero J (2013) Microbial diversity and the presence of algae in halite endolithic communities are correlated to atmospheric moisture in the hyper-arid zone of the Atacama Desert. Environ Microbiol 17(2):299–315

Rosen (1994) Paleoclimate and basin evolution of playa systems. In: Rosen MR (ed) Geological Society of America Special Paper 289. Geological Society of America, Boulder, pp 1–18

Rothschild L, Mancinelli R (2001) Life in extreme environments. Nature 409:1092–1101

Rothschild L, Giver L, White M, Mancinelli R (1994) Metabolic activity of microorganisms in evaporites. J Phycol 30:431–438

Ruff S, Farmer J (2016) Silica deposits on Mars with features resembling hot spring biosignatures at El Tatio in Chile. Nature 7:13554

Ruff S, Farmer J, Calvin W, Herkenhoff K, Johnson J, Morris R, Rice M, Arvidson R, Bell J, Christensen P, Squyres S (2011) Field reconnaissance geologic mapping of the columbia hills, mars based on MER spirit and MRO HiRISE observations. JGR 116:E00F23. https://doi.org/10.1029/2010JE003767

Rundel P (1978) Ecological relationships of desert fog zone lichens. Bryologist 81(2):277–293. https://doi.org/10.2307/3242189

Schlesinger H, Pippen J, Wallenstein M, Hofmockel K, Klepeis D, Mahall B (2003) Community composition and photosynthesis by photoautotrophs under quartz pebbles, southern Mojave Desert. Ecology 84(12):3222–3231. https://doi.org/10.1890/02-0549

Schulze-Makuch D, Wagner D, Kounaves S, Mangelsdorf K, Devine K, Vera JP, Schmitt-Kopplin-P, Grossart HP, Parro V, Kaupenjohann M, Galy A, Schneider B, Airo A, Frösler J, Davila A, Arens F, Cáceres L, Cornejo F, Carrizo D, Dartnell L, DiRuggiero J, Flury M, Ganzert L, Gessner M, Grathwohl P, Guan L, Heinz J, Hess M, Keppler F, Maus D, McKay CP, Meckenstock R, Montgomery W, Oberlin E, Probst A, Sáenz J, Sattler T, Schirmack J, Sephton M, Schloter M, Valenzuela J, Vestergaard G, Wörmer L, Zamorano P (2018) Transitory microbial habitat in the hyperarid Atacama Desert. PNAS 115(11):2670–2675. https://doi.org/10.1073/pnas.1714341115

Shaw P, Bryant R (2011) Pans, playas and salt lakes. In: Thomas DSG (ed) Arid zone geomorpholgy. University of Oxford, London, UK

Shen J, Wyness A, Claire M, Zerkle A (2021) Spatial variability of microbial communities and salt distributions across a latitudinal aridity gradient in the Atacama Desert. Microb Ecol 82:442–458. https://doi.org/10.1007/s0024-020-01672-w

Sirisena K, Ramirez S, Steele A, Glamoclija M (2018) Microbial diversity of hypersaline sediments from Lake Lucero playa in White Sands National Monument, New Mexico, USA. Microb Ecol 76:404–418. https://doi.org/10.1007/s00248-018-1142-z

Sirisena K, Ramirez S, Steele A, Glamoclija M (2019) Microbial diversity of hypersaline sediments from Lake Lucero playa in White Sands National Monument, New Mexico, USA. Microb Ecol 76:404–418. https://doi.org/10.1007/s00248-018-1142-z

Skelley AM, Aubrey AD, Willis PA, Amashukeli X, Ehrenfreund P, Bada JL, Grunthaner FJ, Mathies RA (2007) Organic amine biomarker detection in the Yungay region of the Atacama Desert with the Urey instrument. J Geophys Res Biogeosci 112(G4)

Smith P, Tamppari L, Arvidson R, Bass D, Blaney D, Boynton W, Carswell A, Catling D, Clark B, Duck T, DeJong E, Fisher D, Goetz W, Gunnlaugsson H, Hecht M, Hipkin V, Hoffman J, Hviid S, Keller H, Kounaves S, Lange C, Lemmon M, Madsen M, Markiewicz W, Marshall J, McKay CP, Mellon M, Ming D, Morris R, Pike W, Renno N, Staufer U, Stoker C, Taylor P, Whiteway J, Zent A (2009) H_2O at the phoenix landing site. Science 325:58–61. https://doi.org/10.1126/science.1172339

Soil Survey Staff (1999) Soil taxonomy: A basic system of soil classification for making and interpreting soil surveys, 2nd edn. U.S. Department of Agriculture NRCS

Squyres S, Grotzinger J, Arvidson R, Bell J, Calvin W, Christensen R, Clark B, Crisp J, Farrand H, Herkenhoff K, Johnson J, Klingelhofer G, Knoll A, McLennan S, McSween H, Morris R, Rice J, Rider R, Soderblom L (2004) In situ evidence for an ancient aqueous environment at Meridiani Planum, Mars. Science 306:1709–1714. https://doi.org/10.1126/science.1104559

Squyres S, Arvidson R, Ruff S, Gellert R, Morris R, Ming D, Crumpler L, Farmer J, Des Marais D, Yen A, McLennan S, Calvin W, Bell J, Clark B, Wang A, McCoy T, Schmidt M, de Souza P (2008) Detection of silica-rich deposits on Mars. Science 320:1063. https://doi.org/10.1126/science.1155429

Stern J, Sutter B, Freissinet C, Navarro-González R, McKay CP, Archer P, Buch A, Brunner A, Coll P, Eigenbrode J, Faíren A, Franz H, Glavin D, Kashyap S, McAdam A, Ming D, Steele A, Szopa C, Wray P, Martín-Torres F, Zorzano M, Conrad P, Mahaffy P, MSL Science Team (2015) Evidence for indigenous nitrogen in sedimentary and aeolian deposits from the Curiosity rover investigations at Gale Crater, Mars. Proc Natl Acad Sci USA 112:4245–4250

Steven B, Gallegos-Graves L, Belnap J, Kuske C (2013) Dryland soil microbial communities display biogeographic patterns associated with soil depth and soil parent material. FEMS Microbiol Ecol 86:101–113. https://doi.org/10.1111/1574-6941.12143

Stoertz G, Ericksen GE (1974) Geology of salars in Northern Chile. US Geological Survey, Professional paper, no. 811

Stomeo F, Makhalanyane T, Valverde A, Pointing S, Stevens M, Cary SC, Tuffin M, Cowan D (2012) Abiotic factors influence microbial diversity in permanently cold soil horizons of a maritime-associated Antarctic Dry Valley. FEMS Microbiol Ecol 82(2):326–340. https://doi.org/10.1111/j.1574-6941.2012.01360.x

Stomeo F, Valverde A, Pointing SB, McKay CP, Warren-Rhodes KA, Tuffin M, Seely M, Cowan D (2013) Hypolithic and soil microbial community assembly along an aridity gradient in the Namib Desert. Extremophiles 17:329–337

Sun H (2013) Endolithic microbial life in extreme cold climate: snow is required, but perhaps less is more. Biology 2:693–701. https://doi.org/10.3390/biology2020693

Sutter B, McAdam A, Mahaffy P, Ming D, Edgett K, Rampe E, Eigenbrode J, Franz H, Friessinet C, Grotzinger J, Steele A, House C, Archer P, Malespin C, Navarro-González R, Stern J, Bell J, Calef F, Gellert R, Glavin D (2017) Evolved gas analyses of sedimentary rocks and eolian sediment in Gale Crater, Mars: results of the Curiosity rover's sample analysis at Mas instrument from Yellowknife Bay to the Namib Dune. JGR Planets 122:2574–2609. https://doi.org/10.1002/2016JE005225

Szopa C, Freissinet C, Glavin D, Millan M, Buch A, Franz H, Summons R, Sumner D, Sutter B, Eigenbrode J, Williams R, Navarro-González R, Guzman M, Malespin C, Teinturier S, Mahaffy P, Cabane M (2020) First detections of dichlorobenzene isomers and trichloromethylpropane from organic matter indigenous to Mars mudstone in Gale Crate, Mars: results from the sample analysis at Mars instrument onboard the Curiosity rover. Astrobiology 20(2)

Szynkiewicz A, Moore C, Glamoclija M, Bustos D, Pratt L (2010) Origin of coarsely crystalline gypsum domes in a saline playa environment at white sands national monument, New Mexico. JGR 115:F02021. https://doi.org/10.1029/2009JF001592

Thornwaite CW (1948) An approach toward a rational classification of climate. Geographical Rev 38(1):55–94

Traci C, Streten-Joyce C, Dalton R, Nussear K, Gibb K, Christian K (2010) Microclimate and limits to photosynthesis in a diverse community of hypolithic cyanobacteria in northern Australia. Environ Microbiol 12(3):592–607. https://doi.org/10.1111/j.1462-2920.2009.02098

Treiman A, Bish D, Vaniman D, Chipera S, Blake D, Ming D, Morris R, Bristow T, Morrison S, Baker M, Rampe E, Downs R, Filiberto J, Glazner A, Gellert R, Thompson L, Schmidt M, Le Deit L, Wiens R, McAdam A, Achilles C, Edgett K, Farmer J, Fendrich K, Grotzinger J, Gupta S, Morookian J, Newcombe M, Rice M, Spray J, Stolper E, Sumner D, Vasavada A, Yen A (2016) Mineralogy, provenance and diagenesis of a potassic basaltic sandstone on Mars: CheMin X-ray diffraction of the Windjana sample (Kimberley area, Gale Crater). J Geophys Res Planets 121:75–106. https://doi.org/10.1002/2015JE004932

Turnbull L, Wilcox B, Belnap J, Ravi S, D'Odorico P, Childers D, Gwenzi W, Okin G, Wainwright J, Caylor K, Sankey T (2012) Understanding the role of ecohydrological feedbacks in ecosystem state change in drylands. Ecohydrol. 5:174–183. https://doi.org/10.1002/eco.265

UNEP, United Nations Environment Programme (1997) In: Middleton N, Thompson D (eds) World Atlas Desertification, 2nd edn. Arnold, New York, p 182

UNEP, United Nations Environment Programme (2006) Global Deserts Outlook. UNEP, Nairobi Kenya, p 148

Uritskiy G, Getsin S, Munn A, Gómez-Silva B, Davila A, Glass B, Taylor J, DiRuggiero J (2019) Halophilic microbial community compositional shift after a rare rainfall in the Atacama Desert. ISME J. https://doi.org/10.1038/s41396-019-0468-y

Valverde A, Makhalanyane T, Seely M, Cowan DA (2015) Cyanoabacteria drive community composition and functionality in rock-soil interface communities. Mol Ecol 24:812–821

Van Goethem MW, Makhalanyane TP, Valverde A, Cary SC, Cowan DA (2016) Characterization of bacterial communities in lithobionts and soil niches from Victoria Valley, Antarctica. FEMS Microbiol Ecol 92:fiw051. https://doi.org/10.1093/femsec/fiw051

Vaniman D, Martínez G, Rampe E, Bristow T, Blake D, Yen A, Ming D, Rapin W, Meslin PY, Morookian J, Downs R, Chipera S, Morris R, Morrison S, Treiman A, Achilles C, Roberston K, Grotzinger J, Hazen R, Wiens R, Sumner D (2018) Gypsum, bassanite and anhydrite at Gale crater, Mars. American Mineralogist 103(7):1011–1102

Velland M (2017) The Theory of Ecological Communities. Monographs in Population Biology. University Press, Princeton, USA, p 248

Vikram S, Guerrero L, Makhalanyane T, Le P, Seely M, Cowan D (2016) Metagenomic analysis provides insights into functional capacity in a hyperarid desert soil niche community. Environ Microbiol 18(6):1875–1888

Vogel S (1955) Niedere 'Fensterpflanzen' in der südafrikanischen Wüste. Eine ökologische Schiderung. Beitr Biol Pflanz 31:45–135

Vogel M, Des Marais D, Turk K, Parenteau M, Jahnke L, Kubo M (2009) The role of biofilms in the sedimentology of actively forming gypsum deposits at Guerrero Negro, Mexico. Astrobiol 9(9): 875–893. https://doi.org/10.1089/as5.2008.0325

Walker J, Pace N (2007) Endolithic microbial ecosystems. Ann Rev Microbiol 61:331–347. https://doi.org/10.1146/annurev.micro.61.080706.093302

Walker J, Spear J, Pace N (2005) Geobiology of a microbial endolithic community in the Yellstone geothermal environment. Nature 434:1011–1014

Warren-Rhodes KA, Rhodes KL, Pointing SB et al (2006) Hypolithic cyanobacteria, dry limit of photosynthesis, and microbial ecology in the hyperarid Atacama Desert. Microb Ecol 52:389–398

Warren-Rhodes KA, Dungan J, Piatek J, Stubbs K, Gómez-Silva B, Chen Y, McKay CP (2007) Ecology and spatial pattern of cyanobacterial community island-patches in the Atacama Desert, Chile. JGR 112:G04S15. https://doi.org/10.1029/2006JG000305

Warren-Rhodes, K.A., McKay, C.P., Ng-Boyle, L., Wing, M., Kiekebusch, E., Cowan, D., Stomeo, F., Pointing, S., Kaseke, K., Eckardt, F., Henschel, J., Anisfeld, A., Seely, M. ad K. Rhodes (2013). Physical ecology of hypolithic communities in the central namib desert: the role of fog, rain, rock habitat and light. JGR Biogeosci 118, 1–10. Doi:https://doi.org/10.1002/jgrg.20117.2013.

Warren-Rhodes KA et al (2019) Subsurface microbial habitats in an extreme desert mars-analog environment. Front Microbiol 10:69. https://doi.org/10.3389/fmicb.2019.00069

Washburn AL (1956) Classification of patterned ground and review of suggested origins. GSA Bulletin 67(7):823–866. https://doi.org/10.1130/0016-7606(1956)67(823:COPGAR)2.0.CO.2

Wei S, Lacap-Bugler D, Pérez-Ortega S, de los Rios-Murillo A, Lee C (2014) Characterization of chasmoendolithic community in miers valley, McMurdo dry valleys, Antarctica. Microbiol Ecol 68:351–359. https://doi.org/10.1007/s00248-014-0412-7

Wei S, Lacap-Bugler D, Lau M, Caruso T, Rao S, de los Rios A, Archer S, Chiu J, Higgens C, Van Nostrand J, Zhou J, Hopkins D, Pointing SB (2016) Taxonomic and functional diversity of soil and hypolithic microbial communities in miers valley, mcmurdo dry valleys, antarctica. Front Microbiol 7:1642. https://doi.org/10.3389/fmicb.2016.01642

Weiss M, Preiner M, Xavier J, Zimorski V, Martin W (2016) The physiology and habitat of the last universal common ancestor. PLoS Genet 14(8):e1007518. https://doi.org/10.1038/nmicrobiol.2016.116

Wharton R, Crosby J, McKay CP, Rice J Jr (1995) Paleolakes on Mars. J Paleolimnol 13:167–283

Wierzchos J, Ascaso C (2001) Life, decay and fossilization of endolithic microorganisms from the Ross Desert, Antarctica. Polar Biol. https://doi.org/10.1007/s003000100296

Wierzchos J, Ascaso C (2002) Microbial fossil record of rocks from the Ross Desert, Antarctica: implications in the search for past life on Mars. Int J Astrobiol. https://doi.org/10.1017/S1473550402001052

Wierzchos J, Ascaso C, McKay CP (2006) Endolithic Cyanobacteria in Halite Rocks from the Hyperarid Core of the Atacama Desert. Astrobiology 6:415–422

Wierzchos J, Cámara B, De los Ríos A et al (2011) Microbial colonization of ca-sulfate crusts in the hyperarid core of the atacama desert: implications for the search for life on Mars. Geobiology 9:44–60

Wierzchos J, Davila AF, Sánchez-Almazo IMM et al (2012a) Novel water source for endolithic life in the hyperarid core of the atacama desert. Biogeosciences 9:3071–3098

Wierzchos J, de los Ríos A, Ascaso C (2012b) Microorganisms in desert rocks: the edge of life on earth. Int Microbiol 15:173–183

Wierzchos J, Davila A, Artieda O, Cámara-Gallego B, de los Ríos A, Nealson K, Valea S, García-González T, Ascaso C (2013) Ignimbrite as a substrate for endolithic life in the hyper-arid Atacama Desert: Implications for the search for life on Mars. Icarus 224(2):334–346. https://doi.org/10.1016/j.icarus.2012.06.009

Wierzchos J, DiRuggiero J, Vitek P, Artieda O, Souza-Egipsy V, Skaloud P, Tisza M, Davila A, Vilchez C, Garbayo I, Ascaso C (2015) Adaptation strategies of endolithic chlorophototrophs to survive the hyperarid and extreme solar radiation environment of the Atacama Desert. Front Microbiol 6:934. https://doi.org/10.3389/fmicb.2015.00934

Wierzchos J, Casero C, Artieda O, Ascaso C (2018) Endolithic microbial habitats as refuges for life in polyextreme environment of the Atacama Desert. Curr Opin Microbiol 43:124–131. https://doi.org/10.1016/j.mib.2018.01.003

Wilhelm MB, Davila A, Eignebrode J, Parenteau M, Jahnke L, Liu XL, Summons R, Wray J, Stamos B, O'Reilly S, Williams A (2017) Xeropreservation of functionalized lipid biomarkers in hyperarid soils in the Atacama Desert. Organic Geochem 103:97–104

Wilhelm MB, Davila A, Parenteau M, Jahnke L, Abate M, Cooper G, Taylor Kelly E, Parro V, Villadangos M, Blanco Y, Glass B, Wray J, Eigenbrode J, Simmons R, Warren-Rhodes K (2018) Constraints on the metabolic activity of microorganisms in Atacama surface soils inferred from refractory biomarkers: Implications for Martian habitability and biomarker detection. Astrobiology 18(7):955–966. https://doi.org/10.1089/ast.2017.1705

Williams M, Buck B, Beyene M (2012) Biological soil crusts in the Mojave Desert USA: Micromorphology and pedogenesis. J Soil Sci Soc Am. https://doi.org/10.2136/sssaj2012.0021

Wong K, Lacap D, Lau M, Aitchison J, Cowan D, Pointing SB (2010) Hypolithic colonization of quartz pavement in the high altitude tundra of central Tibet. Microb Ecol 60:730–739

Wray J, Murchie S, Squyres S, Seelos F, Tornabene L (2009) Diverse aqueous environments on ancient Mars revealed in the southern highlands. Geology 37(11):1043–1046. https://doi.org/10.1130/G30331A.1

Yen Y, Ming D, Gellert R, Vaniman D, Clark B, Morris R, Mittlefehldt D, Arvidson R, The Athena and MSL Science teams (2014) Investigation of Martian aqueous processes using multiple alpha particle X-ray spectrometer (APXS) datasets. Eighth Int Mars Conference, July 14–18, 2014, Jet Propulsion Lab, Pasadena CA

Yingst R, Berger J, Cohen B, Hynek B, Schmidt M (2016) Determining best practices in reconnoitering sites for habitability potential on Mars using a semi-autonomous rover: a GeoHeuristic Operational Strategies Test. Acta Astronaut 132:268–281

Yingst R, Bartley J, Chidsey T Jr, Cohen B, Hynek B, Kah L, Minitti M, Vanden Berg M, Williams R, Adams M, Black S, El-Maarry M, Gemperline J, Kronyak R, Lotto M (2020) Is a linear or a walkabout protocol more efficient when using a rover to choose biologically relevant samples in a small region of interest. Astrobiology 20(3):327–347. https://doi.org/10.1089/ast.2019.2090

Yoshikawa K (2003) Origin of the polygons and the thickness of vastitas borealis formation in Western Utopia Planitia on Mars. J Geophys Res Lett 30(12):1603. https://doi.org/10.1029/2003GL017165

Yung C, Chan Y, Lacap D, Pérez-Ortegao S, de los Rios-Murillo A, Lee C, Cary SC, Pointing SB (2014) Characterization of chasmoendolithic community in Miers valley, McMurdo Dry Valleys, Antarctica. Microbiol Ecol 68:351–359. https://doi.org/10.1007/s00248-014-0412-7

Survival under Stress: Microbial Adaptation in Hot Desert Soils

10

Clément Coclet, Don Cowan, and Pedro H. Lebre

Abstract

The combined effects of climate change and anthropogenic-driven soil erosion have led to a progressive increase in aridity and resulting desertification of many of Earth's biomes. Hot deserts are characterized by wide temperature fluctuations, low precipitation levels, and high UV radiation, leading to extreme abiotic conditions that challenge all life. Among these stresses, desiccation, osmotic stress, and oligotrophy have the most impacts on edaphic microbial communities. In response, these communities have developed several adaptative mechanisms that are broadly distributed among taxa and allow them to survive and even thrive under the harsh desert conditions. In this chapter, we discuss the molecular, metabolic, and physiological strategies that enable microbial taxa to survive in hot desert soils. We will focus on both general adaptive mechanisms that are used against multiple stressors, such as sporulation, and stressor-specific adaptions such as desiccation-driven solute accumulation. Understanding the adaptative mechanisms used by microbes to survive in harsh desert-associated conditions will provide useful information on community survival strategies in such ecosystems.

Keywords

Microbial communities · Hot deserts · Abiotic stresses · Desiccation · Molecular and physiological adaptations

Clément Coclet and Pedro H. Lebre contributed equally with all other contributors.

C. Coclet (✉) · D. Cowan · P. H. Lebre
Centre for Microbial Ecology and Genomics (CMEG), University of Pretoria, Pretoria, Gauteng, South Africa

J.-B. Ramond, D. A. Cowan (eds.), *Microbiology of Hot Deserts*, Ecological Studies 244, https://doi.org/10.1007/978-3-030-98415-1_10

10.1 Introduction: Hot Deserts and Their Challenges to Life

Desert biomes represent around 35% of the earth's landmass (Laity 2009; UNEP-WCMC 2007), and this area is likely to increase due to the combined effects of anthropogenically driven desertification and global warming (D'Odorico et al. 2013; Warren 1993). Deserts are present on all the world's continents (Laity 2009), with the largest of these being the "cold" Arctic ($\approx$14 million km^2) and the "hot" North African Sahara ($\approx$9 million km^2) deserts (Tucker et al. 1991). The principal distinction between these two types of biomes is temperature, with cold desert soils reaching temperatures below −60°C (Turner et al. 2009) while the surface soils of hot deserts regularly exceed 65°C (Cowan et al. 2019) and may reach temperatures as high as 80°C (Kieft 2002). Hot deserts are characterized by low levels of precipitation, with precipitation to potential evapotranspiration (P/PET) indices well below the 0.5 threshold defined for semi-arid biomes (UNEP-WCMC 2007; Barrow 1992).

Hot desert ecosystems represent some of the most extreme environments on Earth, due to a combination of stress-inducing abiotic conditions including low water availability, oligotrophy, wide temperature fluctuations, and high levels of UV radiation (Makhalanyane et al. 2015; Lebre et al. 2017). Despite these stressors, a diverse range of microbial taxa can be found in hot desert edaphic niches [(Makhalanyane et al. 2015; Fierer et al. 2012; Pointing and Belnap 2012; Pointing and Belnap 2014); Chaps. 3–9]. Desiccation stress is one of the most extreme challenges to life, since all biological processes require water to operate (Chaplin 2006). The removal of water from the system, either through evaporation or by osmotic processes, can negatively impact the physical stability of microbial cells and their constituents, resulting in cell shrinkage, destruction of membrane surface tension, and the unfolding of protein structures (Potts 1994, 2001). In addition, desiccation leads to accumulation of reactive oxygen species (ROS) such as superoxide (O_2^-) and hydroxyl radicals ($OH^\cdot$) (Imlay and Linn 1988). Cellular iron (Fe^{+2}) plays a crucial role in the production of ROS, reacting with hydrogen peroxide to form ferric ions (Fe^{+3}) and hydroxyl radicals through the Fenton reaction (Billi and Potts 2000). In turn, these compounds accelerate lipid peroxidation, DNA strand breakage via backbone oxidation, and protein side-chain oxidation, resulting in extensive molecular damage and ultimately in cell death (Potts 1994, 2001; Billi and Potts 2000).

Nutrient availability has also been shown to play an important role in driving microbial community structure in hot deserts (Schlesinger et al. 1990; Liu et al. 2018; Bell et al. 2008; Berg and Steinberger 2008). Desert soils are generally considered to be oligotrophic, where limitations on carbon, nitrogen, and/or phosphorus availability may limit microbial community function and biomass (Schlesinger et al. 1990; She et al. 2018; Crawford and Gosz 1982). Similarly, exposure of desert-surface biological systems to solar radiation can also lead to DNA damage through the range of different mechanisms (Rastogi et al. 2010). High energy ionizing UV radiation (UV-B and UV-C; <315 nm) represents less than 1% of solar radiation (Rastogi et al. 2010) but is responsible for extensive DNA lesions through direct interaction with purines and pyrimidines. The generation of photoproducts such as 6-4 pyrimidine-pyrimidone (6-4PPs) and cyclobutane pyrimidine dimers (CPDs) results in changes in DNA conformation (Sinha and Häder 2002; Gallagher and Duker 1989; Lukin and de los Santos C. 2006) and the indirect production of ROS (Santos et al. 2012). These reactions ultimately lead to double-stranded DNA breaks (DSBs) due to replication arrest and the subsequent cleavage of DNA by topoisomerase I (TopI) or other specific endonucleases (Strumberg et al. 2000; Ohnishi et al. 2009; Batista et al. 2009).

Microbial communities inhabiting hot desert soils are continuously or intermittently exposed to the abiotic stresses described above. Consequently, microorganisms exposed to these conditions have developed complementary molecular, metabolic, and physiological strategies that ensure cellular integrity while maintaining the ability to respond rapidly to environmental changes (Fig. 10.1) (Lebre et al. 2017). In the following sections, we discuss how microorganisms have adapted to survive the abiotic stresses imposed by the hot desert macro- and microenvironment.

10.2 Adaptation to Desiccation and Osmotic Stresses

10.2.1 Sporulation

Sporulation is a common physiological strategy adopted by many organisms to survive extreme desiccation, in which cells differentiate into a reversible dormant state that can persist for a long time in the absence of water (Lennon and Jones 2011; Higgins and Dworkin 2012; Setlow 2014). Sporulation is therefore seen as an important survival mechanism for hot desert edaphic (exposed soil) communities (Makhalanyane et al. 2015). During spore formation, cells modify the peptidoglycan cross-linkage of the cell wall to create a denser structure that leads to the formation of the compact shell (Setlow 2014). These compact cells have low water content owing to the accumulation of minerals such as manganese ions (Mn^{2+}), dipicolinic acid (DPA), and small acid-soluble proteins (SASPs) (Higgins and Dworkin 2012; García 2011; Kaplan-levy et al. 2010). Spores are a common feature of several bacterial phyla, including Actinobacteria (Chen and Alexander 1973) and Firmicutes, such as *Bacillus* and *Paenibacillus* species (Bueche et al. 2013; McHugh et al. 2017; Schulze-Makuch et al. 2018; Connon et al. 2007). The capacity of these phyla to sporulate might, in part, explain their existence in a wide range of different environments, including desert ecosystems, by facilitating survival under desiccating conditions (Pointing et al. 2007).

Cyanobacteria, which are highly abundant in microbial communities residing on the ventral surfaces of translucent rocks in desert pavement, termed hypolithons (Li et al. 2005; Gao and Garcia-Pichel 2011), can form spore-like structures called akinetes under conditions of nutrient starvation or desiccation (see reference (Kaplan-levy et al. 2010) for a comprehensive review). Akinetes formed by *Anabaena cylindrica* have been shown to be able to survive for long periods of

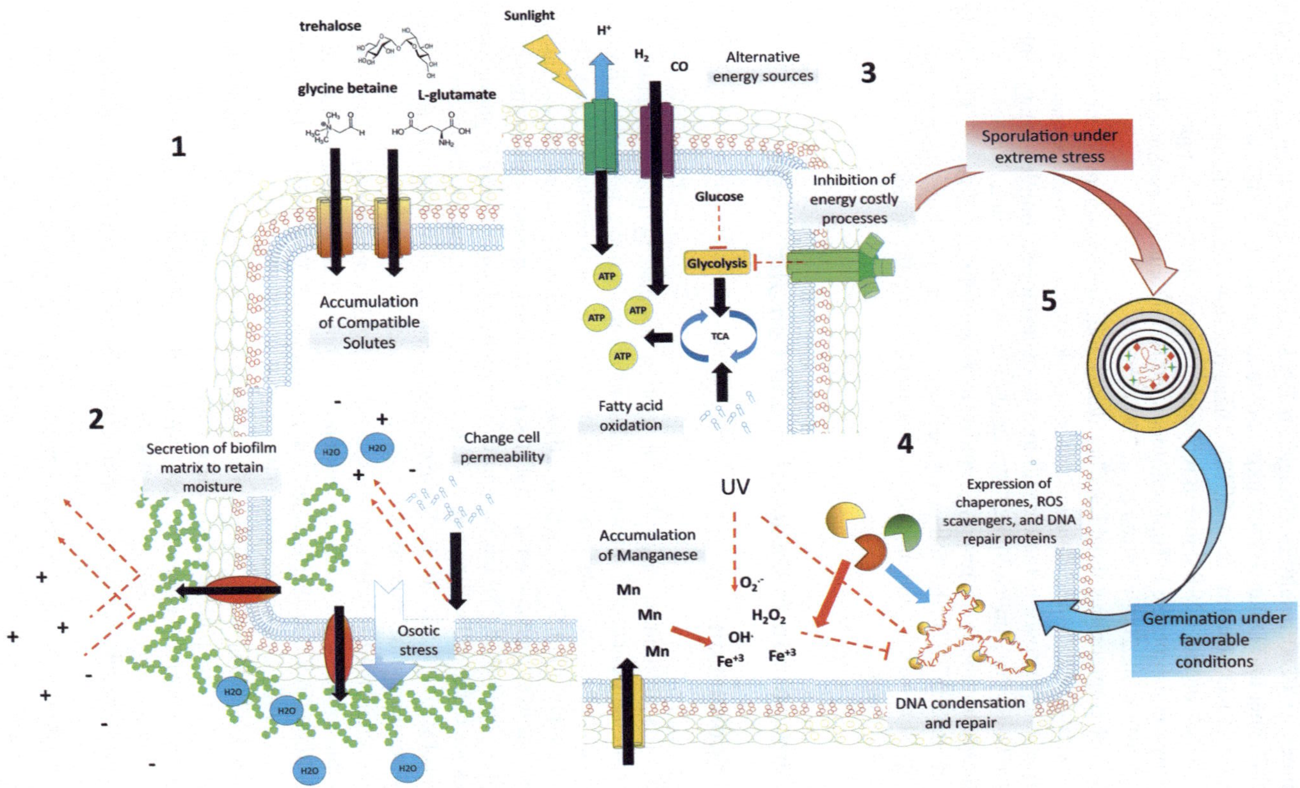

Fig. 10.1 Multifactorial model of molecular and physiological mechanisms for survival of microorganisms in hot deserts. These include (**1**) accumulation of compatible solutes that help stabilize the membrane and other macromolecules in the cell in the absence of water, as well as serving as alternative energy

sources; (**2**) secretion of EPS and other molecules to form a biofilm matrix that retains moisture and temperature, while changing cell membrane permeability to prevent water loss and ion influx (ions represented as + and – in the figure); (**3**) inhibition of energy-costly processes while using alternative energy sources for energy production, including the use of rare gases and solar energy; (**4**) expression of heat-shock, DNA repair, and ROS-scavenging proteins that mitigate the detrimental effects of ROS, as well as the accumulation of manganese (Mn), which also scavenges ROS; and (**5**) differentiation into spores in extreme conditions when maintenance through other adaptive processes is no longer possible. Blue and red arrows represent positive and negative interactions, respectively desiccation, maintaining germination capability even after 5 years in a dark and desiccated environment (Olsson-Francis et al. 2009). Four genes (i.e., *hep1*, *devR*, *hetR*, and *argL*) have been shown to be involved in the differentiation of akinetes from photosynthetically active cyanobacterial cells (Zhang et al. 2006). The *hep1* gene encodes for an ABC transporter required for the deposition of polysaccharides in the envelope of akinetes, while *devR* encodes a response regulator of a two-component system (Leganés 1994). Another gene affecting akinete development in *Nostoc ellipsosporum* is *argL*, which encodes an N-acetylglutamate semialdehyde dehydrogenase, an enzyme involved in L-arginine biosynthesis (Leganés et al. 1998).

10.3 Membrane Modifications

As the main barrier between the intracellular and external environment, the cellular membrane is severely de-stabilized during desiccation and osmotic stress (Van De Mortel and Halverson 2004). Consequently, microorganisms modify the composition of their lipid bilayer in order to preserve the physical membrane structure under these abiotic stresses (Van De Mortel and Halverson 2004; Halverson and Firestone 2000). Under desiccation and high temperature stress conditions, the ratio of saturated to unsaturated fatty acids is increased (Brown et al. 2000; Mutnuri et al. 2005; Scherber et al. 2009), resulting in a higher packing of the lipid bilayer that inhibits water loss. Upon re-wetting, the degree of membrane saturation is reduced by the expression of fatty acid desaturases (LeBlanc et al. 2008; Fierer et al. 2003). In addition, the accumulation of cyclopropane fatty acids has been shown to assist intracellular pH homeostasis by reducing membrane permeability to protons (Shabala and Ross 2008; Kocharunchitt et al. 2014). Under osmotic stress, cells accumulate negatively charged phospholipids, including phosphatidylglycerol lipids and cardiolipin, on the membrane surface (Brown et al. 2000; Romantsov et al. 2009). These lipids play a dual role in osmoprotection by inhibiting the formation of non-bilayer structures (Brown et al. 2000; Romantsov et al. 2009) and by positively affecting osmosensing proteins such as the osmolyte–H^+ symporter ProP (Lebre et al. 2017; Romantsov et al. 2009).

10.4 Accumulation of Compatible Solutes

In addition to changes in membrane fluidity, the accumulation of small molecules also contributes to the maintenance of intracellular homeostasis and macromolecular stability during desiccation and osmotic stress (Potts 1994; Chen and Murata 2002; Santos and Costa 2002; Pade and Hagemann 2015). While extreme halophiles typically use a "salting-in" strategy, by importing inorganic ions or osmolytes into the cell to regulate the osmotic pressure between the intracellular and extracellular environment ((Harding et al. 2016), desiccation-tolerant bacteria use a "salting-out" process, with the accumulation of neutral organic solutes (compatible solutes) that replace water as stabilizers of protein and membrane structure (Lebre et al. 2017). To date, a large number of these solutes have been characterized, including sugars such as trehalose, sucrose, and amino acid derivatives such as glycine betaine (Lebre et al. 2017; Santos and Costa 2002; Riedel and Lehner 2007; Oren 2008). Notably, different compatible solutes are associated with the osmotic response of specific phyla. For instance, glycine betaine is the most widely used compatible solute for the majority of prokaryotes and eukaryotes (Sleator and Hill 2002), while glucosylglycerol has mainly been reported as a compatible solute in cyanobacteria (Welsh 2000; Klähn and Hagemann 2011).

Many of these compounds are also by-products or intermediates in general metabolic pathways, and therefore, these biosynthetic pathways are widespread in microbial communities in many of Earth's terrestrial biomes, including desert ecosystems (LeBlanc et al. 2008; Oren 2008; Cytryn et al. 2007; Li et al. 2012). However, compounds such as ectoine are energetically expensive to synthesize (Oren 1999), and therefore compatible solute acquisition rather than *de novo* biosynthesis is often found to be overexpressed under desiccation stress (Bremer and Krämer 2019). The TreBCR and TreA–TreF pathways, which are involved in the transport of trehalose into the bacterial cell, are upregulated in desiccated *S. enterica* cells (Li et al. 2012). Transcriptomics and proteomics studies have also shown the upregulation of several compatible solute importers in response to desiccation (Gunasekera et al. 2008; Thi Le et al. 2016), including the glycine betaine/proline uptake system encoded by *proUVWX* (Sleator and Hill 2002), which is also used as a biomarker for osmotic stress resistance (Zhou et al. 2013; Shi et al. 2019).

10.5 Protein Protection

Late embryogenesis abundant (LEA) proteins are a family of hyper-hydrophilic proteins that accumulate in response to cellular dehydration. Originally identified as plant proteins associated with seed desiccation tolerance, LEA proteins have been identified in a wide range of organisms, including microorganisms (Danyluk et al. 1994; Artus et al. 1996; Sasaki et al. 2014). LEA proteins are thought to stabilize proteins through a similar mechanism to compatible solutes, by providing a hydrophilic environment around macromolecules that compensates for the loss of water during dehydration (Battaglia et al. 2008; Chakrabortee et al. 2007). These proteins have since been associated with the drought response mechanisms in a range of higher plants, lower eukaryotes, and prokaryotes (Battaglia et al. 2008; Chen et al. 2019; Hundertmark and Hincha 2008; Ikeno and Haruyama 2013; Magwanga et al. 2018; Jin et al. 2019). Based on sequence identity, LEA proteins can be clustered into at least seven families (Battaglia et al. 2008), all of which have a common glycine-rich and highly hydrophilic backbone that remains in an unstructured, coiled state during normal hydrated conditions (Battaglia et al. 2008; Garay-Arroyo et al. 2000). Proteins with a high degree of homology to the LEA family 3 are found in a range of prokaryotes that are well adapted to desiccation, including *Deinococcus radiodurans* and *Haemophilus influenza* (Battaglia et al. 2008; Slade and Radman

2011). In addition, LEA family 2, also referred to as LEA 14 proteins, are found in both bacteria and archaea (Mertens et al. 2018). Members from this family contain a unique water stress and hypersensitive response (WHy) domain that has been shown to be overexpressed during desiccation (Jaspard and Hunault 2014; Ciccarelli and Bork 2005). A recent study has suggested that WHy domain containing proteins might be involved in a broader stress response physiology (Anderson et al. 2015). While the exact role that LEA proteins play in drought tolerance is still a matter of some debate, it has been shown that *D. radiodurans* mutants lacking family 3 LEA-like proteins are extremely sensitive to dehydration (Battista et al. 2001). Several studies have also highlighted the role of LEA proteins as osmoprotectants, where they act by either binding directly to protein surfaces (thereby compensating for water loss) or re-ordering water molecules around the proteins to limit further water loss (Chakrabortee et al. 2007; Wise and Tunnacliffe 2004).

10.6 Biofilm Formation and EPS Production

Biofilms are a common feature of microbial life (Flemming and Wingender 2010), with biofilm-coated communities (also referred to as biocrusts; see Chap. 3 of this book for more details on the microbial ecology of hot desert BSCs) covering around 70 % of the desert soil surface (Cowan et al. 2019; Pointing and Belnap 2014; Belnap 1995; Chamizo et al. 2016; Warren-Rhodes et al. 2019; Tracy et al. 2010; Schlesinger et al. 2003). Biofilms are extracellular matrices mainly composed of extracellular polymeric substances (EPS), which include exopolysaccharides, proteins, nucleic acids, and lipids (Flemming et al. 2016; Tamaru et al. 2005). Exopolysaccharides are the main components of biofilms and can be diverse in their physical and chemical properties (Flemming and Wingender 2010). The majority of these exopolysaccharides are polyanionic, due to the presence of either uronic acids (d-glucuronic acid being the commonest) or ketal-linked pyruvate, and inorganic residues, such as phosphate or sulfate [(Anderson et al. 2015); see (Flemming et al. 2016) for a detailed review of EPS substances and functions]. Due to their polyanionic nature, EPS exhibit hydrophilic properties that contribute to water retention during desiccation and rapid absorption during the infrequent periods of rehydration (Tamaru et al. 2005). By maintaining a higher matrix water potential in the biofilm environment immediately surrounding the cell, the EPS also increase nutrient availability during soil desiccation by allowing greater solute diffusion (Kieft 2002; Roberson and Firestone 1992). Cyanobacteria produce large amounts of EPS, and are the primary source of EPS genes in biofilms (Varin et al. 2012). The production and secretion of EPS by these taxa not only benefits the species responsible for biofilm production, but also increases the desiccation tolerance of other microorganisms in the biofilm community (Knowles and Castenholz 2008). Bacteria can also be protected against osmotic stress by the hydrated microenvironment created by the surrounding EPS (Flemming and Wingender 2010). EPS are not unique to bacteria, since they can be produced by the methanogenic archaeon *Methanosarcina barkeri* (Anderson et al. 2012). Some of the most abundant EPS producers are microalgae (in particular, diatoms) (Cooksey and Wigglesworth-Cooksey 1995).

Information regarding the genes encoding the proteins that synthesize these EPS is limited (see Marvasi et al. (2010) for a detailed review), especially in the microbiomes of extreme environments such as hot deserts. The over-induction of the *yjbEFGH* and *wca* operons, involved in EPS production by *E. coli,* may be an adaptive response to osmotic stress (Ionescu and Belkin 2009). Several genes involved in biofilm formation in *Acinetobacter baumannii,* such as the membrane transporter *ompA* and the *pgaABCD* adhesion locus, have been shown to contribute to desiccation tolerance of this bacterium (Selasi et al. 2016). In addition, Yoshimura et al. (2007) reported that the expression of Group 2 sigma factor gene (*sigJ*) in the cyanobacterium *Nostoc* sp HK-01 was increased during dehydration, resulting in the upregulation of genes related to exopolysaccharide biosynthesis. Other sigma factors, such as *AlgU* and *RpoS,* were also found to be essential for the tolerance of *Pseudomonas fluorescens* against desiccation and hyperosmolarity stresses, due in part to the activation of EPS biosynthesis pathways (Schnider-Keel et al. 2001; Sarniguet et al. 1995).

10.7 Adaptation to Oligotrophy

Most hot desert soils are considered to be oligotrophic environments in which energy availability, especially in the form of readily accessible carbon and nitrogen sources, is extremely limited (Kieft 2002; Dommergues et al. 1978; Morita 1988). In addition, metabolic activity in hot desert soils (and therefore the capacity to acquire energy) is limited by the extreme levels of desiccation experienced throughout the year (Lebre et al. 2017). To cope with this limitation, microorganisms downregulate many energy-intensive processes such as flagellar motility (Breeuwer et al. 2003; Burgess et al. 2016; Katoh et al. 2004; Crowe et al. 2002), as well as metabolic pathways that depend on water, such as photosynthesis (Billi and Potts 2002). Photosynthesis also generates ROS as by-products, and therefore, its downregulation is also a preservation strategy to prevent DNA damage (Bär et al. 2002; Gruzdev et al. 2012). However, the downregulation of major anabolic pathways does not mean that microbial communities are totally dormant in hot desert soils. A recent study using transcriptomic analysis (León-Sobrino et al. 2019) has shown that microbial assemblages are metabolically active in hyper-arid soils, despite the low water availability in the immediate environment. Crucially, non-photosynthetic microbiota in desiccated desert soils were shown to express many growth-related and major catabolic pathway genes, in addition to desiccation-tolerance gene markers. Other studies have shown the induction of pathways involved in aerobic respiration (such as glycolysis and the TCA cycle) during desiccation (Burgess et al. 2016; Katoh et al. 2004; Rajeev et al. 2013). For instance, taxa such as *Bradyrhizobium japonicum* and *Salmonella enterica* have been shown to overexpress genes involved in fatty acid metabolism as a response to desiccation stress. One such example is 3-ketoacyl-CoA thiolase, which is involved in the degradation of long chain fatty acids into acetyl-coA (Katoh et al. 2004; Crowe et al. 2002). Fatty acid oxidation generates more ATP per atom than glucose, and is therefore a viable alternative source of energy in arid and hyper-arid environments while freeing glucose for conversion into trehalose, which plays a major role in tolerance to desiccation (Crowe et al. 1998, 2002).

Anoxygenic phototrophy, where light is harvested by bacteriochlorophylls (BChl) and rhodopsins to generate a membrane potential and drive ATP generation, has also been identified as a possible adaptive mechanism for energy metabolism under desiccated conditions and in oligotrophic niches (Leung et al. 2020). Anoxygenic phototrophy allows bacteria to perform photosynthesis without the requirement of water as an electron donor (Larkum et al. 2018), a potential advantage for organisms in low water activity environments. Another potential advantage of anoxygenic phototrophy is the ability to use inorganic and organic compounds, including Fe^{2+}, H_2, HS , NO^{2-}, and organic central metabolism intermediates, as electrons donors to generate an electron gradient across the cell membrane (Fischer et al. 2016). Anoxygenic phototrophic bacteria are found across a broad range of taxa, including thermophiles present in poly-extreme environments (Yurkov and Hughes 2017; Madigan 2017; Madigan and Jung 2009; Tahon and Willems 2017; Guerrero et al. 2017). In particular, rhodopsin-light harvesting systems, which involve the activity of a proton-pump rhodopsin encoded by a single gene, are assumed to be simple enough for dissemination via lateral gene transfer, and are present in approximately 50% of the metagenome-assembled genomes (MAGs) derived from desert environments (Leung et al. 2020; Pushkarev et al. 2018; Finkel et al. 2013). Microorganisms that use rhodopsin as an energy transduction mechanism under anoxygenic conditions do not have the required machinery for autotrophic CO_2 assimilation, and are therefore photoheterotrophic (Larkum et al. 2018). By comparison, BChl-dependent light harvesting systems have been documented in a limited number of phyla in desert soil biomes (Leung et al. 2020), such as Proteobacteria isolated from Antarctic continental soils (Tahon and Willems 2017). Considering the prevalence of Proteobacteria in hot desert soils (Bachar et al. 2012; Lester et al. 2007), it is expected that BChl-dependent anoxygenic phototrophic bacteria should also be functionally important in these niches (Makhalanyane et al. 2015). A recent study (Tang et al. 2018) showed that anoxygenic phototrophic bacteria containing BChl-a, which is able to absorb near-infrared light (NIR), have a positive effect on the growth and biodiversity of biological soil crusts communities in the Hopq Desert in China. This suggests that anoxygenic photosynthesis, in addition to being a survival mechanism under desiccation stress, might be an essential process for the maintenance and active growth of microbial communities in desert ecosystems (Tang et al. 2018).

More recently, studies have pointed to the viability of rare gas harvesting and oxidation (lithoheterotrophy), particularly of dihydrogen, carbon monoxide, and methane, as a mechanism for growth and survival of microbial communities in extremely oligotrophic desert ecosystems (Leung et al. 2020). Soil communities have long been known to consume trace amounts of atmospheric H_2 (Schmidt 1969; Greening et al. 2015), but recent studies suggest that this process may be more ecologically important than previously considered, particularly in extreme soil biotopes which have limited capacity for photoautotrophy (Ji et al. 2017; Ortiz et al. 2020). The biological consumption of atmospheric H_2 is primarily attributed to high-affinity group 5 [NiFe]-hydrogenases, first discovered in *Streptomyces* spp. (Meredith et al. 2014; Piché-Choquette and Constanta 2019). These have been now identified in numerous microorganisms belonging to soil-dominant taxa such as Actinobacteria, Chloroflexi, and Alphaproteobacteria (Greening et al. 2015; Greening et al. 2016; Islam et al. 2019; Schäfer et al. 2013; Constant et al. 2011). Candidate genera from the phylum Actinobacteria, Candidatus Eremiobacteraeota (desert bacterial phylum), and Candidatus Dormibacteraeota (dormant bacterial phylum) have been particularly associated with the processes of H_2-dependent CO_2 fixation and energy generation from H_2 oxidation (Ji et al. 2017). Originally identified in remote Antarctic soils (Ji et al. 2017), microbial genomes encoding group 5 [NiFe]-hydrogenase genes have now been identified in hot desert soils (the Atacama Desert) (Lynch et al. 2014), suggesting that trace gas metabolism, as a source of fixed carbon and energy, may have much wider ecological significance than originally proposed. It has been recently noted that H_2 oxidation is a hydrogenic (water generating) reaction: offering the intriguing prospect that this energy-generating reaction may also contribute metabolic water to desiccated soil microbial communities (Lynch et al. 2014).

CO consumption has been shown to occur in desert soils (Ji et al. 2017). The enzyme responsible for CO oxidation, carbon monoxide dehydrogenase (CODH), is widely distributed among phyla that are prevalent in desert environments, including Proteobacteria (King and Weber 2007), Actinobacteria (Ji et al. 2017), and Chloroflexi (Islam et al. 2019). Additionally, CODH has been shown to be upregulated when organic carbon is exhausted (Schäfer et al. 2013; Cordero et al. 2019; Patrauchan et al. 2012). Importantly, two isolates from extreme halophilic niches, the euryarchaeotal *Halorubrum* str. BV1 and the proteobacterial *Alkalilimnicola ehrlichii* MLHE-1, were shown to consume CO at very low water activities (King 2017), reinforcing the suggestion that CO consumption is a viable energy and carbon source for desert communities, even during long periods of desiccation.

By contrast, methane consumption is considered to have a more limited role in survival and maintenance of desert microbial communities, possibly due to its stronger dependence on water content, relative to other rare gas oxidation processes (Leung et al. 2020; Fischer et al. 2016). Low water potential is known to severely inhibit microbial methane uptake and oxidation (King 2017).

10.8 Adaptations to UV Radiation

Among the extreme abiotic stresses in desert niches, solar exposure and associated ionizing radiation (IR), in the form of high-intensity UV radiation, have severe detrimental effects on DNA and other macromolecules (Lebre et al. 2017). In particular, UV irradiation can induce the formation of free radicals and other ROS that cause oxidative damage to DNA and proteins, including DNA double-strand breaks (DSBs) (Rastogi et al. 2010). Therefore, the abilities of cells to protect and quickly repair DNA damage are seen as crucial for the survival of desert soil microbial communities.

The main two-step mechanism by which UV-tolerant microorganisms are thought to be able to repair UV-induced damage is termed "extended synthesis-dependent strand annealing" (ESDSA) (Zahradka et al. 2006). In the first step, which is *polA*-dependent, overlapping fragments are used as templates for single-strand extension from the DSBs, resulting in fragments with long overhangs. After several rounds of re-annealing and extension of the fragments, *recA* drives the maturation of the re-annealed fragments through homologous recombination. Importantly, the ESDSA pathway relies on there being multiple copies of the cell's chromosome (polyploidy) to guarantee that at least one of the overlapping regions contains the correct DNA code (Zahradka et al. 2006). Polyploidy has been described for both bacterial and archaeal taxa that are routinely found in desert ecosystems, including cyanobacteria (Griese et al. 2011) and haloarchaea (Zerulla and Soppa 2014), and is associated with both UV and desiccation tolerance (Soppa 2014). Additionally, several proteins of unknown function that are common to *Deinococcus* species could assist the ESDSA pathway (Liu et al. 2003; Sghaier et al. 2008; Makarova et al. 2007; Tanaka et al. 2004). These proteins, named Ddr (DNA damage response) and Ppr (pleiotropic protein promoting DNA repair) proteins, are thought to stabilize the long single-stranded overhangs or to enhance the activity and expression of RecA (Lim et al. 2019). The production of small proteins, termed alpha/beta-type SASPs, during sporulation is also known to protect cells from ROS-induced damage and UV radiation (Sasaki et al. 2014; Mertens et al. 2018; Mason et al. 1988; Mason and Setlow 1984; Connors et al. 1986) by binding to DNA and altering its photochemical properties (Fairhead et al. 1994; Setlow and Setlow 1993).

Proteins are also particularly vulnerable to UV-driven ROS damage, especially via amino acid oxidation which leads to protein denaturation and loss of activity (Daly 2009; Fredrickson et al. 2008; Fagliarone et al. 2017; Ezraty et al. 2017). Studies on the resistance of *Deinococcus* species to ROS-induced damage have shown that resistance to IR exposure is related to protein oxidation rather than DNA damage (Daly et al. 2007; Krisko and Radman 2010). A growing body of evidence has shown that UV resistance is linked to Mn^{2+} concentrations in the cell (Daly 2009). Mn^{2+} acts as a protectant due to its ability to scavenge ROS, as well as to bind to peptide complexes, thus physically protecting them from ROS-mediated oxidation (Daly et al. 2010; Singh et al. 2019; Culotta and Daly 2013). Accumulation of Mn^{2+} has been described in several desert soil taxa, including the cyanobacterium *Synechocystis sp.* PCC 6803 (Paulino-Lima et al. 2016; Keren et al. 2002) and *D. radiodurans*, which can accommodate as much as 300-fold higher concentrations of Mn^{2+} compared to UV-sensitive taxa (Sghaier et al. 2008). Mn^{2+} is also bound as a cofactor in the active sites of catalases and peroxide dismutases (SOD), both of which are ROS-scavenging enzymes commonly associated with tolerance to oxidative stress (Rajeev et al. 2013; Liu et al. 2003). Notably, the formation of rock varnishes in deserts, which are dark layered coats of clay minerals mixed with manganese also involved in UV protection (Potter and Rossman 1979), has been linked to fungal and bacterial oxidation of Mn^{2+} (Kuhlman et al. 2006; Lang-Yona et al. 2018; Dorn and Oberlander 1981; Nagy et al. 1991). Since this oxidation is also a mechanism by which microorganisms scavenge ROS inside the cell, rock varnishes are postulated to provide microbial communities with protection against both UV radiation and desiccation (Kuhlman et al. 2008).

10.9 Temperature Adaptations

Temperatures at or near the surface of desert soils in hot deserts regularly exceed 65°C (Cowan et al. 2019) and therefore can be very challenging for the survival of microbial communities. In addition to driving water loss via evaporation, high temperatures lead to increased membrane fluidity which may result in cell leakage and rupture (Los and Murata 2004), and loss of activity of thermosensitive proteins via heat denaturation (Matsuura et al. 2015).

One of the principal mechanisms of adaptation to elevated temperatures is the expression of specific proteins that stabilize other macromolecules. These include the shock-response chaperones *groES/EL* and *dnaJ-dnaK*, which are induced in the filamentous cyanobacterium *Microcoleus vaginatus* during heat shock (Rajeev et al. 2013). Another class of chaperones important in maintaining protein homeostasis is termed the small heat-shock proteins (sHsps), which are induced in response to a number of stressors, including elevated temperatures and low pH (Voth and Jakob 2017). Thermotolerance in cyanobacteria is dependent on the activity of sHsps, as well as other families of heat-shock proteins such as Hsp100 and Hsp90 (Rajaram et al. 2014). The model mechanism for the activity of these chaperones involves the unfolding of the protein structure upon activation, thereby exposing intrinsically disordered (ID) regions in the chaperones that interact with other unfolding proteins, thereby stabilizing them (Voth and Jakob 2017).

In addition to its role as an osmoprotectant, the EPS biofilm matrix has also been proposed to confer protection to microorganisms against extreme temperature fluctuations (Costa et al. 2018; Yin et al. 2019). A study on the thermophilic acidophile *Sulfolobus* sp. showed an increase in biofilm formation at higher temperatures (Koerdt et al. 2010), while EPS from cold-adapted bacteria such as *Bacteriovorax* and *Colwellia* GW185 was shown to increase survivability at low temperatures (Williams et al. 2009), as well as to protect cells from freeze–thaw cycles (Caruso et al. 2018). Similarly, accumulation of the compatible solute trehalose has been associated with increased thermotolerance in several bacteria and fungi (Welsh 2000). For instance, increases in trehalose concentration strongly correlated to the increased heat resistance of *Streptomyces griseus* spores (McBride and Ensign 1987). In another study, *Rhizobium etli* mutants lacking the trehalose-6-P-synthase gene (*otsA*) were unable to synthesize trehalose and were sensitive to high temperatures as a result (Reina-Bueno et al. 2012). Under heat shock, trehalose increases melting temperatures of bound proteins and stabilizes partially folded proteins (Baptista et al. 2008; Singer and Lindquist 1998), while also inhibiting the formation of denatured protein aggregates (Hottiger et al. 1994). Additionally, it has been suggested that trehalose might act synergistically with chaperones to increase the thermostability of proteins (Jain and Roy 2009).

10.10 Concluding Remarks

Recent studies focusing on the survival of microbial life in extreme environments have revealed a broad array of adaptive strategies by which microorganisms survive (and thrive) under challenging conditions. Many of these adaptive mechanisms are universal responses that are independent of the type of stress (Kieft 2002), i.e., occur as part of a broader response strategy to common stressors such as the production of ROS or changes in osmotic gradients across the cell membrane. Recently, new advanced "omics" approaches have enabled researchers to investigate more holistic models of hot desert ecosystem functioning. The reader is referred to Chap. 2 of this book for a comprehensive description of the novel methodologies used or that could be used in hot desert microbial ecology studies. For instance, a recent study combining metagenomics and functional characterization has revealed the presence of novel high-affinity rare gas sequestration proteins that could drive microbial subsistence under extreme nutrient and water deprivation (Ortiz et al. 2020). A recent viromics study (Van Goethem et al. 2019) has also highlighted the potential role of viruses in the horizontal transfer of sporulation genes and auxiliary metabolic genes (AMGs) across hosts under extreme conditions, thus suggesting that viruses might be important for the survivability of communities in extreme biomes. Given the concerns about climate change, and given the desertification that accompanies such changes, it is timely that we attempt to comprehend the role and responses of desert communities to inhospitable environments.

References

Anderson KL, Apolinario EE, Sowers KR (2012) Desiccation as a long-term survival mechanism for the archaeon Methanosarcina barkeri. Appl Environ Microbiol. 78:1473–1479. https://doi.org/10.1128/AEM.06964-11

Anderson D, Ferreras E, Trindade M, Cowan D (2015) A novel bacterial water Hypersensitivity-like protein shows in vivo protection against cold and freeze damage. FEMS Microbiol Lett. 362:1–8. https://doi.org/10.1093/femsle/fnv110

Artus NN, Uemura M, Steponkus PL, Gilmour SJ, Lin C, Thomashow MF (1996) Constitutive expression of the cold-regulated Arabidopsis thaliana COR15a gene affects both chloroplast and protoplast freezing tolerance. Proc Natl Acad Sci U S A. 93:13404–13409. https://doi.org/10.1073/pnas.93.23.13404

Bachar A, Soares MIM, Gillor O (2012) The effect of resource islands on abundance and diversity of bacteria in arid soils. Microb Ecol. 63:694–700. https://doi.org/10.1007/s00248-011-9957-x

Baptista RP, Pedersen S, Cabrita GJM, Otzen DE, Cabral JMS, Melo EP (2008) Thermodynamics and mechanism of cutinase stabilization by trehalose. Biopolymers. 89:538–547. https://doi.org/10.1002/bip.20926

Bär M, Von Hardenberg J, Meron E, Provenzale A (2002) Modelling the survival of bacteria in drylands: the advantage of being dormant. Proc R Soc B Biol Sci. 269:937–942. https://doi.org/10.1098/rspb.2002.1958

Barrow CJ. World atlas of desertification (United nations environment programme). Edward Arn. In: Middleton N, Thomas D, editors. Land Degradation & Development. Edward Arn. London; 1992. pp. 249–249.

Batista LFZ, Kaina B, Meneghini R, Menck CFM (2009) How DNA lesions are turned into powerful killing structures: insights from UV-induced apoptosis. Mutat Res - Rev Mutat Res. 681:197–208. https://doi.org/10.1016/j.mrrev.2008.09.001

Battaglia M, Olvera-Carrillo Y, Garciarrubio A, Campos F, Covarrubias AA (2008) The enigmatic LEA proteins and other hydrophilins. Plant Physiol. 148:6–24. https://doi.org/10.1104/pp.108.120725

Battista JR, Park MJ, McLemore AE (2001) Inactivation of two homologues of proteins presumed to be involved in the desiccation tolerance of plants sensitizes Deinococcus radiodurans R1 to desiccation. Cryobiology. 43:133–139. https://doi.org/10.1006/cryo.2001.2357

Bell C, McIntyre N, Cox S, Tissue D, Zak J (2008) Soil microbial responses to temporal variations of moisture and temperature in a Chihuahuan Desert grassland. Microb Ecol. 56:153–167. https://doi.org/10.1007/s00248-007-9333-z

Belnap J (1995) Surface disturbances: their role in accelerating desertification. Environ Monit Assess. 37:39–57. Available: papers2://publication/uuid/BEABF50C-E2C4-49E1-8C54-A3B0772A4B0E

Berg N, Steinberger Y (2008) Role of perennial plants in determining the activity of the microbial community in the Negev Desert ecosystem. Soil Biol Biochem. 40:2686–2695. https://doi.org/10.1016/j.soilbio.2008.07.019

Billi D, Potts M (2000) Life without water: responses of prokaryotes to desiccation. In: Environmental stressors and gene responses. Elsevier, Amsterdam, pp 181–192

Billi D, Potts M (2002) Life and death of dried prokaryotes. Res Microbiol. 153:7–12

Breeuwer P, Lardeau A, Peterz M, Joosten HM (2003) Desiccation and heat tolerance of Enterobacter sakazakii. J Appl Microbiol. 95:967–973. https://doi.org/10.1046/j.1365-2672.2003.02067.x

Bremer E, Krämer R (2019) Responses of microorganisms to osmotic stress. Annu Rev Microbiol. 73:313–334. https://doi.org/10.1146/annurev-micro-020518-115504

Brown GR, Sutcliffe IC, Bendell D, Cummings SP (2000) The modification of the membrane of Oceanomonas baumannii(T) when subjected to both osmotic and organic solvent stress. FEMS Microbiol Lett. 189:149–154. https://doi.org/10.1016/S0378-1097(00)00269-X

Bueche M, Wunderlin T, Roussel-Delif L, Junier T, Sauvain L, Jeanneret N et al (2013) Quantification of endospore-forming firmicutes by quantitative PCR with the functional gene spo0A. Appl Environ Microbiol. 79:5302–5312. https://doi.org/10.1128/AEM.01376-13

Burgess CM, Gianotti A, Gruzdev N, Holah J, Knøchel S, Lehner A et al (2016) The response of foodborne pathogens to osmotic and desiccation stresses in the food chain. Int J Food Microbiol. 221:37–53. https://doi.org/10.1016/j.ijfoodmicro.2015.12.014

Caruso C, Rizzo C, Mangano S, Poli A, di Donato P, Finore I et al (2018) Production and biotechnological potential of extracellular polymeric substances from sponge-associated Antarctic bacteria. Appl Environ Microbiol. 84:1–18. https://doi.org/10.1128/AEM.01624-17

Chakrabortee S, Boschetti C, Walton LJ, Sarkar S, Rubinsztein DC, Tunnacliffe A (2007) Hydrophilic protein associated with desiccation tolerance exhibits broad protein stabilization function. Proc Natl Acad Sci U S A. 104:18073–18078. https://doi.org/10.1073/pnas.0706964104

Chamizo S, Cantón Y, Rodríguez-Caballero E, Domingo F (2016) Biocrusts positively affect the soil water balance in semiarid ecosystems. Ecohydrology. 9:1208–1221. https://doi.org/10.1002/eco.1719

Chaplin M (2006) Do we underestimate the importance of water in cell biology? Nat Rev Mol Cell Biol. 7:861–866. https://doi.org/10.1038/nrm2021

Chen M, Alexander M (1973) Survival of soil bacteria during prolonged desiccation. Soil Biol Biochem. 5:213–221. https://doi.org/10.1016/0038-0717(73)90004-7

Chen THH, Murata N (2002) Enhancement of tolerance of abiotic stress by metabolic engineering of betaines and other compatible solutes. Curr Opin Plant Biol. 5:250–257. https://doi.org/10.1016/S1369-5266(02)00255-8

Chen Y, Li C, Zhang B, Yi J, Yang Y, Kong C et al (2019) The role of the Late Embryogenesis-Abundant (LEA) protein family in development and the abiotic stress response: a comprehensive expression analysis of potato (Solanum Tuberosum). Genes (Basel). 10:1–16. https://doi.org/10.3390/genes10020148

Ciccarelli FD, Bork P (2005) The WHy domain mediates the response to desiccation in plants and bacteria. Bioinformatics. 21:1304–1307. https://doi.org/10.1093/bioinformatics/bti221

Connon SA, Lester ED, Shafaat HS, Obenhuber DC, Ponce A (2007) Bacterial diversity in hyperarid atacama desert soils. J Geophys Res Biogeosciences. 112:1–9. https://doi.org/10.1029/2006JG000311

Connors MJ, Mason JM, Setlow P (1986) Cloning and nucleotide sequencing of genes for three small, acid-soluble proteins from Bacillus subtilis spores. J Bacteriol. 166:417–425. https://doi.org/10.1128/jb.166.2.417-425.1986

Constant P, Chowdhury SP, Hesse L, Pratscher J, Conrad R (2011) Genome data mining and soil survey for the novel group 5 [NiFe]-hydrogenase to explore the diversity and ecological importance of presumptive high-affinity H 2-oxidizing bacteria. Appl Environ Microbiol. 77: 6027–6035. https://doi.org/10.1128/AEM.00673-11

Cooksey KE, Wigglesworth-Cooksey B (1995) Adhesion of bacteria and diatoms to surfaces in the sea: a review. Aquat Microb Ecol. 9:87–96. https://doi.org/10.3354/ame009087

Cordero PRF, Bayly K, Man Leung P, Huang C, Islam ZF, Schittenhelm RB et al (2019) Atmospheric carbon monoxide oxidation is a widespread mechanism supporting microbial survival. ISME J. 13:2868–2881. https://doi.org/10.1038/s41396-019-0479-8

Costa OYA, Raaijmakers JM, Kuramae EE (2018) Microbial extracellular polymeric substances: ecological function and impact on soil aggregation. Front Microbiol. 9:1–14. https://doi.org/10.3389/fmicb.2018.01636

Cowan DA, Hopkins DW, Jones BE, Maggs-Kölling G, Majewska R, Ramond J-B (2019) Microbiomics of Namib Desert habitats. Extremophiles.:1–28. https://doi.org/10.1007/s00792-019-01122-7

Crawford CS, Gosz JR (1982) Desert ecosystems: their resources in space and time. Environ Conserv. 9:181–195

Crowe JH, Carpenter JF, Crowe LM (1998) The role of vitrification in anhydrobiosis. Annu Rev Physiol. 60:73–103. https://doi.org/10.1146/annurev.physiol.60.1.73

Crowe JH, Oliver AE, Tablin F (2002) Is there a single biochemical adaptation to anhydrobiosis? Integr Comp Biol. 42:497–503. https://doi.org/10.1093/icb/42.3.497

Culotta VC, Daly MJ (2013) Manganese complexes: diverse metabolic routes to oxidative stress resistance in prokaryotes and yeast. Antioxidants Redox Signal. 19:933–944. https://doi.org/10.1089/ars.2012.5093

Cytryn EJ, Sangurdekar DP, Streeter JG, Franck WL, Chang WS, Stacey G et al (2007) Transcriptional and physiological responses of Bradyrhizobium japonicum to desiccation-induced stress. J Bacteriol. 189:6751–6762. https://doi.org/10.1128/JB.00533-07

D'Odorico P, Bhattachan A, Davis KF, Ravi S, Runyan CW (2013) Global desertification: Drivers and feedbacks. Adv Water Resour. 51:326–344. https://doi.org/10.1016/j.advwatres.2012.01.013

Daly MJ (2009) A new perspective on radiation resistance based on Deinococcus radiodurans. Nat Rev Microbiol. 7:237–245. https://doi.org/10.1038/nrmicro2073

Daly MJ, Gaidamakova EK, Matrosova VY, Vasilenko A, Zhai M, Leapman RD et al (2007) Protein oxidation implicated as the primary determinant of bacterial radioresistance. PLoS Biol. 5:769–779. https://doi.org/10.1371/journal.pbio.0050092

Daly MJ, Gaidamakova EK, Matrosova VY, Kiang JG, Fukumoto R, Lee DY et al (2010) Small-molecule antioxidant proteome-shields in Deinococcus radiodurans. PLoS One. 5:10–15. https://doi.org/10.1371/journal.pone.0012570

Danyluk J, Houde M, Rassart É, Sarhan F (1994) Differential expression of a gene encoding an acidic dehydrin in chilling sensitive and freezing tolerant gramineae species. FEBS Lett. 344: 20–24. https://doi.org/10.1016/0014-5793(94)00353-X

Dommergues YR, Belser LW, Schmidt EL (1978) Limiting factors for microbial growth and activity in soil. In: Advances in microbial ecology. Springer, Boston, MA, pp 49–104

Dorn RI, Oberlander TM (1981) Microbial origin of desert varnish. Science (80-). 213:1245–1247. https://doi.org/10.1126/science.213.4513.1245

Ezraty B, Gennaris A, Barras F, Collet JF (2017) Oxidative stress, protein damage and repair in bacteria. Nat Rev Microbiol. 15:385–396. https://doi.org/10.1038/nrmicro.2017.26

Fagliarone C, Mosca C, Ubaldi I, Verseux C, Baqué M, Wilmotte A et al (2017) Avoidance of protein oxidation correlates with the desiccation and radiation resistance of hot and cold desert strains of the cyanobacterium Chroococcidiopsis. Extremophiles. 21:981–991. https://doi.org/10.1007/s00792-017-0957-8

Fairhead H, Setlow B, Waites WM, Setlow P (1994) Small, acid-soluble proteins bound to DNA protect Bacillus subtilis spores from killing by dry heat. Appl Environ Microbiol. 60:2647–2649. https://doi.org/10.1128/aem.61.7.2787-2790.1995

Fierer N, Schimel JP, Holden PA (2003) Influence of drying-rewetting frequency on soil bacterial community structure. Microb Ecol. 45:63–71. https://doi.org/10.1007/s00248-002-1007-2

Fierer N, Leff JW, Adams BJ, Nielsen UN, Bates ST, Lauber CL et al (2012) Cross-biome metagenomic analyses of soil microbial communities and their functional attributes. Proc Natl Acad Sci U S A. 109:21390–21395. https://doi.org/10.1073/pnas.1215210110

Finkel OM, Béjà O, Belkin S (2013) Global abundance of microbial rhodopsins. ISME J. 7:448–451. https://doi.org/10.1038/ismej.2012.112

Fischer WW, Hemp J, Johnson JE (2016) Evolution of oxygenic photosynthesis. Annu Rev Earth Planet Sci. 44:647–683. https://doi.org/10.1146/annurev-earth-060313-054810

Flemming H-C, Wingender J (2010) The biofilm matrix. Nat Rev Microbiol. 8:623–633. https://doi.org/10.1038/nrmicro2415

Flemming HC, Wingender J, Szewzyk U, Steinberg P, Rice SA, Kjelleberg S (2016) Biofilms: an emergent form of bacterial life. Nat Rev Microbiol. 14:563–575. https://doi.org/10.1038/nrmicro.2016.94

Fredrickson JK, Li SMW, Gaidamakova EK, Matrosova VY, Zhai M, Sulloway HM et al (2008) Protein oxidation: Key to bacterial desiccation resistance? ISME J. 2:393–403. https://doi.org/10.1038/ismej.2007.116

Gallagher PE, Duker NJ (1989) Formation of purine photoproducts in a defined human dna sequence. Photochem Photobiol. 49:599–605. https://doi.org/10.1111/j.1751-1097.1989.tb08430.x

Gao Q, Garcia-Pichel F (2011) Microbial ultraviolet sunscreens. Nat Rev Microbiol. 9:791–802. https://doi.org/10.1038/nrmicro2649

Garay-Arroyo A, Colmenero-Flores JM, Garciarrubio A, Covarrubias AA (2000) Highly hydrophilic proteins in prokaryotes and eukaryotes are common during conditions of water deficit. J Biol Chem. 275:5668–5674. https://doi.org/10.1074/jbc.275.8.5668

García AH (2011) Anhydrobiosis in bacteria: from physiology to applications. J Biosci. 36:939–950. https://doi.org/10.1007/s12038-011-9107-0

Greening C, Constant P, Hards K, Morales SE, Oakeshott JG, Russell RJ et al (2015) Atmospheric hydrogen scavenging: from enzymes to ecosystems. Appl Environ Microbiol. 81:1190–1199. https://doi.org/10.1128/AEM.03364-14

Greening C, Biswas A, Carere CR, Jackson CJ, Taylor MC, Stott MB et al (2016) Genomic and metagenomic surveys of hydrogenase distribution indicate H 2 is a widely utilised energy source for microbial growth and survival. ISME J. 10:761–777. https://doi.org/10.1038/ismej.2015.153

Griese M, Lange C, Soppa J (2011) Ploidy in cyanobacteria. FEMS Microbiol Lett. 323:124–131. https://doi.org/10.1111/j.1574-6968.2011.02368.x

Gruzdev N, Mcclelland M, Porwollik S, Ofaim S, Pinto R, Saldinger-Sela S (2012) Global transcriptional analysis of dehydrated Salmonella enterica serovar Typhimurium. Appl Environ Microbiol. 78:7866–7875. https://doi.org/10.1128/AEM.01822-12

Guerrero LD, Vikram S, Makhalanyane TP, Cowan DA (2017) Evidence of microbial rhodopsins in Antarctic Dry Valley edaphic systems. Environ Microbiol. 19:3755–3767. https://doi.org/10.1111/1462-2920.13877

Gunasekera TS, Csonka LN, Paliy O (2008) Genome-wide transcriptional responses of Escherichia coli K-12 to continuous osmotic and heat stresses. J Bacteriol. 190:3712–3720. https://doi.org/10.1128/JB.01990-07

Halverson LJ, Firestone MK (2000) Differential effects of permeating and nonpermeating solutes on the fatty acid composition of Pseudomonas putida. Appl Environ Microbiol. 66:2414–2421. https://doi.org/10.1128/AEM.66.6.2414-2421.2000

Harding T, Brown MW, Simpson AGB, Roger AJ (2016) Osmoadaptative strategy and its molecular signature in obligately halophilic heterotrophic protists. Genome Biol Evol. 8: 2241–2258. https://doi.org/10.1093/gbe/evw152

Higgins D, Dworkin J (2012) Recent progress in Bacillus subtilis sporulation. FEMS Microbiol Rev. 36:131–148. https://doi.org/10.1111/j.1574-6976.2011.00310.x

Hottiger T, De Virgilio C, Hall MN, Boller T, Wiemken A (1994) The role of trehalose synthesis for the acquisition of thermotolerance in yeast: II. Physiological concentrations of trehalose increase the thermal stability of proteins in vitro. Eur J Biochem. 219:187–193. https://doi.org/10.1111/j.1432-1033.1994.tb19929.x

Hundertmark M, Hincha DK (2008) LEA (Late Embryogenesis Abundant) proteins and their encoding genes in Arabidopsis thaliana. BMC Genomics. 9:1–22. https://doi.org/10.1186/1471-2164-9-118

Ikeno S, Haruyama T (2013) Boost protein expression through co-expression of LEA-like peptide in Escherichia coli. PLoS One. 8. https://doi.org/10.1371/journal.pone.0082824

Imlay J, Linn S. DNA damage and oxygen radical toxicity. Science (80-). 1988 [cited 20 Dec 2013]. Available: http://www.sciencemag.org/content/240/4857/1302.short

Ionescu M, Belkin S (2009) Overproduction of exopolysaccharides by an Escherichia coli K-12 rpoS mutant in response to osmotic stress. Appl Environ Microbiol. 75:483–492. https://doi.org/10.1128/AEM.01616-08

Islam ZF, Cordero PRF, Feng J, Chen YJ, Bay SK, Jirapanjawat T et al (2019) Two Chloroflexi classes independently evolved the ability to persist on atmospheric hydrogen and carbon monoxide. ISME J. 13:1801–1813. https://doi.org/10.1038/s41396-019-0393-0

Jain NK, Roy I (2009) Effect of trehalose on protein structure. Protein Sci. 18:24–36. https://doi.org/10.1002/pro.3

Jaspard E, Hunault G (2014) Comparison of amino acids physico-chemical properties and usage of late embryogenesis abundant proteins, hydrophilins and WHy domain. PLoS One. 9:1–16. https://doi.org/10.1371/journal.pone.0109570

Ji M, Greening C, Vanwonterghem I, Carere CR, Bay SK, Steen JA et al (2017) Atmospheric trace gases support primary production in Antarctic desert surface soil. Nature. 552:400–403. https://doi.org/10.1038/nature25014

Jin X, Cao D, Wang Z, Ma L, Tian K, Liu Y et al (2019) Genome-wide identification and expression analyses of the LEA protein gene family in tea plant reveal their involvement in seed development and abiotic stress responses. Sci Rep. 9:1–15. https://doi.org/10.1038/s41598-019-50645-8

Kaplan-levy RN, Hadas O, Summers ML, Sukenik A (2010) Dormancy and Resistance in Harsh. Environments. 21:189–202. https://doi.org/10.1007/978-3-642-12422-8

Katoh H, Asthana RK, Ohmori M (2004) Gene expression in the cyanobacterium Anabaena sp. PCC7120 under desiccation. Microb Ecol. 47:164–174. https://doi.org/10.1007/s00248-003-1043-6

Keren N, Kidd MJ, Penner-Hahn JE, Pakrasi HB (2002) A light-dependent mechanism for massive accumulation of manganese in the photosynthetic bacterium Synechocystis sp. PCC 6803. Biochemistry. 41:15085–15092. https://doi.org/10.1021/bi026892s

Kieft TL (2002) Hot desert soil microbial communities. Encycl Environ Microbiol. 86:1576–1586

King GM (2017) Water potential as a master variable for atmosphere-soil trace gas exchange in arid and semi-arid ecosystems. In: Steven B (ed) The biology of arid soils. De Gruyter, Berlin, pp 31–45

King GM, Weber CF (2007) Distribution, diversity and ecology of aerobic CO-oxidizing bacteria. Nat Rev Microbiol. 5:107–118. https://doi.org/10.1038/nrmicro1595

Klähn S, Hagemann M (2011) Compatible solute biosynthesis in cyanobacteria. Environ Microbiol. 13:551–562. https://doi.org/10.1111/j.1462-2920.2010.02366.x

Knowles EJ, Castenholz RW (2008) Effect of exogenous extracellular polysaccharides on the desiccation and freezing tolerance of rock-inhabiting phototrophic microorganisms. FEMS Microbiol Ecol. 66:261–270. https://doi.org/10.1111/j.1574-6941.2008.00568.x

Laity JJ (2009) In: Sons JW (ed) Deserts and desert environments, vol 3. Wiley-Blackwell, Chichester, UK

UNEP-WCMC. A spatial analysis approach to the global delineation of dryland areas of relevance to the CBD Programme of Work on Dry and Subhumid Lands. 2007.

Warren A (1993) Desertification as a global environmental issue. GeoJ 31:11–14. https://doi.org/10.1007/BF00815896

Tucker CJ, Dregne HE, Newcomb WW (1991) Expansion and contraction of the Sahara desert from 1980 to 1990. Science (80-). 253:299–300

Turner J, Anderson P, Lachlan-Cope T, Colwell S, Phillips T, Kirchgaessner A et al (2009) Record low surface air temperature at Vostok station, Antarctica. J Geophys Res Atmos. 114:1–14. https://doi.org/10.1029/2009JD012104

Makhalanyane TP, Valverde A, Gunnigle E, Frossard A, Ramond JB, Cowan DA (2015) Microbial ecology of hot desert edaphic systems. FEMS Microbiol Rev. 39:203–221. https://doi.org/10.1093/femsre/fuu011

Lebre PH, De Maayer P, Cowan DA (2017) Xerotolerant bacteria: Surviving through a dry spell. Nat Rev Microbiol. 15:285–296. https://doi.org/10.1038/nrmicro.2017.16

Pointing SB, Belnap J (2012) Microbial colonization and controls in dryland systems. Nat Rev Microbiol. 10:551–562. https://doi.org/10.1038/nrmicro2831

Pointing SB, Belnap J (2014) Disturbance to desert soil ecosystems contributes to dust-mediated impacts at regional scales. Biodivers Conserv. 23:1659–1667. https://doi.org/10.1007/s10531-014-0690-x

Potts M (1994) Desiccation tolerance of prokaryotes. Microbiol Rev. 58:755–805. https://doi.org/10.1128/mmbr.58.4.755-805.1994

Potts M (2001) Desiccation tolerance: a simple process? Trends Microbiol. 9:553–559. https://doi.org/10.1016/S0966-842X(01)02231-4

Schlesinger WH, Reynolds JF, Cunningham GL, Huenneke LF, Jarrell WM, Virginia R et al (1990) Biological Feedbacks in Global Desertification. Science (80-). 247:1043–1048

Liu P, Klose M, Conrad R (2018) Temperature effects on structure and function of the methanogenic microbial communities in two paddy soils and one desert soil. Soil Biol Biochem. 124:236–244. https://doi.org/10.1016/j.soilbio.2018.06.024

She W, Bai Y, Zhang Y, Qin S, Feng W, Sun Y et al (2018) Resource availability drives responses of soil microbial communities to short-term precipitation and nitrogen addition in a desert shrubland. Front Microbiol. 9:1–14. https://doi.org/10.3389/fmicb.2018.00186

Rastogi RP, Richa KA, Tyagi MB, Sinha RP (2010) Molecular mechanisms of ultraviolet radiation-induced DNA damage and repair. J Nucleic Acids. 2010. https://doi.org/10.4061/2010/592980

Sinha RP, Häder DP (2002) UV-induced DNA damage and repair: a review. Photochem Photobiol Sci. 1:225–236. https://doi.org/10.1039/b201230h

Lukin M, de los Santos C. (2006) NMR structures of damaged DNA. Chem Rev. 106:607–686. https://doi.org/10.1021/cr0404646

Santos AL, Gomes NCM, Henriques I, Almeida A, Correia A, Cunha Â (2012) Contribution of reactive oxygen species to UV-B-induced damage in bacteria. J Photochem Photobiol B Biol. 117:40–46. https://doi.org/10.1016/j.jphotobiol.2012.08.016

Strumberg D, Pilon AA, Smith M, Hickey R, Malkas L, Pommier Y (2000) Conversion of topoisomerase I cleavage complexes on the leading strand of ribosomal DNA into 5-′-Phosphorylated DNA double-strand breaks by replication runoff. Mol Cell Biol. 20:3977–3987. https://doi.org/10.1128/mcb.20.11.3977-3987.2000

Ohnishi T, Mori E, Takahashi A (2009) DNA double-strand breaks: Their production, recognition, and repair in eukaryotes. Mutat Res - Fundam Mol Mech Mutagen. 669:8–12. https://doi.org/10.1016/j.mrfmmm.2009.06.010

Lennon JT, Jones SE (2011) Microbial seed banks: the ecological and evolutionary implications of dormancy. Nat Rev Microbiol. 9:119–130. https://doi.org/10.1038/nrmicro2504

Setlow P (2014) Germination of spores of Bacillus species: what we know and do not know. J Bacteriol. 196:1297–1305. https://doi.org/10.1128/JB.01455-13

McHugh TA, Compson Z, Gestel N van, Hayer M, Ballard L, Haverty M, et al. Climate controls prokaryotic community composition in desert soils of the southwestern United States. FEMS Microbiol Ecol. 2017;93: 1–13. doi:https://doi.org/10.1093/femsec/fix116

Schulze-Makuch D, Wagner D, Kounaves SP, Mangelsdorf K, Devine KG, De Vera JP et al (2018) Transitory microbial habitat in the hyperarid Atacama Desert. Proc Natl Acad Sci U S A. 115: 2670–2675. https://doi.org/10.1073/pnas.1714341115

Pointing SB, Warren-Rhodes KA, Lacap DC, Rhodes KL, McKay CP (2007) Hypolithic community shifts occur as a result of liquid water availability along environmental gradients in China's hot and cold hyperarid deserts. Environ Microbiol. 9:414–424. https://doi.org/10.1111/j.1462-2920.2006.01153.x

Li WJ, Chen HH, Zhang YQ, Kim CJ, Park DJ, Lee JC et al (2005) Citricoccus alkalitolerans sp. nov., a novel actinobacterium isolated from a desert soil in Egypt. Int J Syst Evol Microbiol. 55:87–90. https://doi.org/10.1099/ijs.0.63237-0

Olsson-Francis K, de la Torre R, Towner MC, Cockell CS (2009) Survival of Akinetes (resting-state cells of cyanobacteria) in low earth orbit and simulated extraterrestrial conditions. Orig Life Evol Biosph. 39:565–579. https://doi.org/10.1007/s11084-009-9167-4

Zhang CC, Laurent S, Sakr S, Peng L, Bédu S (2006) Heterocyst differentiation and pattern formation in cyanobacteria: a chorus of signals. Mol Microbiol. 59:367–375. https://doi.org/10.1111/j.1365-2958.2005.04979.x

Leganés F (1994) Genetic evidence that hepA gene is involved in the normal deposition of the envelope of both heterocysts and akinetes in Anabaena variabilis ATCC 29413. FEMS Microbiol Lett. 123:63–67. https://doi.org/10.1016/0378-1097(94)90275-5

Leganés F, Fernández-Piñas F, Wolk CP (1998) A transposition-induced mutant of Nostoc ellipsosporum implicates an arginine-biosynthetic gene in the formation of cyanophycin granules and of functional heterocysts and akinetes. Microbiology. 144:1799–1805

Van De Mortel M, Halverson LJ (2004) Cell envelope components contributing to biofilm growth and survival of Pseudomonas putida in low-water-content habitats. Mol Microbiol. 52:735–750. https://doi.org/10.1111/j.1365-2958.2004.04008.x

Mutnuri S, Vasudevan N, Kastner M, Heipieper HJ (2005) Changes in fatty acid composition of Chromohalobacter israelensis with varying salt concentrations. Curr Microbiol. 50:151–154. https://doi.org/10.1007/s00284-004-4396-2

Scherber CM, Schottel JL, Aksan A (2009) Membrane phase behavior of Escherichia coli during desiccation, rehydration, and growth recovery. Biochim Biophys Acta - Biomembr. 1788:2427–2435. https://doi.org/10.1016/j.bbamem.2009.08.011

LeBlanc JC, Gonçalves ER, Mohn WW (2008) Global response to desiccation stress in the soil actinomycete Rhodococcus jostii RHA1. Appl Environ Microbiol. 74:2627–2636. https://doi.org/10.1128/AEM.02711-07

Shabala L, Ross T (2008) Cyclopropane fatty acids improve Escherichia coli survival in acidified minimal media by reducing membrane permeability to H+ and enhanced ability to extrude H+. Res Microbiol. 159:458–461. https://doi.org/10.1016/j.resmic.2008.04.011

Kocharunchitt C, King T, Gobius K, Bowman JP, Ross T (2014) Global genome response of Escherichia coli O157:H7 Sakai during dynamic changes in growth kinetics induced by an abrupt downshift in water activity. PLoS One. 9. https://doi.org/10.1371/journal.pone.0090422

Romantsov T, Guan Z, Wood JM (2009) Cardiolipin and the osmotic stress responses of bacteria. Biochim Biophys Acta - Biomembr. 1788:2092–2100. https://doi.org/10.1016/j.bbamem.2009.06.010

Santos H, Costa MS (2002) Compatible solutes of organisms that live in hot saline environments. Environ Microbiol. 4:501–509

Pade N, Hagemann M (2015) Salt acclimation of cyanobacteria and their application in biotechnology. Life. 5:25–49. https://doi.org/10.3390/life5010025

Riedel K, Lehner A (2007) Identification of proteins involved in osmotic stress response in Enterobacter sakazakii by proteomics. Proteomics. 7:1217–1231. https://doi.org/10.1002/pmic.200600536

Oren A (2008) Microbial life at high salt concentrations: phylogenetic and metabolic diversity. Saline Syst 4:1–13. https://doi.org/10.1186/1746-1448-4-2

Sleator RD, Hill C (2002) Bacterial osmoadaptation: the role of osmolytes in bacterial stress and virulence. FEMS Microbiol Rev. 26:49–71. https://doi.org/10.1016/S0168-6445(01)00071-7

Welsh DT (2000) Ecological significance of compatible solute accumulation by micro- organisms: from single cells to global climate. FEMS Microbiol Rev. 24:263–290. https://doi.org/10.1016/S0168-6445(99)00038-8

Li H, Bhaskara A, Megalis C, Lou TM (2012) Transcriptomic analysis of salmonella desiccation resistance. Foodborne Pathog Dis. 9:1143–1151. https://doi.org/10.1089/fpd.2012.1254

Oren A (1999) Bioenergetic aspects of halophilism. Microbiol Mol Biol Rev. 63:334–348. Available: http://www.ncbi.nlm.nih.gov/pubmed/10357854

Thi Le P, Makhalanyane TP, Guerrero LD, Vikram S, van De Peer Y, Cowan DA (2016) Comparative metagenomic analysis reveals mechanisms for stress response in hypoliths from extreme hyperarid deserts. Genome Biol Evol. 8:2737–2747. https://doi.org/10.1093/gbe/evw189

Zhou A, He Z, Qin Y, Lu Z, Deng Y, Tu Q et al (2013) StressChip as a high-throughput tool for assessing microbial community responses to environmental stresses. Environ Sci Technol. 47: 9841–9849. https://doi.org/10.1021/es4018656

Shi Z, Yin H, Van Nostrand JD, Voordeckers JW, Tu Q, Deng Y et al (2019) Functional gene array-based ultrasensitive and quantitative detection of microbial populations in complex communities. mSystems. 4:1–16. https://doi.org/10.1128/msystems.00296-19

Sasaki K, Christov NK, Tsuda S, Imai R (2014) Identification of a novel LEA protein involved in freezing tolerance in wheat. Plant Cell Physiol. 55:136–147. https://doi.org/10.1093/pcp/pct164

Magwanga RO, Lu P, Kirungu JN, Lu H, Wang X, Cai X et al (2018) Characterization of the late embryogenesis abundant (LEA) proteins family and their role in drought stress tolerance in upland cotton. BMC Genet. 19:1–31. https://doi.org/10.1186/s12863-017-0596-1

Slade D, Radman M (2011) Oxidative stress resistance in deinococcus radiodurans. Microbiol Mol Biol Rev. https://doi.org/10.1128/mmbr.00015-10

Mertens J, Aliyu H, Cowan DA (2018) LEA proteins and the evolution of the WHy domain. Appl Environ Microbiol. 84:1–8. https://doi.org/10.1128/AEM.00539-18

Wise MJ, Tunnacliffe A (2004) POPP the question: what do LEA proteins do? Trends Plant Sci. 9: 13–17. https://doi.org/10.1016/j.tplants.2003.10.012

Warren-Rhodes KA, Lee KC, Archer SDJ, Cabrol N, Ng-Boyle L, Wettergreen D et al (2019) Subsurface microbial habitats in an extreme desert Mars-analog environment. Front Microbiol. 10:1–11. https://doi.org/10.3389/fmicb.2019.00069

Tracy CR, Streten-Joyce C, Dalton R, Nussear KE, Gibb KS, Christian KA (2010) Microclimate and limits to photosynthesis in a diverse community of hypolithic cyanobacteria in northern Australia. Environ Microbiol. 12:592–607. https://doi.org/10.1111/j.1462-2920.2009.02098.x

Schlesinger WH, Pippen JS, Wallenstein MD, Hofmockel KS, Klepeis DM, Mahall BE (2003) Community composition and photosynthesis by photoautotrophs under quartz pebbles, southern Mojave Desert. Ecology. 84:3222–3231. https://doi.org/10.1890/02-0549

Tamaru Y, Takani Y, Yoshida T, Sakamoto T (2005) Crucial role of extracellular polysaccharides in desiccation and freezing tolerance in the terrestrial cyanobacterium Nostoc commune. Appl Environ Microbiol. 71:7327–7333. https://doi.org/10.1128/AEM.71.11.7327-7333.2005

Roberson EB, Firestone MK (1992) Relationship between desiccation and exopolysaccharide production in a soil Pseudomonas sp. Appl Environ Microbiol. 58:1284–1291. https://doi.org/10.1128/aem.58.4.1284-1291.1992

Varin T, Lovejoy C, Jungblut AD, Vincent WF, Corbeil J (2012) Metagenomic analysis of stress genes in microbial mat communities from Antarctica and the high Arctic. Appl Environ Microbiol. 78:549–559. https://doi.org/10.1128/AEM.06354-11

Marvasi M, Visscher PT, Casillas ML (2010) Exopolymeric substances (EPS) from Bacillus subtilis: polymers and genes encoding their synthesis. FEMS Microbiol Lett. 313:1–9. https://doi.org/10.1111/j.1574-6968.2010.02085.x

Selasi GN, Nicholas A, Jeon H, Na SH, Il KH, Kim YJ et al (2016) Differences in biofilm mass, expression of biofilm-associated genes, and resistance to desiccation between epidemic and sporadic clones of carbapenem-resistant acinetobacter baumannii sequence type 191. PLoS One. 11:1–11. https://doi.org/10.1371/journal.pone.0162576

Yoshimura H, Okamoto S, Tsumuraya Y, Ohmori M (2007) Group 3 sigma factor gene, sigJ, a key regulator of desiccation tolerance, regulates the synthesis of extracellular polysaccharide in cyanobacterium Anabaena sp. strain PCC 7120. DNA Res. 14:13–24. https://doi.org/10.1093/dnares/dsm003

Schnider-Keel U, Lejbølle KB, Baehler E, Haas D, Keel C (2001) The sigma factor AlgU (AlgT) controls exopolysaccharide production and tolerance towards desiccation and osmotic stress in the biocontrol agent pseudomonas fluorescens CHA0. Appl Environ Microbiol. 67:5683–5693. https://doi.org/10.1128/AEM.67.12.5683-5693.2001

Sarniguet A, Kraus J, Henkels MD, Muehlchen AM, Loper JE (1995) The sigma factor σs affects antibiotic production and biological control activity of Pseudomonas fluorescens Pf-5. Proc Natl Acad Sci U S A. 92:12255–12259. https://doi.org/10.1073/pnas.92.26.12255

Morita RY (1988) Bioavailability of energy and its relationship to growth and starvation survival in nature. Can J Microbiol. 34:436–441. https://doi.org/10.1139/m88-076

León-Sobrino C, Ramond JB, Maggs-Kölling G, Cowan DA (2019) Nutrient acquisition, rather than stress response over diel cycles, drives microbial transcription in a hyper-arid Namib desert soil. Front Microbiol. 10:1–11. https://doi.org/10.3389/fmicb.2019.01054

Rajeev L, Da Rocha UN, Klitgord N, Luning EG, Fortney J, Axen SD et al (2013) Dynamic cyanobacterial response to hydration and dehydration in a desert biological soil crust. ISME J. 7: 2178–2191. https://doi.org/10.1038/ismej.2013.83

Leung PM, Bay SK, Meier DV, Chiri E, Cowan DA, Gillor O et al (2020) Energetic basis of microbial growth and persistence in desert ecosystems. mSystems. 5:1–14. https://doi.org/10.1128/msystems.00495-19

Larkum AWD, Ritchie RJ, Raven JA (2018) Living off the sun: chlorophylls, bacteriochlorophylls and rhodopsins. Photosynthetica. 56:11–43. https://doi.org/10.1007/s11099-018-0792-x

Yurkov V, Hughes E (2017) Aerobic anoxygenic phototrophs: four decades of mystery. In: Modern topics in the phototrophic prokaryotes. Springer, Cham, pp 193–214. https://doi.org/10.1007/978-3-319-46261-5

Madigan MT (2017) Anoxygenic phototrophic bacteria from extreme environments. Photosynth Res. 76:157–171. https://doi.org/10.1007/978-3-319-46261-5_13

Madigan MT, Jung DO (2009) An overview of purple bacteria: systematics, physiology, and habitats. In: The purple phototrophic bacteria. Springer, Dordrecht, pp 1–15. https://doi.org/10.1007/978-1-4020-8815-5_1

Tahon G, Willems A (2017) Isolation and characterization of aerobic anoxygenic phototrophs from exposed soils from the Sør Rondane Mountains. East Antarctica. Syst Appl Microbiol. 40:357–369. https://doi.org/10.1016/j.syapm.2017.05.007

Pushkarev A, Inoue K, Larom S, Flores-uribe J, Singh M, Konno M, et al. Discovered using functional metagenomics. Nature. 2018; 1. Available: http://www.nature.com/articles/s41586-018-0225-9

Lester ED, Satomi M, Ponce A (2007) Microflora of extreme arid Atacama Desert soils. Soil Biol Biochem Biochem. 39:704–708

Tang K, Jia L, Yuan B, Yang S, Li H, Meng J et al (2018) Aerobic anoxygenic phototrophic bacteria promote the development of biological soil crusts. Front Microbiol. 9:1–12. https://doi.org/10.3389/fmicb.2018.02715

Schmidt U (1969) Molecular hydrogen in the atmosphere. Tellus. 26:78–90. https://doi.org/10.1029/ja074i024p05807

Ortiz M, Bosch J, Coclet C, Johnson J, Lebre P, Salawu-rotimi A et al (2020) Microbial nitrogen cycling in antarctic soils. Microorganisms. 8

Meredith LK, Rao D, Bosak T, Klepac-Ceraj V, Tada KR, Hansel CM et al (2014) Consumption of atmospheric hydrogen during the life cycle of soil-dwelling actinobacteria. Environ Microbiol Rep. 6:226–238. https://doi.org/10.1111/1758-2229.12116

Piché-Choquette S, Constanta P (2019) Molecular hydrogen, a neglected key driver of soil biogeochemical processes. Appl Environ Microbiol. 85:1–19. https://doi.org/10.1128/AEM.02418-18

Schäfer C, Friedrich B, Lenza O (2013) Novel, oxygen-insensitive group 5 [NiFe]-hydrogenase in Ralstonia eutropha. Appl Environ Microbiol. 79:5137–5145. https://doi.org/10.1128/AEM.01576-13

Lynch RC, Darcy JL, Kane NC, Nemergut DR, Schmidt SK (2014) Metagenomic evidence for metabolism of trace atmospheric gases by high-elevation desert actinobacteria. Front Microbiol. 5:1–13. https://doi.org/10.3389/fmicb.2014.00698

Patrauchan MA, Miyazawa D, LeBlanc JC, Aiga C, Florizone C, Dosanjh M et al (2012) Proteomic analysis of survival of rhodococcus jostii RHA1 during carbon starvation. Appl Environ Microbiol. 78:6714–6725. https://doi.org/10.1128/AEM.01293-12

Zahradka K, Slade D, Bailone A, Sommer S, Averbeck D, Petranovic M et al (2006) Reassembly of shattered chromosomes in Deinococcus radiodurans. Nature. 443:569–573. https://doi.org/10.1038/nature05160

Zerulla K, Soppa J (2014) Polyploidy in haloarchaea: advantages for growth and survival. Front Microbiol. 5:1–8. https://doi.org/10.3389/fmicb.2014.00274

Soppa J (2014) Polyploidy in archaea and bacteria: about desiccation resistance, giant cell size, long-term survival, enforcement by a eukaryotic host and additional aspects. J Mol Microbiol Biotechnol. 24:409–419. https://doi.org/10.1159/000368855

Liu Y, Zhou J, Omelchenko MV, Beliaev AS, Venkateswaran A, Stair J et al (2003) Transcriptome dynamics of deinococcus radiodurans recovering from ionizing radiation. Proc Natl Acad Sci U S A. 100:4191–4196. https://doi.org/10.1073/pnas.0630387100

Sghaier H, Ghedira K, Benkahla A, Barkallah I (2008) Basal DNA repair machinery is subject to positive selection in ionizing-radiation-resistant bacteria. BMC Genomics. 9:1–7. https://doi.org/10.1186/1471-2164-9-297

Makarova KS, Omelchenko MV, Gaidamakova EK, Matrosova VY, Vasilenko A, Zhai M et al (2007) Deinococcus geothermalis: The pool of extreme radiation resistance genes shrinks. PLoS One. 2. https://doi.org/10.1371/journal.pone.0000955

Tanaka M, Earl AM, Howell HA, Park MJ, Eisen JA, Peterson SN et al (2004) Analysis of Deinococcus radiodurans's transcriptional response to ionizing radiation and desiccation reveals novel proteins that contribute to extreme radioresistance. Genetics. 168:21–33. https://doi.org/10.1534/genetics.104.029249

Lim S, Jung JH, Blanchard L, De Groot A (2019) Conservation and diversity of radiation and oxidative stress resistance mechanisms in Deinococcus species. FEMS Microbiol Rev. 43:19–52. https://doi.org/10.1093/femsre/fuy037

Mason JM, Fajardo-Cavazos P, Setlow P (1988) Levels of mRNAs which code for small, acid-soluble spore proteins and their LacL gene fusions in sporulating cells of Bacillus subtilis. Nucleic Acids Res. 16:6567–6583

Mason JM, Setlow P (1984) Expression of Bacillus megaterium and Bacillus subtilis Small Acid-Soluble Spore Protein Genes During Stationary-Phase Growth of Asporogenous B. subtilis Mutants. J Bacteriol. 157:931–933

Setlow B, Setlow P (1993) Binding of small, acid-soluble spore proteins to DNA plays a significant role in the resistance of Bacillus subtilis spores to hydrogen peroxide. Appl Environ Microbiol. 59:3418–3423. https://doi.org/10.1128/aem.59.10.3418-3423.1993

Krisko A, Radman M (2010) Protein damage and death by radiation in Escherichia coli and Deinococcus radiodurans. Proc Natl Acad Sci U S A. 107:14373–14377. https://doi.org/10.1073/pnas.1009312107

Singh N, Savanur MA, Srivastava S, D'Silva P, Mugesh G (2019) A manganese oxide nanozyme prevents the oxidative damage of biomolecules without affecting the endogenous antioxidant system. Nanoscale. 11:3958–3967. https://doi.org/10.1039/c8nr09397k

Paulino-Lima IG, Fujishima K, Navarrete JU, Galante D, Rodrigues F, Azua-Bustos A et al (2016) Extremely high UV-C radiation resistant microorganisms from desert environments with different manganese concentrations. J Photochem Photobiol B Biol. 163:327–336. https://doi.org/10.1016/j.jphotobiol.2016.08.017

Potter RM, Rossman GR (1979) The manganese- and iron-oxide mineralogy of desert varnish. Chem Geol. 25:79–94. https://doi.org/10.1016/0009-2541(79)90085-8

Kuhlman KR, Fusco WG, La Duc MT, Allenbach LB, Ball CL, Kuhlman GM et al (2006) Diversity of microorganisms within rock varnish in the Whipple Mountains, California. Appl Environ Microbiol. 72:1708–1715. https://doi.org/10.1128/AEM.72.2.1708-1715.2006

Lang-Yona N, Maier S, Macholdt DS, Müller-Germann I, Yordanova P, Rodriguez-Caballero E et al (2018) Insights into microbial involvement in desert varnish formation retrieved from metagenomic analysis. Environ Microbiol Rep. 10:264–271. https://doi.org/10.1111/1758-2229.12634

Nagy B, Nagy LA, Rigali MJ, Jones WD, Krinsley DH, Sinclair NA (1991) Rock varnish in the Sonoran Desert: microbiologically mediated accumulation of manganiferous sediments. Sedimentology. 38:1153–1171. https://doi.org/10.1111/j.1365-3091.1991.tb00376.x

Kuhlman KR, Venkat P, La Duc MT, Kuhlman GM, McKay CP (2008) Evidence of a microbial community associated with rock varnish at Yungay, Atacama Desert, Chile. J Geophys Res Biogeosciences. 113:1–14. https://doi.org/10.1029/2007JG000677

Los DA, Murata N (2004) Membrane fluidity and its roles in the perception of environmental signals. Biochim Biophys Acta - Biomembr. 1666:142–157. https://doi.org/10.1016/j.bbamem.2004.08.002

Matsuura Y, Takehira M, Joti Y, Ogasahara K, Tanaka T, Ono N et al (2015) Thermodynamics of protein denaturation at temperatures over 100°C: CutA1 mutant proteins substituted with hydrophobic and charged residues. Sci Rep. 5:1–9. https://doi.org/10.1038/srep15545

Voth W, Jakob U (2017) Stress-activated chaperones: a first line of defense. Trends Biochem Sci. 42:899–913. https://doi.org/10.1016/j.tibs.2017.08.006

Rajaram H, Chaurasia AK, Apte SK (2014) Cyanobacterial heat-shock response: role and regulation of molecular chaperones. Microbiol (United Kingdom). 160:647–658. https://doi.org/10.1099/mic.0.073478-0

Yin W, Wang Y, Liu L, He J (2019) Biofilms: the microbial "protective clothing" in extreme environments. Int J Mol Sci. 20. https://doi.org/10.3390/ijms20143423

Koerdt A, Gödeke J, Berger J, Thormann KM, Albers SV (2010) Crenarchaeal biofilm formation under extreme conditions. PLoS One. 5. https://doi.org/10.1371/journal.pone.0014104

Williams HN, Turng BF, Kelley JI (2009) Survival response of Bacteriovorax in surface biofilm versus suspension when stressed by extremes in environmental conditions. Microb Ecol. 58: 474–484. https://doi.org/10.1007/s00248-009-9499-7

McBride MJ, Ensign JC (1987) Effects of intracellular trehalose content on Streptomyces griseus spores. J Bacteriol. 169:4995–5001. https://doi.org/10.1128/jb.169.11.4995-5001.1987

Reina-Bueno M, Argandoña M, Nieto JJ, Hidalgo-García A, Iglesias-Guerra F, Delgado MJ et al (2012) Role of trehalose in heat and desiccation tolerance in the soil bacterium Rhizobium etli. BMC Microbiol. 12:1–17. https://doi.org/10.1186/1471-2180-12-207

Singer MA, Lindquist S (1998) Multiple effects of trehalose on protein folding in vitro and in vivo. Mol Cell. 1:639–648. https://doi.org/10.1016/S1097-2765(00)80064-7

Van Goethem MW, Swenson TL, Trubl G, Roux S, Northen TR (2019) Characteristics of wetting-induced bacteriophage blooms in biological soil crust. MBio. 10:1–15. https://doi.org/10.1128/mBio.02287-19

The Response of Soil Microbial Communities to Hydration and Desiccation Cycles in Hot Desert Ecosystems

11

Adam Štovícek and Osnat Gillor

Abstract

Vast regions of the Earth's surface are arid, characterized by sparse vegetation and dominated by dry and barren soil surfaces. Microbial life in these soils must survive long periods of severe desiccation that are interspersed by infrequent rainfall events. The soil microbial communities must utilize any available water, even the mild wetting of the soil surface by morning dew or incoming fog that accommodates a swift response by the phototrophs. In contrast, major rain events demand drastic modifications of the soil microbiome activity, structure, and interactions. In this chapter, we review the microbial response to hydration and desiccation cycles that characterize desert ecosystems and the resulting dynamics in microbial community composition and activity. These wetting cycles are reflected in terrestrial microbiome ecological functions and provide new insights into the mechanisms that sustain the microbial communities in arid soils.

Keywords

Arid · Rain · Flood · Soil · Microbiome

A. Štovícek (✉)
Department of Microbiology, Nutrition and Dietetics, Czech University of Life Sciences Prague, Prague, Czech Republic
e-mail: stovicek@af.czu.cz

O. Gillor (✉)
Zuckerberg Institute for Water Research, Blaustein Institutes for Desert Research, Ben-Gurion University of the Negev, Midreshet Ben Gurion, Israel
e-mail: gilloro@bgu.ac.il

J.-B. Ramond, D. A. Cowan (eds.), *Microbiology of Hot Deserts*, Ecological Studies 244, https://doi.org/10.1007/978-3-030-98415-1_11

11.1 Introduction to Desert Terminology

Hot Desert Definition Hot deserts are characterized by extreme fluctuation in temperature and precipitation patterns. Primary production in hot deserts is controlled by water availability (Noy-Meir 1973). They are usually oligotrophic and occasionally hypersaline with high levels of ultraviolet radiation. The commonly used Köppen-Geiger climate classification is based on the threshold values and seasonality of monthly air temperature and precipitation. Hot deserts are defined as areas with low mean annual precipitation that mainly (>70% of the time) occurred in summer or winter and with a mean summer temperature of above 22 °C (Beck et al. 2018). According to this definition, hot deserts make up around 20% of the earth's land surface (Beck et al. 2018; Peel et al. 2007).

Aridity Definition Arid climates are often defined by an aridity index (AI), a quantitative indicator of the degree of water deficiency at a given location (Stadler 1998). AI calculations are not straightforward as they present the interplay between rainfall, temperature, and evaporation depicting long-term climatic water deficiency. Therefore, they are traditionally calculated as the ratio between P (precipitation) and PET (potential evapotranspiration) (Cherlet et al. 2018). Alternative aridity measures were suggested, calculating the ratio between the season (winter and/or summer) rainfall and the mean annual temperature; or the growing season precipitation and the total amount of water required for a given crop (Stadler 1998). Regardless of the formulation, the aridity measures are applied at continental and subcontinental levels and used in relation with the distribution of natural vegetation and crops (Cherlet et al. 2018).

Soil Water Potential The degree of soil "dryness" is expressed as water potential, which is a compound effect of soil matrix potential, solute potential, and pressure potential. Water storage is a function of the distribution of soil pore-size governed by the soil texture and structure. Clay-rich soils have the greatest total water holding capacity because they have the largest pore space (Saxton et al. 1986). However, total water holding capacity is governed by the water storage and redistribution capacity in a given soil type in response to differences in potential energy required to move or extract water from a given soil (Hong et al. 2018).

Three moisture states are used to describe soil water potential: saturation, field capacity, and permanent wilting point, all related to the potential energy required to extract water from the soil (Chandler et al. 2017). Field capacity represents the soil water content retained against the force of gravity by the matric potential of soil pores. Decreases in water content correspond to decrease in soil matric potential, and as a result, water is held more strongly to mineral surfaces due to water molecules cohesive forces, and between water and mineral particles adhesive or capillary forces (O'Geen 2013). The water potential is measured in pressure units (Mega Pascals), with 0 MPa being a saturated equilibrium, and negative values arise as soil dries resulting in suction or tension on the soil aqueous phase. Field capacity represents the soil water content retained against the force of gravity by matric forces at a tension of about −0.033 MPa. As the water content decreases, soil matric potential decreases, becoming more negative, and the water hold more strongly to particles surfaces due to cohesive forces between the water molecules and adhesive forces between the water and mineral particles (Saxton et al. 1986). At less than −1.5 MPa, the soil is too dry for agronomic plants to uptake water to their roots, defined as the permanent wilting point (O'Geen 2013). Desert soils, during prolonged draught, could reach −6 MPa (Monson and Smith 1982).

Desert Soil Type and Salinity Desert soils are typically highly saline (depending on localized draining conditions) and alkaline (pH higher than 7) buffered by high levels of calcium carbonate. Alkalinity affects the availability of phosphates and other micronutrients, whose solubility drops with increasing pH. In a comprehensive review, Makhalanyane et al. (2015) present a summary of hot deserts soil characteristics from around the globe. This summary suggests that, regardless of the desert soil type, the pH is mostly above 7, the organic carbon is low (generally below 1%), and so is the nitrogen content (below 0.1%) pointing to alkaline, oligotrophic soils in most desert ecosystems (Makhalanyane et al. 2015).

Rain, Fog, and Dew Climatological Description The three main precipitation modes in hot desert ecosystems are rain, fog, and dew. *Rain* events are highly variable in scale and time, with extended periods of drought in between. The rainy season can be very short (<20 days annually), with a significant amount of precipitation ranging from <5 mm to >20 mm of rain in a single pulse, and concentrated in a few rain events (Noy-Meir 1973; Peel et al. 2007). The composition of total dissolved ions (TDI) in desert rain was shown to be highly variable depending on dust load and draught duration, yet in the Negev Desert rain TDI lower by 1.5–7 times than dew (Kidron and Starinsky 2012).

Both *fog and dew* may be a significant water source in deserts as this water source is more reliable than rain and may be available for over 200 days a year and accumulate to as much as 150 mm of additional moisture (Belnap et al. 2016). Fog and dew occur in coastal desert regions, like the eastern edge of the Atacama Desert, North Saini, Baja California, or the western edge of the Sahara and Namib deserts. At night, surfaces cool through radiant heat loss to below water vapor condensation temperature (dew point), hence water is deposited directly on the cold surfaces in the form of dew. Alternatively, the supplemental moisture may come in the form of droplets suspended in the air near ground level, called radiation fog. Fog forms when heat loss through radiation cools the ground and chills the air in the lowest few meters to below the dew point temperature (Eckardt et al. 2013).

11.2 Response to Hydration in Desert Ecosystems

For most parts, the desert soil communities are subjected to extremely low water potential, which is composed of low matrix potential in addition to high solute potential if the soil is saline (Kieft et al. 1987). Desert soil matrix potential is low during the summer months, dropping to -10 MPa, but during the rain event, this matrix potential can quickly raise to 0.1 MPa (Scanlon 1994).

11.2.1 Rhythm of Plant Primary Production

The main limiting factors of primary production in arid soils are moisture and nitrogen availability (Austin et al. 2004). During the year, desert organisms experience a severe water shortage intercepted by brief and highly variable precipitation pulses (Noy-Meir 1973) that trigger well-documented pulses of vegetative growth (Noy-Meir 1974; Ogle and Reynolds 2004). The emerging annual vegetation focuses on expedient and rapid production of as a large number of seeds as possible (Noy-Meir 1974). In addition, the boost in vegetation triggers increased metabolic activity powered by the extensive carbon fixation (Huxman et al. 2004) and succeeded by mineralization of the newly available resources (Austin et al. 2004). Most of the freshly generated biomass is attributed to annuals that dry shortly after the rain, leaving behind dormant seeds ready to germinate at the next rain event (Huxman et al. 2004). Moreover, a large proportion of desert vascular plants biomass perish during the dry season (Khadka et al. 2018). Theoretically, this would allow for cycling of carbon and other nutrients contained by the plants, but due to the habitat disconnect and fragmentation in the desiccated soil, much of the carbon is unavailable to microorganisms, leaving physical weathering as the main degradation pathway throughout the dry season (Moorhead and Reynolds 1991). A subsequent seasonal rain increases habitat connectivity and allows soil microbes access to carbon from decaying plant matter (Ebrahimi and Or 2015).

11.2.2 Carbon Dioxide Pulse

In soil ecosystems, the respiration and moisture relationship depend on the specific physiological responses of microbial communities to water stress (Bailey et al. 2019). In arid soils, hydration that follows a long drought results in a sudden pulse-like event of rapidly increasing soil CO_2 efflux named the "Birch effect" (Birch 1958). The Birch effect is variable depending on the previous history of rainfall, soil moisture, and microbial activity (Fig. 11.1). Yet, it was shown that a stronger soil respiration response to hydration occurs at the first rain events after a prolonged dry period (Moyano et al. 2013). The precise mechanisms causing this CO_2 spike are not fully clear, and the proposed mechanisms include both biotic and abiotic processes. Several hypotheses have been proposed, including death of microbial cells (Borken and Matzner 2009), osmotic shock-triggered release and subsequent assimilation of low weight organic molecules (Fierer and Schimel 2003), soil aggregate disruption (Denef et al. 2001), and cell-independent enzymatic activity (Fraser et al. 2016). Regardless of the mechanism, rain events result not only in the sudden emission of CO_2 but also in the sudden availability of nutrients coupled with the temporary increase in the soil water content. Soil aqueous phase enables the dispersal of nutrients to the cells triggering an increase in net metabolic activity (Blazewicz et al. 2014; Fierer and Schimel 2003; Huxman et al. 2004).

The origins of the CO_2 spike were debated for decades and were proposed to originate from either physico-chemical, biochemical and/or biological sources (Fraser et al. 2016; Kim and Or 2017; Moyano et al. 2013). Soil physico-chemical processes may involve rainwater entering dry soil aggregates, pressurizing the air inside, and disrupting them. As a result, previously protected soil organic matter is exposed to the reactivated microbial community that quickly devours it (Denef et al. 2001). An alternative proposed mechanism is photodegradation, a water-independent process that was found to be responsible for degrading plant material in semi-arid ecosystems (Austin and Vivanco 2006). This decomposition pathway is independent of water availability and results in the accumulation of decomposition products in the soil that could be utilized by microorganisms upon hydration (Moyano et al. 2013).

The biochemical processes include extracellular enzyme activity that decomposes organic matter under dry conditions in soil pores at low water potential that prohibit cellular metabolism but not enzyme activity. The enzymatic degradation releases chemically labile substrates that are inaccessible to microorganisms due to the dry conditions and therefore accumulate during desiccation (Steinweg et al. 2012). Hydration increases the connectivity between soil pores, and thus accumulated

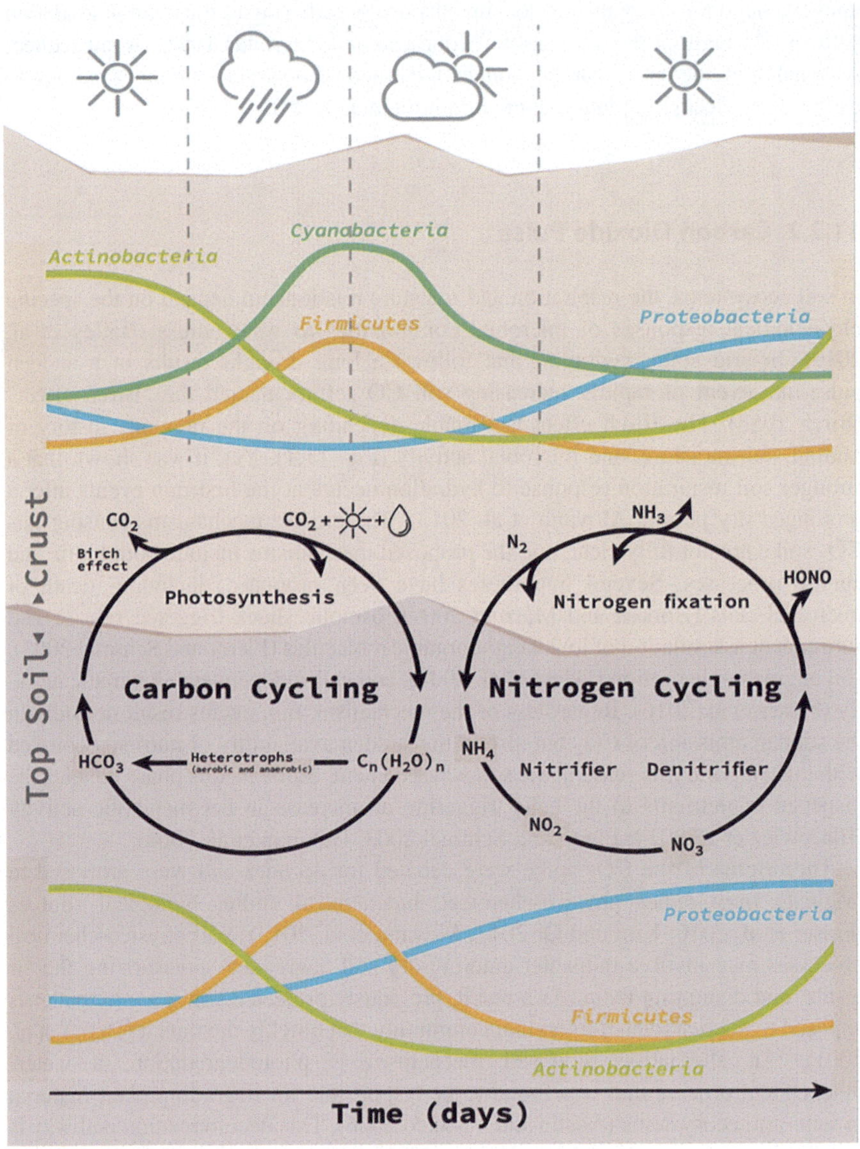

Fig. 11.1 Key microbial taxa, functions, and biogeochemical interactions in desert soil surface (biocrust) and the next horizon (topsoil). Hydration and desiccation mediated changes in the relative abundance of soil taxa are illustrated in conjugation with carbon and nitrogen cycling. The schematic illustration is adapted from field (Placella et al. 2012; Šťovíček et al. 2017), lab (Angel and Conrad 2013), and model (Kim and Or 2016, 2017) reports

nutrients diffuse and become available to the resuscitated microbial community that uses them for growth and respiration (Fraser et al. 2016).

Biological processes are linked to the Birch Effect either through active and/or passive metabolite release. Soil microbes were shown to lower their solute potential in response to desiccation by synthesizing or importing osmolytes. These low molecular weight organic solutes include C- and N-containing compounds such as some sugars (e.g., trehalose), ammonium compounds (e.g., betaine) and nucleic acid derivatives (e.g., ectoine) (Warren 2014). Following hydration, soil microbes face strong negative solute potential and need to dispose of the accumulated intracellular osmolytes to avoid uncontrolled influx of water. The release of the osmolytes from dead cells provide readily available metabolites that are quickly assimilated during hydration of the resuscitating cells, providing substrated that suport the respiration pulse (Fierer and Schimel 2003). Concomitantly, hydration was proposed to invoke cell lysis due to the sudden increase of turgor pressure, resulting in cytoplasmatic membrane rupture and the release of intracellular content (Borken and Matzner 2009). In addition to direct respiration, the osmolytes could trigger priming of the bulk soil carbon sources (Moyano et al. 2013).

11.2.3 Nitrogen Pulsing

Slow gas exchange was often measured on wet desert soil, combined with the increase in metabolic activity, typically suggesting that oxygen is depleted from some soil microhabitats. The decreased oxygen availability forces microbes to switch to respiring other available substrates, such as nitrates. This results in the transformation of nitrates into nitrites and eventually even nitrogen, which escape the system (Austin et al. 2004). This process reduces the nitrogen reserves in the soil, thus limiting both microbial and plant primary production. These subsequently rely on a new spike in nitrogen fixation occurring primarily in the biological soil crust (Elbert et al. 2012; Ramond et al. 2018).

The vast majority of studies describing desert soil dynamics concentrate on hydration events that follow a long drought (Blazewicz et al. 2014; Kieft et al. 1987; Placella et al. 2012). However, these reports ignored the following desiccation. This presents only part of the hydration-desiccation cycle experienced by microorganisms in arid soil environments. The gradual drying that follows rain events is at best described as a return to the initial microbial community (Bell et al. 2008; Clark et al. 2008; Placella et al. 2012) yet, the patterns and dynamics of the soil bacterial communities remain fragmentary.

11.3 Bacterial Composition in Desert Soils

11.3.1 Biological Soil Crust

The surface of desert soil is typically covered by a biological soil crust (biocrust) that refers to soil surface cemented by extracellular polymeric substances of microbial origin (Belnap et al. 2016). Biocrust can vary in structure, thickness, and species composition, but commonly it contains a high percentage of photosynthetic algae and/or cyanobacteria (Belnap et al. 2016). Biocrust structure is influenced not only by composition but also by water availability, radiation, temperature, and soil texture, resulting in several types of crusts (Belnap 2003). The microbiology of hot desert biocrusts is described in Chap. 3. The water sources of the photosynthetic microorganisms residing in the biocrust are rain and dew. Dew is an ephemeral source of hydration that is available for a few hours, mostly at night, while day-time dew can be available for only a few minutes before it evaporates (Hill et al. 2015; Kidron and Starinsky 2012; Oren et al. 2019). Nevertheless, in arid and hyper-arid environments, it is more reliable than rain as there is a handful of rainy days but over 200 days of dew annually (Hill et al. 2015). The dew allows the biocrust to fix enough carbon and nitrogen to support not only the associated heterotrophic microorganisms, but also the heterotrophs below, by leeching the organic carbon and nitrogen during wetting events (Belnap 2003).

Biocrust Primary Producers Biocrusts are the most important primary producer in deserts. Unlike most vascular plants, it can benefit from dew, which may be a more reliable and available water source throughout the year in some deserts (Hagemann et al. 2016; Raanan et al. 2016; Uclés et al. 2016). This allows crust photosynthesis throughout the year as opposed to the short and unpredictable rainy season. Biocrusts are characterized in accordance with the composition of photosynthetic microorganisms they contain: Smooth crust is formed in hyper-arid regions that are too dry to support the growth of mosses and lichens, hence this crust is dominated by cyanobacteria. Rugose crusts are formed in arid hot regions, could reach up to 3 cm thick, and may contain very low lichen moss cover. Cool deserts are dominated by pinnacled crusts, with a thickness of up to 15 cm, that are usually dominated by lichens (>40% cover). Finally, rolling crusts are found in both hot and cold deserts, where the soil freezes occasionally, they reach up to 5 cm thick and are partially covered by mosses and lichen (Belnap 2003).

Biocrust Community Composition In addition to the primary producers, biocrusts in arid environments support several populations from five different taxa,including *Alpha-Proteobacteria*, *Bacteroidota*, *Chloroflexota*, *Gemmatimonadota*, and *Acidobacteriota* (Karaoz et al. 2018; Meier et al. 2021). The most diverse phylum *Actinobacteria* and specifically the class *Rubrobacteraceae* was abundant in both the biocrust and topsoil (Meier et al. 2021; Šťovíček et al. 2017). The *Rubrobacteraceae* cells are well adapted to survival under extreme desiccation conditions without entering dormancy, by encoding genetic potential for a mixotrophic lifestyle.

These adaptations include rhodopsin and high-affinity hydrogenase, exclusive reliance on the manganese-based catalase and superoxide dismutase for combating reactive oxygen species (ROS), DNA-binding rubrerythrin and dedicated polymerases that can repair UV-induced double strand breaks in the DNA (Meier et al. 2021).

11.3.2 Topsoil Community Composition

Bacteria Arid soils harbor a large proportion of *Actinobacteria* and *Proteobacteria* (Delgado-Baquerizo et al. 2018; Makhalanyane et al. 2015; Chap. 4) typical to both cold and hot deserts (Fierer et al. 2012). These similarities between the two desert types are likely due to similarities in AI and productivity rates (Delgado-Baquerizo et al. 2018). Members of the *Actinobacteria* phylum appear to be ubiquitous in desert soils. Recent bacterial community analyses conducted in the Negev desert determined that members of the *Actinobacteria* represented over 40% of the total bacterial community in arid soils (Bachar et al. 2010; Makhalanyane et al. 2015). It was suggested that the relative abundance of soil *actinobacteria* increases at extreme conditions. Several studies have shown *Actinobacteria* to be more resilient to low levels of soil moisture compared to the general bacterial population. *Proteobacteria* are also abundant in desert soil, including members of the alpha, beta, and gamma orders (Bachar et al. 2010; Makhalanyane et al. 2015). This phylum is functionally important in nutrient-limited arid environments due to its role in carbon and nitrogen cycling in both the biocrust and topsoil (Fierer et al. 2007). *Bacteroidetes* are also common in desert soils, which is surprising given their anaerobic phenotype (Fierer et al. 2007; Makhalanyane et al. 2015). Other bacterial lineages identified in desert environments include *Firmicutes*, mainly the *Clostridia* class, in addition to *Bacilli* and *Paenibacilli* that can form endospores, which facilitate survival under desiccating conditions, but were shown to germinate rapidly upon hydration (Karaoz et al. 2018).

Fungi Fungi, along with heterotrophic and phototrophic bacteria, algae, and mosses, are prevalent components of biocrusts in arid and semi-arid areas. They can occur as free-living fungi or in symbiosis with cyanobacteria or algae to form lichens. The fungal compositions and their critical roles in biocrust biogeochemistry have been studied in desert ecosystems around the world, including the Negev Desert (Grishkan and Kidron 2013), Chihuahuan, and Sonoran Desert (Bates et al. 2012), Arabian Desert (Abed et al. 2013), and the Tengger Desert (Zhang et al. 2016). These studies showed that Ascomycota are the dominant fungi in biocrusts and in the rhizosphere of dryland perennials (Rudgers et al. 2018). In some areas, the biocrusts fungal communities are dominated by melanin-containing species with large multicellular conidia (Grishkan et al. 2015). It was hypothesized that in arid environments, perennial plants are often nitrogen-limited but can gain access to nitrogen by networking with nitrogen-fixing biocrusts through fungal loop that link resources that would otherwise be disconnected in space or time (Rudgers et al. 2018). However, this hypothesis requires further testing.

11.4 Soil Bacteria Response to Hydration-Desiccation Cycles in Desert Ecosystems

11.4.1 The Response of Desert Soil Bacteria to Desiccation

The desert-dwelling microorganisms developed several remarkable adaptations to desiccation that allow for a population response to changing environmental conditions (Fowler and Winstanley 2018). Chap. 10 specifically describes the resistance mechanisms developed by microorganisms, at the cellular and community levels, to cope with hot deserts' extreme environmental conditions. Certain bacteria from the phylum *Firmicutes* can produce resilient spores capable of surviving long periods of desiccation. Other bacterial species, like Azotobacter, can form cysts that are similar to spores but are comparatively less resistant (Karaoz et al. 2018; Meier et al. 2021). Some taxa, including *Deinococcaceae*, *Rubrobacteridae,* and *Thermoleophilidae*, were shown to develop an intricate strategy to survive desiccation and related oxidation stress not by producing spores or cysts, but by evolving efficient protective and repair mechanisms coping with desiccation induced damage (Baubin et al. 2019; Jin et al. 2019; Meier et al. 2021). These adaptations help microbial species thrive during a dry season or survive until the next hydration.

Dormancy Importance in Desert Soil Dormancy is increasingly seen as an important evolutionary bet-hedging strategy (Shoemaker and Lennon 2018). It has been estimated that as much as 90% of soil microbes, regardless of the biome or ecosystem, are usually inactive (Joergensen and Wichern 2018; Nannipieri 2020). The majority of soil-dwelling organisms are persisting in an inactive form waiting for favorable growth conditions (Shoemaker and Lennon 2018). In desert environments, the microorganisms persist through long periods of drought, subjected to extensive stress caused by high salinity, extensive radiation, and lack of nutrients. Bacteria persist through dormancy, including quiescence, latency, persistence, sporulation, or encystation (Nannipieri 2020). Only sporulation and encystation are defined as physiological states, while the rest of the terms are loosely defined as periods of reduced physiological activity in different contexts. The common trait of dormancy is the ability to switch between active and inactive states (Shoemaker and Lennon 2018. The inactive, resilient form should enable the cells to persist through unfavorable conditions and subsequently reseed an active population (Fowler and Winstanley 2018). This distinguishes dormant organisms from oligotrophic slow-growing microorganisms.

Detrimental Effects of Desiccation A loss of cellular water in microorganisms could be manifested by cell shrinkage, change in shape or a change in color (darkening) due to pigment oxidation (Potts et al. 2005). Moreover, desiccation may result in an increase of intracellular salt concentration, crowding of macromolecules, changes in biophysical properties, such as fluidity (increased viscosity), acquisition of static charge, or damage to external structures such as pili (Potts 1994). These changes impose stress on the cells, resulting in hyperosmolarity and hyper-oxidation by reactive oxygen species (ROS) that have a detrimental effect on: (I) proteins, resulting in their aggregation and misfolding; (II.) nucleic acids (DNA and RNA) that are exposed to ROS damage; and (III.) the membranes that lose fluidity and elasticity (Potts 2001; Potts et al. 2005).

Chemical and Biochemical Desiccation Effectors Several chemical reactions occur during reduced water availability in a cell, including the Maillard reaction that occurs in the presence of sugars and aminoacids at low water activity and high temperature, which could severely damage essential cell components. In addition, Haber–Weiss reaction and Fenton reaction (a catalytic process that converts hydrogen peroxide to a hydroxyl free radical) involve metal (typically iron) and produce ROS that further damage the cells' DNA and protein (Meier et al. 2021).

The Haber-Weis reaction is catalyzed by metals and occurs between hydrogen peroxide and superoxide radicals, resulting in hydroxyl radicals (Imlay 2008). Hydroxyl radicals and their derivatives are among the most reactive oxygen species known (Potts 1994). For protection from hydroxyl radicals damage, aerobic organisms have developed an array of enzymes that actively scavenge for both reactants of the Haber–Weis reaction. These are broadly grouped as catalases, removing hydrogen peroxide and superoxide dismutase, thus neutralizing superoxide radicals. During desiccation, cells are exposed to high rates of ROS damage and therefore accumulate high concentrations of both catalase and superoxide dismutases which remain active even throughout a prolonged desiccation protecting the cells from oxygen damage (Du et al. 2014; Tang et al. 2007).

It was recently shown that the acute water scarcity in arid lands also results in some microbes engaging in light-independent carbon fixation using trace gases, such as dihydrogen (H_2) and carbon monoxide (CO) to maintain the cell's integrity in dormant states. Furthermore, trace gas oxidation could potentially support primary production, as oxidation of atmospheric H_2 has been shown to sustain CO_2 fixation in cold (Ji et al. 2017) and hot (Bay et al. 2020; Meier et al. 2021) desert soils. This minimalistic mode of energy production was reported to be inversely proportional to water availability and was suggested to serve predominantly to maintain and repair cell damage, rather than support active growth (Leung et al. 2020).

Passive Desiccation Stress Effector Desiccation tolerant bacteria have developed remarkable mechanisms of coping with desiccation induced damage. Many xerophiles are Gram-positive, non-spore forming species (Potts et al. 2005). Alongside enzymes that help protect the cells against desiccation damage, they accumulate passive stress effectors that do not require energy input maintaining function even through prolonged desiccation (Yu et al. 2015). Common passive stress effectors in xerophiles include (I) manganese Mn(II) ions (Lang-Yona et al. 2018);

(II) non-reducing disaccharides (trehalose and sucrose) (Azua-Bustos et al. 2014); and (III) hydrophilins (short unstructured hydrophilic proteins) (Jiang et al. 2017), that accumulate in high concentration in drying cells.

The Role of Manganese in Desiccation-Related ROS Scavenging Xerophilic bacteria often accumulate Mn(II) ions, increasing the intracellular Mn(II) to Fe ratio helping to protect the cells against desiccation induced oxidation damage. Mn(II) is a cofactor for superoxide dismutase, which helps neutralize damaging ROS species. Mn(II) also acts as a part of non-proteinaceous manganese antioxidants (i. e., phosphates, lactates, and trehalose complexes), which neutralize oxidative stress (Aguirre and Culotta 2012). Furthermore, the Mn(II) ions, unlike Fe, do not partake in Haber-Weiss and Fenton reactions, which generate ROS in the first place. Xerophilic bacteria, such as Deinococcus, Lactobacillus, and cyanobacteria, have been reported to accumulate an unusually high concentration of Mn(II) (Daly et al. 2004), which can be linked to their capacity to withstand oxidative stress (Paulino-Lima et al. 2016). Indeed, bacterial communities from extreme arid environments also actively partake in Mn(II) metabolism (Grote and Krumbein 1992; Hungate et al. 1987).

The Effect of Trehalose and Hydrophilins as Desiccation Protectants Trehalose has been extensively studied as a passive stress effector in xerophyles. Together with Mn(II) ions, it accumulates in desiccating cells of xerophilic bacteria and yeast (Aguirre and Culotta 2012). Trehalose and Mn(II) co-occur with hydrophilins, which is a diverse group of proteins that share some common features, including at least 6% glycine content, small amino acids (such as alanine and serine) and a high hydrophilicity index over 1.0 (Slade and Radman 2011), however, their function has not yet been fully established. Both trehalose and hydrophilins have been shown to protect membrane structure, stabilize protein folding, circumvent protein aggregation during desiccation, maintain enzymatic activity at low water potential, and help remove detrimental ROS (Lang-Yona et al. 2018; Mertens et al. 2018; Yu et al. 2015).

11.4.2 The Response of Desert Soil Bacteria to Hydration

Desert soil microorganisms face numerous stresses during a hydration event due to the sudden shift from highly desiccated to hydrated conditions. The sudden difference in the cell turgor pressure, exerted by the osmotic flow of water into the cell, may lead to cell bursting (plasmoptysis), the release of cellular content and finally cell death (Halverson et al. 2000). Bacteria have developed various defense mechanisms to survive this sudden and unexpected osmotic stress, including robust cell walls that can withstand an increase of osmotic pressure (Kieft et al. 1987; Pointing and Belnap 2012; Potts et al. 2005), or the release of osmolytes to counter the sudden pressure difference (Halverson et al. 2000). While releasing osmolytes, microorganisms can decrease their available amino acid pool by 25% and their low molecular weight sugars by 20% (Halverson et al. 2000). The released nutrients could become available to neighboring cells alongside other previously inaccessible substrates (Fierer and Schimel 2003).

Changes in Soil Microbial Communities in Response to Hydration During a rain event, newly available resources (water and nutrients) trigger dramatic community changes (Fig. 11.1) (Blazewicz et al. 2014; Tecon and Or 2017). The reactivated communities consume the substrate, deplete oxygen, and release secondary metabolites (Kim and Or 2016; Tecon and Or 2017). Previously inactive microbes take advantage of the increased substrate availability (Or et al. 2007; Wang and Or 2012) as the access of water creates a continuous film around soil particles, allowing migration of chemicals and cells (Wang and Or 2012). This allows the interaction of previously separated organisms that could be manifested by substrate-mediated competition (Kim and Or 2016), microbicidal secondary metabolite mediated competition (Charlop-Powers et al. 2014), predator-prey interactions (Rodriguez-Zaragoza et al. 2005), or phage bloom (Van Goethem et al. 2016).

The microbial community responds quickly to a rain event that alters their composition (Fig. 11.1). Actinobacteria were shown to dominate the desert soils around the globe (Chap. 4) like the Negev (Angel and Conrad 2013; Baubin et al. 2021; Šťovíček et al. 2017), Atacama (Fernández-Martínez et al. 2019), or Namib (Makhalanyane et al. 2015). During a rain event, the relative abundance of Actinobacteria in both the topsoil and biocrust rapidly decreased from over 40% to less than 5% (Angel and Conrad 2013; Šťovíček et al. 2017; Baubin et al. 2021). The Firmicutes quickly germinate and dominate the community for a short while [Fig. 11.1 (Šťovíček et al. 2017)]. In the biocrust, the cyanobacteria take advantage of the water availability to photosynthesize and proliferate, correspondingly their relative abundance increase (Angel and Conrad 2013; Baubin et al. 2021). The proteobacteria, a dominant group within the community, are not rapid responders, but following the rain event, their abundance slowly grows (Fig. 11.1). The mechanisms that underlie these changes were modeled and suggested to result from trophic interactions mediated by aqueous phase connectivity (Ebrahimi and Or 2015; Kim and Or 2016, 2017). Hydration was shown to mediate competition in otherwise segregated communities, thus reducing the overall richness and evenness while altering the community composition (Šťovíček et al. 2017; Kim and Or 2017; Baubin et al. 2021).

Substrate Mediated Competition Hydration increases the connectivity of soil microhabitats allowing for substrate diffusion across larger distances, and consequently more microbes have access to the substrate pool (Dechesne et al. 2008; Wang and Or 2013). Numerous seasonally opportunistic microbial taxa, such as *Bacilli* (Karaoz et al. 2018), take advantage of the plentiful water and substrate and germinate, increasing the competition pressure over nutrients (Fig. 11.1). This substrate competition was shown to have the potential to drive hydration-related dynamics between copiotrophic and oligotrophic desert soil communities (Bauer et al. 2018).

Secondary Metabolites Mediated Competition The increased connectivity allows the diffusion of secondary metabolites through the soil profile (Wang and Or 2013). This could result in higher rates of dilution of the secondary metabolites, possibly below their active concentration, but may also enable microbicidal compounds to reach and influence a higher number of cells. The net outcome of this tradeoff is likely specific to the local topology and the resistance of the community members to the toxic compounds. However, little research is available and predictions are limited to local measurements combined with substrate diffusion models (Kim and Or 2016).

Predator–Prey Interaction Seasonal rain events allow the movement of predators scavenging for their prey) (Dechesne et al. 2008). Both prokaryotic (e.g., Bdellovibrio and Myxobacteria) or eukaryotic (e.g., flagellates, amoeba, nematodes, and ciliates) predators were detected in desert soil (Pérez et al. 2016), including the Negev (Rodriguez-Zaragoza et al. 2005) and were shown to activate upon hydration (Darby et al. 2006). Flagellate movement was shown to require a very high level of water content, typically near to field capacity (Wang and Or 2013), offering a short window of opportunity for predators to act before the water content drops. Mixobacteria and amoeba move by gliding, a motion that is likely more important in desert soils as it requires lower water content (Raanan et al. 2016; Rodriguez-Zaragoza et al. 2005). Moreover, these taxa could benefit from EPS to "lubricate a movement path" (Kidron et al. 2020; Roberson and Firestone 1992).

Viral Interaction Previous studies point towards an unusually high proportion of lysogenic viruses in desert soil metagenomes (Kuzyakov and Mason-Jones 2018; Prestel et al. 2013; Chap. 6). The life cycle of lysogenic viruses may offer an advantage in the challenging desert environment, allowing viruses to persist under the hosts' protection during desiccation periods (Zablocki et al. 2016, 2017). The lytic phase may be activated upon hydration, where water-filled pores convey an increased habitat connectivity and allow dispersal of viral particles to new hosts. Viruses could convey new functions to their hosts, providing competitive advantages and resulting in mutualistic relationships (Prestel et al. 2013). However, the role of viruses in desert soils is undoubtedly complex and hitherto poorly understood.

11.4.3 Pulse Theory and Grime's Hypothesis

In desert ecosystems, the ecosystem has to adjust to cycles between periods of resource and water scarcity and the short pulses of water and nutrient availability (Austin et al. 2004). This pulse is accompanied by a CO_2 pulse as well as large community changes with microbial groups resuscitating in as little as 1 hour after the wetting event (Placella et al. 2012). The periods of low productivity, where harsh conditions limit nutrient acquisition and availability might either increase or decrease competition, depending on the interactions between of growth and the survival component of fitness and resource dynamics (Goldberg and Novoplansky 1997). Resources' availability in deserts is dictated by water availability, thus dividing desert organisms' existence between pulse and interpulse periods. Pulse is a period when resources are temporarily available, resulting in growth and resource accumulation, while interpulse is a period when resources are unavailable (Noy-Meir 1973). It is important to note that these are periods whose perceived length depends on any individual's physiological characteristics. Pulse periods are governed by competition that affects population dynamics, whereas interpulse is dominated by population persistence. This is true only if the resource availability during interpulse is driven by abiotic processes (e.g., drainage and evaporation of water) and if survival during interpulse is not positively correlated with growth during pulse periods (Goldberg and Novoplansky 1997; Grime 1977). This suggests that during a rain period, desert soil microbial communities might be subjected to more intense competition (Grime 1977).

Water availability in desert is predominantly determined by abiotic processes, such as evaporation or drainage. This confirms the first assumption of the Grime hypothesis. It has also been previously shown that desert soil has distinct populations connected with the dry and wet periods with an anticorrelated response to a hydration event (Fierer et al. 2012; Placella et al. 2012; Barnard et al. 2013; Blazewicz et al. 2014). This may be an evidence supporting the second assumption of Grime's hypothesis and confirming its validity for desert soil systems. The measurements of in situ diversity following a wetting event have described a higher degree of coexistence reflected in increased diversity under drier conditions (Carson et al. 2010), validating Grime hypothesis.

11.5 Conclusion

Rain events in desert environments significantly impact the entire ecosystem, triggering noticeable regrowth and bloom of annuals and perennials alike. It has profound effects on the soil bacterial communities shaping their composition and activity. The effect of hydration was described under controlled conditions, field conditions, and through mathematical modeling, focusing on the sudden increase in water availability and the resulting bacterial dynamics. The bacterial community quickly responds to water access as well as severe desiccation, and we have described the mechanisms involved. However, the response of corresponding members of the desert soil microbiome is poorly understood and trophic interactions were seldom explored. We briefly reviewed the role of predation and viral bloom may play in shaping the microbial community during rain events and subsequent desiccation. A global overview of the soil community is lacking and would form the next frontier in our understanding of the response of the desert soil microbiome to hydration-desiccation cycles.

References

Abed RMM, Al-Sadi AM, Al-Shehi M, Al-Hinai S, Robinson MD (2013) Diversity of free-living and lichenized fungal communities in biological soil crusts of the Sultanate of Oman and their role in improving soil properties. Soil Biol Biochem 57:695–705. https://doi.org/10.1016/j.soilbio.2012.07.023

Aguirre JD, Culotta VC (2012) Battles with iron: manganese in oxidative stress protection. J Biol Chem 287:13541–13548. https://doi.org/10.1074/jbc.R111.312181

Angel R, Conrad R (2013) Elucidating the microbial resuscitation cascade in biological soil crusts following a simulated rain event: microbial resuscitation in biological soil crusts. Environ Microbiol. https://doi.org/10.1111/1462-2920.12140

Austin AT, Vivanco L (2006) Plant litter decomposition in a semi-arid ecosystem controlled by photodegradation. Nature 442:555–558. https://doi.org/10.1038/nature05038

Austin AT, Yahdjian L, Stark JM, Belnap J, Porporato A, Norton U, Ravetta D, Schaeffer SM (2004) Water pulses and biogeochemical cycles in arid and semiarid ecosystems. Oecologia 141:221–235. https://doi.org/10.1007/s00442-004-1519-1

Azua-Bustos A, Zúñiga J, Arenas-Fajardo C, Orellana M, Salas L, Rafael V (2014) Gloeocapsopsis AAB1, an extremely desiccation-tolerant cyanobacterium isolated from the Atacama Desert. Extremophiles 18:61–74. https://doi.org/10.1007/s00792-013-0592-y

Bachar A, Al-Ashhab A, Soares MIM, Sklarz MY, Angel R, Ungar ED, Gillor O (2010) Soil microbial abundance and diversity along a low precipitation gradient. Microb Ecol 60:453–461. https://doi.org/10.1007/s00248-010-9727-1

Bailey VL, Pries CH, Lajtha K (2019) What do we know about soil carbon destabilization? Environ Res Lett 14:083004. https://doi.org/10.1088/1748-9326/ab2c11

Barnard RL, Osborne CA, Firestone MK (2013) Responses of soil bacterial and fungal communities to extreme desiccation and rewetting. ISME J 7:2229–2241. https://doi.org/10.1038/ismej.2013.104

Bates ST, Nash TH, Garcia-Pichel F (2012) Patterns of diversity for fungal assemblages of biological soil crusts from the southwestern United States. Mycologia 104:353–361. https://doi.org/10.3852/11-232

Baubin C, Farrell AM, Šťovíček A, Ghazaryan L, Giladi I, Gillor O (2019) Seasonal and spatial variability in total and active bacterial communities from desert soil. Pedobiologia 74:7–14. https://doi.org/10.1016/j.pedobi.2019.02.001

Baubin C, Ran N, Siebner H, Gillor O (2021) The response of desert biocrust bacterial communities to hydration-desiccation cycles (preprint). Microbiology. https://doi.org/10.1101/2021.04.04.438350

Bauer MA, Kainz K, Carmona-Gutierrez D, Madeo F (2018) Microbial wars: competition in ecological niches and within the microbiome. Microb Cell 5:215–219. https://doi.org/10.15698/mic2018.05.628

Bay SK, McGeoch MA, Gillor O, Wieler N, Palmer DJ, Baker DJ, Chown SL, Greening C (2020) Soil bacterial communities exhibit strong biogeographic patterns at fine taxonomic resolution. mSystems 5:e00540-20. https://doi.org/10.1128/mSystems.00540-20

Beck HE, Zimmermann NE, McVicar TR, Vergopolan N, Berg A, Wood EF (2018) Present and future Köppen-Geiger climate classification maps at 1-km resolution. Sci Data 5:180214. https://doi.org/10.1038/sdata.2018.214

Bell C, McIntyre N, Cox S, Tissue D, Zak J (2008) Soil microbial responses to temporal variations of moisture and temperature in a Chihuahuan Desert grassland. Microb Ecol 56:153–167. https://doi.org/10.1007/s00248-007-9333-z

Belnap J (2003) The world at your feet: desert biological soil crusts. Front Ecol Environ 1:181–189. https://doi.org/10.1890/1540-9295(2003)001[0181:TWAYFD]2.0.CO;2

Belnap J, Weber B, Büdel B (2016) Biological soil crusts as an organizing principle in drylands. In: Weber B, Büdel B, Belnap J (eds) Biological soil crusts: an organizing principle in drylands, ecological studies. Springer International Publishing, Cham, pp 3–13. https://doi.org/10.1007/978-3-319-30214-0_1

Birch HF (1958) The effect of soil drying on humus decomposition and nitrogen availability. Plant Soil 10:9–31. https://doi.org/10.1007/BF01343734

Blazewicz SJ, Schwartz E, Firestone MK (2014) Growth and death of bacteria and fungi underlie rainfall-induced carbon dioxide pulses from seasonally dried soil. Ecology 95:1162–1172. https://doi.org/10.1890/13-1031.1

Borken W, Matzner E (2009) Reappraisal of drying and wetting effects on C and N mineralization and fluxes in soils. Glob Chang Biol 15:808–824. https://doi.org/10.1111/j.1365-2486.2008.01681.x

Carson JK, Gonzalez-Quinones V, Murphy DV, Hinz C, Shaw JA, Gleeson DB (2010) Low pore connectivity increases bacterial diversity in soil. Appl Environ Microbiol 76:3936–3942. https://doi.org/10.1128/AEM.03085-09

Chandler DG, Seyfried MS, McNamara JP, Hwang K (2017) Inference of soil hydrologic parameters from electronic soil moisture records. Front Earth Sci 5:25. https://doi.org/10.3389/feart.2017.00025

Charlop-Powers Z, Owen JG, Reddy BVB, Ternei MA, Brady SF (2014) Chemical-biogeographic survey of secondary metabolism in soil. Proc Natl Acad Sci 111:3757–3762. https://doi.org/10.1073/pnas.1318021111

Cherlet M, Hutchinson C, Reynolds J, Hill J, Sommer S, von Maltitz G (2018) World atlas of desertification rethinking land degradation and sustainable land management. Publication Office of the European Union, Luxembourg

Clark JS, Campbell JH, Grizzle H, Acosta-Martìnez V, Zak JC (2008) Soil microbial community response to drought and precipitation variability in the Chihuahuan Desert. Microb Ecol 57:248–260. https://doi.org/10.1007/s00248-008-9475-7

Daly MJ, Gaidamakova EK, Matrosova VY, Vasilenko A, Zhai M, Venkateswaran A, Hess M, Omelchenko MV, Kostandarithes HM, Makarova KS, Wackett LP, Fredrickson JK, Ghosal D (2004) Accumulation of Mn(II) in Deinococcus radiodurans facilitates gamma-radiation resistance. Science 306:1025–1028. https://doi.org/10.1126/science.1103185

Darby BJ, Housman DC, Zaki AM, Shamout Y, Adl SM, Belnap J, Neher DA (2006) Effects of altered temperature and precipitation on desert protozoa associated with biological soil crusts. J Eukaryotic Microbiology 53:507–514. https://doi.org/10.1111/j.1550-7408.2006.00134.x

Dechesne A, Or D, Smets BF (2008) Limited diffusive fluxes of substrate facilitate coexistence of two competing bacterial strains: substrate diffusion and bacterial competition. FEMS Microbiol Ecol 64:1–8. https://doi.org/10.1111/j.1574-6941.2008.00446.x

Delgado-Baquerizo M, Oliverio AM, Brewer TE, Benavent-González A, Eldridge DJ, Bardgett RD, Maestre FT, Singh BK, Fierer N (2018) A global atlas of the dominant bacteria found in soil. Science 359:320–325. https://doi.org/10.1126/science.aap9516

Denef K, Six J, Paustian K, Merckx R (2001) Importance of macroaggregate dynamics in controlling soil carbon stabilization: short-term effects of physical disturbance induced by dry–wet cycles. Soil Biol Biochem 33:2145–2153. https://doi.org/10.1016/S0038-0717(01)00153-5

Du F, Shi H, Zhang X, Xu X (2014) Responses of reactive oxygen scavenging enzymes, proline and malondialdehyde to water deficits among Six secondary successional seral species in loess plateau. PLoS One 9:e98872. https://doi.org/10.1371/journal.pone.0098872

Ebrahimi A, Or D (2015) Hydration and diffusion processes shape microbial community organization and function in model soil aggregates: hydration affects microbial activity in soil aggregates. Water Resour Res 51:9804–9827. https://doi.org/10.1002/2015WR017565

Eckardt FD, Soderberg K, Coop LJ, Muller AA, Vickery KJ, Grandin RD, Jack C, Kapalanga TS, Henschel J (2013) The nature of moisture at Gobabeb, in the central Namib Desert. J Arid Environ 93:7–19. https://doi.org/10.1016/j.jaridenv.2012.01.011

Elbert W, Weber B, Burrows S, Steinkamp J, Büdel B, Andreae MO, Pöschl U (2012) Contribution of cryptogamic covers to the global cycles of carbon and nitrogen. Nat Geosci 5:459–462. https://doi.org/10.1038/ngeo1486

Fernández-Martínez MÁ, dos Santos Severino R, Moreno-Paz M, Gallardo-Carreño I, Blanco Y, Warren-Rhodes K, García-Villadangos M, Ruiz-Bermejo M, Barberán A, Wettergreen D, Cabrol N, Parro V (2019) Prokaryotic community structure and metabolisms in shallow subsurface of Atacama Desert playas and alluvial fans after heavy rains: repairing and preparing for next dry period. Front Microbiol 10:1641. https://doi.org/10.3389/fmicb.2019.01641

Fierer N, Schimel JP (2003) A proposed mechanism for the pulse in carbon dioxide production commonly observed following the rapid rewetting of a dry soil. Soil Sci Soc Am J 67(3): 798–805. https://doi.org/10.2136/sssaj2003.7980

Fierer N, Bradford MA, Jackson RB (2007) Toward an ecological classification of soil bacteria. Ecology 88:1354–1364. https://doi.org/10.1890/05-1839

Fierer N, Leff JW, Adams BJ, Nielsen UN, Bates ST, Lauber CL, Owens S, Gilbert JA, Wall DH, Caporaso JG (2012) Cross-biome metagenomic analyses of soil microbial communities and their functional attributes. Proc Natl Acad Sci 109:21390–21395. https://doi.org/10.1073/pnas.1215210110

Fowler AC, Winstanley HF (2018) Microbial dormancy and boom-and-bust population dynamics under starvation stress. Theor Popul Biol 120:114–120. https://doi.org/10.1016/j.tpb.2018.02.001

Fraser FC, Corstanje R, Deeks LK, Harris JA, Pawlett M, Todman LC, Whitmore AP, Ritz K (2016) On the origin of carbon dioxide released from rewetted soils. Soil Biol Biochem 101:1–5. https://doi.org/10.1016/j.soilbio.2016.06.032

Goldberg D, Novoplansky A (1997) On the relative importance of competition in unproductive environments. J Ecol 85:409. https://doi.org/10.2307/2960565

Grime JP (1977) Evidence for the existence of three primary strategies in plants and its relevance to ecological and evolutionary theory. Am Nat 111:1169–1194. https://doi.org/10.1086/283244

Grishkan I, Kidron GJ (2013) Biocrust-inhabiting cultured microfungi along a dune catena in the western Negev Desert, Israel. Eur J Soil Biol 56:107–114. https://doi.org/10.1016/j.ejsobi.2013.03.005

Grishkan I, Jia R-L, Kidron GJ, Li X-R (2015) Cultivable microfungal communities inhabiting biological soil crusts in the Tengger Desert, China. Pedosphere 25:351–363. https://doi.org/10.1016/S1002-0160(15)30003-5

Grote G, Krumbein WE (1992) Microbial precipitation of manganese by bacteria and fungi from desert rock and rock varnish. Geomicrobiol J 10:49–57. https://doi.org/10.1080/01490459209377903

Hagemann M, Henneberg M, Felde VJMNL, Berkowicz SM, Raanan H, Pade N, Felix-Henningsen P, Kaplan A (2016) Cyanobacterial populations in biological soil crusts of the Northwest Negev Desert, Israel – effects of local conditions and disturbance. FEMS Microbiol Ecol 93(6):fiw228. https://doi.org/10.1093/femsec/fiw228

Halverson LJ, Jones TM, Firestone MK (2000) Release of intracellular solutes by four soil bacteria exposed to dilution stress. Soil Sci Soc Am J 64:1630. https://doi.org/10.2136/sssaj2000.6451630x

Hill AJ, Dawson TE, Shelef O, Rachmilevitch S (2015) The role of dew in Negev Desert plants. Oecologia 178:317–327. https://doi.org/10.1007/s00442-015-3287-5

Hong B, Lin Q, Yu S, Chen Y, Chen Y, Chiang P (2018) Urbanization gradient of selected pharmaceuticals in surface water at a watershed scale. Sci Total Environ 634:448–458. https://doi.org/10.1016/j.scitotenv.2018.03.392

Hungate B, Danin A, Pellerin NB, Stemmler J, Kjellander P, Adams JB, Staley JT (1987) Characterization of manganese-oxidizing (MnII→MnIV) bacteria from Negev Desert rock varnish: implications in desert varnish formation. Can J Microbiol 33:939–943. https://doi.org/10.1139/m87-165

Huxman TE, Snyder KA, Tissue D, Leffler AJ, Ogle K, Pockman WT, Sandquist DR, Potts DL, Schwinning S (2004) Precipitation pulses and carbon fluxes in semiarid and arid ecosystems. Oecologia 141:254–268. https://doi.org/10.1007/s00442-004-1682-4

Imlay JA (2008) Cellular defenses against superoxide and hydrogen peroxide. Annu Rev Biochem 77:755–776. https://doi.org/10.1146/annurev.biochem.77.061606.161055

Ji M, Greening C, Vanwonterghem I, Carere CR, Bay SK, Steen JA, Montgomery K, Lines T, Beardall J, van Dorst J, Snape I, Stott MB, Hugenholtz P, Ferrari BC (2017) Atmospheric trace gases support primary production in Antarctic desert surface soil. Nature. https://doi.org/10.1038/nature25014

Jiang S, Wang J, Liu X, Liu Y, Guo C, Zhang L, Han J, Wu X, Xue D, Gomaa AE, Feng S, Zhang H, Chen Y, Ping S, Chen M, Zhang W, Li L, Zhou Z, Zuo K, Li X, Yang Y, Lin M (2017) DrwH, a novel WHy domain-containing hydrophobic LEA5C protein from Deinococcus radiodurans, protects enzymatic activity under oxidative stress. Sci Rep 7:9281. https://doi.org/10.1038/s41598-017-09541-2

Jin M, Xiao A, Zhu L, Zhang Z, Huang H, Jiang L (2019) The diversity and commonalities of the radiation-resistance mechanisms of Deinococcus and its up-to-date applications. AMB Expr 9:138. https://doi.org/10.1186/s13568-019-0862-x

Joergensen RG, Wichern F (2018) Alive and kicking: why dormant soil microorganisms matter. Soil Biol Biochem 116:419–430. https://doi.org/10.1016/j.soilbio.2017.10.022

Karaoz U, Couradeau E, da Rocha UN, Lim H-C, Northen T, Garcia-Pichel F, Brodie EL (2018) Large blooms of *Bacillales* (*Firmicutes*) underlie the response to wetting of cyanobacterial biocrusts at various stages of maturity. mBio 9:e01366–e01316. https://doi.org/10.1128/mBio.01366-16

Khadka J, Yadav N, Granot G, Grafi G (2018) Seasonal growth of Zygophyllum dumosum Boiss.: summer dormancy is associated with loss of the permissive epigenetic marker dimethyl H3K4 and extensive reduction in proteins involved in basic cell functions. Plan Theory 7:59. https://doi.org/10.3390/plants7030059

Kidron GJ, Starinsky A (2012) Chemical composition of dew and rain in an extreme desert (Negev): cobbles serve as sink for nutrients. J Hydrol 420–421:284–291. https://doi.org/10.1016/j.jhydrol.2011.12.014

Kidron GJ, Wang Y, Herzberg M (2020) Exopolysaccharides may increase biocrust rigidity and induce runoff generation. J Hydrol 588:125081. https://doi.org/10.1016/j.jhydrol.2020.125081

Kieft TL, Soroker E, Firestone MK (1987) Microbial biomass response to a rapid increase in water potential when dry soil is wetted. Soil Biol Biochem 19:119–126. https://doi.org/10.1016/0038-0717(87)90070-8

Kim M, Or D (2016) Individual-based model of microbial life on hydrated rough soil surfaces. PLoS One 11:e0147394. https://doi.org/10.1371/journal.pone.0147394

Kim M, Or D (2017) Hydration status and diurnal trophic interactions shape microbial community function in desert biocrusts. Biogeosciences 14:5403–5424. https://doi.org/10.5194/bg-14-5403-2017

Kuzyakov Y, Mason-Jones K (2018) Viruses in soil: Nano-scale undead drivers of microbial life, biogeochemical turnover and ecosystem functions. Soil Biol Biochem 127:305–317. https://doi.org/10.1016/j.soilbio.2018.09.032

Lang-Yona N, Maier S, Macholdt DS, Müller-Germann I, Yordanova P, Rodriguez-Caballero E, Jochum KP, Al-Amri A, Andreae MO, Fröhlich-Nowoisky J, Weber B (2018) Insights into microbial involvement in desert varnish formation retrieved from metagenomic analysis. Environ Microbiol Rep 10:264–271. https://doi.org/10.1111/1758-2229.12634

Leung PM, Bay SK, Meier DV, Chiri E, Cowan DA, Gillor O, Woebken D, Greening C (2020) Energetic basis of microbial growth and persistence in desert. Ecosystems mSystems 5:e00495-19. https://doi.org/10.1128/mSystems.00495-19

Makhalanyane TP, Valverde A, Gunnigle E, Frossard A, Ramond J-B, Cowan DA (2015) Microbial ecology of hot desert edaphic systems. FEMS Microbiol Rev 39:203–221. https://doi.org/10.1093/femsre/fuu011

Meier DV, Imminger S, Gillor O, Woebken D (2021) Distribution of Mixotrophy and desiccation survival mechanisms across microbial genomes in an arid biological soil crust community. mSystems 6:e00786-20. https://doi.org/10.1128/mSystems.00786-20

Mertens J, Aliyu H, Cowan DA (2018) LEA proteins and the evolution of the why domain. Appl Environ Microbiol 84:e00539-18. https://doi.org/10.1128/AEM.00539-18

Monson RK, Smith SD (1982) Seasonal water potential components of Sonoran Desert plants. Ecology 63:113–123. https://doi.org/10.2307/1937037
Moorhead DL, Reynolds JF (1991) A general model of litter decomposition in the northern Chihuahuan Desert. Ecol Model 56:197–219. https://doi.org/10.1016/0304-3800(91)90200-K
Moyano FE, Manzoni S, Chenu C (2013) Responses of soil heterotrophic respiration to moisture availability: an exploration of processes and models. Soil Biol Biochem 59:72–85. https://doi.org/10.1016/j.soilbio.2013.01.002
Nannipieri P (2020) Soil is still an unknown biological system. Appl Sci 10:3717. https://doi.org/10.3390/app10113717
Noy-Meir I (1973) Desert ecosystems: environment and producers. Annu Rev Ecol Syst 4:25–51. https://doi.org/10.1146/annurev.es.04.110173.000325
Noy-Meir I (1974) Desert ecosystems: higher trophic levels. Annu Rev Ecol Syst 5:195–214. https://doi.org/10.1146/annurev.es.05.110174.001211
O'Geen AT (2013) Soil water dynamics. Nat Educ Knowl 4:9
Ogle K, Reynolds JF (2004) Plant responses to precipitation in desert ecosystems: integrating functional types, pulses, thresholds, and delays. Oecologia 141:282–294. https://doi.org/10.1007/s00442-004-1507-5
Or D, Smets BF, Wraith JM, Dechesne A, Friedman SP (2007) Physical constraints affecting bacterial habitats and activity in unsaturated porous media – a review. Adv Water Resour 30: 1505–1527. https://doi.org/10.1016/j.advwatres.2006.05.025
Oren N, Raanan H, Kedem I, Turjeman A, Bronstein M, Kaplan A, Murik O (2019) Desert cyanobacteria prepare in advance for dehydration and rewetting: the role of light and temperature sensing. Mol Ecol 28:2305–2320. https://doi.org/10.1111/mec.15074
Paulino-Lima IG, Fujishima K, Navarrete JU, Galante D, Rodrigues F, Azua-Bustos A, Rothschild LJ (2016) Extremely high UV-C radiation resistant microorganisms from desert environments with different manganese concentrations. J Photochem Photobiol B Biol 163:327–336. https://doi.org/10.1016/j.jphotobiol.2016.08.017
Peel MC, Finlayson BL, McMahon TA (2007) Updated world map of the Köppen-Geiger climate classification. Hydrol Earth Syst Sci Discuss 4:439–473. https://doi.org/10.5194/hessd-4-439-2007
Pérez J, Moraleda-Muñoz A, Marcos-Torres FJ, Muñoz-Dorado J (2016) Bacterial predation: 75 years and counting!: bacterial predation. Environ Microbiol 18:766–779. https://doi.org/10.1111/1462-2920.13171
Placella SA, Brodie EL, Firestone MK (2012) Rainfall-induced carbon dioxide pulses result from sequential resuscitation of phylogenetically clustered microbial groups. Proc Natl Acad Sci 109: 10931–10936. https://doi.org/10.1073/pnas.1204306109
Pointing SB, Belnap J (2012) Microbial colonization and controls in dryland systems. Nat Rev Microbiol 10:551–562. https://doi.org/10.1038/nrmicro2831
Potts M (1994) Desiccation tolerance of prokaryotes. Microbiol Rev 58:755–805
Potts M (2001) Desiccation tolerance: a simple process? Trends Microbiol 9:553–559. https://doi.org/10.1016/S0966-842X(01)02231-4
Potts M, Slaughter SM, Hunneke F-U, Garst JF, Helm RF (2005) Desiccation tolerance of prokaryotes: application of principles to human cells. Integr Comp Biol 45:800–809. https://doi.org/10.1093/icb/45.5.800
Prestel E, Regeard C, Salamitou S, Neveu J, DuBow MS (2013) The bacteria and bacteriophages from a Mesquite flats site of the Death Valley desert. Antonie Van Leeuwenhoek 103:1329–1341. https://doi.org/10.1007/s10482-013-9914-4
Raanan H, Felde VJMNL, Peth S, Drahorad S, Ionescu D, Eshkol G, Treves H, Felix-Henningsen P, Berkowicz SM, Keren N, Horn R, Hagemann M, Kaplan A (2016) Three-dimensional structure and cyanobacterial activity within a desert biological soil crust: biological soil crust structure and activity. Environ Microbiol 18:372–383. https://doi.org/10.1111/1462-2920.12859
Ramond J-B, Woodborne S, Hall G, Seely M, Cowan DA (2018) Namib Desert primary productivity is driven by cryptic microbial community N-fixation. Sci Rep 8:6921. https://doi.org/10.1038/s41598-018-25078-4

Roberson EB, Firestone MK (1992) Relationship between desiccation and exopolysaccharide production in a soil pseudomonas sp. Appl Environ Microbiol 58:1284–1291
Rodriguez-Zaragoza S, Mayzlish E, Steinberger Y (2005) Vertical distribution of the free-living amoeba population in soil under desert shrubs in the Negev Desert, Israel. Appl Environ Microbiol 71:2053–2060. https://doi.org/10.1128/AEM.71.4.2053-2060.2005
Rudgers JA, Dettweiler-Robinson E, Belnap J, Green LE, Sinsabaugh RL, Young KE, Cort CE, Darrouzet-Nardi A (2018) Are fungal networks key to dryland primary production? Am J Bot 105:1783–1787. https://doi.org/10.1002/ajb2.1184
Saxton KE, Rawls WJ, Romberger JS, Papendick RI (1986) Estimating generalized soil-water characteristics from texture. Soil Sci Soc Am J 50:1031–1036. https://doi.org/10.2136/sssaj1986.03615995005000040039x
Scanlon BR (1994) Water and heat fluxes in desert soils: 1. Field studies. Water Resour Res 30: 709–719. https://doi.org/10.1029/93WR03251
Shoemaker WR, Lennon JT (2018) Evolution with a seed bank: the population genetic consequences of microbial dormancy. Evol Appl 11:60–75. https://doi.org/10.1111/eva.12557
Slade D, Radman M (2011) Oxidative stress resistance in Deinococcus radiodurans. Microbiol Mol Biol Rev 75:133–191. https://doi.org/10.1128/MMBR.00015-10
Stadler SJ (1998) Aridity indices. In: Encyclopedia of hydrology and lakes. Springer, Dordrecht, pp 78–83. https://doi.org/10.1007/1-4020-4497-6_20
Steinweg JM, Dukes JS, Wallenstein MD (2012) Modeling the effects of temperature and moisture on soil enzyme activity: linking laboratory assays to continuous field data. Soil Biol Biochem 55:85–92. https://doi.org/10.1016/j.soilbio.2012.06.015
Šťovíček A, Kim M, Or D, Gillor O (2017) Microbial community response to hydration-desiccation cycles in desert soil. Sci Rep 7:45735. https://doi.org/10.1038/srep45735
Tang D, Shi S, Li D, Hu C, Liu Y (2007) Physiological and biochemical responses of Scytonema javanicum (cyanobacterium) to salt stress. J Arid Environ 71:312–320. https://doi.org/10.1016/j.jaridenv.2007.05.004
Tecon R, Or D (2017) Biophysical processes supporting the diversity of microbial life in soil. FEMS Microbiol Rev 41:599–623. https://doi.org/10.1093/femsre/fux039
Uclés O, Villagarcía L, Cantón Y, Domingo F (2016) Partitioning of non rainfall water input regulated by soil cover type. Catena 139:265–270. https://doi.org/10.1016/j.catena.2015.02.018
Van Goethem MW, Makhalanyane TP, Valverde A, Cary SC, Cowan DA (2016) Characterization of bacterial communities in lithobionts and soil niches from Victoria Valley, Antarctica. FEMS Microbiol Ecol 92:fiw051. https://doi.org/10.1093/femsec/fiw051
Wang G, Or D (2012) A hydration-based biophysical index for the onset of soil microbial coexistence. Sci Rep 2:881. https://doi.org/10.1038/srep00881
Wang G, Or D (2013) Hydration dynamics promote bacterial coexistence on rough surfaces. ISME J 7:395–404. https://doi.org/10.1038/ismej.2012.115
Warren CR (2014) Response of osmolytes in soil to drying and rewetting. Soil Biol Biochem 70: 22–32. https://doi.org/10.1016/j.soilbio.2013.12.008
Yu LZ-H, Luo X-S, Liu M, Huang Q (2015) Diversity of ionizing radiation-resistant bacteria obtained from the Taklimakan Desert: diversity of radiation-resistant bacteria. J Basic Microbiol 55:135–140. https://doi.org/10.1002/jobm.201300390
Zablocki O, Adriaenssens EM, Cowan D (2016) Diversity and ecology of viruses in Hyperarid Desert soils. Appl Environ Microbiol 82:770–777. https://doi.org/10.1128/AEM.02651-15
Zablocki O, Adriaenssens EM, Frossard A, Seely M, Ramond J-B, Cowan D (2017) Metaviromes of extracellular soil viruses along a Namib Desert aridity gradient. Genome Announc 5 (e01470–16):e01470–e01416. https://doi.org/10.1128/genomeA.01470-16
Zhang T, Jia R-L, Yu L-Y (2016) Diversity and distribution of soil fungal communities associated with biological soil crusts in the southeastern Tengger Desert (China) as revealed by 454 pyrosequencing. Fungal Ecol 23:156–163. https://doi.org/10.1016/j.funeco.2016.08.004

Hot Desert Microbiology: Perspectives in a Warming World 12

Jean-Baptiste Ramond and Don A. Cowan

Abstract

Deserts are the most dominant terrestrial environments as they cover over a third of the Earth's emerged surface. These arid ecosystems further influence global biogeochemical cycling particularly via the emission of dust. These dust clouds can travel thousands of kilometers and fertilize very distant environments as well as intensify global warming. This is concerning as desert surfaces are expanding with climate change. This concluding chapter therefore briefly discusses possible novel research avenues that desert microbial ecologist could follow in the context of climate change.

Deserts biomes experience a very wide range of macro-climatic conditions. They may be either hot (e.g., the Sahara Desert) or cold (the Antarctic McMurdo Dry Valleys), may be coastal (e.g., the Atacama and Namib Deserts) or inland (e.g., Mojave and Gobi Deserts) and may be low or high altitude. However, the critical common feature shared by desert biomes is a general deficiency in water availability. To be considered a desert, an environment must present an aridity index (AI), defined as the ratio of Precipitation (P) over Potential Evapotranspiration (PET) [AI=P/PET], below 0.65. The AI allows a further sub-classification of the aridity

J.-B. Ramond (✉)
Department of Molecular Genetics and Microbiology, Pontificia Universidad Católica de Chile, Santiago, Chile

Centre for Microbial Ecology and Genomics, Department of Biochemistry, Genetics and Microbiology, University of Pretoria, Pretoria, South Africa
e-mail: jbramond@bio.uc.cl

D. A. Cowan
Centre for Microbial Ecology and Genomics, Department of Biochemistry, Genetics and Microbiology, University of Pretoria, Pretoria, South Africa

J.-B. Ramond, D. A. Cowan (eds.), *Microbiology of Hot Deserts*, Ecological Studies 244, https://doi.org/10.1007/978-3-030-98415-1_12

status of drylands from dry-subhumid ($0.5 < AI < 0.65$) to hyperarid ($AI < 0.05$). Using these AI-based definitions, drylands/deserts represent the most dominant biome on Earth, covering approximately 40% of the planet's terrestrial surface and being the only biome present on the six continents. To obtain a better understanding of the climatic and geomorphological features that give rise to hot deserts globally, the reader is referred to Chap. 1 of this book.

Given the scale of dryland coverage on terrestrial Earth, a comprehensive understanding of how these dominant ecosystems function is highly relevant at the planetary scale. The chapters compiled in this book provide a comprehensive "*microbes-eye view*" on how these fragile ecosystems are driven by microbially-mediated processes. This is further emphasized by the fact that our species, *Homo sapiens,* the most widespread (and devastating; IPCC 2021) ecosystem engineer on the planet, has a high socioeconomic dependence on dryland ecosystems. More than 2 billion humans, mostly from poor and developing countries are potentially impacted by dryland expansions (desertification processes), according to the United Nations Decade for Deserts and the fight against Desertification (UNDDD) 2010–2020 reports (http://www.un.org/en/events/desertification_decade/whynow.shtml).

Deserts have a substantial impact on the overall functioning of planet Earth (Pointing and Belnap 2014; Kok et al. 2017). Annually, billions of tons of soil-derived dust are emitted from deserts and dispersed at very large scales, with significant impacts on local biogeochemistry and climate (Herut et al. 2002; Jickells et al. 2005; Kellogg and Griffin 2006; Bristow et al. 2010; Gonzalez-Martin et al. 2014; Pointing and Belnap 2014; Kok et al. 2017; Šikoparija 2020). Remarkably, the phosphorous-deficient Amazon Basin in South America, which is the most productive region in the world, receives a significant amount of phosphorous (P) from a relatively small area in the southern Sahara in Chad, the Bodélé Depression of around 10,800 km^2; circa the surface of Jamaica or Lebanon (Bristow et al. 2010). Some 6.5 Tg of Fe and 0.12 Tg of P [$1\ Tg = 10^9\ kg$] are emitted each year from the Bodélé Depression, fertilizing the otherwise oligotrophic Atlantic Ocean or the other P-starved Amazon environments (Bristow et al. 2010). Similarly, the dust aerosols emitted by Asian deserts have been shown to cross the Pacific Ocean and even reach the eastern coast of the north America (Kellogg and Griffin 2006), a journey of over 15,000 km. Desert dust aerosols may even participate in global warming (Kok et al. 2017), a cause for concern given that global desert surface areas are increasing with climate change (e.g., Huang et al. 2016, 2017). Desert soil stabilization via the restauration of microbial-dominated biological soil crusts clearly represents a relevant strategy that could be implemented to mitigate the impact of climate change at both the desert biome and the global Earth system scales (e.g., Chap. 3; Bowker 2007; Tucker et al. 2020). Given that BSCs represent N and C fixation hubs in such environments, such a strategy would have a beneficial effect on hot desert primary production, as outlined in Chaps. 3 and 7.

In a world hit by the worst pandemic of modern record, it is also noteworthy that cross-continental desert dust transport events represent potential human and plant health hazards as exogenous pathogens (particularly spore-forming microorganisms) may hitch-hike on dust particles and colonize/affect new hosts (Griffin 2007; Gonzalez-Martin et al. 2014; Zhang et al. 2016; Salawu-Rotimi et al. 2021). In this context, the recent advances in microbial aerobiology—particularly meta'omics' (e.g., Archer and Pointing 2020; Archer et al. 2021; Maki et al. 2021; Chap. 2)—represents a key future avenue of research for desert microbial ecology.

In addition to water deficiency, a range of other abiotic stresses are imposed on desert (micro)biota. These may include (hyper)oligotrophy, high soil salinity, high UV irradiation, and high daily and seasonal temperature fluctuations (Noy-Meir 1973). As a result, microorganisms are often referred to as the main driving forces of hot desert biogeochemical cycling (Pointing and Belnap 2012; Makhalanyane et al. 2015; Cowan et al. 2020). Chap. 7 summarizes the data indicating how microbial communities from different desert biotopes perform key steps of the C, N, and P cycles. There is even a growing body of evidence demonstrating that desert BSCs are crucial for the cycling of S (e.g., Qi et al. 2021; Zhang et al. 2021). Indeed, as clearly described in the various chapters of this book—and contrary to historical belief[1]—these depauperate ecosystems are populated by a wide array of taxa (e.g., Chaps. 3–9). Diverse, active, and highly adapted microbial communities have successfully colonized the various niches present in hot deserts, from exposed desert pavements to cryptic refuge niches (such as hypo/endoliths; Chaps. 3–9; Cowan et al. 2020). Even the harshest of desert ecosystems, such as in the most hyperarid regions of the Atacama and Namib Deserts, are colonized by a wide range of active microbial taxa (Gunnigle et al. 2014, 2017; Schulze-Makuch et al. 2018; León-Sobrino et al. 2019; Chaps. 4 and 9). The unique and specialized adaptations of microorganisms to the polyextreme conditions imposed by desert environments—many of which are described in Chap. 10—are exemplified by the recent discovery of the microbial metabolic capacity for harvesting atmospheric trace gasses such as H_2 and CO, effectively unlimited resources in otherwise depauperate environments, for energy and biomass production (e.g., Ji et al. 2017; Jordaan et al. 2020). The observation that aerobic H_2 oxidation is "hydrogenic" (water-producing: Ortiz et al. 2021) offers the intriguing and potentially paradigm-changing possibility that desert soil microbial communities generate their own water! The extent to which metabolic hydrogenesis contributes to the water budgets of desert soil microbiomes, and whether this process is capable of supporting basal cellular metabolism (or even higher desiccation-sensitive metabolic functions such as photosynthesis and cell division) is currently unknown.

Nevertheless, this recent discovery strongly suggests that, within the many desert niches described in this book, new metabolic processes may remain to be discovered,

[1] In 1903, when Captain Scott first discovered the Dry Valleys in Antarctica, he wrote "*It is worthy to record, too, that we have seen no living thing, not even a moss or a lichen; all that we did find, far inland amongst the moraine heaps, was the skeleton of a Weddell seal, and how that came there is beyond guessing. It is certainly a valley of the dead; even the great glacier which once pushed through it has withered away*" (Scott 1907). Yet this Mars-like environment unarguably harbors diverse and active microbial communities (e.g., Lee et al. 2012; Chan et al. 2013; Ortiz et al. 2020; Canini et al. 2021; Chap. 10).

thanks to microbial metabolic ingenuity and plasticity when confronted with environmental extremes. The latest high-resolution and high-throughput methods used to study (desert) environmental microbial communities, which are described in Chap. 2, could be central to such discoveries. It is an exciting prospect that such discoveries could lead to the future development of new biotechnologies and even improvements in desert farming (Marasco et al. 2012; Bull et al. 2016; Chap. 8).

However, the world's climate is clearly changing and, in general, desert regions are predicted to become hotter and drier (Huang et al. 2016, 2017). For the first time, the most recent IPCC report clearly states that climate change is an anthropogenic phenomenon: "*It is unequivocal that human influence has warmed the atmosphere, ocean and land. Widespread and rapid changes in the atmosphere, ocean, cryosphere and biosphere have occurred.*" and "*Each of the last four decades has been successively warmer than any decade that preceded it since 1850*" (IPCC 2021). This global temperature increase is clearly shown in Fig. 12.1, a 20-year comparison of global temperature anomalies in 2001, compared to 2021. Furthermore, 50 °C and above temperatures are now measured at high frequencies globally (Fig. 12.2; Di Luca et al. 2020) and are not limited to drylands. In June 2021, a record high 49.5 °C was measured in British Columbia (Canada) at a latitude of 50.2333° N!

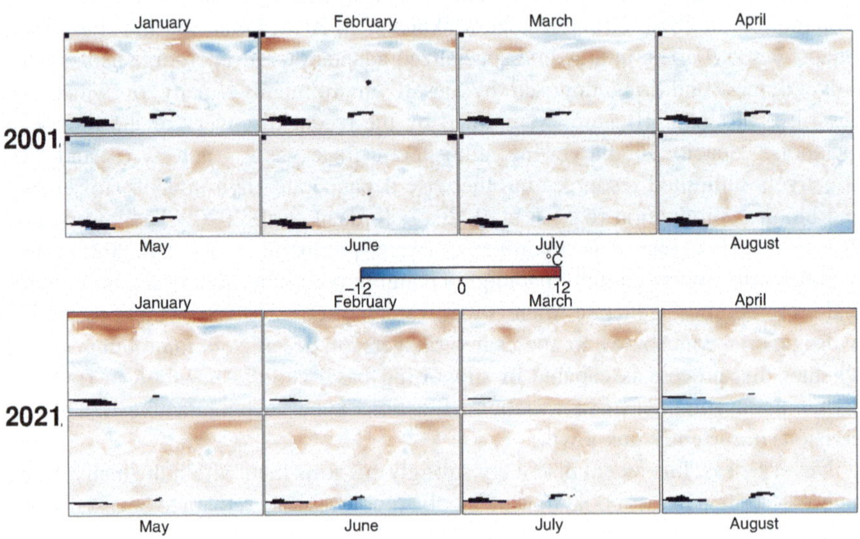

Fig. 12.1 Maps of the monthly global temperature anomalies on Earth from January to August in 2001 and 2021. Imagery produced by the NASA Earth Observations team based on data provided by the NASA Goddard Institute for Space Studies (GISS). Downloaded from https://neo.sci.gsfc.nasa.gov/ on the 6th of October 2021. The maps depict how much warmer (i.e., redder) or colder (i.e., bluer) a region may be in a given month compared to the norm for that same month in the same region from 1951 to 1980. These maps do not depict absolute temperature but instead show temperature anomalies, or how much it has changed. The source data for these images is 2 × 2 degrees—or 180 × 90 pixels

Fig. 12.2 Photograph of a bus stop in the suburbs of Madrid (Spain) on the 14th of August 2021 indicating a temperature of 50 °C (!). *Photo courtesy Mr. José Luis Corbacho*

In heating steadily warming world, can hot desert microbial communities maintain their essential ecosystem functions? This is a critical question, the answer to which requires both a qualitative and quantitative understanding of the functions of desert soil microbiomes, and how such functions may change with changes in water availability and temperature regimes. These two factors are intimately interconnected: increased temperatures will elevate rates of evapotranspiration and decrease mean desert soil moisture contents. In habitats where water-stress is the dominant driver of microbial community structure and function (Makhalanyane et al. 2015), it is confidently predicted that increased mean temperatures will negatively impact microbial communities and the processes they mediate (Neilson et al. 2017; Jansson and Hofmockel 2020). Temperature rises in African and North America deserts have already been shown to reduce the photosynthetic rates in lichens (Maphangwa et al. 2012) and mosses (Grote et al. 2010), and to significantly decrease biological soil crust cover (~ 44% in 4 years) at two semiarid sites in Spain (Maestre et al. 2013).

However, climate change and the associated regional warming, despite being global phenomena (IPCC 2014, 2021), will have locally specific impacts (Faramarzi et al. 2013; IPCC 2014). Different deserts and the different habitats of each desert harbor unique microbial assemblages that may react differently to thermal and xeric impacts; i.e., may exhibit different degrees of structural and functional resilience (Caruso et al. 2011; Johnson et al. 2017; Chaps. 10 and 11). Consequently, a global effort is necessary to evaluate the impacts of increasing temperatures in each of the Earth's hot deserts, by cataloging and understanding the responsiveness of the different microbial communities, and the way in which such responses affect ecosystem services. This is particularly relevant, given that climate models often lack microbially-mediated data (Treseder et al. 2012; Jansson and Hofmockel 2020). Apart from higher seasonal temperatures (Fig. 12.1), climate change is also predicted to lead to hydrological cycle discrepancies in hot deserts, i.e., longer droughts, and with fewer but more intensive precipitation events (Faramarzi et al. 2013; IPCC 2014, 2021). In this context, the studies summarized in Chap. 11 provide critical information on how hot desert microbial communities may react to changes in water inputs.

Our conclusion, supported by the comprehensive chapters contributing to this text, is that the past decades of research on desert soil ecosystems using the latest and most sophisticated *omics* technologies have done much to help us understand the structure and functional capacity of desert soil microbiomes. We also conclude that there remain huge gaps in our understanding of these fascinating systems, particularly relating to the quantitative aspects of soil microbiome function and their responsiveness to the abiotic variables to which they are exposed. The next two decades of research on desert soil microbiomics may be even more exciting that the last!

References

Archer SD, Pointing SB (2020) Anthropogenic impact on the atmospheric microbiome. Nat Microbiol 5(2):229–231

Archer S, Lee K, Caruso T, Leung M, Tong X, Hinchliffe G, Maki T, Santl-Temkiv T, Warren-Rhodes K, Gomez-Silva B et al. (2021) Diverse recruitment to a globally structured atmospheric microbiome

Bowker MA (2007) Biological soil crust rehabilitation in theory and practice: an underexploited opportunity. Restor Ecol 15(1):13–23

Bristow CS, Hudson-Edwards KA, Chappell A (2010) Fertilizing the Amazon and equatorial Atlantic with west African dust. Geophys Res Lett 37(14):L14807

Bull AT, Asenjo JA, Goodfellow M, Gómez-Silva B (2016) The Atacama Desert: technical resources and the growing importance of novel microbial diversity. Annu Rev Microbiol 70: 215–234

Canini F, Geml J, D'Acqui LP, Buzzini P, Turchetti B, Onofri S, Ventura S, Zucconi L (2021) Fungal diversity and functionality are driven by soil texture in Taylor Valley, Antarctica. Fungal Ecol 50:101041

Caruso T, Chan Y, Lacap DC, Lau MC, McKay CP, Pointing SB (2011) Stochastic and deterministic processes interact in the assembly of desert microbial communities on a global scale. ISME J 5(9):1406–1413

Chan Y, Van Nostrand JD, Zhou J, Pointing SB, Farrell RL (2013) Functional ecology of an Antarctic dry valley. Proc Natl Acad Sci 110(22):8990–8995

Cowan DA, Hopkins DW, Jones BE, Maggs-Kölling G, Majewska R, Ramond JB (2020) Microbiomics of Namib Desert habitats. Extremophiles 24(1):17–29

Di Luca A, de Elía R, Bador M, Argüeso D (2020) Contribution of mean climate to hot temperature extremes for present and future climates. Weather Clim Extremes 28:100255

Faramarzi M, Abbaspour KC, Vaghefi SA, Farzaneh MR, Zehnder AJ, Srinivasan R, Yang H (2013) Modeling impacts of climate change on freshwater availability in Africa. J Hydrol 480: 85–101

Gonzalez-Martin C, Teigell-Perez N, Valladares B, Griffin DW (2014) The global dispersion of pathogenic microorganisms by dust storms and its relevance to agriculture. Adv Agron 127:1–41

Griffin DW (2007) Atmospheric movement of microorganisms in clouds of desert dust and implications for human health. Clin Microbiol Rev 20(3):459–477

Grote EE, Belnap J, Housman DC, Sparks JP (2010) Carbon exchange in biological soil crust communities under differential temperatures and soil water contents: implications for global change. Glob Chang Biol 16(10):2763–2774

Gunnigle E, Ramond JB, Frossard A, Seeley M, Cowan D (2014) A sequential co-extraction method for DNA, RNA and protein recovery from soil for future system-based approaches. J Microbiol Methods 103:118–123

Gunnigle E, Frossard A, Ramond JB, Guerrero L, Seely M, Cowan DA (2017) Diel-scale temporal dynamics recorded for bacterial groups in Namib Desert soil. Sci Rep 7(1):1–12

Herut B, Collier R, Krom MD (2002) The role of dust in supplying nitrogen and phosphorus to the Southeast Mediterranean. Limnol Oceanogr 47(3):870–878

Huang J, Yu H, Guan X, Wang G, Guo R (2016) Accelerated dryland expansion under climate change. Nat Clim Chang 6(2):166–171

Huang J, Li Y, Fu C, Chen F, Fu Q, Dai A, Shinoda M, Ma Z, Guo W, Li Z, Zhang L (2017) Dryland climate change: recent progress and challenges. Rev Geophys 55(3):719–778

IPCC (2021) Summary for policymakers. In: Masson-Delmotte V, Zhai P, Pirani A, Connors SL, Péan C, Berger S, Caud N, Chen Y, Goldfarb L, Gomis MI, Huang M, Leitzell K, Lonnoy E, Matthews JBR, Maycock TK, Waterfield T, Yelekçi O, Yu R, Zhou B (eds) Climate change 2021: the physical science basis. Contribution of working group I to the sixth assessment report of the intergovernmental panel on climate change. Cambridge University Press, Cambridge. In Press

IPCC (Intergovernmental Panel on Climate Change) (2014) Climate change 2014: Synthesis report. Contribution of working groups I, II and III to the fifth assessment report of the intergovernmental panel on climate change. Core writing team, pp. 1–151

Jansson JK, Hofmockel KS (2020) Soil microbiomes and climate change. Nat Rev Microbiol 18(1): 35–46

Ji M, Greening C, Vanwonterghem I, Carere CR, Bay SK, Steen JA, Montgomery K, Lines T, Beardall J, Van Dorst J, Snape I (2017) Atmospheric trace gases support primary production in Antarctic desert surface soil. Nature 552(7685):400–403

Jickells TD, An ZS, Andersen KK, Baker AR, Bergametti G, Brooks N, Cao JJ, Boyd PW, Duce RA, Hunter KA, Kawahata H (2005) Global iron connections between desert dust, ocean biogeochemistry, and climate. Science 308(5718):67–71

Johnson RM, Ramond JB, Gunnigle E, Seely M, Cowan DA (2017) Namib Desert edaphic bacterial, fungal and archaeal communities assemble through deterministic processes but are influenced by different abiotic parameters. Extremophiles 21(2):381–392

Jordaan K, Lappan R, Dong X, Aitkenhead IJ, Bay SK, Chiri E, Wieler N, Meredith LK, Cowan DA, Chown SL, Greening C (2020) Hydrogen-oxidizing bacteria are abundant in desert soils and strongly stimulated by hydration. Msystems 5(6):e01131–e01120

Kellogg CA, Griffin DW (2006) Aerobiology and the global transport of desert dust. Trends Ecol Evol 21(11):638–644

Kok JF, Ridley DA, Zhou Q, Miller RL, Zhao C, Heald CL, Ward DS, Albani S, Haustein K (2017) Smaller desert dust cooling effect estimated from analysis of dust size and abundance. Nat Geosci 10(4):274–278

Lee CK, Barbier BA, Bottos EM, McDonald IR, Cary SC (2012) The inter-valley soil comparative survey: the ecology of Dry Valley edaphic microbial communities. ISME J 6(5):1046–1057

León-Sobrino C, Ramond JB, Maggs-Kölling G, Cowan DA (2019) Nutrient acquisition, rather than stress response over diel cycles, drives microbial transcription in a hyper-arid Namib Desert soil. Front Microbiol 10:1054

Maestre FT, Escolar C, de Guevara ML, Quero JL, Lázaro R, Delgado-Baquerizo M, Ochoa V, Berdugo M, Gozalo B, Gallardo A (2013) Changes in biocrust cover drive carbon cycle responses to climate change in drylands. Glob Chang Biol 19(12):3835–3847

Makhalanyane TP, Valverde A, Gunnigle E, Frossard A, Ramond JB, Cowan DA (2015) Microbial ecology of hot desert edaphic systems. FEMS Microbiol Rev 39(2):203–221

Maki T, Lee KC, Pointing SB, Watanabe K, Aoki K, Archer SD, Lacap-Bugler DC, Ishikawa A (2021) Desert and anthropogenic mixing dust deposition influences microbial communities in surface waters of the western Pacific Ocean. Sci Total Environ 719:148026

Maphangwa KW, Musil CF, Raitt L, Zedda L (2012) Experimental climate warming decreases photosynthetic efficiency of lichens in an arid south African ecosystem. Oecologia 169(1): 257–268

Marasco R, Rolli E, Ettoumi B, Vigani G, Mapelli F, Borin S, Abou-Hadid AF, El-Behairy UA, Sorlini C, Cherif A, Zocchi G (2012) A drought resistance-promoting microbiome is selected by root system under desert farming. PLoS One 7(10):e48479

Neilson JW, Califf K, Cardona C, Copeland A, Van Treuren W, Josephson KL, Knight R, Gilbert JA, Quade J, Caporaso JG, Maier RM (2017) Significant impacts of increasing aridity on the arid soil microbiome. MSystems 2(3):e00195–e00116

Noy-Meir I (1973) Desert ecosystems: environment and producers. Annu Rev Ecol Syst 4(1):25–51

Ortiz M, Bosch J, Coclet C, Johnson J, Lebre P, Salawu-Rotimi A, Vikram S, Makhalanyane T, Cowan D (2020) Microbial nitrogen cycling in Antarctic soils. Microorganisms 8(9):1442

Ortiz M, Leung PM, Shelley G, Jirapanjawat T, Nauer PA, Von Goethem MW, Bay SK, Islam ZF, Jordaan K, Vikram M, Hogg ID, Makhalanyane TP, Chown SL, Grinter R, Cowan DA, Greening C (2021) Multiple energy sources and metabolic strategies sustain microbial diversity in Antarctic desert soils. Proc Nat Acad Sci USA 118:e2025322118

Pointing, S.B. and Belnap, J. (2012) Microbial colonization and controls in dryland systems. Nature Reviews Microbiology 10(8):551–562

Pointing SB, Belnap J (2014) Disturbance to desert soil ecosystems contributes to dust-mediated impacts at regional scales. Biodivers Conserv 23(7):1659–1667

Qi J, Liu Y, Wang Z, Zhao L, Zhang W, Wang Y, Li X (2021) Variations in microbial functional potential associated with phosphorus and sulfur cycling in biological soil crusts of different ages at the Tengger Desert. China Applied Soil Ecology 165:104022

Salawu-Rotimi A, Lebre PH, Vos HC, Fister W, Kuhn N, Eckardt F, Cowan DA (2021) Gone with the wind: microbial communities associated with dust from emissive farmlands. Microb Ecol 82:859–869

Schulze-Makuch D, Wagner D, Kounaves SP, Mangelsdorf K, Devine KG, de Vera JP, Schmitt-Kopplin P, Grossart HP, Parro V, Kaupenjohann M et al (2018) Transitory microbial habitat in the hyperarid Atacama Desert. Proc Natl Acad Sci 115(11):2670–2675

Scott RF (1907) The voyage of the discovery, vol 1. Charles Scribner's Sons, New York

Šikoparija B (2020) Desert dust has a notable impact on aerobiological measurements in Europe. Aeolian Res 47:100636

Treseder KK, Balser TC, Bradford MA, Brodie EL, Dubinsky EA, Eviner VT, Hofmockel KS, Lennon JT, Levine UY, MacGregor BJ, Pett-Ridge J (2012) Integrating microbial ecology into ecosystem models: challenges and priorities. Biogeochemistry 109(1):7–18
Tucker C, Antoninka A, Day N, Poff B, Reed S (2020) Biological soil crust salvage for dryland restoration: an opportunity for natural resource restoration. Restor Ecol 28:S9–S16
Zhang X, Zhao L, Tong DQ, Wu G, Dan M, Teng B (2016) A systematic review of global desert dust and associated human health effects. Atmos 7(12):158
Zhang XC, Li JY, Liu JL, Yuan CX, Li YN, Liu BR, Yan XF (2021) Temporal shifts in cyanobacterial diversity and their relationships to different types of biological soil crust in the southeastern Tengger Desert. Rhizosphere 17:100322

The manufacturer's authorised representative in the EU is Springer Nature Customer Service Centre GmbH, Europaplatz 3, 69115 Heidelberg, Germany. If you have any concerns regarding our products, please contact ProductSafety@springernature.com

Printed and bound by CPI Group (UK) Ltd, Croydon, CR0 4YY
23/09/2026
02221546-0013